Philosophy of Logic and Mathematics

Publications of the Austrian Ludwig Wittgenstein Society
New Series

―

Volume 27

Philosophy of Logic and Mathematics

Proceedings of the 41st International Wittgenstein Symposium

Edited by
Gabriele M. Mras, Paul Weingartner, and
Bernhard Ritter

DE GRUYTER

ISBN 978-3-11-076347-8
e-ISBN (PDF) 978-3-11-065788-3
e-ISBN (EPUB) 978-3-11-065454-7
ISSN 2191-8449

Library of Congress Control Number 2019939345

Bibliographic information published by the Deutsche Nationalbibliothek
The Deutsche Nationalbibliothek lists this publication in the Deutsche Nationalbibliografie; detailed bibliographic data are available on the Internet at http://dnb.dnb.de.

© 2021 Walter de Gruyter GmbH, Berlin/Boston
This volume is text- and page-identical with the hardback published in 2019.
Printing and binding: CPI books GmbH, Leck

www.degruyter.com

Contents

Gabriele M. Mras, Paul Weingartner, and Bernhard Ritter
Preface —— IX

Part I: Philosophy of Logic

Eric Snyder and Stewart Shapiro
Link's Revenge: A Case Study in Natural Language Mereology —— 3

Gerhard Schurz
Universal Translatability: An Optimality- Based Justification of (Classical) Logic —— 37

Gila Sher
Invariance and Necessity —— 55

Itala M. Loffredo D'Ottaviano and Hércules de Araújo Feitosa
Translations Between Logics: A Survey —— 71

Jan Woleński
On the Relation of Logic to Metalogic —— 91

Edi Pavlović and Norbert Gratzl
Free Logic and the Quantified Argument Calculus —— 105

Gabriel Sandu
Dependencies Between Quantifiers Vs. Dependencies Between Variables —— 117

Wolfgang Kienzler
Three Types and Traditions of Logic: Syllogistic, Calculus and Predicate Logic —— 133

Günther Eder
Truth, Paradox, and the Procedural Conception of Fregean Sense —— 153

Christoph C. Pfisterer
Wittgenstein and Frege on Assertion —— 169

Maria van der Schaar
Assertions and Their Justification: Demonstration and Self-Evidence —— 183

Elena Dragalina-Chernaya
Surprises in Logic: When Dynamic Formality Meets Interactive Compositionality —— 197

Part II: Philosophy of Mathematics

Paolo Mancosu and Benjamin Siskind
Neologicist Foundations: Inconsistent Abstraction Principles and Part-Whole —— 215

William Tait
What Hilbert and Bernays Meant by "Finitism" —— 249

Juliet Floyd
Wittgenstein and Turing —— 263

Charles Parsons
Remarks on Two Papers of Paul Bernays —— 297

Richard Zach
The Significance of the Curry-Howard Isomorphism —— 313

Felix Mühlhölzer
Reductions of Mathematics: Foundation or Horizon? —— 327

Jan von Plato
What Are the Axioms for Numbers and Who Invented Them? —— 343

Part III: Wittgenstein

Mathieu Marion and Mitsuhiro Okada
Following a Rule: Waismann's Variation —— 359

Michael Potter
Propositions in Wittgenstein and Ramsey —— 375

Jean-Yves Béziau
An Unexpected Feature of Classical Propositional Logic in the *Tractatus* —— 385

Janusz Kaczmarek
Ontology in *Tractatus Logico-Philosophicus*: A Topological Approach —— 397

Franz Berto
Adding 4.0241 to TLP —— 415

Štefan Riegelnik
Understanding Wittgenstein's Wood Sellers —— 429

Susan Edwards-McKie
On the Infinite, In-Potentia: Discovery of the Hidden Revision of *Philosophical Investigations* and Its Relation to TS 209 Through the Eyes of Wittgensteinian Mathematics —— 441

Richard Heinrich
Incomplete Pictures and Specific Forms: Wittgenstein Around 1930 —— 457

Oliver Feldmann
„Man kann die Menschen nicht zum Guten führen" – Zur Logik des moralischen Urteils bei Wittgenstein und Hegel —— 473

Esther Ramharter
Der Status mathematischer und religiöser Sätze bei Wittgenstein —— 485

Richard Raatzsch
Gutes Sehen —— 499

Timm Lampert
Wittgenstein's Conjecture —— 515

Index of Names —— 535

Index of Subjects —— 539

Gabriele M. Mras, Paul Weingartner, and Bernhard Ritter
Preface

The title of these Proceedings raises questions about the very idea that logic and mathematics are genuine philosophical subjects. Whilst such questions might be explored with particular reference to any number of distinct historical periods, from Ancient Greece onwards, particular focus is given here to the period from the end of the 19th century to the present. That said, Aristotle and Plato are, inevitably, invoked in the task of clarifying the philosophical status of logic and mathematics.

According to Aristotle (cf. Met. 981b 28 – 982b 10), metaphysics (also referred to as "first philosophy") aims to identify the first principles (axioms) that underpin our foundational explanations, knowledge and reasoning (cf. Post. An. I, 2). The assumptions that these truths should be foundational and as few as possible in number is an important precursor to the axiomatic method and the axiomatic scientific system first built by Frege for logic and then by Cantor, Peano, Russell, and Hilbert for mathematics. Whilst Aristotle holds that highly general principles and axioms, which have their roots in philosophy, underlie every science, Plato stresses that the roots of rational philosophy lie in mathematics, particularly in geometry. "Nobody who has not studied geometry should enter my house" was written above the entrance of his school. Without the appropriate preparatory engagement with mathematics, especially geometry, a philosopher is unable to appreciate the independent nature of an eternal truth, and its maximally general nature, open to multiple substitutions (cf. Rep. 524d – 527d; Tim. 53d – 54b). Unsurprisingly, Aristotelian and Platonic legacies, and their inherent tensions, pervade much of the contemporary investigations in these Proceedings.

Frege, in his *Begriffsschrift – eine der arithmetischen nachgebildeten reinen Formelsprache des Denkens*, formulates what we now call *propositional* and *predicate logic*; although there were earlier alternative formulations by De Morgan, Boole, and Schröder. The *Begriffsschrift* provides the building blocks of all contemporary logic courses: the view of sentential connectives "and", "or", "if – then" as determined by the combination of truth values of the sentences so connected. This new Concept Script also provides a theory of quantification that matches best the requirement to provide a structure for various kinds of propositions. Frege was not concerned, though, with logic for its own sake. His goal to represent inferential

---

Gabriele M. Mras, Vienna University of Economics, gabriele.mras@wu.ac.at
Paul Weingartner, University of Salzburg, paul.weingartner@sbg.ac.at
Bernhard Ritter, University of Klagenfurt, bernhard.ritter@aau.at

relations was a means to establish a "secure foundation" for mathematics, or rather arithmetic.

There are, at least, two main achievements in Frege's conception of predicates as functional expressions: 1. a formalization of inferences involving relational terms that could not be handled by syllogistic logic; 2. a structure, suggested by the function and argument analysis, that delivers a radical shift in our understanding of generality. The quantifier variable view of generality makes it possible to represent inferences involving multiple quantifiers and relational terms of the kind needed to formalize the content of a sentence such as "There are infinitely many prime numbers". It also becomes possible to reduce the relation of "concept-subordination" to the inference rule of the material implication. Frege's calculus thus provides for the first time a unified system of propositional and predicate logic. His attempt to provide arithmetic with a secure foundation in purely logical terms must, however, be considered a failure.

The difficulties of providing a set-theoretical foundation for arithmetic did lead to a re-evaluation of the concept of a set, not least because of various well-known paradoxes that result from the unrestricted introduction of comprehension and abstraction principles. The serious limitations of this project, discovered by Russell, Tarski, and Turing, were laid down in a much more general way by Gödel in his famous incompleteness theorem (cf. Gödel 1931). Thus even the restricted sense of constructive methods adopted by Brouwer's intuitionism and Hilbert's finitist proofs must be considered as inadequate. Neologicism emerged in the 1980s as a reaction to this negative outcome. Neologicism is an attempt to preserve the key tenets of Frege's project, namely the analyticity of arithmetic knowledge, by means of a weakened understanding of those basic laws Frege used to establish logicism.

Parts I and II of these Proceedings deal with the history of logic, the question of the nature of logic, the relation of logic and mathematics, modal or alternative logics (many-valued, relevant, paraconsistent logics) and their relations, including translatability, to classical logic in the Fregean and Russellian sense, and, more generally, the aim or aims of philosophy of logic and mathematics. Also explored are several problems concerning the concept of definition, non-designating terms, the interdependence of quantifiers, and the idea of an assertion sign.

One part of the Proceedings of the Wittgenstein Symposium is traditionally open to papers on any aspect of Wittgenstein's philosophy, whether or not their authors directly address the particular theme of the conference. The contributions concerned with Wittgenstein's investigations into the philosophy of logic and mathematics pursue issues relating to logical necessity, the undeniability of the law of the excluded middle, and the source of self-evidence, often characterized in the literature as the 'rule-following considerations'. Additionally, they examine Wittgensteinian attitudes towards the very idea of set-theory as a possible foundation for arithmetic. In some sense, whilst it is possible to reduce larger mathematical areas to more restricted or privileged ones, such as set theory, what, if anything might such reduction achieve or reveal? If we follow Wittgenstein in resisting the idea that such 'reductions' illuminate supposedly foundational matters, where does this leave us? Does it mean that there are or can be radically divergent systems and practices that would, accordingly, be no less as justified (or unjustified) as our own? Proceeding along these lines, one might even conclude that what is and what is not evident, or necessary, or possible, entirely depends on us – hardly a philosophically satisfactory position and most likely not Wittgenstein's.

The volume also includes a number of contributions on specific issues concerning Wittgenstein's views on moral and religious judgements. A common theme running through almost all of these discussions is Wittgenstein's concern with the question of whether or not there are limits, or conditions, to sense and, if so, whether it is possible to determine them.

Acknowledgment: We would like to thank all the authors for their contributions and kind cooperation. We also like to thank Beate Cemper, Thomas Hainscho, Bernd Nussbaumer, and Joaquín Padilla Montani for a number of Word-to-LaTeX transcriptions as well as suggestions on how to improve the code. Lastly we would like to thank Jana Habermann and Nancy Christ from De Gruyter.

Bibliography

Aristotle (Met.): *Metaphysics, Books 1–9.* Cambridge (MA), London: Harvard University Press, Heinemann, 1975.
Aristotle (Post. An.): *Posterior Analytics.* Oxford University Press, New York, 1993² (1975).
Frege, Gottlob (1879): *Begriffsschrift – eine der arithmetischen nachgebildeten reinen Formelsprache des Denkens.* Halle: Nebert.
Gödel, Kurt (1931): "Über formal unentscheidbare Sätze der Principia Mathematica und verwandter Systeme I." In *Monatshefte für Mathematik und Physik,* Vol. 38, 173 – 198.
Plato (Tim.): *Timaeus.* In: *Timaeus – Critias – Cleitophon – Menexenus – Epistles.* Cambridge (MA), London: Harvard University Press, 1999⁹ (1929).
Plato (Rep): *The Republic.* Shorey, Paul (transl.). London: Heinemann; Cambridge (MA): Harvard University Press, 1970.

Part I: **Philosophy of Logic**

Eric Snyder and Stewart Shapiro
Link's Revenge: A Case Study in Natural Language Mereology

Abstract: Most philosophers are familiar with the metaphysical puzzle of the statue and the clay. A sculptor begins with some clay, eventually sculpting a statue from it. Are the clay and the statue one and the same thing? Apparently not, since they have different properties. For example, the clay could survive being squashed, but the statue could not. The statue is recently formed, though the clay is not, etc. Godehart Link 1983's highly influential analysis of the count/mass distinction recommends that English draws a distinction between uncountable "stuff" and countable "things". There are two mereological relations, related in specific ways. Our primary question here is whether an empirically adequate account of the mass/count distinction really does require distinguishing "things" from "stuff", and thus postulating two corresponding mereological relations, or if instead positing only one sort of entity and corresponding mereological relation is sufficient, as other semantic theories would have it. This question is meant to be one of what we call *natural language mereology*. We are asking about the mereological commitments of English, or perhaps competent speakers of English, and not about ultimate reality as such. There is no pretense that we will definitively solve the *metaphysical* puzzle of the statue and clay.

1 Introduction

Most philosophers are familiar with the metaphysical puzzle of the statue and the clay. A sculptor begins with some clay, eventually sculpting a statue from it. Are the clay and the statue one and the same thing? Apparently not, since they have different properties. For example, the clay could survive being squashed, but the statue could not. The statue is recently formed, though the clay is not, etc.

LEIBNIZ'S LAW is the thesis that identical things have the same properties. It follows from this that the statue and the clay are distinct, since they have different properties: one can survive being squashed, the other cannot; one is recently formed, the other is not, etc.

Many metaphysical puzzles then follow: When exactly does the clay become the statue? How can two co-located objects exist concurrently, etc.

Eric Snyder, Smith College, Northampton (MA), USA
Stewart Shapiro, The Ohio State University, Columbus (OH), USA, shapiro.4@osu.edu
DOI 10.1515/9783110657883-2

Given the persistence of this puzzle, it would be rather surprising if the relatively nascent science of natural language semantics required taking a stance on the issue. Yet that is precisely what Godehart Link 1983's highly influential analysis of THE COUNT/MASS DISTINCTION recommends. In particular, Link argues on the basis of examples like (1) that English draws a distinction between uncountable "stuff" and countable "things".

(1) This ring is new, but the gold in the ring is old.

Link's argument is straightforward: The ring and the gold constituting it have different properties, even though the ring is completely constituted by the gold. It follows, from Leibniz's Law, that the two are distinct. To quote Link 1983: 128 directly:[1]

> Our guide in ontological matters has to be language itself, it seems to me. So if we have, for instance, two expressions a and b that refer to entities occupying the same place at the same time but have different sets of predicates applying to them, then the entities referred to are simply not the same. From this it follows that my ring and the gold making up my ring are different entities.

If providing an empirically adequate semantics for nouns requires postulating such a distinction, and if "our guide to ontological matters has to be language itself", then it would appear that we have little choice but to accept that the ring and the gold are in fact different, as are the statue and the clay.

More to the point, since 'ring' is a count noun and 'gold' is a mass noun, Link infers that the denotations of the two nouns must be different *sorts* of things. Very roughly, rings are "things", while gold is "stuff".

More technically, Link distinguishes between ATOMIC INDIVIDUALS and MASS QUANTITIES. Atomic individuals are countable entities serving as the denotations of singular count nouns like 'ring'. Plural nouns like 'rings' then denote PLURALITIES, or mereological sums of atoms. Atoms are related to pluralities, and pluralities to other pluralities, via the INDIVIDUAL PARTHOOD relation, represented as '$\sqsubseteq_i$', and defined as (2), where '$x \sqcup_i y$' is the individual sum of x and y.

(2) $\forall x, y.\ x \sqsubseteq_i y \leftrightarrow x \sqcup_i y = y$

Ordering the atoms via individual parthood results in an atomic join semilatice structure like the following, where arrows represent $\sqsubseteq_i$.

---

[1] See also Link 1998.

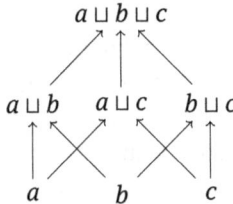

Call this THE COUNT DOMAIN.

Mass quantities, in contrast, are related via MATERIAL PARTHOOD, represented as '$\sqsubseteq_m$', and defined similarly to (2). Ordering mass quantities via material parthood results in a different, but similarly structured, semilattice. Call this THE MASS DOMAIN.

The count and mass domains are related via a homomorphism, a function h, mapping pluralities to mass quantities. If a is an individual (or a sum of individuals), then $h(a)$ is the material stuff it is (or they are) made of. So if a is the notorious ring, then $h(a)$ is the gold constituting it. Since h is a homomorphism, it preserves the material constitution of atomic individuals: if $a \sqsubseteq_i b$, then $h(a) \sqsubseteq_m h(b)$. For example, if Link's ring a is an individual part of his collection of rings b, then the gold constituting Link's ring is a material part of that collection.

If Link's ring and the gold constituting are located in different, though related, domains, then it is hardly surprising that they have different properties. After all, they would be *different entities*. As Link acknowledges, the resulting ontology is anything but parsimonious, from a purely metaphysical perspective.

> [L]et a and b denote two atoms in A. Then there are two more individuals to be called below $a + b$ and $a \oplus b$, $a + b$ is still a singular object in A, the *material fusion of a and b*; $a \oplus b$ is the *individual sum* or *plural object* of a and b. The theory is such that $a + b$ constitutes, but is not identical with, $a \oplus b$. This looks like a wild Platonistic caprice strongly calling for Occam's Razor. Language, however, seems to function that way. Take for a, b two rings recently made out of some old Egyptian gold. Then *the rings*, $a \oplus b$, are new, *the stuff*, $a + b$, is old.

There are two ways of interpreting Link's proposal concerning ontology. On the first, ontologies which attempt to reduce "things" to "stuff", or vice versa, by appeal to, say, basic facts about physics are making some kind of methodological mistake—the guide to reality is language, not science.

On the second, more plausible, interpretation, Link's semantics is not meant to be an account of ultimate reality as such. Rather, it is intended to be of a piece with what Strawson 1959 calls DESCRIPTIVE METAPHYSICS, or what Bach 1986 calls NATURAL LANGUAGE METAPHYSICS, if those are different (see Pelletier 2011). On this approach, Link's semantics exposes the ontological commitments of English itself, or perhaps of competent English speakers. The language functions *as if* "things" and "stuff" are different. Whether or not one sort of entity is "really" reducible to the other, in some metaphysically loaded sense, is a separate matter.

Our primary question here is whether an empirically adequate account of the mass/count distinction really does require distinguishing "things" from "stuff", and thus postulating two corresponding mereological relations, or if instead positing only one sort of entity and corresponding mereological relation is sufficient, as other semantic theories would have it.

This question is meant to be one of what we call NATURAL LANGUAGE MEREOLOGY. We are asking about the mereological commitments of English, or perhaps competent speakers of English, and not about ultimate reality as such. Thus, to return to our original example, there is no pretense that we will definitively solve the *metaphysical* puzzle of the statue and clay.

Nevertheless, our question is important because, as we will see, Link's argument for distinguishing "things" from "stuff" leads to an apparent dilemma. On the one hand, it is easy to generate examples similar to Link's, but which do not involve a mixture of count nouns and mass nouns. For example, consider (3), due originally to Susan Rothstein 2010, Rothstein 2017.

(3) The bricks of the wall are old, but the wall is new.

Here we have a singular count noun ('wall') and a plural noun ('bricks'). We may assume that the bricks completely constitute the wall, and yet they have different properties.

Now consider another example due to Rothstein, which involves only mass nouns.

(4) The gold in the jewelry is old, but the jewelry is new.

Again, we may assume that the gold in the jewelery completely constitutes the jewelry, and yet they have different properties.

The operative general principle, which we dub LINK'S MORAL, appears to be that if something completely constitutes something else but the two have different properties, then they must be different sorts of things, belonging to completely different domains. Link's original argument would not only justify positing distinct domains and corresponding mereological relations for count and mass nouns, but also corresponding distinctions between singular and plural nouns, and also mass nouns like 'gold' and so-called "object mass nouns" like 'jewelry'. In fact, we will argue in §3 that it vindicates postulating *indefinitely many* sorts of entities, not just "things" and "stuff".

This might be reasonably taken to show that we should instead adopt what we call a ONE-DOMAIN ANALYSIS, as opposed a TWO-DOMAIN ANALYSIS such as Link's. One-domain analyses postulate only one sort of entity and one corresponding mereological relation in capturing the characteristic semantic differences between count and mass nouns.

These labels —"one-domain analysis" and "two-domain analysis"— are borrowed from Chierchia 1998 and Rothstein 2010, Rothstein 2017. To quote Rothstein 2017: 91f. directly:

> Link (1983) proposes that homogeneous and non-homogeneous singular predicates have their denotations in different domains, reflecting the fact that they denote different kinds of entities. [Mass nouns] have their denotations in a non-atomic domain, and denote non-atomic Boolean semilattices. [Singular count nouns] have their denotation in an atomic domain and denote sets of atoms ...
>
> Link's model captures the distinction between objects and stuff as an ontological distinction between two different kinds of things. It posits two different semantic domains representing two different kinds of entities related by ... material constitution.

Thus, the intended effect of Link's Moral is that if x completely constitutes y despite x and y having different properties, then x and y are in different (but related) "domains", with different corresponding mereological relations, in precisely this sense.

As we will see in §2, both one-domain analyses and two-domain analyses purport to capture the key semantic differences between count nouns and mass terms. So the primary argument for two-domain analyses cannot be that postulating a sortal distinction between "things" and "stuff" does a better job at that. Nor can it be that two-domain analyses better track a brute metaphysical intuition that "things" and "stuff" are *fundamentally* different, in some metaphysically loaded sense, at least not if "our guide to ontological matters has to be language itself".

Now, one-domain analyses typically assume that the nominal domain consists of "stuff" which may or may not be "packaged" into countable bits—the *atoms*—in context. Hence, on such analyses, Link's ring and the gold constituting it stand in the *same* parthood-relation to each other. They are, to put it bluntly, *the same stuff*. In other words, 'the ring' and 'the gold in the ring' are coextensional, in which case it would appear that they cannot have different properties, after all.

And the same holds of the wall and the bricks constituting it, the jewelry and the gold constituting it, and many more pairs, in fact. In essence, if we reject Link's Moral in favor of a one-domain theory, then we are immediately saddled with explaining the observation which led Link to postulate separate domains in the first place. We call this predicament LINK'S REVENGE.

The goal of §4 is to survey two seemingly plausible responses to Link's Revenge on behalf of one-domain analyses. In particular, we will attempt to fill in the missing details of two suggestions from Rothstein 2010. On both suggestions, what examples like (1), (3), and (4) reveal is not that natural language sortally distinguishes "things" from "stuff", but rather that there is some kind of *intensionality* associated with the accompanying noun phrases in those examples, thus explaining why there is a failure of substitutivity. And though both explications appear initially very plausible, we will ultimately see that neither is acceptable without significant challenges.

Ultimately, our goal in this paper is not to adjudicate between one-domain and two-domain analyses, or between the two possibilities sketched in §4. Rather, it is to raise the apparent dilemma already sketched, and to survey some possible resolutions, assuming that Link's Moral is rejected. The challenges facing the two suggestions sketched here are only intended to illustrate the difficulty of adequately addressing Link's Revenge.

Nevertheless, finding a potential resolution to Link's Revenge is important, for two reasons. First, with the exception of Link, the predominant theories of the count/mass distinction within linguistic semantics are one-domain analyses. Examples include Krifka 1989, Gillon 1992, Chierchia 1998, Chierchia 2010, and Rothstein 2010, Rothstein 2017. Thus, *some* plausible response to Link's Revenge is in order. Secondly, the question of how many mereological relations are needed to account for the count/mass distinction is, we take it, of primary importance to natural language mereology. Thus, finding a plausible resolution to Link's Revenge would go some way towards giving a definitive answer to that question, namely 'one'.

The rest of the paper is laid out as follows. In §2, we sketch the characteristic differences between count and mass nouns, and show how both one-domain and two-domain analyses purport to account for those differences. In §3, we look closer at Link's original argument for a two-domain analysis, and show how it seemingly leads to an explosion of nominal domains and corresponding mereological relations, thus leading to the adoption of a one-domain analysis. We then sketch two particularly natural responses to Link's Revenge, along with their difficulties, in §4. We conclude the paper in §5, where we summarize the paper and suggest some alternative ways out of Link's Revenge not considered here.

2 Single-Domain and Double-Domain Theories

The count/mass distinction is typically presented as a series of characteristic contrasts.[2] For example, whereas count nouns can occur with cardinality modifiers such as 'two', mass nouns (usually) cannot.

(5) Mary bought two {rings/??golds}.

Similarly, whereas count nouns cannot usually occur in the singular with classifiers like 'piece of' or 'kilo of', mass nouns usually can.

(6) Mary bought three pieces of {??ring/gold}.

Also, while count nouns can occur with distributive determiners like 'every', mass nouns (usually) cannot.

(7) Mary bought every {ring/??gold}.

Similarly, whereas mass nouns are typically acceptable with modifiers like 'little' or 'much', count nouns are instead typically acceptable only with modifiers like 'several' or 'many'.

(8) a. Mary bought several {rings/??gold(s)}
 b. Mary bought little {??ring/gold}.

Finally, and relatedly, whereas count nouns are typically acceptable with reciprocal distributive predicates like 'stacked on top of each other', mass nouns typically are not.

(9) The {rings were/??gold was} stacked on top of each other.

Both single domain and double domain theories purport to explain contrasts like (5)–(9), but they do so in different ways. Link explains these differences in terms of the semantic properties of the nouns involved. Specifically, whereas mass nouns are CUMULATIVE, singular count nouns are not.

(10) **Cumulativity:** $\forall P. \forall x, y.\ P(x) \wedge P(y) \rightarrow P(x \sqcup y)$

---

[2] See Rothstein 2017.

For example, if x and y are both quantities of water, then the sum of x and y is also a quantity of water. In contrast, if x is a chair y is a chair, the sum of x and y is not a chair.

Furthermore, whereas some mass nouns are (apparently) DIVISIVE, count nouns are not.

(11) **Divisiveness**: $\forall P. \forall x. \exists y, z.\ P(x) \rightarrow [P(y) \wedge P(z) \wedge y \sqsubseteq x \wedge z \sqsubseteq x \wedge \neg y \circ z]$

In other words, entities satisfying divisive predicates can always be "split" into smaller, non-overlapping parts which also satisfy that predicate. This implies that not only do the denotations of count nouns and mass nouns belong to different domains, those domains are structurally different: whereas count nouns form an *atomic* semilattice structure, mass nouns instead form an *atomless*, or GUNKY, semilattice structure.[3] Assuming cardinality modifiers count atoms, and that distributive expressions "distribute" down to atoms, it's little wonder we see contrasts like (5)–(9).

In contrast, Chierchia 1998's highly influential single-domain analysis explains contrasts like (5)–(9) through the nature of the proposed denotations for count and mass nouns. Specifically, count and mass nouns both denote sets of atoms forming semilattice structures like the following:

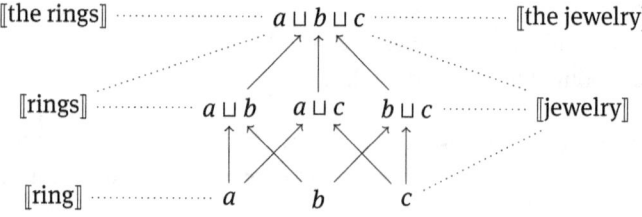

Thus, singular count nouns such as 'ring' denote atoms, plural nouns such as 'rings' denote proper sums of atoms, and mass nouns such as 'jewelry' denote the closure of the atoms under sum-formation. Moreover, as on Link's analysis, 'the' is a maximality operator, and so 'the rings' and 'the jewelry' will denote the same maximal sum, namely the rings, assuming (for simplicity) that they are the only pieces of jewelry. Incidentally, this will also be the denotation of 'the gold in the rings', given that the rings are completely constituted by gold.

---

[3] We are following the exegesis of Rothstein 2017 and others (like Landman 2012) here. In fact, Link 1983: 131 appears to be neutral on the atomicity of mass nouns: "In addition to the domain of individuals, E, there is a set D which is endowed with a join operator "⊔" making D into a complete, but *not necessarily atomic*, join-semilattice" (emphasis added).

As a result, only one mereological relation is needed on Chierchia's analysis. Moreover, the explanation of contrasts like (5)–(9) falls out from the nature of the denotations assumed. Specifically, mass nouns literally neutralize the singular/-plural contrast, in virtue of denoting *both* atoms and pluralities. As such, they cannot be pluralized, since they are already plural. Further, because their denotations do not include just atoms, they cannot be counted or occur acceptably with distributive expressions. Lastly, they require classifiers like 'piece of' or 'quantity of' to be counted, as the latter partition mass entities into countable atoms (see §4.1.2).

But why think that count and mass nouns should be analyzed homogeneously in this manner? Chierchia 1998: 348 explains:

> The main argument in favor of the present view of mass nouns is one of economy. The structure revealed by plurals suffices to account for the properties of mass nouns. Why hypothesize two different domains when all that is needed to account for mass nouns can be found in the familiar atomic domain of count objects? The intuition that a mass noun like *furniture* means something subtly but deeply different from a count counterpart like *pieces of furniture* is an optical illusion, a gestalt effect due to the different groupings of their denotations.

In other words, if it is possible to explain the characteristic contrasts between count and mass nouns without positing distinct domains of "things" and "stuff", then one ought to do so, all else being equal.

Hence, the argument for double-domain theories such as Link's cannot simply be a metaphysical hunch that "things" and "stuff" are by their very nature different sorts of things, and that the count/mass distinction is tracking this difference.[4] At least that cannot be the argument if we are engaged in *natural language mereology*, in which case positing distinct domains would be legitimate only if providing an empirically adequate account of that distinction required doing so. Yet this is precisely what *both* single and double-domain analyses purport to do.

To be clear, Link's argument for two domains does *not* rely on antecedent metaphysical intuitions. Rather, it relies only on the observation that a ring and the gold constituting it can have different properties, in which case it would appear that single-domain theories such as Chierchia's are empirically inadequate. After all, if 'the ring' and 'the gold in the ring' both denote *the same maximal sum*, then shouldn't they have the same properties?

---

[4] See Pelletier 1975 for relevant discussion here.

3 Link's Moral

Let's return to Link's argument for double-domains. Assuming (1) is true, the referent of 'this ring' and the referent of 'the gold in the ring' must have different properties.

(1) This ring is new, but the gold in the ring is old.

Specifically, on Link's semantics, the denotations of 'the ring' and 'the gold in the ring' are given in (12a,b), where 'σ' is Link's maximality-operation, and A and Q range over atomic entities and mass quantities, respectively.

(12) a. 〚the ring〛 = $\sigma x \in A : \mathrm{ring}(x)$
 b. 〚the gold in the ring〛 = $\sigma x \in Q : \mathrm{gold}(x) \wedge x \sqsubseteq_m \mathrm{the\text{-}ring}$

Thus, 'the ring' refers to some unique atomic ring, while 'the gold in the ring' refers to the maximal quantity of gold which stands in the material-part of relation to that ring. Assuming that the ring is completely constituted by the gold, both are the same "stuff". So, if 'the ring' and 'the gold in the ring' both referenced that "stuff", their referents would be identical, thus violating Leibniz's Law. Upholding the latter, we conclude that atoms and mass quantities constitute different domains altogether.

Link's argument seemingly relies on a more general principle, which we call *Link's Moral*:

(13) If x and y have different properties, yet one completely constitutes the other, then x and y are different sorts of things. That is, x and y belong to different domains.

While this would certainly justify Link's double-domain analysis, it would also appear to justify even more domains and corresponding mereological relations. For example, consider again (3).

(3) The bricks of the wall are old, but the wall is new.

Here we have two count nouns: 'the bricks of the wall' and 'the wall'. On Link's analysis, 'the wall' denotes a specific atom, while 'the bricks of the wall' denotes the maximal sum of atomic bricks which stand in the individual-part of relation to the wall ('*' is Link's pluralization operator, where '*P' denotes the closure of the P-atoms under sum-formation).

(14) a. ⟦the wall⟧ = $\sigma x \in A$: wall(x)
   b. ⟦the bricks of the wall⟧ = $\sigma x \in A$: *brick(x) ∧ $x \sqsubseteq_i$ the-wall

Now, suppose that the wall is completely constituted by the bricks. Since these have different properties, they must be different sorts of things, and thus constitute different domains, by Link's Moral. The trouble, of course, is that the bricks and the wall are supposed to located in the *same* domain, namely that of atomic entities.

Now reconsider (4).

(4) The gold in the jewelry is old, but the jewelry is new.

Here we have two mass nouns, 'the gold in the jewelry' and 'the jewelry'. On Link's analysis, these too can be coextensional.

(15) a. ⟦the jewelry⟧ = $\sigma x \in Q$: jewelry(x)
   b. ⟦the gold in the jewelry⟧ = $\sigma x \in Q$: gold(x) ∧ $x \sqsubseteq_m$ the-jewelry

In particular, assuming the jewelry is completely constituted by the gold in it, we are led to conclude that the jewelry and the gold form different domains, by Link's Moral. The trouble, once again, is that the denotations of mass nouns are supposed to be located in the *same* domain, namely that of mass quantities.

In the next section, we consider certain responses available to Link in light of these two examples. To anticipate, it might be reasonably thought that Link 1984, Link 1998's theory of *groups* could be used to explain how the wall and the bricks, and the jewelry and the gold, do in fact constitute separate domains. Thus, groups represent a seemingly plausible way of retaining Link's Moral in light of Rothstein's examples.

Despite this possibility, maintaining Link's Moral in full generality would appear to *massively* overgenerate domains and corresponding mereological relations. Consider (16), due to Oliver & Smiley 2001.

(16) a. Russell and Whitehead were logicians.
   b. The molecules of Russell and Whitehead were logicians.

'Be logicians' is a DISTRIBUTIVE PREDICATE, meaning that it applies to all parts of a given plurality. In Link's semantics, 'Russell and Whitehead' denotes the sum of Russell and Whitehead, while 'the molecules of Russell and Whitehead' denote atoms which are individual parts of the aforementioned sum. Since parthood is transitive, Link's semantics would thus appear to predict that (16a) entails (16b), contrary to fact. Intuitively, the problem is that because mereological sums do not have unique decompositions, we cannot semantically distinguish the plurality of Russell and Whitehead from its proper parts.[5]

According to Oliver and Smiley,[6] examples like this reveal that mereological analyses of plurals, such as Link's, are fundamentally misguided. However, this objection neglects an important property of atoms, as they appear in various semantic treatments, namely that they are *property-relative*. In other words, when doing semantics, it never makes sense to talk about atoms *full-stop*, but only atoms of a certain kind, atoms of a given property P.

Here, for example, is the definition of ATOMICITY in (17a) from Krifka 1989, along with the accompanying definitions of ATOMIC PREDICATE in (17b) and ATOMIC PARTHOOD in (17c), where S restricts the P-atoms to a certain sort.

(17) a. $\forall x. \forall P.\ \text{ATOM}_S(x, P) \leftrightarrow P(x) \wedge \neg \exists y.\ y \sqsubset_S x \wedge P(y)$
   b. $\forall P.\ \text{ATOMIC}_S(P) \leftrightarrow \forall x.\ P(x) \rightarrow \text{ATOM}_S(x, P)$
   c. $\forall x, y.\ x \sqsubseteq_{At,S} y \leftrightarrow x \sqsubseteq_S y \wedge \text{ATOM}_S(x, S)$

Thus, x is an atomic-P relative to sort S just in case it has no proper parts which are also Ps; P is an atomic predicate relative to S just in case every member of it's extension is a P-atom in S; and x is an atomic part of y relative S just in case x is an S-part of y and x is atomic in S.

In a Link-style semantics, singular count nouns are atomic predicates in this sense, and distributive predicates apply to atomic parts as defined in (17c). In both (16a,b), the relevant P-atoms will be individual logicians, including both Russell and Whitehead. Relative to the property of being a molecule of Russell or Whitehead, on the other hand, the relevant P-atoms will be the molecules belonging to either Russell or Whitehead. Hence, neither Russell nor Whitehead are atoms relative to this property, and none of the atomic-molecules have the property of being logicians. In other words, it simply does not follow on Link's analysis that (16a) entails (16b).

Nevertheless, (16) does seemingly represent a problem for *Link's Moral*. Again, Russell and Whitehead are completely constituted by their molecules, presumably. In other words, the correct denotations for 'Russell and Whitehead' and 'the molecules of Russell and Whitehead' are presumably those in (18).

(18) a. $[\![\text{Russell and Whitehead}]\!] = r \sqcup w$
   b. $[\![\text{the molecules of Russell and Whitehead}]\!] = \sigma x \in A :\ {}^*\texttt{molecule}(x)$
      $\wedge\ x \sqsubseteq_i r \sqcup w$

---

[5] For similar arguments, see Rayo 2002 and McKay 2006.
[6] See also Oliver & Smiley 2013.

There are two issues. First, because 'molecule' is a count noun, 'the molecules of Russell and Whitehead' must be located in the domain of atomic entities. Likewise with Russell and Whitehead, of course. Hence, the mereological relation holding between the molecules and the sum of Russell and Whitehead should be that of individual-part of. On the other hand, any physical mass quantities standing in the material-part of relation to Russell and Whitehead – their blood, their hair, etc – will also be constituted by those molecules. And these too can have different properties.

(19) The molecules of Russell's hair are old, but Russell's hair is (comparatively) new.

This should not be surprising on Link's analysis, given that molecules belong to the count domain, and Russell's hair to the mass domain. What is surprising, however, is that the former *constitutes* the latter.

(20) a. ⟦Russell's hair⟧ = $\sigma x \in Q$: hair-belonging-to-Russell(x)
 b. ⟦the molecules of Russell's hair⟧ = $\sigma x \in A$: *molecule(x) $\wedge$
    $x \sqsubseteq_?$ Russell's-hair

But then it is hard to see what this constitution-relation could be. It cannot be that of individual-part of, since Russell's hair would then be in the atomic domain, and thus countable. Conversely, it cannot be that of material-part of, since the molecules constituting it would be mass quantities, and thus non-countable.

Perhaps one can set this aside. A more pressing issue is that since the molecules and the sum consisting of Russell and Whitehead have different properties, they must constitute completely different domains by Link's Moral. We can reproduce this argument for practically *any* atomic entity or sum of atomic entities. That is, for any atom or sum of atoms referenced by a definite noun phrase like 'Mary', 'that chair', or 'these people', the molecules constituting them will have different properties than the things referenced. Thus, if Link's Moral held in full generality, we would have *indefinitely many* domains and corresponding mereological relations, not just the two Link hypothesizes.

To be clear, the problem isn't merely that this would require positing more domains than just the two Link originally hypothesizes. Semanticists regularly posit a variety of different sorts of entities, including, for example, events, kinds, degrees, numbers, times, and locations. A seemingly plausible justification is that natural language regularly makes category distinctions corresponding to these, as witnessed in the distinction between various kinds of predicates, nouns and verbs, measure phrases, tenses, locatives, ... And something similar might be said with respect to the count/mass distinction, perhaps. But nothing remotely similar can be said for the sorts of additional sortal distinctions that would be required to maintain Link's Moral in light of examples like (16) and (19).

In short, while Link's Moral would certainly vindicate his two domain analysis, it also appears to massively overgenerate domains and corresponding mereological relations. Thus, unless some principled reason can be given for restricting Link's Moral to just the count and mass domains, or unless some other general background principle can be found which would have the same effect, the right response would appear to be rejecting Link's Moral. In that case, however, we would have no obvious reason for adopting a double-domain analysis, opting instead for a single-domain analysis.

But now we have come full circle. Again, if all there is, at least with respect to the nominal domain, is "stuff" which is "packaged" in context into countable "things", as one-domain theories suggest, then how can it be that that Link's ring and the gold constituting it can have different properties? Likewise for the wall and the bricks constituting it, the jewelry and the gold constituting it, Russell and Whitehead and the molecules constituting them, etc.

If a single domain and corresponding mereological relation is all that is needed to adequately model the count/mass distinction, then clearly some response to Link's Revenge is in order. But what?

4 Two Avenues of Response

From here on, we assume that some kind of one-domain analysis is correct. The question, then, is how to make sense of examples like (1), (3), and (4) if the pairs of definite noun phrases in those examples refer to the same "stuff", and so are coextensional.

(1) This ring is new, but the gold in the ring is old.
(3) The bricks of the wall are old, but the wall is new.
(4) The gold in the jewelry is old, but the jewelry is new.

Commenting on these examples in a footnote, Rothstein 2010: 365, fn. 10 suggests a possible way out:

> One possible solution is to treat *wall* analogously to *deck* [as in *deck of cards*], justifying this by the plausible assumption that walls are greater than the sums of bricks that compose them... [This] is a version of the problem that occurs in the mass domain too... *This jewelry is new, but the gold it is made of is old*. The mass entity in jewelry cannot be equated with

the mass entity in gold since they have different properties, even though they are apparently identical. This implies that generally 'artefact' predicates like *jewelry* involve a packaging or perspective function as part of their lexical meaning, so that [*gold*] and [*jewelry*] can be identified as the same spatiotemporal entity but presented under different perspectives or guises and with different properties.

Because Rothstein's comments here are only meant to be suggestive, she does not elaborate on how exactly this suggestion should be carried out, or indeed how might be used in response to Link's Revenge. Thus, our task will be to fill in these missing details.

We will do so by appealing to two theories which appear particularly well suited to implement Rothstein's proposal. On both, the truth of examples like (1), (3), and (4) is to be explained through some kind of *intensionality* associated with the noun phrases involved in those examples: 'the ring' and 'the gold in the ring', 'the wall' and 'the bricks of the wall', and 'the jewelry' and 'the gold in the jewelry'. Where they differ is the *source* of that intensionality, corresponding to different views of the semantic function of those noun phrases.

On the first suggestion, the noun phrases are *referential* expressions (type e) referring to *groups* in the sense of Link 1984, Link 1998. Just as a deck of cards can consist completely of cards but have its own identity in virtue of representing the cards as a unified whole, likewise the wall and the jewelry have their own identity beyond the the bricks or the gold in virtue of representing them as unified wholes. Assuming with Link that groups are intensional, and so cannot be identified by their members, it would be hardly surprising if the wall qua group of bricks and the bricks qua sums of atoms can have different properties despite consisting of the same atomic constituents, namely bricks. Likewise for Link's ring, the jewelry, and the gold constituting them. This is this THE GROUP-FORMING STRATEGY.

On the second suggestion, inspired by Landman 1989b, the noun phrases involved are not referential expressions. Rather, they are intensional generalized quantifiers (of hyperintensional type $\langle\langle e, p\rangle, p\rangle$, where p is the type of propositions), expressing properties of restricted properties. These restricted properties can be thought of as representing the bricks, the wall, etc. through different aspectual "guises". They are given either explicitly through aspectual phrases like 'as a group of bricks' and 'qua sums of atoms', or else contextually when no overt aspectual phrases are available. This guarantees that the different noun phrases involved will express different second-order properties in different contexts, thus explaining how (1), (3), and (4) can be true. This is THE ASPECT-RESTRICTION STRATEGY.

On both strategies, then, the moral of Link's original example (1), and others like it, is that the meanings of definite noun phrases like 'the ring', 'the gold in the ring', etc. cannot be identified with the "stuff" they denote.

We will consider both strategies in what follows, while pointing out their apparent challenges. Though we will conclude these challenges are substantial, this should not be taken as an indictment on Rothstein's suggestion, as there could be alternatives to those considered here which do not face those problems. Rather, the challenges considered here are only intended to illustrate how difficult answering Link's Revenge really is.

4.1 The Group-Forming Strategy

Consider (21), due originally to Link 1984, Link 1998.

(21) The red cards and the blue cards are shuffled.

As Link observes, (21) is ambiguous between COLLECTIVE INTERPRETATION according to which the red cards and the blue cards are shuffled together, and a DISTRIBUTIVE INTERPRETATION on which each deck of cards is individually shuffled. Now, given Link 1983's original analysis, 'the red cards' and 'the blue cards' should both denote maximal sums of atomic cards.

(22) a. ⟦the red cards⟧ = $\sigma x \in A : \text{red}(x) \wedge {}^*\text{card}(x)$]
  b. ⟦the blue cards⟧ = $\sigma x \in A : \text{blue}(x) \wedge {}^*\text{card}(x)$]
  c. ⟦the red cards and the blue cards⟧ = [$\sigma x \in A : \text{red}(x) \wedge {}^*\text{card}(x)$] ⊔ [$\sigma x \in A : \text{blue}(x) \wedge {}^*\text{card}(x)$]

Since 'and' denotes the join-operation, 'the red cards and the blue cards' will thus denote the maximal sum consisting of the red cards and the blue cards. Moreover, since cumulative predicates apply to sums, while distributive predicates apply to all parts of sums, including their atomic parts, the prediction is that the collective interpretation should be true if the maximal sum in (22c) has the property of being shuffled, while the distributive interpretation should be true if each atomic card within that maximal sum has the same property, i.e. each individual card is *itself* shuffled. Clearly, that is not the intended interpretation, and, it seems, we cannot get the one on which the red cards are shuffled and the blue cards are shuffled separately.

To overcome this difficulty, Link develops a theory of *groups*. Groups are like pluralities (sums) in that they are inherently plural, (typically) having more than one atomic constituent.[7] Linguistically, however, they are canonically referenced by different sorts of phrases. Whereas pluralities are prototypically referenced by conjunctive noun phrases like 'John and Mary', groups are prototypically referenced by GROUP NOUNS like 'deck' in 'deck of cards'. Both can be referenced, however, by definite plurals such as 'the red cards', as evidenced by examples like (21).

However, the crucial difference between pluralities and groups, on Link's analysis, is that whereas pluralities are inherently extensional, and so are identified by their atomic parts, groups are inherently intensional, and hence cannot be so identified. Consider (23), due to Landman 1989a.

(23) a. The judges are on strike.
     b. The hangmen are on strike.

Suppose we happen to live in a small town where the judges moonlight as the hangmen. As Landman observes, a prisoner sentenced to die would be ill-advised to infer (23b) from (23a) in such a situation. After all, it could be that the judges *qua* judges are on strike, while the judges *qua* hangmen are not.

Link models this sort of intensionality by introducing a distinction between PURE ATOMS and IMPURE ATOMS. Pure atoms are ordinary atomic entities like John, Mary, that table, etc. They are the sorts of things which when summed together form pluralities. Impure atoms, on the other hand, are groups formed from pluralities in the following manner. There are two operations, ↑ (GROUP-FORMATION) mapping pluralities to impure atoms, and a converse operation ↓ (MEMBER-SPECIFICATION) mapping impure atoms to the pluralities from which they are formed. ↑ is one-to-one but not onto, whereas ↓ is onto but not one-to-one. Thus, whereas every plurality forms a group, it needn't follow that every group corresponds to a unique plurality. In the case of (23), for instance, the judges and the hangmen form different groups, despite being formed from the same plurality.

Crucially, impure atoms are sortally distinct from pure atoms, i.e. they form a separate domain. The picture is roughly as follows, where dots represent group-formation.

---

[7] We say "typically" here because it is in principle possible to have a group having only one member, thanks to the idempotency of ⊔ (Krifka 1989), i.e. $x \sqcup x = x$ for any x.

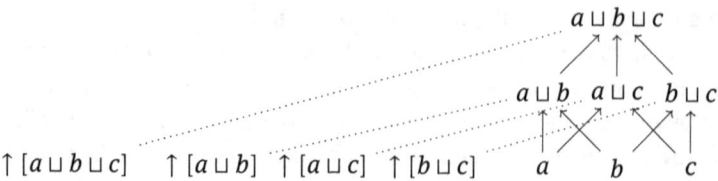

We begin with some pure atoms, a, b, c, and form pluralities from them through the sum-operation. We then form groups from these pluralities through the ↑-operation, and these now serve as atomic entities within a new semilattice structure consisting only of impure atoms. These can then be summed to form pluralities of groups, and the process can be iterated to form groups of groups, groups of groups of groups, etc.[8]

With groups in hand, it is easy to see how (21) can have both a collective and distributive interpretation. On the cumulative interpretation, 'be shuffled' applies to the plurality consisting of the two maximal sums, and thus the totality of red and blue cards. The distributive interpretation arises instead if 'be shuffled' applies to the two decks as groups:

(24) a. $\text{shuffled}([\sigma x \in A : \text{red}(x) \wedge {}^*\text{card}(x)] \sqcup [\sigma x \in A : \text{blue}(x) \wedge {}^*\text{card}(x)])$ (collective)
b. $\text{shuffled}(\uparrow [\sigma x \in A : \text{red}(x) \wedge {}^*\text{card}(x)] \sqcup \uparrow [\sigma x \in A : \text{blue}(x) \wedge {}^*\text{card}(x)])$ (distributive)

Since groups are atoms, the distributive interpretation will be true just in case each deck is shuffled, as desired.

4.1.1 Extending Groups

The question here is whether Link's theory of groups can be extended so as to capture the truth of (3) and (4), and without multiplying domains unnecessarily.

(3) The bricks of the wall are old, but the wall is new.
(4) The gold in the jewelry is old, but the jewelry is new.

---

[8] See Landman 1989a. We'd want the process to be cumulative, in the sense that, for example, there is a group consisting of some groups plus some individuals.

The basic idea is simple enough: if we allow that the wall is a group whose members are the bricks constituting it, and that the jewelry is similarly a group whose members are the rings, bracelets, etc. constituting it, then since groups have different properties from the pluralities forming them, it's hardly surprising that (3) and (4) can be true. What's more, because groups are presumably needed to model semantic phenomena like (21) anyway, even for one-domain analyses, this postulation of separate domains is independently justified. This is *the Group-Forming Strategy*.

Let's begin with (3). Again, the basic idea is to analyze 'the wall' as referencing a group formed from a sum of pure atomic bricks, as suggested in (25), where 'IA' is the domain of impure atoms, or groups, at a given level.[9]

(25) a. ⟦the wall⟧ = $\sigma x \in IA$: wall(x)
b. ⟦the bricks of the wall⟧ = $\sigma x \in A$: *brick(x) ∧ $x \sqsubseteq$ ↓ [the-wall]

Thus, 'the wall' will refer to the unique impure atom whose only constituents are pure bricks, while 'the bricks of the wall' will refer to the maximal plurality of pure bricks. The mereological relation holding between them is that of *member-specification*: the pure bricks are group-members of the wall. Nevertheless, they are distinct, just as the deck of red cards is distinct from the maximal sum of red cards. And just as decks of cards and the pluralities of cards constituting them can have different properties, e.g. (26) can only be true of the plurality and not the group,

(26) The red cards are stacked on top of each other.

likewise the wall qua group of bricks may have properties different from the bricks qua components of the wall, as revealed by e.g. (3).

Something similar can be said about (4), it seems. So-called OBJECT MASS NOUNS such as 'jewelry', 'silverware', and 'furniture' serve as major motivations for one-domain analyses.[10] That's because, unlike e.g. 'gold' or 'water', 'jewelry', 'silverware', and 'furniture' denote collections of apparently *countable* entities, e.g. rings and bracelets, forks and knives, and chairs and sofas.[11] Hence, to recall the quote from Chierchia 1998 in §2: "The intuition that a mass noun like *furniture* means something subtly but deeply different from a count counterpart like *pieces of furniture* is an optical illusion, a gestalt effect due to the different groupings of

---

[9] The fact that groups iterate in way resembling sets means that we are flirting with (Russell's) paradox. See Snyder & Shapiro (ms.) for details plus a possible solution.
[10] The label 'object mass nouns' comes from Rothstein 2010, Rothstein 2017.
[11] See Barner & Snedeker 2005.

their denotations." According to Chierchia, this suggests that we ought view the denotations of mass nouns and count nouns alike: both denote atoms, though the atoms of non-object mass nouns such as 'gold' and 'water' are typically "vague", and thus far less easily identifiable than those of e.g. 'furniture'.

Suppose Chierchia is right. Then the analogy between the bricks of the wall and the gold in the jewelry becomes apparent: just as we can view the wall as an impure atom whose sub-constituents are pure atomic bricks, we can likewise view the jewelry as an impure atom whose pure atomic sub-constituents are pieces of jewelry such as rings and bracelets. And as before, these may have different properties, even if the jewelry is completely constituted by the rings and the bracelets:

(27) The rings and the bracelets of the jewelry are stacked on top of each other, but the jewelry itself is not.

This suggests the analysis in (28).[12]

(28) a. ⟦the jewelry⟧ = $\sigma x \in IA$: jewelry(x)
b. ⟦the rings and the bracelets of the jewelry⟧ = $\sigma x \in A$: $\exists y, z \in A$: *ring(y) ∧ *bracelet(z) ∧ $x = y \sqcup z \wedge x \sqsubseteq \downarrow$ [the-jewelry]

Thus, as with the wall, the jewelry may be viewed as a group whose group-members are rings and bracelets, and so the relevant mereological relation holding between them will be that of group-membership.

Now consider the remaining examples considered above.

(1) This ring is new, but the gold in the ring is old.
(4) The gold in the jewelry is old, but the jewelry is new.
(19) The molecules of Russell's hair are old, but Russell's hair is (comparatively) new.

As before, the idea would be to view the ring and the jewelry as a group whose group-members are "vague" (pure) gold-atoms, and likewise to view Russell's hair as a group consisting of "vague" (pure) molecule-atoms. Hence, the only semantically significant difference between examples like (3) and (27) and those like (1), (4), and (19) would be that the pure atoms in latter cases are not as easily identifiable as those in the former cases.

Finally, consider (29a), modeled on an example from Pelletier 2011.

---

[12] For a similar analysis of cumulative conjunction, see Krifka 1990.

(29) a. The snow is new, but the water constituting the snow is old, and the hydrogen and oxygen molecules constituting the water are even older.
b. The art installation is new, though the walls of the art installation are old, and the bricks of the walls are even older.
c. The art exhibit is new, though the jewelry of the art exhibit is old, and the gold in the jewelry is even older.

As Pelletier rightly notes, (29a) is puzzling for Link's original analysis, since it would appear to require that the hydrogen and oxygen molecules stand in a mereological relation to the water, and that the water to stands in the *same* mereological relation the snow, despite 'molecule' being count, and 'water' and 'snow' being mass. And as (29b,c) reveal, similar examples can be reproduced for the other sorts of nouns considered here.

On the present suggestion, these examples might be seen as witness to *group iteration*. As mentioned above, groups iterate in such a way that we can have groups of groups, groups of groups of groups, etc. Thus, for (29a), we might view the pure atoms as the "vague" hydrogen and oxygen molecules. These constitute the first "level" of impure atoms, namely the "vague" water-atoms. And these in turn form the second "level" of impure atoms, namely the "vague" snow atoms. And something similar can be said about (29b,c), of course. In short, it would appear that adopting a one-domain analysis along with Link's theory of groups is sufficient to explain how all of the variations on Link's original example can be true.

4.1.2 Challenges for the Group-Forming Strategy

Despite its apparent advantages, the Group-Forming Strategy faces two significant challenges. The first is raised by Rothstein 2010: 365.

> One possible solution is to treat *wall* analogously to *deck*, justifying this by the plausible assumption that walls are greater than the sums of bricks that compose them. However, against this is the intuition that while *deck* is defined as a set of cards, *wall* denotes a set of entities that are objects in their own right, rather than being an expression that classifies bricks...

Expressions such as 'deck' (as in 'deck of cards') are called "group nouns" because their function, intuitively, is to combine with a noun to denote groups of things having that property. Group nouns are a subclass of English CLASSIFIERS, or expressions whose function is to combine with a noun to produce a countable or measurable predicate. Rothstein 2017 organizes these into two kinds. The first are COUNTING CLASSIFIERS:

Category	Example Classifier	Example Classifier Phrase
Unit Classifier	'unit', 'item'	'item of clothing'
Apportioning Classifier	'grain', 'quantity'	'grain of rice'
Container Classifier	'box', 'cup'	'box of books'
Group Classifier	'group', 'deck'	'deck of cards'
Arrangement Classifier	'row', 'pile'	'row of cabbage'

Opposed to counting classifiers are MEASURING CLASSIFIERS.

Category	Example Classifier	Example Classifier Phrase
Lexical Measure	'kilo', 'liter'	'kilo of cocaine'
Container Measure	'bottle', 'glass'	'glass of water'
'-ful' Measure	'pocketful', 'busful'	'pocketful of sand'
'-worth' Measure	'dollarsworth', 'poundsworth'	'ten dollarsworth of nickels'

Thus, Rothstein's argument can be summarized as follows. If 'wall' were analogous to 'deck', it would be a classifier, specifically a counting classifier. But whereas nothing is a deck *outright*, but only a deck *of* something, something can be a wall outright, independent of whether it consists of bricks, cardboard, etc. In other words, there is hardly any plausibility to the claim that 'wall' is a classifier, and so the analogy between 'deck of cards' as denoting a group whose group-members are (pure) atomic-cards, and 'wall of bricks' as denoting a group whose (pure) atomic-bricks, collapses. And the same complaint could be leveled at the suggestion that 'the ring', 'the jewelry', and 'Russell's hair' are group-referring expressions.

One may reasonably question the apparently operative presumption here, namely that 'the wall', 'the ring', etc. are plausibly understood as a group-denoting term only if they are appropriately analogous to group classifiers. As we have seen, Link 1984, Link 1998 argues that postulating groups is necessary to account for ambiguities like (21).

(21) The red cards and the blue cards are shuffled.

Notice that group classifiers allow for the same ambiguity.

(30) The deck of red cards and the deck of blue cards are shuffled.

On Link's account, presumably, the distributive interpretation arises if the definite plurals here reference groups, while the collective interpretation arises through applying member-specification (↓) to those groups, thus returning the totality of cards in both decks. In other words, group classifiers are not required to generate these sorts of distributive/collective ambiguities. Moreover, insofar as groups are needed to explain such ambiguities, it would seem that 'the red cards' *is* appropriately analogous to 'the deck of red cards'. If so, then why not think that 'the wall', 'the ring', etc. are too?

Setting this aside, there appears to be a more direct, but related, challenge to the Group-Forming Strategy. Clearly, we can use group classifiers to talk about the bricks as a group; witness 'that group of bricks'. Presumably, this refers to the same group which would result through applying group-formation to 'the bricks of the wall'. But now consider (31).

(31) That group of bricks is old, but the wall is new.

In other words, we can reformulate the same kind of problematic example using the hypothesized referent of 'the bricks of the wall'. But since 'the wall' is, by hypothesis, coreferential with 'that group of bricks', it appears that appealing to groups will not help explain how (3) can be true.

Worse yet, we can easily produce similar examples for the other problem cases. Consider (32), for instance.

(32) The quantity of gold in the jewelry is old, but the jewelry is new.

Arguably, apportioning classifiers like 'quantity' serve a semantic function similar to group classifiers,[13] but with mass nouns. In other words, they partition uncountable "stuff" into countable, unified portions, much like how 'group of bricks' partitions pluralities of bricks into a countable, unified whole. Hence, 'the quantity of gold in the jewelry' plausibly references the same group of (pure) gold-atoms hypothesized as the referent of 'the jewelry' in (4). If so, then once again appealing to groups will not help explain how the gold in the jewelry and the jewelry itself can have different properties.

A different kind of challenge for the Group-Forming Strategy concerns the source of intensionality supposed for the examples discussed. Consider again Landman 1989a's (23).

(23a) The judges are on strike.

---

13 See Scontras 2014.

(23b) The hangmen are on strike.

To repeat, the proposed explanation for why (23a) does not entail (23b) is that because groups are intensional, and so cannot be identified by atomic parts of the sums from which they are formed, 'the judges' and 'the hangmen' can reference different groups despite extensionally consisting of the same individuals. Hence the failure of substitutivity in (23a,b).

As Landman notes, the trouble is that exactly similar examples can be produced using definite *singular* noun phrases like 'the judge' and 'the hangmen'. Thus, consider Landman's (33).

(33) a. The judge is on strike.
 b. The hangman is on strike.

As before, if John happens to moonlight as both judge and hangman in our small town, one would be ill-advised to infer (33b) from (33a): in his capacity as judge, John may have good reasons to be on strike, even if he feels compelled to carry out his duties as a hangman. Yet the claim that 'the judge' and 'the hangmen' reference a *group* seems far less intuitive.

Thus, Landman reasonably concludes that the sort of intensionality witnessed in both sorts of examples is better located in the meaning of noun phrase itself, not in the *sorts* of things referenced, i.e. groups. Thus, we are led to the same problematic conclusion: identifying the referents of the various definite plurals above with groups will not explain how the wall and the bricks constituting it, the jewelry and the gold constituting it, etc. can have different properties.

All of this suggests that a different kind of explanation is in order. In the next subsection, we will consider a solution modeled from Landman's own theory of examples like (23) and (33). On that account, the intensionality witnessed is not a function of the sorts of things referenced, but rather of the meanings of the definite noun phrases involved. In effect, substitutivity fails because we are restricting the properties expressed to different aspectual "guises". As a result, the meanings expressed are different, even if the things referenced are extensionally the same. Hence we call it "the Aspect-Restriction Strategy".

4.2 The Aspect-Restriction Strategy

Unlike the Group-Forming Strategy, the Aspect-Restriction Strategy locates the source of intensionality witnessed in examples like (3) and (4) in the meanings of the component noun phrases.

(3) The bricks of the wall are old, but the wall is new.
(4) The gold in the jewelry is old, but the jewelry is new.

This is in keeping with a second suggestion of Rothstein 2010: 355, elaborated as follows:

> The mass entity in *jewelery* cannot be equated with the mass entity in *gold* since they have different properties, even though they are apparently identical. This implies that generally 'artifact' predicates like *jewelery* involve a packaging or perspective function as part of their lexical meaning, so that [the lexical meanings of *gold* and *jewelry*] can be identified as the same spatiotemporal entity but presented under different perspectives or guises and with different properties. But if this kind of lexical packaging is needed anyway in the mass domain, then the problem of the wall and the sum of bricks that makes it up can be solved at the level of [the lexical meanings of *wall* and *brick*], in which case [the lexical meaning of *brick*] will not include the sum of bricks presented as a wall.

In other words, if the source of intensionality can be located within the lexical meanings of the component nouns, then there is no need to appeal to groups to explain the failure of substitutivity witnessed in (3) and (4).

How might this suggestion be spelled out? A natural place to look would be Landman's analysis of groups. As mentioned, Landman argues that the kind of intensionality witnessed in (34) should be located in the meanings of the component nouns, rather than groups, precisely because it exists for both definite plural and definite singular noun phrases alike.

(34) a. The {judge is/judges are} on strike.
 b. The {hangman is/hangmen are} on strike.

Thus, unlike Link 1984, Link 1998, Landman models groups *extensionally*, using sets. Pluralities correspond to sets of entities, and groups correspond to sets of sets of entities—the group formed from a plurality is the singleton of the corresponding set. The same operations are available relating pluralities and groups, namely ↓ and ↑, only now both are *bijective*: for every plurality there is a unique corresponding group, and vice versa

In other words, the source of intensionality witnessed in (34) is to be located in the meaning of the component noun 'judge'. The basic idea is that we are not interpreting 'the judge' and 'the hangman' in (33a,b) as singular terms referring to the same individual, but rather as properties of that individual considered under a certain *aspect*.

Supposing that John happens to be the lone judge and hangman in our small town, (33a,b) can be respectively paraphrased as (35a,b).

(35) a. As a judge, John is on strike.
   b. As a hangman, John is on strike.

John's CHARACTER is the set of properties John possesses, and the semantic function of aspectual phrases like 'as a judge' is to restrict these properties to a certain aspect, corresponding to the different functional roles he plays. These may be thought of as John under different "perspectives" or "guises" in that when evaluating (35a,b), we are considering not merely the properties of John as such, but rather the properties of John *qua judge* or John *qua hangman*. Since he may have different properties when considered under different aspects of his character, substitutivity fails in (33a,b) even though, intuitively, both noun phrases describe the same individual.

To develop (or, better, to start developing) this plan, Landman adopts a hyperintensional semantics, with two basic types: e (individuals) and p (propositions). He then introduces an aspectual-operator $\upharpoonright$, which takes an entity and a property (type $\langle e, p \rangle$) and returns a property of properties (type $\langle \langle e, p \rangle, p \rangle$).

Accordingly, (35a,b) can be represented as (36a,b), respectively, where 'on-strike' also expresses a property (type $\langle e, p \rangle$).

(36) a. $j \upharpoonright$ judge(on-strike)
   b. $j \upharpoonright$ hangman(on-strike)

Thus, (35a) will be true if being on strike is among the properties John has in his role as judge, and similarly for (35b), in his role as hangman. The important thing to note is that (36a,b) are *not* equivalent, simply because '$j \upharpoonright$ judge' and '$j \upharpoonright$ hangman' can express different second-order properties, corresponding to different aspects of John's character.

Of course, not just *any* set of properties will represent an aspect of John's character. Certain conditions must be imposed, and Landman lays down several. The first guarantees that what we may call John's HAECCEITY—the property of being identical to John—is in the set.

(37) a. $j \upharpoonright$ judge($\lambda x.\, x = j$)
   b. John qua judge is still John.

The second guarantees that the restricting property is among the set.

(38) a. $j \upharpoonright$ judge(judge)
   b. John qua judge is a judge.

The third guarantees that restricting John's character to his haecceity does no restricting: it returns *all* of John's properties.

(39) a. $[j \upharpoonright \lambda x. x = j] = \lambda P. P(j)$
b. John qua John is John.

The next five govern the internal logic of property restriction.

(40) a. $[j \upharpoonright \text{judge}(P) \wedge j \upharpoonright \text{judge}(Q)] \rightarrow j \upharpoonright \text{judge}(P \wedge Q)$
b. $[j \upharpoonright \text{judge}(P) \wedge P \rightarrow Q] \rightarrow j \upharpoonright \text{judge}(Q)$
c. $\neg \exists P. j \upharpoonright \text{judge}(P \wedge \neg P)$
d. $\forall P. j \upharpoonright \text{judge}(P \vee \neg P)$
e. $j \upharpoonright \text{judge}(P) \rightarrow \text{judge}(j)$

Jointly, these tell us that restricted properties are ULTRAFILTERS of properties which include the haecceity of the individual whose properties are being restricted and the property doing the restricting itself.

4.2.1 Extending Aspect-Restriction

It is relatively straightforward to extend Landman's theory to the examples of interest. Because noun phrases more generally are interpreted as second-order properties, we can interpret 'the bricks of the wall' in (3) and 'the gold in the jewelry' in (4) similarly as restricted terms, where these contextually-determined restrictions intuitively represent different "perspectives" or "guises".

Following Rothstein 2010's suggestion, we assume that the source of this aspectual relativity is the component nouns, e.g. 'brick' or 'gold'. This leads to a completely general, and seemingly plausible, answer to Link's Revenge.

Let's begin with (3). Suppose a wall was just constructed out of some old bricks. In such a scenario, it seems plausible that the bricks are understood under the guise of components of the wall, whereas the wall itself is understood under the guise of a unified structure. Hence, we might plausibly paraphrase (3) as (41a), formalized as (41b).[14]

---

14 Following Ladusaw 1982, we have been assuming that 'of' denotes the parthood relation between entities (hyperintensional type $\langle e, \langle e, p \rangle \rangle$), as suggested in (i).

(i) $\llbracket \text{of} \rrbracket = \lambda x \lambda y. y \sqsubseteq x$

However, now that noun phrases are interpreted at the level of second-order properties, a type-mismatch arises when we combine 'of' with 'the bricks' and 'the wall'. To remedy this, one could

(41) a. As components of the wall, the bricks of the wall are old, though as a structure, the wall is new.
b. $\sigma x \in A :$ bricks-of-the-wall$(x) \upharpoonright \lambda y.\, y \sqsubseteq$ the-wall$(\text{old}) \wedge \sigma x \in A :$ wall$(x) \upharpoonright$ structure(new)

According to (41b), we are considering the bricks of the wall under their guise as components of the wall, whereas we are considering the wall under its guise as a unified structure, independent of its component parts. And just as John can be trustworthy under certain guises, e.g. being a judge, while being corrupt under others, e.g. being a hangman, despite still being John, similarly the bricks of the wall under the guise of components can be old, while the wall under the guise of unified structure can be new, despite both being the same material stuff.

A similar analysis is available for (4). In one plausible scenario, for instance, some jewelry was recently made from some ancient Egyptian gold. In such a scenario, (4) is plausibly true because we are considering the gold in its role as materially constituting the jewelry, while considering the jewelry itself as a unified collection, or artifact to follow Rothstein 2010, independent of its material components. Thus, we might paraphrase (4) as (42a), analyzed as (42b).

(42) a. As the material constituting the jewelry, the gold in the jewelry is old, though as an artifact, the jewelry itself is new.
b. $\sigma x \in A :$ gold-in-the-jewelry $\upharpoonright \lambda y.\, y \sqsubseteq$ the-jewelry(old) $\wedge \sigma x \in A :$ jewelry$(x) \upharpoonright$ artifact(new)

According to (42b), we are considering the gold in the jewelry under its guise as materially constituting the gold, and we are considering the jewelry itself as an artifact, independent of its material constitution. And as with John under his different roles and the bricks and the wall under their different roles, these too can have different properties.

It is easy to see that similar analyses are available for the other examples discussed above. What's more, the present analysis does not share the problems mentioned in §4.1.2 for the Group-Forming-Strategy. For example, because 'that group of bricks' is also a definite noun phrase, Landman's theory implies that it

---

raise the type of 'of' accordingly. Alternatively, one could define a hyperintensional analog of Partee 1986's LOWER type-shifting operation, as in (i).

(ii) LOWER* $= \lambda \mathcal{Q}_{\langle\langle e,p\rangle,p\rangle}.\, \sigma x \in A[\forall P_{\langle e,p\rangle}.\, P(x) \leftrightarrow \mathcal{Q}(P)]$

We will remain neutral on this issue in what follows, largely ignoring the semantic contribution of 'of' in the noun phrases of interest.

too should be contextually restricted to a certain guise, presumably one similar to that suggested in (41a).

(31) That group of bricks is old, but the wall is new.

Likewise for (32), of course.

(32) The quantity of gold in the jewelry is old, but the jewelry is new.

In short, extending Landman's analysis of group-like phenomena to the cases of interest affords a completely general, and independently motivated, response to Link's Revenge.

4.2.2 Challenges to Aspect-Restriction

Despite these apparent advantages over the Group-Forming Strategy, the Aspect-Restriction Strategy faces its own challenges. We will consider two of them here.

First, because the Strategy relies crucially on Landman's analysis of groups, it is only as adequate as Landman's analysis itself. However, it has been charged that Landman's principles governing aspectual restriction, though initially plausible, lead to inconsistency.[15]

Suppose that John has two jobs: he works as a judge during the day, and as a hangman at night. Also, suppose that being a hangman implies being a non-judge. Now, by (38) John qua judge is a judge.

(38a) $j \upharpoonright \text{judge}(\text{judge})$

Also, by (40e), this implies that John is a judge.

(40e) $j \upharpoonright \text{judge}(P) \rightarrow \text{judge}(j)$

Finally, by (40b), if John qua judge is John and if this implies that John is a judge, then in fact John is a judge.

(40b) $[j \upharpoonright \text{judge}(P) \land P \rightarrow Q] \rightarrow j \upharpoonright \text{judge}(Q)$

---

15 See Szabo 2003 and Asher 2011.

But by exactly similar reasoning, if John qua non-judge (i.e. hangman) is not a judge and this implies that John is not a judge, then he is indeed not a judge. Thus, we have that John is both a judge and not a judge, which is obviously inconsistent.

There are different potential responses available here. Most obviously, one could deny that being a hangman implies being a non-judge. After all, one wouldn't normally infer from the fact that John is both a judge and a hangman that John has contradictory properties. Furthermore, the argument will not go through if we instead adopt the seemingly more plausible assumption that, *as a hangman*, John is not a judge. In that case, it does not follow by (40e) that John qua hangman is a non-judge, but only that he is a hangman. Hence the need to assume that being a judge and being a hangman are mutually exclusive properties. And there are doubtless other potential responses available.[16]

Nevertheless, there is a related, but more pressing, concern for present purposes: the proposed extension of Landman's analysis does not appear capable of actually solving our original puzzle. Consider (43a), which is intuitively true.

43. a. As bricks, the bricks of the wall are old, but as a wall, the wall itself is new.
    b. $\sigma x \in A : \text{bricks-of-the-wall}(x) \upharpoonright \text{bricks(old)} \wedge$
       $\sigma x \in A : \text{wall}(z) \upharpoonright \text{wall(new)}$

This makes sense on the semantics under consideration since, after all, we are considering the bricks and the wall under different aspectual guises.

But now consider (44), which is also seemingly true.

(44) a. As the bricks of the wall, the bricks of the wall are old, but as the wall, the wall itself is new.
     b. $\sigma x \in A : \text{bricks-of-the-wall}(x) \upharpoonright \lambda y. y = \text{the-bricks-of-the-wall(old)}$
        $\wedge \sigma x \in A : \text{wall}(z) \upharpoonright \lambda w. w = \text{the-wall(new)}$

Recall that by (39a), restricting entities to their haecceities returns all *unrestricted* properties of the entity in question.

(39a) $[j \upharpoonright \lambda x. x = j] = \lambda P. P(j)$

For example, restricting the bricks of the wall to their haecceity returns the set of their unrestricted properties.

---

[16] For example, Szabo 2003 considers a weaker formulation of Landman's principles. Alternatively, one might consider a stronger interpretation of the conditional.

(45) $σx ∈ A$: bricks-of-the-wall$(x) \upharpoonright λy. y$ = the-bricks-of-the-wall
 = $λP_{\langle e,p \rangle}. P$(the-bricks-of-the-wall)

But now the problem should be apparent: (44b), and thus (44a), is equivalent to (3) but *without* any aspectual restrictions. Since (44a) seems true, it seems that aspectually restricting the offending noun phrases will not explain how the bricks and the wall can have different properties.

Worse, similar examples are easily formulated for the other examples considered above.

(46) a. As the gold in the jewelry, the gold in the jewelry is old, but as the jewelry itself, the jewelry is new.
b. As the gold in the ring, the gold in the ring is old, but as the ring itself, the ring is new.

As with (44a), (46a,b) are plausibly true, despite being equivalent to (4) and (1) *without* aspectual restrictions. One potential moral here is that restricting properties to a haecceity should not return a set of unrestricted properties. After all, as Landman 1989b: 733 observes, (39a) "is not absolutely necessary, but very convenient". Still, (39a) does seem a particularly natural constraint on aspectual restriction: being John is a property John has, no matter how we view him, presumably. How, then, could restricting John's properties to that of being John return anything other than John under *no* aspectual guises?

5 Conclusion

We began with Link 1983's claim that because Link's ring and the gold constituting it have different properties despite existing in the same place and time, they must be different sorts of things, constituting different domains with different corresponding mereological relations. We then observed that the general underlying principle supporting Link's contention would appear to massively overgenerate domains and corresponding mereological relations. This led to adopting a single-domain analysis, which in turn required an explanation of Link's original observation. We then considered two initially plausible explanations, inspired by Rothstein 2010, concluding that neither is without significant challenges.

What should we conclude from all of this? Perhaps the upshot is that we ought to adopt a different theory of aspectual restriction. Other available theories include those of Jäger 2003, Szabo 2003, and Asher 2011. An altogether different option would be to view the culprit noun phrases as referring to different sorts of intensional, aspectually anchored objects, perhaps along the lines of Fine 1982's QUA OBJECTS. And a third option would be to analyze them as referring instead to INDIVIDUAL CONCEPTS, i.e. functions from worlds to entities, perhaps restricted to what Aloni 2001 calls CONCEPTUAL COVERS. We will not pursue these alternatives further here. Suffice it to say that though numerous possibilities exist, determining which is most suitable for the various examples considered here constitutes an important, but arguably at least book-long, project.

Nevertheless, *if* some such analysis could be made to work, and so we had a satisfactory response to Link's Revenge, the upshot would be that one-domain analyses are adequate to explain the count/mass distinction. Hence, given the methodological orientation of natural language mereology assumed here, we would conclude that only one domain is needed to account for that phenomena, and thus that natural language presupposes just one mereological relation, at least with regard to that phenomena.

This is not to say that natural language presupposes only one mereological relation *more generally*, however. There are still further domains or sorts to take into consideration, including kinds, events, numbers, degrees, etc. Assuming that at least some of these are genuinely distinct, the question remains whether we should view the mereological relations ordering entities within them as distinct, or whether instead we should view natural language as committed to a single mereological relation operating over different sorts.

On the one hand, if these different domains are just that—domains of a certain relation—then it would appear true merely by definition of 'relation' that we have different mereological relations. On the other hand, consider again Krifka 1989's definition of atomic parthood in (17c), where 'S' ranges over sorts.

(17c) $\forall x, y.\ x \sqsubseteq_{At,S} y \leftrightarrow x \sqsubseteq_S y \wedge \text{ATOM}_S(x, S)$

This suggests that we have just one domain, subdivided into different sorts, each of which is ordered by a *single* mereological relation $\sqsubseteq$. Indeed, this is how Krifka himself captures various well known similarities between the meanings of nouns and verbs. However, whether these are really just notational variants, or whether they instead represent substantially different empirical claims, is something else we must leave for future research.

Bibliography

Aloni, M. (2001): *Quantification Under Conceptual Covers*. University of Amsterdam, PhD thesis, dissertation.
Asher, N. (2011): *Lexical Meaning in Context: A Web of Words*. Cambridge University Press.
Bach, E. (1986): "Natural Language Metaphysics." In: Marcus, R. B. and et al (eds.): *Logic, Methodology, and the Philosophy of Science VII*, 573 – 595.
Barner, D. and Snedeker, J. (2005): "Quantity Judgments and Individuation: Evidence That Mass Nouns Count." In: *Cognition*, Vol. 97, No. 1, 41 – 66.
Chierchia, G. (1998): "Reference to Kinds Across Languages." In: *Natural Language Semantics*, Vol. 6, 339 – 405.
Chierchia, G. (2010): "Mass Nouns, Vagueness and Semantic Variation." In: *Synthese*, Vol. 174, No. 1, 99–149.
Fine, K. (1982): "Acts, Events and Things." In: Leinfellner, Werner et al. (eds.): *Language and Ontology: Proceedings of the Sixth International Wittgenstein Symposium*. Wien: Hölder-Pichler-Tempsky, 97 – 105.
Gillon, B. S. (1992): "Towards a Common Semantics for English Count and Mass Nouns." *Linguistics and philosophy*, Vol. 15, No. 6, 597 – 639.
Jäger, G. (2003): "Towards an Explanation of Copula Effects." In: *Linguistics and Philosophy*, Vol. 26, No. 5, 557 – 593.
Krifka, M. (1989): "Nominal Reference, Temporal Constitution, and Quantification in Event Semantics." In: von Bentham, J., Bartsch, R., and von Emde-Boas, P. (eds.): *Semantics and Contextual Expressions*. Dordrecht: Foris Publications, 75 – 115.
Krifka, M. (1990): "Boolean and Non-Boolean *and*." In: Kalman, L. and Polos, L., (eds.): *Papers from the Second Symposium on Logic and Language*. Budapest: Akadémmiai Kiadó.
Ladusaw, W. A. (1982): "Semantic Constraints on the English Partitive Construction." In: *Proceedings of WCCFL 1*, Vol. 1, 231 – 242.
Landman, F. (1989a): "Groups, i." In: *Linguistics and Philosophy*, Vol. 12, No. 5, 559 – 605.
Landman, F. (1989b): "Groups, ii." *Linguistics and Philosophy*, Vol. 12, No. 6, 723 – 744.
Landman, F. (2012): *Structures for Semantics*. Vol. 45. Springer Science & Business Media.
Link, G. (1983): "The Logical Analysis of Plurals and Mass Terms: A Lattice-Theoretic Approach." In: Bäuerle, R., Schwarze, C., and von Stechow, A. (eds.): *Meaning, Use, and Interpretation of Language*, 303 – 323.
Link, G. (1984): "Hydras: On the Logic of Relative Constructions With Multiple Heads." In: *Varieties of Formal Semantics*, 302 – 323.
Link, G. (1998): *Algebraic Semantics in Language and Philosophy*. SCLI Publications.
McKay, T. (2006): *Plural Predication*. Oxford: Oxford University Press.
Oliver, A. and Smiley, T. (2001): "Strategies for a Logic of Plurals." In: *The Philosophical Quarterly*, Vol. 51, No. 204, 289 – 306.
Oliver, A. and Smiley, T. (2013): *Plural Logic: Revised and Enlarged*. Oxford: Oxford University Press.
Partee, B. (1986): "Noun Phrase Interpretation and Type-Shifting Principles." In: Groenendijk, J., de Jongh, D., and Stokhof, M. (eds): *Studies in Discourse Representation Theory and the Theory of Generalized Quantifiers*. Dordrecht: Foris Publications.
Pelletier, F. J. (1975): "Non-Singular Reference: Some Preliminaries." In: *Philosophia*, Vol. 5, 451 – 465.

Pelletier, F. J. (2011): "Descriptive Metaphysics, Natural Language Metaphysics, Sapir-Whorf, and all that Stuff: Evidence from the Mass-Count Distinction." In: *The Baltic International Yearbook of Cognition, Logic and Communication*, Vol. 6, No. 1, 7.

Rayo, A. (2002): "Word and Objects." In *Noûs*, Vol. 36, No. 3, 436 – 464.

Rothstein, S. (2010): "Counting and the Mass/Count Distinction." In: *Journal of semantics*, Vol. 27, No. 3, 343 – 397.

Rothstein, S. (2017): *Semantics for Counting and Measuring*. Cambridge: Cambridge University Press.

Scontras, G. (2014): *The Semantics of Measurement*. PhD thesis, Harvard University.

Snyder, E. and Shapiro, S. (unpubl.): "Superplurals, Groups, and Paradox."

Strawson, P. F. (1959): *Individuals: An Essay in Descriptive Metaphysics, London*, 1969.

Szabo, Z. G. (2003): "On Qualification." In: *Philosophical Perspectives*, Vol. 17, No. 1, 385 – 414.

Gerhard Schurz
Universal Translatability: An Optimality-Based Justification of (Classical) Logic

Abstract: In order to prove the validity of logical rules, one has to assume these rules in one's metalogic. But how is a non-circular justification of a logical system possible? The question becomes especially pressing insofar in present time a variety of *non-classical* alternatives to classical logics have been developed. Is the threatening situation of an epistemic circle or infinite regress unavoidable? The situation seems hopeless. Yet, in this paper I suggest a positive solution to the problem based on the fact that logical systems are *translatable* into each other. I propose a translation method based on introducing additional concepts into the language of classical logic. Based on this method I demonstrate that all finite multi-valued logics – and I conjecture all non-classical logics – can be translated into classical logic. If this argument is correct, it would show that classical logic is *optimal* in the following sense: by using it we cannot lose, because if another logic turns out to have advantages for certain purposes, we can translate and thus embed it into classical logic. This optimality argument does not exclude that there can be other, non-classical logics that are likewise optimal in the explained sense.

1 Introduction

In other writings (Schurz 2008a, Schurz 2018, Schurz 2019) I defend a 'modernized' version of an internalist and foundation-oriented epistemology. Within this epistemological framework the class of 'basic' beliefs that are considered as 'immediately evident' or not in need of further justification is *minimal* (consisting only of analytic and introspective beliefs); moreover circular justifications are rejected because they are demonstrably epistemically worthless (see sec. 2). In such an epistemological framework the 'epistemic load' that has to be carried by deductive, inductive or abductive reasoning becomes high. Therefore the justification of the truth-conduciveness of these inferences – in the strict or at least high probability sense – acquires central importance. In other writings I have studied the problem of justifying inductive inferences, i.e., Hume's problem (Schurz 2018, Schurz 2019). In contrast, this paper is devoted to the problem of finding a non-circular justification for a system of logic. Thus the primary question of this paper will be: how can

Gerhard Schurz, DCLPS, University of Düsseldorf, Germany

we justify the rules of classical logic? Thereby I will focus on the justification of systems of propositional logic.

2 The Justification of (Classical) Logic and the Problem of Circularity

According to (what I call) the naïve answer to our question, the justification of deduction rules of classical propositional logic is *unproblematic*: it simply *follows* from the semantic definition of the logical operations in terms of truth tables. Based on them we can prove that the logical rules are valid, i.e., truth-preserving under all possible (truth-value) valuations, or in all 'possible worlds'. The problem of this answer is that the semantic proofs are implicitly *circular*. They lead into an *infinite regress* of meta-levels because in the proof of the validity of logical rule rules one needs these rules again in one's meta-logic. This is so because in a meta-logical proof the cognitive content of the truth tables has to be 'verbalized' by means of implication relations between the truth value of the compound formula and that of its components. Here are two examples:

Example 1. *Semantic proof of the simplification rule $p \land q/p$:*

(1) $True(p \land q)$	Premise
(2) $True(p \land q) \to (True(p) \land True(q))$	From $\land$'s truth-table
(3) $True(p) \land True(q)$	From (1), (2) by Modus Ponens
(4) $True(p)$	From (3) by the simplification rule

Step 2 is based on the truth table of conjunction telling us that in the only 'possible world' or line in the truth table in which "$p \land q$" is true (premise 1), both p and q are true. In step 3 we infer from this fact that "p" is true by employing the simplification rule at the meta-level. Note that implicitly also a universal generalization step is involved: from the fact that the inference holds for an arbitrary valuation we infer that it holds for all valuations.

Example 2. *Semantic proof of the validity of Modus Ponens $p \to q, p/q$:*

(1) $True(p \to q)$	Premise
(2) $True(p)$	Premise
(3) $True(p \to q) \to (True(p) \to True(q))$	From $\to$'s truth table
(4) $True(p) \to True(q)$	From (1), (3) and Modus Ponens
(5) $True(q)$	From (2), (4) and Modus Ponens

In this example we proved the validity of Modus Ponens in the object language by using it twice at the meta-level.

Experienced logicians are accustomed to meta-language circularities of this sort. The question we have to ask here is: are these circularities harmful?, and if yes, in which respect? Of course, one may legitimately argue that these circularities give us a deeper semantic understanding of these logical operations. However, if these circular semantic proofs are used epistemically, in order to lend additional justification to the logical rules, they are indeed harmful, for the reason that with help of rule-circular arguments, intuitively rather irrational rules can be 'pseudo-justified'. Here is an example of a circular justification of the irrational rule of 'Modus Morons' that goes back to Susan Haack (Haack 1976: 115); I present her circular 'proof' in a slightly different (in my eyes more straightforward) way:

Example 3. *Circular 'proof' of the 'validity' of the invalid rule of 'Modus Morons'* $p \to q, q/p$:

(1) $True(p \to q)$ *Premise*
(2) $True(q)$ *Premise*
(3) $True(p \to q) \to (True(p) \to True(q))$ *From $\to$'s truth table*
(4) $True(p) \to True(q)$ *From (1), (3) and Modus Ponens*
(5) $True(p)$ *From (2), (4) and Modus Morons*

Example 3 shows clearly that rule-circular arguments cannot convey epistemic support, because the rules of 'Modus Morons' is obviously invalid. Even more drastic examples of circular 'justifications' of obviously irrational rules are given in Achinstein 1974 and Schurz 2019, sec. 2.4 and 3.3.

Is the threat of a justification circle or regress unavoidable in logic? *No*, it is not. But it tells us that semantic explications of logical concepts and rules *cannot stop* the justificational regress. At some meta-language level we must stop the regress by assuming the principles of classical logic *as given*. Technically this is done by assuming an *axiomatic system*: a system of axioms and rules from which (hopefully) all other logically valid theorems can be derived. Epistemologically this means that we consider the axiom and rules of classical logic as *basic* in the explained sense, that is, as immediately evident and not in need of further justification.

Our next question is of course: is this epistemically satisfactory? What justifies us in considering the principles of classical logic as basic? The *traditional* answer to this question argues that there is a crucial difference between the problem of justifying *induction* and the problem of justifying *deduction*. We can easily imagine possible worlds in which induction fails. But we can hardly imagine possible worlds in which (classical) logic fails, because we presuppose our logic already in the *representation* of these worlds. For this reason, deductive logic is basic and needs no justification. In this sense, Kant and Wittgenstein said that "logic is transcendental".

Maybe this traditional justification was satisfactory two centuries ago, but in contemporary philosophy it is no longer convincing, because it assumes that possible worlds must be represented by means of *classical* logic. However, in present days there is a variety of *non-classical* logics that deviate from the principles of classical logic in certain respects. The best known examples are multi-valued logics that assume that there are more than two truth values, for example, the values "true", "false", and "undetermined".

What we have in present day is a situation of *logical pluralism*. But what does this epistemologically mean? Given that deductive logic is part of the 'deepest' basis, or of the 'innermost' core of our system of belief, logical pluralism seems to be a *threat for rationality* insofar alternative logical systems seem to be cognitively *incommensurable*. However that may be (we return to this question below), given this situation our problem terminates in the following question: *How can one justify classical logic, or a system of logic at all, in view of the situation of logical pluralism?*

3 Optimality Justifications of Logics by Means of Universal Translation Arguments

There is a kind of higher-order justification that doesn't lead into a circle or infinite regress: an epistemic optimality justification. An optimality justification does not attempt to demonstrate that a given epistemic method or system is strictly or probabilistically reliable, in the sense of leading to the truth in all or most cases. It pursues a more modest epistemic goal, namely to demonstrate that a given method (or system) is epistemically *optimal* among all competing methods (of a given kind, e.g., induction or deduction) that are *cognitively accessible* to the given epistemic agent. My paradigm case of an optimality justification is the justification of meta-induction. In other writings I have proved that a certain method of meta-induction is predictively optimal in the long run among all prediction methods that are accessible to the forecaster, even in possible worlds in which the success rates of the competing prediction methods are permanently changing (Schurz 2008b, Schurz 2019). This universal optimality result provides us with a weak a priori justification of meta-induction that can stop the justification regress for the problem of justifying induction.

Note that the restriction of the optimality claim to *cognitively accessible* methods is crucial for the possibility of a universal optimality argument. The possibility of the optimality proof rests on the fact that the strategy of meta-induction has a universal learning ability: it tracks the success rates of all accessible methods and incorporates these methods into its own strategy as soon as they are sufficiently successful, by predicting a weighted average of the predictions of accessible methods, using weights that are a delicately chosen function of their so far achieved success rates. What the optimality justification of meta-induction implies is that by using the optimal method *on top* of all other cognitively accessible methods one can only gain but never lose anything (in the sense of Reichenbach's idea of a 'best alternative' justification; see Reichenbach 1949, sec. 91). The optimality justification of meta-induction does not exclude that there are 'demonic worlds' in which all methods of prediction fail, it only entails that *if* there is are successful prediction methods in a given environment, then meta-induction will be certainly among them.

In the next section I intend to apply the method of optimality justifications to the problem of justifying rules of deduction, in particular to the deductive system of classical propositional logic. Of course, the epistemic goal of deductive logic is different from that of induction; it is not prediction but, more basically, obtaining a coherent *representation* of the facts of the world. But there is also a commonality: as in the area of induction, an optimal logical strategy must operate at the meta-level and must possess means to incorporate alternative logical methods into itself. My crucial idea is to use (ideally universal) *translation functions* between logical systems for this purpose. Thus, my epistemic account is based on the fact that logical systems are translatable into each other. The existence of translation functions between different logical systems is nothing new in the logical literature, but what I will show is that a logical translation function can be constructed that meets the requirements of *semantic*, i.e. meaning-preserving *translations* (which is not the case for the standard translation functions studied in the literature; see below). My major epistemic argument will be that the existence of a meaning-preserving translation from non-classical logics into classical logic L shows that by using classical logic L one can only gain but never loose, because everything expressible in the non-classical logic can also be expressed in the classical logic. In the next section I will carry out my idea for the case of the translation of multi-valued logics into classical (bivalent) logic.

4 Translating Three-Valued (or Multi-Valued) Logic Into Classical Logic

In what follows the indexed letter L_i varies over logical systems, and $\mathcal{L}_i$ designates the language of such a logic. L_2 denotes the classical (bivalent) propositional *logic* (consisting of its logical axioms, theorems and its valid inferences, the latter being denotes as $\vDash_{L_2}$). The language $\mathcal{L}_2$ contains ¬, ∧, ∨ as primitive propositional connectives (material implication '⊂' and equivalence '≡' being defined in the usual way). Languages are identified with the set of their well-formed formulas. I use

$p_1, p_2, \ldots, q, r \ldots$ as (propositional) variables,

$A, B, \ldots, S, \ldots$ as schematic letters for arbitrary formulas (i.e., sentences), and

$\Gamma, \Delta, \ldots$ designate arbitrary sets of formulas.

L_3 is Łukasiewicz' three-valued logic (Łukasiewicz 1920) with the truth-values true (t), false (f) and u for 'undetermined'. The language $\mathcal{L}_3$ has four basic truth-functional connectives ¬, ∧, ∨ and → (the conditional → not being definable in terms of the other three connectives). As usual one assumes a linear ordering among the truthvalues of a (finite) multi-valued logic; in our case the ordering is

$f < u < t$; or represented as ranks: −1, 0, +1.

Based on this ordering, Łukasiewicz' three-valued truth-tables for the four connectives are easily explained. The truth value of ¬p is the inverse of p's truth value, that of $p \vee q$ is the maximum and that of $p \wedge q$ the minimum of the truth values of p and q. Finally, $p \rightarrow q$'s truth value equals true, undecided or false, respectively, if the rank difference between q's truth value and p's truth value is not smaller than 0 / minus 1/ minus 2.

p	¬p	p	q	p∧q	p∨q	p → q
t	f	t	t	t	t	t
u	u	t	u	u	t	u
f	t	t	f	f	t	f
		u	t	u	t	t
		u	u	u	u	t
		u	f	f	u	u
		f	t	f	t	t
		f	u	f	u	t
		f	f	f	f	t

The notion of logical truth and validity in multi-valued logics is defined analogously as in bivalent logics. Let

$\mathcal{P}$ be the denumerable set of propositional variables and

$val_3: \mathcal{P} \to \{t, u, f\}$ range over trivalent truth-valuations over the (propositional) variables that are recursively extended to arbitrary complex formulas of $\mathcal{L}_3$ by way of the above truth tables.

Then an $\mathcal{L}_3$-formula A is logically true in L_3, in short $\models_3 A$ iff $val_3(A) = t$ for all (possible) trivalent valuations, and A follows from a formula set Γ in L_3, in short $\Gamma \models_3 A$, iff all trivalent valuations making all formulas in Γ true make A true.

It is well known that some typical theorems and meta-theorems of classical L_2 are not among the theorems of L_3. Here are some examples.
Some theorems of L_3: $p \to (q \to p)$, $(\neg q \to \neg p) \leftrightarrow (p \to q)$, $(p \lor q) \leftrightarrow (p \to q) \to q$.
Some non-theorems of L_3: $p \lor \neg p$, $\neg(p \land \neg p)$, $(p \lor q) \leftrightarrow (\neg p \to q)$.
Deduction theorem ($\Gamma, A \models B$ iff $\Gamma \models A \to B$) fails for L_3.

In general, the notion of validity in multi-valued logics is defined by assuming a subset $Des \subset Val$ of designated (truth-) values (Val being the set of all truth-values) and defining a formula A as valid in L_{Val} if all L_{Val}-valuations convey to A a designated value. The triple $<Val, Des, \{t_c : c \in C\}>$ (with $\{t_c : c \in C\}$ being the set of truth-tables for a set of connectives C) is also called a Val-valued *logical matrix*.

Turning back to Łukasiewicz' three-valued logic L_3 I explain now my major idea: the sentences of L_3 can be *translated* into sentences of L_2 – in a strict sense of translation – by introducing the following three additional propositional operators into $\mathcal{L}_2$:

the operators of "being true" (T), "being false" (F) and "being undetermined" (U), *as understood in three-valued logics*.

The crucial point of this step is this: even if S is a sentence of the three-valued logic, the sentences $T(S)$, $F(S)$ and $U(S)$ are nevertheless two-valued, obeying the following truth table:

p	T(p)	U(p)	F(p)
t	t	f	f
u	f	t	f
f	f	f	t

We do not need to add the three truth-value operators T, U and F explicitly to the language $\mathcal{L}_3$ because these operators are definable within L_3 as follows. That S is undetermined can be asserted via the formula

$\text{UNDET}(S) =_{def} (S \lor \neg S) \to \neg(S \lor \neg S)$.

In L_3 it holds that $U(S) \leftrightarrow \text{UNDET}(S)$, i.e., the two have the same truth table. Likewise the formulas

$\text{TRUE}(S) =_{def} S \wedge \neg\text{UNDET}(S)$ and $\text{FALSE}(S) =_{def} \neg S \wedge \neg\text{UNDET}(S)$

have the same three-valued truth table as $T(S)$ and $F(S)$, respectively.

By adding the Łukasiewicz-operators T, U, F to our classical language $\mathcal{L}_2$ we obtain the extended classical language $\mathcal{L}_{2.Luk}$ whose formulas are still evaluated bivalently and whose basic logical laws are still the classical laws of L_2. Of course, within L_2 the operators T, U and F figure as intensional (non-bivalently-truth-functional) operators, similar as the operators of modal logic.

Based on the truth tables of these three operators, every semantic rule of three-valued logic can be translated into a set of corresponding axioms formulated in the expanded language of classical logic. Łukasiewicz' three-valued truth table for negation is represented by three axioms:

$T(\neg A) \leftrightarrow F(A) \qquad U(\neg A) \leftrightarrow U(A) \qquad F(\neg A) \leftrightarrow T(A)$.

The truth table for the conjunction is represented by:

$T(A \wedge B) \leftrightarrow T(A) \wedge T(B) \qquad F(A \wedge B) \leftrightarrow F(A) \vee F(B)$
$U(A \wedge B) \leftrightarrow (U(A) \wedge \neg F(B)) \vee (U(B) \wedge \neg F(A))$

Similarly for the disjunction:

$T(A \vee B) \leftrightarrow T(A) \vee T(B) \qquad F(A \vee B) \leftrightarrow F(A) \wedge F(B)$
$U(A \vee B) \leftrightarrow (U(A) \wedge \neg T(B)) \vee (U(B) \wedge \neg T(A))$.

and for the (three-valued, non-material) implication:

$T(A \to B) \leftrightarrow F(A) \vee (U(A) \wedge \neg F(B)) \vee (T(A) \wedge T(B))$
$U(A \to B) \leftrightarrow (T(A) \wedge U(B)) \vee ((U(A) \wedge F(B))$
$F(A \to B) \leftrightarrow T(A) \wedge F(B)$

Finally we add the trivalent truth-value axiom:

$T(S) \dot\vee U(S) \dot\vee F(S)$ ("$\dot\vee$" for exclusive disjunction).

The set of these axiom schemata forms the axiom system Ax_{Luk} of Łukasiewicz' logic in the expanded language of classical logic $\mathcal{L}_{2.Luk}$.

We show now how all $\mathcal{L}_3$-formulas can be translated into $\mathcal{L}_{2.Luk}$. Our translation function is based on the

truth view of assertion: asserting a sentence S means to assert that it is true.
We assume the assertion view holds not only for L_2 but also for L_3 (and, as we think, for all propositional languages). Thus the translation functions "trans$_{3\to 2}$" (from $\mathcal{L}_3$ into $\mathcal{L}_2$) is as follows:

trans$_{3\to 2}(S) = T(S)$, for all $S \in \mathcal{L}_3$.

Since this translation is meaning-preserving, it is of course injective, thus:

trans$_{2\to 3}(T(S)) = S$.

We now show that the translation trans$_{3\to 2}$ preserves meaning and L_3-logical truth (or validity) in a precise sense. For this purpose, we have to introduce *some terminology*. In what follows, O_i ranges over the three trivalent truth-value operators, T, U and F. $\mathcal{P}(S) = p_1, \ldots, p_{k(S)}$ denotes the set of variables occurring in sentence S. We speak of the "p_i" as "unboxed variables" and of the statements "$O_i p_j$" as "boxed variables". For $\mathcal{P}$ a set of unboxed variables, $O\mathcal{P} = \bigcup_{p \in P}\{Tp, Up, Fp\}$ denotes the corresponding set of boxed variables. If $\mathcal{P}$ is the (denumerable) set of unboxed variables common to $\mathcal{L}_3$ and $\mathcal{L}_{2.Luk}$, then following from the non-truthfunctional nature of the operators O_i, truth-valuations over $\mathcal{L}_{2.Luk}$ are defined over the set of elementary formulas $P \cup O\mathcal{P}$. Let Val$_3(\mathcal{P})$ be the set of all trivalent valuations over $\mathcal{P}$ and Val$_3(\mathcal{L}_3)$ the set of (recursively extended) trivalent valuations over sentences of $\mathcal{L}_3$. Moreover, let Val$_{2.Luk}(O\mathcal{P})$ be the set of all bivalent valuations over $O\mathcal{P}$ satisfying the axiom $T(S) \dot\vee U(S) \dot\vee F(S)$ and let Val$_{2.Luk}(\mathcal{L}_{2.Luk})$ be the set of all (recursively extended) bivalent truth-valuations over formulas of the expanded language satisfying the axioms of Ax_{Luk} for $\neg$, $\wedge$, $\vee$ and $\to$. Then:

Theorem 1. *(1.1) There is a bijective correspondence f between trivalent and bivalent valuations, $f : \text{Val}_3(\mathcal{L}_3) \leftrightarrow \text{Val}_{2.Luk}(\mathcal{L}_2)$ ("$f : \leftrightarrow$" for bijection), such that a trivalent valuation verifies a sentence S exactly iff its bivalent counterpart valuation verifies $T(S)$, i.e. $\forall S \in \mathcal{L}_3$ and $val_3 \in \text{Val}_3(\mathcal{L}_3)$: $val_3 \models S$ iff $f(val_3) =_{def} val_2 \models T(S)$.*
(1.2) A sentence is logically true in L_3 exactly if its translation is a logical consequence of the corresponding axiomatic system Ax_{Luk} in the expanded language of L_2, and likewise for valid inferences. Thus formally:
  $\models_{L_3} S$ *iff* $Ax_{Luk} \models_{L_2} T(S)$, *and*
  Prem $\models_{L_3} S$ *iff* $Ax_{Luk}, \{T(P) : P \in \text{Prem}\} \models_{L_2} T(S)$.

5 Discussion and Generalization of the Proposed Translation Method

In the next three subsections of this section we clarify what we think has been achieved by our translation method by means of three subsections. In the last subsections we discuss generalizations of our results.

5.1 Meaning Preservation

The translation $trans_{3\to 2}$ together with the axiom system Ax_{Luk} strictly preserves the semantic meaning of the trivalent operators, which are part of $\mathcal{L}_{2.Luk}$ as well as of $\mathcal{L}_3$. It also preserves the meaning of the (propositional) variables 'as good as possible' in the following sense. Of course, the meaning of "p" cannot be strictly the same in L_3 and L_2, because in L_3 p has three and in L_2 two truth values. However, the meaning of the more *fine-grained* propositions $T(p)$, $O(p)$ and $U(p)$ is strictly the same in L_3 and in L_2.

This brings me to my central thesis: *every proposition that can be expressed in L_3 can also be expressed in L_2*. Asserting a sentence S in L_3 is expressed by asserting $T(S)$ in L_2; moreover by applying the Ax_{Luk}-equivalences (that are valid in L_2 as well as in L_3), the semantic composition of S in L_3 is fully reflected by the Ax_{Luk}- equivalences of $T(S)$ in L_2.

5.2 Comparison with the Literature

Fact 5.1 distinguishes my account from the translation functions between logics studied in previous literature. In the latter work, translations are usually not accompanied by expansions of the (classical) language, on the cost that these translation functions do not preserve meaning and semantic composition of statements; they only preserve the consequence operation. One example are the abstract 'translations functions' studied by Jerábek 2012. These translation functions map the formulas of the language $\mathcal{L}$ of a propositional logic L into formulas of a language $\mathcal{L}'$ of a logic L', such that if $A \vdash_L B$, then $f(A) \vdash_{L'} f(B)$ (where f need neither be injective nor surjective); if the other direction holds as well, the translation is called conservative. Given an enumeration of all $\mathcal{L}$-formulas and the nth formula A_n of $\mathcal{L}$, Jerabek's translation into the language $\mathcal{L}_2$ of classical logic, $t(A_n)$, is roughly speaking defined as $X \vee (q_n \wedge Y)$, where X is the disjunction of translations of all premises with indices $< n$ that entail A_n, q_n is a new variable and Y is the conjunction of all translations of implications $C \to D$ with indices $< n$ such that $\{C, A_n\}$ entails D (Jerábek 2012: 669). Jerábek 2012 theorem 2.6 proves that classical logic is 'translation-universal' in the sense that every finitary deductive system in countably many formulas can be conservatively translated into classical logic ($\mathcal{L}_2, \vdash_2$); moreover many other non-classical propositional logics are universal in this sense. The result is technically impressive, but obviously, Jerabek's 'translation' function neither preserves the meaning or semantic composition of formulas nor even their syntactic structure; it is constructed just for the purpose to preserve the consequence operation.

Another abstract translation function has been proposed by Rutz 1972, who translates formulas of an n-valued logic into k-tuples of formulas of the classical logic. His translation is based on the abstract idea that every number n has a binary representation by means of $\log_2(n)$ binary digits. Again, this translation does not preserve semantic meaning nor syntactic structure.

An example of a semantic 'translation' or embedding of non-classical into classical logic is the bivalent reduction of multi-valued logic proposed by Suszko 1977. Given a standard multi-valued logic with a subset Des $\subset$ Val of designated truth values, Suszko proposed to translate the disjunction (or set) of the designated truth-values into the bivalent value "true" and the disjunction (or set) of the non-designated truth-values into "false". Suszko's translation is useful for many purposes (cf. Béziau 1999). However, Suszko's translation does not preserve the semantic meaning of the propositional connectives: they become intensional under Suszko's bivalent semantics (Malinowski 1993: 79, Wansing/Shramko 2008). For example, both p and $\neg p$ may have the truth-value false; thus the law of excluded middle, $p \vee \neg p$, is not longer valid in Suszko's bivalent semantics.

5.3 Bridge Axioms Between L_3 and L_2.

So far, the bivalent truth-value of the boxed variables ($O_i p$) are independent from the bivalent truth-value of the unboxed variables. In fact, for every $S \in \mathcal{L}_3$, the truth value of $T(S)$ only depends on the truth values of the boxed but not of the unboxed variables. However, for semantic coherence between L_3 and L_2 we require the following bridge axioms:

$T(S) \to S$ and $F(S) \to \neg S$,

or in words, a trivalently true (or false) sentence is also bivalently true (or false, respectively), while for undetermined sentences their bivalent truth value is left open. Philosophically speaking our translation does not prescribe whether a trivalently undetermined statement should be bivalently classified as true or false; this may depend on the particular content of p. And this has to be so: the *converse implications must not hold*. Otherwise the translation would not be conservative and the translated three-valued logic would collapse into two valued logic. For example from $S \to T(S)$ we could derive $\vDash_{L_2} p \vee \neg p$ and thus $\vDash_{L_2} T(p \vee \neg p)$, and by employing the inverse translation $\vDash_{L_3} p \vee \neg p$; moreover by application of the axioms Ax_{Luk} (that are also valid in L_3) we would obtain from this $\vDash_{L_3} T(p) \vee F(p)$.

5.4 Generalizations

It is rather obvious that the same translation strategy applies to all many-valued logics representable by means of a matrix of *finitely* many truth values. Thus, if an n-valued logic L_n is based on a matrix $< Val_n, Des_k, \{t_c : c \in C\} >$ with $|Val_n| = n$, $|Des_k| = k < n$ and $C = \{\neg, \wedge, \vee, \to\}$, then we introduce the n intensional operators $O_1, \ldots, O_n$ for the n truth values, the equivalence axioms for $\neg, \wedge, \vee$ and $\to$ describing the truth tables in terms of these n operators and the n-valent truth-value axiom $O_1(S) \dot\vee \ldots \dot\vee O_n(S)$, and prove the translation theorem in the same way as above.

It is worth emphasizing that also *para-consistent* logics can be characterized by means of finite truth value matrices, including the value "both true and false" (Priest 1979, Priest 2013, sec. 3.6). Thus also many paraconsistent logics are covered by this generalization. Moreover, also logics with consequence relations based on Malinowski's (1993) quasi-matrices that are not Suszko-reducible are translatable by our strategy. These consequence relations require both the semantic preservation of designated truth values from premise to conclusion and of antidesignated values from conclusion to premise (Wansing/Shramko 2008: 412). Thus for these consequence relations the translation theorem applies in the following form: $A \vDash_{Quasi} B$ iff $D(A) \vDash_{Ax.Quasi} D(B)$ and $AD(B) \vDash_{Ax.Quasi} AD(A)$, where "$D(S)$" and "$AD(S)$" are the $\mathcal{L}_2$-expressions asserting that S has one of the finitely designated (or antidesignated, respectively) truth values.

However, I want to propose an even stronger thesis, namely:

Conjecture: a similar translation strategy applies to all kinds of non-classical logics, even those not characterizable by finite matrices.

My reason for this conjecture is the following: almost all non-classical logics known to me *use classical logic in their meta-language* in which they describe the (correct and complete) semantics of their non-classical principles. Therefore there should exist ways to translate the non-classical principles into classical logic, by introducing intensional operators into $\mathcal{L}_2$ that correspond to the semantic concepts used in the classical meta-language describing the semantics of these logics. Further elaborations of this conjecture and investigations of its tenability against objections are work for the future.

6 Epistemological Conclusions and Discussion of Possible Objections

6.1 Epistemological Conclusion

If our argument is correct, it shows that every non-classical logic can be represented within classical logic, because everything expressible in the former can be expressed in the latter without loss of meaning – by expanding the classical language with appropriate operators and axioms for them. Since the basic laws and rules of classical logic are still valid in this expanded system, this argument gives us an *optimality justification* of classical logic: By using classical logic our conceptual representation system *can only gain but can never lose*, because if another logic has advantages for certain purposes, we can translate and thus embed it into classical logic. What is furthermore achieved by this result is *epistemic commensurability* and thus a refutation of cognitive relativism at the most fundamental level: different logical frameworks are not incommensurable (in the sense of Kuhn 1962), because they are translatable into each other.

6.2 First Objection: The Possibility of Inverse Translation

An apparent 'objection' to my account points out that there is also the possibility of an inverse translation relation, of classical logic into three-valued logic (or more generally, into a non-classical logic), by expanding the non-classical language with intensional operators for bivalent truth-values and corresponding axioms. Indeed, it can be shown that such inverse translation are often possible (because of space limitations we cannot demonstrate this here). Yet this 'objection' is not really an objection to my optimality thesis, because *optimality does not entail dominance*. That a logic (or method) is optimal among a class of logics (or methods) only means that it has maximal value (according to a certain evaluation method). This does not exclude that there may be other logics (or methods) that are likewise maximally good. The stronger property of being *better* than all other logics (or method) in a given class is called *dominance*.

6.3 Logical Pluralism

We have just seen that optimality does not imply dominance. Thus the defender of a three-valued logic may argue that his system is optimal, too, since he can translate every bivalent system into his trivalent logic. Even if it were the case that optimality in our sense holds for a very broad class of logics (as this is the case for the abstract translations studied by Jeřábek 2012), this fact would not undermine the epistemic force of the optimality justification. Rather, it would establish a sort of 'universal commensurability', a situation of *logical pluralism in harmony*: all alternative logics would be 'primus inter pares' because all of them would be translation-universal. However, it is not at all clear whether translation-universality (in our sense) is indeed a property of many logics; this is an open question and its investigation is work for the future.

So far our evaluation of logics was only based on their logical representation power, and logics that are intertranslatable are 'on par' in this respect. This evaluation perspective does not exclude, however, that there may be specific reasons why particular logics are preferred for particular purposes. For example, many people prefer classical logic because it is psychologically more natural. Other people have argued that classical logic is unnatural because it allows irrelevant inferences such as "ex falso quodlibet" and "verum ex quodlibet". Some of these people argue that paraconsistent logics can avoid these problems. On the other hand, for many people the paraconsistent assumption that a proposition can be both true and false is itself highly unnatural, and the paradoxes of relevance are better solved within a Grice-inspired theory of relevance (Schurz 1991, Schurz/Weingartner 2010). The *upshot* of these considerations is that no general epistemic justification of logical systems can be obtained by way of *intuitions* of this sort, because these intuitions are highly subjective. In contrast, optimality justifications based on universal translation relations are perfectly objective and thus epistemically preferable.

6.4 The Meta-Level Objection

In sect. 5.4 we based our generalized conjecture on the argument that almost all non-classical logics use classical logic in their metalanguage in which they formulate their semantics. A perfectly legitimate objection against this assumption may be that this is not generally true. There are some proponents of non-classical logic – for example adherents of intuitionistic logic – that insist on meta-level coherence and use their non-classical logic also at the meta-level. First of all, note that my conjecture is only based the assumption that there *exists* a classical semantics for the non-classical logic, not that its adherent has to accept it. For example, intuitionistic logic has a correct and complete classical Kripke frame semantics (Moschowakis 2018, sec. 5.1), although Brouwer would presumably not be happy with it.

However, there is a deeper objection lurking behind the question of meta-level reflection: in what kind of meta-logic is the proof of the translation theorem carried out? Prima facie, I would answer: in classical logic (because this is what I intuitively assume). However, the proponent of say an intuitionistic logic could object that (s)he accepts only intuitionistically valid meta-logical proof steps, which means that this proof must not employ the classical versions of double negation or indirect proof, by which one can infer A from $\neg\neg A$. My answer to this latter objection is twofold:

Answer 1: Upon closer inspection, the logical rules that are assumed in the meta-logical proof of the translation theorem are extremely weak. Since the axioms Ax_{Luk} are valid in both L_2 and L_3, all what is needed to carry out the proof is the rule of *replacement of logical equivalents*, which formulated as sequent rule says: $\Gamma \models A \leftrightarrow B / \Gamma \models C \leftrightarrow C[B/A]$, where "$\leftrightarrow$" is the equivalence connective of the non-classical logic. Thus roughly speaking the meta-logical proof is acceptable also in all non-classical logics accepting the replacement of equivalents.

Answer 2: Replacement of equivalent is valid in almost all but not all logics, for example, not in hyperintensional logics (cf. Leitgeb 2018). More generally speaking, there may be cases of a logic L^* translatable into classical logic L_2 by means of a meta-logical proof that employs rules not accepted by the non-classical logic L^*. What should we infer epistemologically from such a situation? Of course, for the adherent of L_2 the meta-logical proof is still acceptable and she can reasonably argue that she has a way of translating logic L^* into her system. However, what we do not have established in this situation is a *full commensurability* of the non-classical logic L^* with classical logic, because the defender of L^* (at the meta-level) will not accept the possibility of such a translation.

As a further step we could try to apply the translation function at the meta-level and translate the meta logical proof of the $L^* \rightarrow L_c$-translation theorem into the meta-language of L^*. The meta-level adherent of L^* would then have to accept the translation of this proof – because it holds by the axiomatic characterizations of the operators T_2 and F_2 which are stipulated 'by definition' – and, thus, she would have to accept the assertion "the $L^* \rightarrow L_c$-translation theorem is True$_2$". Even if this is the case, she may diagnose this result as useless, because she rejects the bivalent truth-operator "T_2" as being without clear meaning; in particular, the result doesn't entail that the translation theorem is valid in the L^*-meta-logic. Thus we better try to establish commensurability also at the meta-level. Given that the meta-logical proof of the translation thesis employs only extremely weak proof steps, there is legitimate hope that it should be acceptable by all or at least almost all non-classical logicians, and if not, that there are appropriate weakenings of the translation thesis that are acceptable by non-classical logicians. The investigation of this hopeful conjectures is again work for the future.

Bibliography

Achinstein, Peter (1974): "Self-Supporting Inductive Arguments." In: *The Justification of Induction*. Swinburne, R. (ed.). Oxford: Oxford University Press, 134 – 138.
Béziau, Jean-Yves (1999): "A Sequent Calculus for Łukasiewicz's Three-Valued Logic Based on Susko's Bivalent Semantics." In: *Bulletin of the Section of Logic*, Vol. 28, No. 2, 89 – 97.
Haack, Susan (1976): "The Justification of Deduction." In: *Mind*. LXXXV.337, 112 – 119.
Jeřábek, E. (2012): "The Ubiquity of Conservative Translations." In: *Review of Symbolic Logic*. Vol. 5, No. 4, 666 – 678.
Kuhn, T.S. (1962): *The Structure of Scientific Revolutions*. Chicago: Univ. of Chicago Press (3rd edition 1996).
Leitgeb, H. (2018): "HYPE: A System of Hyperintensional Logic." Appears in *Journal of Philosophical Logic*. (doi.org/10.1007/s10992-018-9467-0).
Łukasiewicz, J. (1920): "O logice trojwartosciowej." In: *Ruch Filozoficny*. Vol. 5, 170 – 171. Translated into English in: Łukasiewicz, J. (1970), *Selected Works*. Borkowski, L. (ed.). Amsterdam: North-Holland and Warsaw.
Malinowski, G. (1993): *Many-Valued Logics*. Oxford: Clarendon Press.
Moschowakis, J. (2018): "Intuitionistic Logic." In: *The Stanford Encyclopedia of Philosophy*. (Winter 2018 Edition) (plato.stanford.edu/archives/win2018/entries/logic-intuitionistic/).
Priest, G. (1979): "Logic of Paradox." In: *Journal of Philosophical Logic*. Vol. 8, 219 – 241.
Priest, G. (2013): "Paraconsistent Logic." In: *Stanford Encyclopedia of Philosophy*. Zalta, E. (ed.), http://plato.stanford.edu.
Reichenbach, H. (1949): *The Theory of Probability*. Berkeley: University of California Press.
Rutz, P. (1972): *Zweiwertige und mehrwertige Logik*. München: Ehrenwirth Verlag.
Salmon, W. C. (1957): "Should We Attempt to Justify Induction?" In: *Philosophical Studies*. Vol. 8, No. 3, 45 – 47.
Schurz, G. (1991): "Relevant Deduction." In: *Erkenntnis*. Vol. 35, 391– 437.
Schurz, G. (2008a): "Third-Person Internalism: A Critical Examination of Externalism and a Foundation-Oriented Alternative." In: *Acta Analytica*. Vol. 23, 9 – 28.
Schurz, G. (2008b): "The Meta-Inductivist's Winning Strategy in the Prediction Game." In: *Philosophy of Science*, Vol. 75, 278 – 305.
Schurz, G. (2018): "Optimality Justifications: New Foundations for Foundation-Oriented Epistemology." In: *Synthese*, Vol. 195, 3877 – 3897.
Schurz, G. (2019): *Hume's Problem Solved: The Optimality of Meta-Induction*. Cambridge (MA): MIT Press.

Schurz, G. and Weingartner, P. (2010): "Zwart and Franssen's Impossibility Theorem Holds for Possible-World-Accounts but not for Consequence-Accounts to Verisimilitude." In: *Synthese*, Vol. 172, 415 – 436.

Suszko, R. (1977): "The Fregean Axiom and Polish Mathematical Logic in the 1920's." In: *Studia Logica*. Vol. 36, 373 – 380.

Wansing, H. and Shramko, Y. (2008): "Suszko's Thesis, Inferential Many-Valuedness, and the Notion of a Logical System." In: *Studia Logica*. Vol. 88, 405 – 429.

Gila Sher
Invariance and Necessity

Abstract: Properties and relations in general have a certain degree of invariance, and some types of properties/relations have a stronger degree of invariance than others. In this paper I will show how the degrees of invariance of different types of properties are associated with, and explain, the modal force of the laws governing them. This explains differences in the modal force of laws/principles of different disciplines, starting with logic and mathematics and proceeding to physics and biology.

1 Introduction

Many philosophers are perplexed by necessity. What is the source of necessity – The world? Our mind (language, concepts, built-in categories)? Is necessity epistemic or metaphysical? Is it possible to explain the necessity of logical laws (logical truths, logical consequences), mathematical laws, physical laws? Are they necessary at all? Do they have the same kind of necessity? Are there many kinds of worldly (metaphysical) necessity or just one kind?

Humeans are deeply disturbed by the thought of necessary physical laws: They view the idea of such laws as an idea of secret, hidden, inexplicable, mystical powers that govern the world, as a remnant of the idea of laws created by God, or by some supernatural forces ...

In this paper I will offer a philosophically systematic explanation of necessity in terms of invariance. This explanation does not purport to say everything there is to say about necessity or laws. But it captures something basic and significant about them, answers some of the above questions, and partly removes some of the worries. And it's down-to-earth in the sense of not invoking any mysterious traits or requiring any mysterious mental capacities.

2 The Idea of Invariance

Invariance in general is a relation: X is invariant under Y. Examples:

Gila Sher, Department of Philosophy, University of California, San Diego, gsher@ucsd.edu

DOI 10.1515/9783110657883-4

1. Logical truths are invariant under changes in (Tarskian) models: They are not affected by such changes. You can replace one model by another, and the logical truths won't "notice", so to speak. If they hold in one model, they hold in all.
2. The laws of physics are invariant under changes of inertial reference frames. They are the same in all such frames. They are indifferent to or don't notice changes in such frames.
3. The laws of universal grammar are invariant under variations in natural language. They hold in all natural languages. They don't distinguish between one natural language and another. They don't even distinguish between actual and (merely) possible natural languages.
4. Logical constants are invariant under all isomorphisms. (Explanation: later on).

3 Fruitfulness and Explanatory Power of Invariance

As these examples suggest, the idea of invariance is very fruitful and has a strong explanatory power. Here are a few citations:

Science
1. Eugene Wigner: "There is a Structure in the laws of nature which we call laws of invariance. This structure is so far-reaching [that] in Some cases ... we guessed [laws of nature] on the basis of the postulate that they fit into the invariance structure. ... [L]aws of nature could not exist without principles of invariance." (Wigner 1967: 29)
2. Jim Woodward: "[E]xplanatory relations must be invariant relations, where a relation is invariant if it remains stable or unchanged as we change various other things. ... [L]aws describe invariant relationships". (Woodward 1997: 26 – 27)

Philosophy
1. Robert Nozick: "Questions about objectiveness depend upon the range of transformations under which something is invariant". (Nozick 2001: 10)
2. Kit Fine: Generality or abstractness are a matter of sensitivity to descriptive differences: More general elements are less sensitive and less general

elements are more sensitive. How do we understand sensitivity? – In terms of invariance. (Paraphrase, Fine 2011: 17)

Mathematics
1. Felix Klein: Developed the Erlangen Program, which offers a characterization of geometries in terms of the transformations under which their notions are invariant.
2. Kronecker: "When the concept of invariants ... is tied ... to the general concept of *equivalence*, then [it] reaches the most general realm of thought". (Cited in Mancosu 2016: 15)

Logic
Tarski: "I suggest that ... we call a notion 'logical' if it is invariant under all possible one-one transformations of the world onto itself". (Tarski 1966/86: 149)

4 Invariance of What?

Different things are invariant under different changes. In this talk I will focus on invariance of *properties* (relations) under certain changes. I will explain necessity in terms of such invariance.

Background Notes. Leaving aside, for the purpose of this paper, many questions concerning objects, properties, and the world, I will start with the following common-sensical picture:

1. The *world* contains: Individuals – level 0; n-place properties of individuals – level 1; properties of properties of level 1 – level 2; ...
2. *Object*: individual or property.
3. An individual can be either *actual* or *non-actual* (counterfactual). The world includes both actual and counterfactual individuals.
4. *Properties* are identified by their actual plus counterfactual extension.
5. The paper focuses on properties of the kind that science, mathematics, and logic are interested in. ("Properties" that lead to paradox, for example, are excluded.) Complete formal systematization: a later task. (Requires a solid idea to systematize).

5 Central Claims Concerning Invariance and Necessity

1. Every property is invariant under some 1–1 [and onto] replacements of individuals.
2. Some properties have a higher degree of invariance than others.
3. The greater the degree of invariance of a given property, the greater the degree of necessity of the laws governing it.
4. This is generalizable to fields of knowledge. It explains why (and in what sense) logical laws have a greater degree of necessity than physical laws, physical laws have a greater degree of necessity than biological laws/regularities, etc.

5.1 Claim 1: *Every property is invariant under some 1–1 [and onto] replacements of individuals.*

Let me begin with an observation about properties. Properties in general are *selective*. They "pay attention" or "are attuned" to *some* features of objects but not others. For example:

The 1st-level property *is-a-human* is not attuned to differences in gender. Gravity (the property of being subject to gravitational forces) is not attuned to differences between living and non-living objects. This introduces the possibility of characterizing, or comparing, properties in terms of what they are attuned and not attuned to. What changes in the world they "notice" and what changes they are "blind" to. This is what my characterization/comparison of properties in terms of invariance will do.

Heuristic notes
1. For ease of explanation I help myself to the language of set-theory and assume bivalence. But this is not essential for the claims made in this paper.
2. For ease of presentation, I think of the total collection of actual plus counterfactual individuals as divisible into domains.
3. In these terms I think of properties as characterized by which objects they hold of in different (actual plus counterfactual) domains.

Notation
D (Domain): a non-empty set of individuals.
r (Replacement Function): any 1–1 function from some domain D onto a domain D' (possibly $D = D'$).

Using this notation, the claim is: $(\forall P)(\exists r)(P$ is invariant under $r)$. r is indexed to some domain D, or a pair of domains, $\langle D, D' \rangle$. When we index it just to D, the understanding is that D' is its range, whatever this is.

The claim itself is trivially true, since every property is invariant under *identity replacements*: functions r that replace each individual in D by itself. But my claim is stronger: properties are commonly invariant under more than just identity r's.

Case 1: 1st-level Properties
Example: *is-a-human*.

Consider:
$D = \{$Obama, Tarski, 1, 2$\}$.
$r : r($Obama$) = $ Trump, $r($Tarski$) = $ Frege, $r(1) = 4$, $r(2) = $ Mt. Everest.
($D' = \{$Trump, Frege, 4, Mt. Everest$\}$.)

Claim: The 1st-level property *is-a-human* is invariant under this r.

Why? Because r replaces each individual that has the property *is-a-human* by an individual that also *has* the property *is-a-human* and each individual that *doesn't have* the property *is-a-human* by an individual that *doesn't have* the property *is-a-human*.

Note that *is-a-human* is not invariant under *all* 1–1 r's. For example, it is not invariant under:

$r' : r'($Obama$) = 1$, $r'($Tarski$) = 2$, $r'(1) = $ Tree1, $r'(2) = $ Tree2.

Because under this r, the property *is-a-human* is changed to (replaced by) the property *is-a-number*. So, *is-a-human* is invariant under *some*, but *not all* r's.

In Fact: *Every* n-place 1st-level property is invariant under *some* r, but *many* 1st-level properties are *not* invariant under *all* r's.

Case 2: 2nd-level properties
Example: IS-A-PROPERTY-OF-MAMMALS. (This is a property of all 1st-level properties that are applicable in principle to Mammals.)

Among the 1st-level properties that have this 2nd-level property are the properties *is-a-human* and *is-a-horse*. Among the 1st-level properties that don't have this 2nd-level property are the properties *is-a-number* and *is-a-tree*.

Explanation: Consider the domain D as above. I.e., $D = \{Obama, Tarski, 1, 2\}$. Let r be a 1-1 function on D where:

r(Obama) = Horse 1, r(Tarski) = Horse 2, $r(1)$ = Tree 1, $r(2)$ = Tree 2.

Claim: IS-A-PROPERTY-OF-MAMMALS is invariant under this r.

Why? Because r induces a replacement of the 1st-level property *is-a-human* by the 1st-level property *is-a-horse*. But the 2nd-level property IS-A-PROPERTY-OF-MAMMALS does not notice this change. From the point of view of IS-A-PROPERTY-OF-MAMMALS there is no difference bet the 1st-level properties *is-a-human* and *is-a-horse*.

Note too that like *is-a-human*, IS-A-PROPERTY-OF-MAMMALS is *not* invariant under every r. For example, it is *not* invariant under r':

r'(Obama) = Tree 1, r'(Tarski) = Tree 2, $r'(1)$ = Snake 1, $r'(2)$ = Snake 2.

r' induces a replacement of *is-a-human* by *is-a-tree*, and *is-a-tree* is not A-PROPERTY-OF-MAMMALS.

In a similar way, any 2nd-level property is invariant under *some* r's, but *many* 2nd-level properties are *not* invariant under *every* r. This applies to properties of any level.

5.2 Claim 2: *Some properties have a higher degree of invariance than others.*

While properties in general are invariant under *some* 1–1 replacements of individuals, *not all* properties are invariant under *the same* 1–1 replacements. This suggests that *some* properties are invariant under *more* 1–1 replacements of individuals than others. "More" can be measured in several ways. As a starting point we understand "more" in the sense of *inclusion*:

P_1 is invariant under more r's than P_2
iff
$\{r: P_1$ is invariant under $r\} \supsetneq \{r: P_2$ is invariant under $r\}$.

Terminology
1. "Degree of invariance of P". The degree of invariance of $P - DI(P) -$ is the class of all r's such that P is invariant under $r : DI(P) =_{Df} \{r : P \text{ is invariant under } r\}$. Clearly, every property has a degree of invariance.
2. "Greater degree of invariance":

$$DI(P_1) > DI(P_2)$$
iff
P_1 is invariant under *more* 1–1 replacements of individuals than P_2 (in the inclusion sense).

Clearly, greater degree of invariance – $DI(X) > DI(Y)$ – is a partial ordering.

Back to claim 2: Some properties have a higher degree of invariance than others. To establish this claim we observe that some properties are invariant under *all* r's. The properties we have discussed so far are not of this kind, but some properties are. For example, *is-identical-to* is invariant under *all* r's. For any r (on any D) and any a, b in D:

$$a = b \rightarrow r(a) = r(b),$$
$$a \neq b \rightarrow r(a) \neq r(b).$$

Clearly, properties that are invariant under *all* r's have a *higher degree* of invariance than other properties. We will say that these properties have a *maximal* degree of invariance.

5.3 Claim 3: *The greater the degree of invariance of a given property, the greater the degree of necessity of its laws.*

Let's look at identity (*is-identical-to*) again. More specifically, let's look at what its *maximal degree of invariance*, by itself, tells us (or determines) about it. (This is informative, because whatever holds of identity due to its maximal invariance holds of *all* properties with maximal invariance.) Let us think of the *idea* of laws *of identity* (intuitively, principles that govern/describe the behavior of identity in all areas, actual and counterfactual, to which it applies), *not* of the *specific* laws of identity. The maximal invariance of identity determines that if a statement/fact is a law of identity, then it holds in *all* domains of individuals, *actual and counterfactual*. Why? Because: Given that identity itself doesn't distinguish between any individuals, its laws cannot be limited to just some domains of individuals. In particular: if the laws of identity hold of *any* individuals, they hold of *all actual and counterfactual* individuals. That is, the laws of identity apply to the *totality* of individuals, *actual and counterfactual*. Conclusion: whatever the laws of identity are, they hold of the totality of individuals, actual and counterfactual. I.e., the laws of identity have a maximal (actual and) counterfactual scope.

Now, given the common understanding of *necessity* in terms of (*actual-*)*counterfactual* scope – namely, for any fact/statement/law X: necessary (X) iff X holds in all (actual-) counterfactual domains – this means that the *laws of identity* (whatever they are) have *maximal necessity*. So:

$$\text{Maximal invariance} \to \text{Maximal necessity.}$$

P is maximally-invariant → Laws of P have maximal actual-counterfactual scope → Laws of P have maximal necessity.

This can be generalized in two directions:

1. The laws of all properties with maximal invariance have maximal necessity.
2. The correlation between degrees of invariance and necessity holds in general, i.e., for laws of all properties (not just for laws of maximally-invariant properties).

Explanations

1. We have seen that due to identity's maximal degree of invariance, its laws, whatever they are, have a maximal degree of necessity. Since this is due just to the maximal degree of invariance of identity, it holds of all properties with a maximal degree of invariance: if they are governed by any laws, they are governed by laws with a maximal degree of necessity. Let us call such properties "maximally-invariant properties", or for short "maximal properties". So: the fact that

$$\text{Degree of invariance of maximal properties}$$
$$>$$
$$\text{Degree of invariance of non-maximal properties,}$$
$$\to:$$
$$\text{Degree of necessity of laws governing maximal properties}$$
$$>$$
$$\text{Degree of necessity of laws governing non-maximal properties.}$$

(The actual-counterfactual scope of the former is greater than that of the latter).

2. The second generalization says that the correlation between degree of invariance and degree of necessity is general. Take, for example, the following properties:

(1) Gravity: x-is-subject-to-gravitational-forces
(2) Evolution: x-is-subject-to-evolutionary-forces.

Intuitively, $DI(1) > DI(2)$.

What leads us to think that this is the case?

> Consider an r that replaces animate objects by inanimate objects.
> (1) – gravity – is invariant under this r, but (2) – evolution – is not.
> →: (1) remains the same in more actual-counterfactual domains than (2).
> →: The actual-counterfactual scope of laws of (1) > the actual-counterfactual scope of laws of (2).

And, in general:

$DI(P_1) > DI(P_2)$
→: P_2 distinguishes between more actual-counterfactual individuals than P_1
→: Actual-counterfactual scope of laws of P_1 > actual-counterfactual scope of laws of P_2
→: $DN(P_1) > DN(P_2)$.

Note: Technically we have to adjust our conception of ">" for pairs of properties which are both not maximally invariant. (Thanks to an anonymous participant in the Wittgenstein Symposium for pointing this out.) For a proposal see Sher, work in progress.

5.4 Claim 4: *The relation between greater property invariance and greater necessity of laws is generalizable to fields of knowledge.*

Take physics, for example. It is quite clear that generally:

DI of maximal properties
>
DI of physical properties.

And, given the systematic connection between invariance and necessity:

DN of laws of maximal properties
>

DN of laws of physical properties.

If we identify a field of knowledge X whose defining properties are maximally invariant, this will establish the claim that

DN of laws of Field X
>
DN of laws of Physics

and explain why this is the case. And if (following the refinement of the definition of ">") we can show that *DI*(the most highly invariant physical properties) > *DI*(the most highly invariant biological properties), we will be able to show that, and explain why, the degree of necessity of physical laws (associated with highly invariant physical properties) is greater than the degree of necessity of biological principles/laws.

6 Application to Logic and Mathematics

We have seen that identity is maximally invariant. It is easy to see that all the properties denoted by the standard logical constants of predicate logic (of any order) have maximal invariance. For example: Consider "$\exists$", viewed as denoting the 2nd-level property of NON-EMPTINESS ["$(\exists x)Px$" says that the (1st-level) property P is not empty].

Clearly, given *any* domain D and a 1–1 replacement r of the Individuals of D, P is replaced by a property P' such that NON-EMPTY (P) iff NON-EMPTY (P'). So: IS-NONEMPTY ($\exists$) is invariant under *all* r's, i.e., is maximally invariant. Similarly, NEGATION ($\sim$), which in predicate-logic contexts ("$\sim Px$") denotes the 2nd-level property of COMPLEMENTATION, is maximally invariant: If r takes P in D to P' in D', then it takes COMPLEMENT(P) in D to COMPLEMENT(P') in D'. So: COMPLEMENTATION($\sim$) has maximal invariance. And the same holds for all the standard logical constants of predicate logic.

From Invariance under r to Invariance under Isomorphisms. Now, if we talk in terms of structures, $\langle D, P \rangle$, $\langle D', P' \rangle$, where D, D' are any domains and P, P' are extensions of properties in D, D', respectively, invariance under all 1–1 r's from D onto D' becomes *invariance under all isomorphisms*. This is the basis for the claim that logical constants are invariant under isomorphisms (in various forms: Mostowski 1957, Lindström 1966, Tarski 1966/86, Sher 1991, and so on). I.e., invariance under isomorphisms is a *criterion* for logical operators (properties), and with a few additional requirements (see Sher 1991), also a criterion for logical constants.

From Logical Constants to Logical Laws. The laws of logic are the laws governing the denotations of the logical constants: the laws of identity, negation (complementation), disjunction (union), existential and universal quantifiers (non-emptiness and universality or empty complement), finite cardinality quantifiers, etc. Since the properties denoted by these constants have a maximal degree of invariance, the relation, established above, between invariance and necessity, determines that the laws governing logical constants/properties have a maximal degree of necessity. Note: this will not change if we add more constants denoting maximal properties to standard logic as logical constants (e.g., MOST, FINITLEY MANY, INDENUMERABLY MANY, IS-SYMMETRIC, IS-WELL-ORDERED, etc.). As a result, the laws governing the resulting logics (their logical truths and logical consequences) will be just as necessary.

We can now close the circle:

Degree of necessity of logical laws
>
Degree of necessity of physical laws
[>
Degree of necessity of biological laws (principles)].

The distinction between types of necessity gives rise to a distinction between types or spaces of possibility:

Space of logical possibility
>
Space of physical possibility
[>
Space of biological possibility].

Ex.: an object that is not subject to gravity is not physically possible, but it is logically possible.

There is much more to say about invariance, necessity, and logic. My most thorough discussions of this topic appears in Sher (1991 and 2016, Part IV). Here I will limit myself to one point, concerning the relation between logic and mathematics:

7 Relation Between Logic and Mathematics

Going back to maximally-invariant properties we notice two interesting things:

1. Not just logical properties, but also many mathematical properties (2nd-level complementation, intersection, inclusion, cardinality properties, ...), are maximally invariant. In fact: the totality of maximally-invariant properties is the totality of mathematical properties, construed (in the large majority of cases) as higher-level properties.
2. Yet: in mathematics itself, mathematical properties are often construed as 1st-level properties, and 1st-level mathematical properties are, for the most part, not maximally-invariant. Examples: Consider the 1st-level properties *is-the-number-1* (= 1), where 1 is an individual, and *is-an-odd-number*. These 1st-level properties are not invariant under all r's. E.g., they are not invariant under any r such that $r(1)$ = Barack Obama. But they are correlated with higher-level maximally-invariant mathematical properties. (Indeed, even mathematical individuals are correlated with higher-level maximally-invariant mathematical properties. For example, the individual 1 is correlated with the 2nd-level cardinality property ONE.) So, what we have is:
    a. Maximally invariant properties include both higher-level mathematical properties and logical properties (which coincide with certain mathematical properties, such as NON-EMPTINESS ($\exists$)).
    b. Maximally-invariant higher-level mathematical properties are systematically correlated with lower-level mathematical properties which are not maximally invariant.

This suggests that there is something in common to logical properties and maximally-invariant mathematical properties, something that also underlies the correlation of non-maximally-invariant mathematical properties with maximally-invariant mathematical properties. Elsewhere I suggested that this common feature is *formality*. Maximal invariance is a criterion of formality, and formality is the common basis of logic and mathematics.

Questions:

1. Is this logicism? No: mathematics is not reducible to logic.
2. Is this identity (mathematics = logic)? No.

What, then, is the relation between logic and mathematics. My answer is that there is a division of labor between mathematics and logic. Mathematics studies formal properties and their laws. Logic builds formal properties into our language as logical-constants, to be used as tools of reasoning, based on the laws of the formal properties they denote: laws of identity, complementation, intersection, inclusion, non-emptiness, finite cardinalities, possibly infinite cardinalities, etc.

One possible objection is: You argue for a close connection between logic and math. But logic is trivial, while mathematics is not. Response: Both logic and mathematics have trivial and non-trivial parts. Examples: both $1 + 1 = 2$ and $Pa \& Qa \rightarrow Pa$ are trivial; both the content of Cantor's (mathematical) theorem (which says that the cardinality of the power set of s is larger than the cardinality of s) and its *logical* derivation from the axioms of ZFC/Z are not trivial.

How do we explain the relation between higher- and lower-level mathematics? In spite of the fact that lower- and higher- level mathematics have different degrees of invariance, there is a systematic connection between them. To understand both the connection and the difference between them, let's note that to understand any field of knowledge, we have to distinguish two questions:

1. What does it study?
2. How does it study it?

What a given field of knowledge studies depends on:

1. What there is to be studied.
2. What the field in question is interested in.

How the given field studies what it is interested in depends on:

1. Humans' cognitive resources.
2. Pragmatic and methodological considerations.

On the present proposal:

1. What does mathematics study? – Formal properties and the laws governing them.
2. How does mathematics study these? – Often, by constructing 1st-level "models" of these properties and laws, i.e., 1st-level structures representing (higher-level) formal properties, and studying these structures (e.g., studying finite cardinality properties by studying numbers (numerical individuals)).
3. Why does mathematics do this? There are many possible explanations. For example: humans may be better at figuring out formal relations when they think in terms of structures of individuals rather than in terms of structures

of properties, properties of properties, and so on. (This is something for cognitive science to investigate.) But the question which is more relevant to us, as philosophers, is: Is it possible, in principle, to adequately study an object of one kind by studying an object of a different kind that systematically represents it? Answer: Yes. (Analogy: designing a big skyscraper, made of steel and concrete by working with a small plastic model of the skyscraper.)

What about necessity? Are the laws of 1st-order arithmetic and set-theory necessary? Answer: Yes. They acquire their necessity from the fact that they represent formal laws, laws that, due to the high degree of invariance of the properties they hold of, are highly – indeed, maximally – necessary.

8 Ramifications for Metaphysical Necessity and Scientific Laws

Necessity is usually viewed as a metaphysical subject-matter. But we explained several basic things about necessity without referring to metaphysics. Why? And what are the degrees of invariance and necessity of metaphysical properties and laws (principles)? Answer: The subject-matter of metaphysics is much less homogeneous, and much more difficult to demarcate, than that of logic, mathematics, physics, or biology. Focusing on invariance, metaphysics deals both with very broad subject matters, such as objects in general, and narrower subject matters with principles such as:

- An object cannot be both all of one color and of another color (at the same time).
- An object cannot have two different temperatures at the same time.

This suggests that metaphysics can be divided into sub-fields: 1. Those whose degree of invariance, hence necessity, is as high as that of logic and mathematics. 2. Those whose degree of invariance, hence, necessity, is lower. Overall, metaphysical invariance, hence necessity, seem to lie in between those of logic/mathematics and physics.

Turning to laws of nature, I mentioned earlier that philosophers of science are especially worried about the necessity of scientific laws, including physical laws. They are also concerned that general philosophy may usurp science. So let me focus on explaining what the philosophical investigation of invariance does, and does not, tell us about the necessity of physical laws. It tells us that the fact that many physical properties have fairly strong degrees of invariance shows that there are objective grounds for modally strong physical laws. But it does not tell us to what extent the world is governed by physical laws and what specific laws it is governed by.

Invariance shows that the world is ready, so to speak, or has an appropriate infrastructure, for being governed by necessary laws. Example: Invariance says that gravity is a candidate for physical laws, while the property of being Mount Everest is not, and it says that if gravity is governed by laws, they have a fairly strong modal force. But it doesn't determine whether such laws would be Newtonian or relativistic. There are also types of necessity that invariance doesn't explain, though they are arguably less philosophically significant than those it does explain. For example: it doesn't explain (1) necessity as fixity or (2) singular necessity (unless they are based on invariance considerations).

1. *Necessity as Fixity*. In mathematics, science, and everyday life, we often identify a property by connecting it to other properties. We often say: "let P be ..." . Say: "Let P be the property of being a set (or a natural number)". Then, this determines the identity of P in advance, and in a sense makes the principles characterizing it necessary (in the given discourse or theory). Similarly, often – either before, or during, or after a given investigation – we identify a property by saying what it is constituted of. Example: "Let rest-energy be energy that does not vanish at 0 speed". Or we can just stipulate: "Let *is-a-dagnet* be the property of being a crazy politician". Then, it is necessary, in the fixity sense, that

(i) Rest energy does not vanish at 0 speed.
(ii) Dagnets are crazy.

This kind of necessity is not explained by invariance (unless there are theoretical reasons, and in particular, reasons involving invariance, for deciding what to hold fixed and how to fix it).

2. *Singular Necessity*. There are a number of fundamental constants in physics: c – the speed of light in a vacuum, h – the Planck constant: a number that the energy of any body must be a multiple of.

These constants are embedded in physically-necessary laws, but they cannot be explained by invariance, unless they follow from considerations of invariance. So if, and to the extent that "Nothing moves faster than c" is required for the invariance of physical laws, its necessity is explained by invariance; otherwise, it is not.

There is much more to say about invariance, including the relation between invariance and abstraction, apriority, and generality. But I leave this for another occasion.

Acknowledgement

I would like to thank the audience at the Wittgenstein Symposium for very helpful comments.

Bibliography

Fine, K. (2011): "What is Metaphysics?" In: T. E. Tahko (ed.). *Contemporary Aristotelian Metaphysics*. Cambridge: Cambridge University Press. Pp. 8–25.

Lindström, P. (1966): "First Order Predicate Logic with Generalized Quantifiers." In: *Theoria*, Vol. 32, 186 – 195.

Mancosu, P. (2016): *Abstraction and Infinity*. Oxford: Oxford University Press.

Mostowski, A. (1957): "On a Generalization of Quantifiers." In: *Fund. Mathematicae*, Vol. 44: 12 – 36.

Nozick, R. (2001): *Invariances: The Structure of the Objective World*. Cambridge: Harvard University Press.

Sher, G. (1991): *The Bounds of Logic: A Generalized Viewpoint*. Cambridge: MIT Press.

Sher, G. (2016): *Epistemic Friction: An Essay on Knowledge, Truth, and Logic*. Oxford: Oxford University Press.

Sher, G.: "Invariance as a Basis for Necessity and Laws". Work in Progress.

Tarski, A. (1966/86): "What are Logical Notions?". In: *History and Philosophy of Logic*, Vol. 7, 1986, 143 – 154.

Wigner, E. (1967): "The Role of Invariance Principles in Natural Philosophy". In: *Symmetries and Reflections*. Bloomington (IN): Indiana University Press, 28 – 37.

Woodward, J. (1997): "Explanation, Invariance, and Intervention." In: *Philosophy of Science*, Vol. 64, Supplement, Proceedings of the 1996 Biennial Meetings of the Philosophy of Science Association. Part II: Symposia Papers, S26–S41.

Itala M. Loffredo D'Ottaviano and Hércules de Araújo Feitosa
Translations Between Logics: A Survey

Abstract: In a series of previous papers we have studied interrelations between logics through the analysis of translations between them. In this paper, providing some brief historical background, we present a general survey of the main questions and problems we have analysed and the results we have obtained. As well as showing the interrelations between our concepts of conservative translation, contextual translation, abstract contextual translation, hypercontextual translation and the concept of isomorphism between logics, we discuss the interrelations among the distinct categories that are constituted by logics and the special types of translations between them.

1 Introduction

For several years the interrelations between logics have been studied by analysing interpretations between them. The first known "translations" concerning classical logic, intuitionistic logic and modal logic were presented by Kolmogorov 1925, Glivenko 1929, Lewis/Langford 1932, Gödel 1933a and Gentzen 1936, some of them developed mainly in order to show the relative consistency of classical logic with respect to intuitionistic logic.

In 1999, da Silva, D'Ottaviano and Sette proposed a very general definition for the concept of translation between logics, logics being characterized as pairs constituted by a set and a consequence operator, and translations between logics being defined as maps that preserve consequence relations (cf. Da Silva et al. 1999). In 2001, we introduced the concept of conservative translation and studied the category whose objects are logics, and whose morphisms are the conservative translations between them (cf. Feitosa/D'Ottaviano 2001). In 2007, Carnielli, Coniglio and D'Ottaviano proposed the concept of contextual translation in order to have a stricter notion of translation and to solve questions related to conservative translations (cf. D'Ottaviano/Feitosa 2007, Carnielli et al. 2009).

Conservative and contextual translations showed themselves, however, to be independent concepts. Recently, with Moreira, we introduced the concept of abstract contextual translation between logics and proved that this new concept is an intermediate concept, relative to the concept of translation, wider than the

---

Itala M. Loffredo D'Ottaviano, University of Campinas, Brazil, itala@cle.unicamp.br
Hércules de Araújo Feitosa, São Paulo State University, Brazil, haf@fc.unesp.br

DOI 10.1515/9783110657883-5

concepts of conservative and contextual translation. We also studied other, stricter, concepts of translations: the conservative-contextual and the hypercontextual translations.

In this paper, providing some brief historical background, we present, without proofs, a general survey of the main questions and problems we have analysed and the results we have obtained. As well as showing the interrelations between our concepts of translation and the concept of isomorphism between logics, we discuss the interrelations among the distinct categories that are constituted by logics and the special types of translations between them.

2 Brief Historical Remarks

The method of studying interrelations between logical systems by the analysis of translations between them was originally introduced by Kolmogorov, in 1925.

The aim of Kolmogorov 1925 "is to explain why" the illegitimate use of the principle of excluded middle in the domain of transfinite arguments "has not yet led to contradictions". Kolmogorov introduces the *general logic of judgements* **B**, and the *special logic of judgements* **H**, that is proved to be equivalent to the classical propositional calculus introduced by Hilbert in 1923 (Hilbert 1923).

It is defined inductively a function k, that associates every formula α of **H** to a formula α^k of **B**, $k : For(\mathbf{H}) \to For(\mathbf{B})$, by adding a double negation in front of every subformula of α. Then, given a set of axioms $\Gamma = \{\alpha_1, ..., \alpha_n\}$, with $\Gamma^k = \{\alpha_1^k, ..., \alpha_n^k\}$, it is proved that[1]:

$$\Gamma \vdash_{\mathbf{H}} \alpha \Rightarrow \Gamma^k \vdash_{\mathbf{B}} \alpha^k.$$

Kolmogorov suggests that a similar result can be extended to quantificational systems and, in general, to all known mathematics, anticipating Gödel's and Gentzen's results on the relative consistency of classical arithmetic with respect to intuitionistic arithmetic (see Feitosa 1997, Feitosa/D'Ottaviano 2001, Carnielli et al. 2009, D'Ottaviano/Feitosa 2011, and D'Ottaviano 2013).

Also related to intuitionism, Glivenko 1929 proves that, if α is a theorem of classical propositional logic **CPL**, then the double negation of α is a theorem of intuitionistic propositional logic **IPL**.

In Gödel 1933a, Gödel proves that, if α is a theorem of **CPL** then, under a specific interpretation g_1, the interpretation of α is a theorem of **IPL**. Gödel shows that this result is also valid relatively to intuitionistic arithmetic **H'** and classical number

---

[1] As usually, the symbol $\Rightarrow$ denotes "if ..., then ..." and the symbol $\Leftrightarrow$ denotes "if, and only if, ...".

theory **PA**:

$$\vdash_{\mathbf{PA}} \alpha \Leftrightarrow \vdash_{\mathbf{H'}} g_1(\alpha).$$

In another short paper of 1933, Gödel introduces an interpretation g_2, that also preserves theoremhood, from **IPL** into a system **G** that is "equivalent" to Lewis' system of strict implication **S4** (see Gödel 1933b and D'Ottaviano/Feitosa 2011).

Yet in 1933, Gentzen published a rigorous and complete paper, with a simpler translation from **CPC** into **IPC**. The aim of Gentzen 1936 is to show that "the applications of the law of double negation in proofs of classical arithmetic can in many instances be eliminated", it "is to a large extent intuitionistically valid".

Gentzen presents a simpler translation g_z from **CPL** into **IPL** and proves that:

$$\vdash_{\mathbf{CPL}} \alpha \Leftrightarrow \vdash_{\mathbf{IPL}} g_z(\alpha).$$

And, as a consequence, he presents a constructive proof of the consistency of elementary classical arithmetic with respect to intuitionistic arithmetic (see Gentzen 1936, Gentzen 1969).

In spite of Kolmogorov, Glivenko, Gödel and Gentzen dealing with interrelations between the studied systems, they are not interested in the meaning of the concept of translation between logics in general. Since then, interpretations between logics have been used to different purposes.

Prawitz/Malmäs 1968 survey these historical papers and is the first paper in which a general definition for the concept of translation between logical systems is introduced. For these two authors, a *translation* from a logical system S_1 into a logical system S_2 is a function $t : For(S_1) \to For(S_2)$, such that:

$$\vdash_{S_1} \alpha \Leftrightarrow \vdash_{S_2} t(\alpha).$$

The system S_1 is then said to be *interpretable* in S_2 by t. Moreover, S_1 is said to be *interpretable with respect to derivability* in S_2 by t if, for every set $\Gamma \cup \{\alpha\} \subseteq$ For (S_1) and $t(\Gamma) = \{t(y) : y \in \Gamma\}$:

$$\Gamma \vdash_{S_1} \alpha \Leftrightarrow t(\Gamma) \vdash_{S_2} t(\alpha).$$

Brown/Suszko 1973 propose a "... framework of the theory of abstract logics" generalizing familiar topological concepts but they are still not interested in translations as such.

Szczerba 1977 also tackles the question but is only concerned with translations between models.

Wójcicki 1988 and Epstein 1990 can be considered the first books with a general systematic study on translations between logics. Both study interrelations between propositional calculi in terms of translations.

For Wójcicki, a logic is a pair (L, C) such that L is a formal language and C is a Tarskian consequence operator in the free algebra of formulas of L. Given two propositional languages S_1 and S_2, with the same set of variables, a mapping $t : S_1 \to S_2$ is a *translation* if, and only if:

(i) There is a formula $\varphi(p_0)$ in S_2 in one variable p_0 such that, for each variable p, $t(p) = \varphi(p)$;

(ii) For each connective μ_i in S_1 of arity k there is a formula φ_i in S_2 in the variables $p_1, \ldots, p_k$, such that $t(\mu_i(\alpha_1, \ldots, \alpha_k)) = \varphi_i(t(\alpha_1), \ldots, t(\alpha_k))$, for $\alpha_1, \ldots, \alpha_k \in S_1$.

A *propositional calculus* is defined as a pair $C = (S, C)$, where C is a consequence operator over the language S.

Finally, for Wójcicki, $C_1 = (S_1, C_1)$ is said to be *translatable* into $C_2 = (S_2, C_2)$ if there is a mapping $t : S_1 \to S_2$, such that for all $\Gamma \cup \{\alpha\} \subseteq S_1$:

$$\alpha \in C_1(\Gamma) \Leftrightarrow t(\alpha) \in C_2(t(\Gamma)).$$

Epstein 1990 defines a *translation* of a propositional logic L into a propositional logic M, in semantical terms, as a map t from the language of L into the language of M such that, for every set $\Gamma \cup \{\alpha\} \subseteq For(L)$:

$$\Gamma \vDash_L \alpha \Leftrightarrow t(\Gamma) \vDash_M t(\alpha).$$

On the other hand, Goguen/Burstall 1984 define a general notion of logic system and his morphisms called institutions, within the framework of category theory (see also Goguen/Burstall 1992, Mossakowski et al. 2005, and Carnielli et al. 2009). Institutions generalize Tarski's notion of truth, by considering (abstract) signatures instead of vocabularies, and abstract (categorial) signature morphisms in the place of translations among vocabularies. In this way, the set of sentences are parameterized by abstract signatures.

We observe that Kolmogorov's, Glivenko's and Gentzen's interpretations are translations in the sense of Prawitz, Wójcicki and Epstein. But Gödel's ones are translations only in Prawitz' sense (see Feitosa 1997 and Feitosa/D'Ottaviano 2001).

3 Translations and Conservative Translations

Da Silva et al. 1999, motivated by D'Ottaviano 1973[2] and Hoppmann 1973[3], and explicitly interested in the study of interrelations between logic systems in general,

---

[2] In this paper variants of Tarskian closure operators characterized by interpretations are studied.
[3] This is apparently the first paper in the literature where the term "translation between general logic systems" is used to mean a function preserving derivability.

propose a general definition for the concept of translation between logics, in order "to single out what seems to be in fact the essential feature of a logical translation": logics are characterized as pairs constituted by an arbitrary set [without the usual requirement of dealing with formulas of a formal language], and a Tarskian consequence operator. Translations between logics are then defined as maps preserving consequence relations.

Definition 3.1. *A Tarskian consequence operator on a set A is a function $C : \mathcal{P}(A) \to \mathcal{P}(A)$ such that, for every $X, Y \subseteq A$:*
(i) $X \subseteq C(X)$;
(ii) $X \subseteq Y \Rightarrow C(X) \subseteq C(Y)$;
(iii) $C(C(X)) \subseteq C(X)$.

Definition 3.2. *A logic $\mathbf{A}$ is a pair (A, C), where the set A is the domain of $\mathbf{A}$ and C is a Tarskian consequence operator on A.*

It is not hard to see that a logic could be defined along the lines of a universal logic, as for instance in Béziau 1995; that is, a logic is basically a pair formed by a set of entities called formulas and a consequence relation, without assuming any properties.

The usual concepts and results on closure spaces are here assumed. The general definition of translation between logics is then presented.

Definition 3.3. *A translation from a logic $\mathbf{A}$ into a logic $\mathbf{B}$ is a map $t : A \to B$ such that, for any $X \subseteq A$:*
$$t(C_A(X)) \subseteq C_B(t(X)).$$

Of course, it is possible to consider logics defined over formal languages.

Definition 3.4. *A logic system defined over L is a pair $\mathbf{L} = (L, C)$, where L is a formal language and C is a structural consequence operator in the free algebra For(L) of the formulas of L.*

We will use the symbol **L** when the system is over a language with particular logical operators, and **A**, **B** for systems over sets only.

If $\mathbf{L}_A$ and $\mathbf{L}_B$ are determined by formal languages with associated syntactic consequence relations $\vdash_A$ and $\vdash_B$, respectively, then t is a translation if, and only if, for $\Gamma \cup \{\alpha\} \subseteq For(L_A)$:
$$\Gamma \vdash_A \alpha \Rightarrow t(\Gamma) \vdash_B t(\alpha).$$

Frequently, when formal languages are considered, it is useful to take translations following a well-defined [syntactical] pattern. This motivates the following definition.

Definition 3.5. *Let L_1 be a language containing only unary and binary connectives and L_2 another language. A translation $t : L_1 \to L_2$ is schematic if there are formulas $\alpha(p)$, $\beta * (p)$ of L_2 [for every unary connective $*$ of L_1] depending just on the propositional variable p, and formulas $y \bowtie (p, q)$ of L_2 [for every binary connective $\bowtie$ of L_1] depending just on propositional variables p, q, such that:*
  (i) $t(p) = \alpha(p)$, for every atomic formula p of L_1;
  (ii) $t(*\varphi) = \beta * (t(\varphi))$;
  (iii) $t(\varphi \bowtie \psi) = y \bowtie (t(\varphi), t(\psi))$.

An initial treatment of a theory of translations between logics is presented by Da Silva et al. 1999, where some connections linking translations between logics and uniformly continuous functions between the spaces of their theories are also investigated.

An important subclass of translations, the conservative translations, is investigated by Feitosa 1997 and Feitosa/D'Ottaviano 2001.

Definition 3.6. *Let A and B be logics. A conservative translation from A into B is a function $t : A \to B$ such that, for every set $X \cup \{x\} \subseteq A$:*

$$x \in C_A(X) \Leftrightarrow t(x) \in C_B(t(X)).$$

In terms of consequence relations, $t : For(L_A) \to For(L_B)$ is a conservative translation when, for every $\Gamma \cup \{\alpha\} \subseteq For(L_A)$:

$$\Gamma \vdash_A \alpha \Leftrightarrow t(\Gamma) \vdash_B t(\alpha).$$

The general notion of translation [Definition 3.3], introduced by da Silva, D'Ottaviano and Sette, accommodates certain maps that seem to be intuitive examples of translations, such as the identity map from intuitionistic into classical logic and the forgetful map from modal logic into classical logic: such cases would be ruled out if the stricter notion of conservative translation, or the general notions of Wójcicki and Epstein, or of contextual translation were imposed.

In particular, the identity function $i : \mathbf{IPL} \to \mathbf{CPL}$, both logics considered in the connectives $\neg, \wedge, \vee, \to$, is a translation, and is a contextual translation, but is not a conservative translation: it suffices to observe that:

$$\nvdash_{\mathbf{IPL}} p \vee \neg p \text{ while } \vdash_{\mathbf{CPL}} p \vee \neg p = i(p \vee \neg p).$$

We observe that translations in the sense of Prawitz/Malmäs 1968 neither coincide with conservative translations, nor with translations in the sense of Definition 3.3, for they preserve only theoremhood, and not derivability.

Translations in the sense of Wójcicki 1988 are particular cases of conservative schematic translations, being derivability-preserving schematic translations in Prawitz and Malmnäs' sense.

Epstein's translations (Epstein 1990) are instances of conservative translations.

We see that the interpretations of Kolmogorov 1925, Glivenko 1929 and Gentzen 1936 are translations in the sense of Prawitz and Malmnäs, Wójcicki and Epstein, and are conservative translations from classical into intuitionistic logic, according to Definition 3.6 (see Feitosa/D'Ottaviano 2001). Such translations are also examples of contextual translations, as we will present in Section 4.

Although Gödel's papers of 1933 and their meaningful extensions by other authors (McKinsey/Tarski 1948, Rasiowa/Sikorski 1953, Solovay 1976, Goldblatt 1978, Boolos 1979a, Boolos 1979b, and Goodman 1984), discussed by D'Ottaviano/Feitosa 2011, are important and relevant, Gödel's interpretations g_1 : **IPL** → **S4** and g_2 : **CPL** → **IPL** are translations only in Prawitz's sense. They do not preserve derivability, even on the propositional level, and hence are not translations in the sense of our definition.

3.1 Some Main Results and Further Research

We have obtained some relevant results to the study of general properties of logic systems from the point of view of translations between them. The proofs can be seen in Feitosa 1997, Feitosa/D'Ottaviano 2001, and D'Ottaviano/Feitosa 2007.

A logic system is *vacuous* if $C(\emptyset) = \emptyset$. A translation $t : \mathbf{L}_1 \to \mathbf{L}_2$ is *literal* for an operator if it maps the same operator from $\mathbf{L}_1$ into $\mathbf{L}_2$.

Proposition 3.7. (i) *If $t : \mathbf{L}_1 \to \mathbf{L}_2$ is a literal translation relative to $\neg$ and $\mathbf{L}_2$ is $\neg$-consistent, then $\mathbf{L}_1$ is $\neg$-consistent.*

(ii) *If there is a recursive and conservative translation from a logic system $\mathbf{L}_1$ into a decidable logic system $\mathbf{L}_2$, then $\mathbf{L}_1$ is also decidable.*

(iii) *Let $\mathbf{L}_1$ be a logic with a set of axioms Λ. If there is a surjective and conservative translation $t : \mathbf{L}_1 \to \mathbf{L}_2$, then $t(\Lambda)$ is a set of axioms for $\mathbf{L}_2$. Additionally, conservative translations preserve non-triviality.*

Theorem 3.8. (i) *A translation $t : A \to B$ is conservative if, and only if, $t^{-1}(C_A(t(X))) \subseteq C_A(X)$, for every $X \subseteq A$.*

(ii) *There is no translation from a non-vacuum system into a vacuum system.*

As a simple consequence of the previous result, there is no recursive conservative translation from first order logic into **CPL**.

A logic **L** *has a deductive implication* if there is a formula $\varphi(p, q)$ depending on two variables such that: $\Gamma, \alpha \vdash \beta$ if, and only if, $\Gamma \vdash \varphi(\alpha, \beta)$.

The next result presents conditions for the preservation of Deduction Metatheorems in the context of deductive implications.

Proposition 3.9. *Let L_1 and L_2 be two logic systems, $\varphi_1(p,q) \in \text{For}(L_1)$ and $\varphi_2(p,q) \in \text{For}(L_2)$. If $t : L_1 \to L_2$ is a conservative translation such that $t(\varphi_1(\alpha, \beta)) = \varphi_2(t(\alpha), t(\beta))$, then always that φ_2 is a deductive implication in L_2, φ_1 is a deductive implication in L_1. If t is surjective and φ_1 is a deductive implication in L_1, then φ_2 is a deductive implication in L_2.*

Feitosa 1997, and Feitosa/D'Ottaviano 2001 prove a necessary and sufficient condition for the existence of conservative translations between two logics, by dealing with the Lindenbaum algebraic structures associated to them. This result provided the authors with a useful method for defining and obtaining conservative translations.

On a logic **A** let's consider the equivalence relation

$$x \sim y \Leftrightarrow C(x) = C(y)$$

and the quotient map

$$Q : A \to A/\sim .$$

As usually, $[x]$ denotes the class of equivalence of x.

The logic $(A/\sim, C/\sim)$ is *co-induced* by **A** and Q if:

$X \subseteq A/\sim$ is closed if, and only if, the set $\{x \in A : [x] \in X\}$ is closed in **A**.

Theorem 3.10. *Let **A** and **B** be logics, with the domain of **B** denumerable; and let $A|_{\sim A}$ and $B|_{\sim B}$ be the logics co-induced by **A**, Q_A and **B**, Q_B, respectively. Then there is a conservative translation $t : A \to B$ if, and only if, there is a conservative translation $t^* : A|_{\sim A} \to B|_{\sim B}$, such that the next diagram commutes:*

$$\begin{array}{ccc} A & \xrightarrow{t} & B \\ Q_A \downarrow & & \downarrow Q_B \\ A|_{\sim A} & \xrightarrow{t^*} & B|_{\sim B} \end{array}$$

Moreover, if such a t^* exists, then it is injective.

We observe that the denumerability of **B** in the hypothesis of the theorem is not necessary if the Choice Axiom is [explicitly] used in the proof.

Based on previous results, D'Ottaviano and Feitosa,[4] dealing with syntactic results, algebraic semantics, and matrix semantics, introduce some conservative translations involving classical logic and the many-valued logics of Łukasiewicz

---

[4] See D'Ottaviano/Feitosa 1999a, D'Ottaviano/Feitosa 1999b, and D'Ottaviano/Feitosa 2000.

and Post; conservative translations involving classical logic, Łukasiewicz' three-valued system L_3, the intuitionistic system I^1 (see Carnielli/Sette 1995); and involving predicate logics.

Queiroz 1997, based on the concept of conservative translation, proposes a general definition for the concept of duality between logics, that allowed the study of new translations and the introduction of new systems (see D'Ottaviano/Feitosa 2007, and D'Ottaviano 2013).

Translations into **CPL** seem to be hard to obtain. Epstein (1990), in particular, proves that under certain circumstances such translations do not exist. However, D'Ottaviano/Feitosa 2006, by using the algebraic semantics associated to the finite-valued Łukasiewicz' logics and to classical propositional calculus, presents a non-constructive proof of the existence of a conservative translation between such logics.

Jeřábek 2012 proves the following general theorem, showing that there are in fact conservative translations from several logics into **CPC**.

Theorem 3.11. *If $L = (L, \vdash)$ is a finitary logic, whose domain L is a denumerable set of formulas, then there is a conservative translation $t : L \to$ **CPC**.*

The above result is valid not only for **CPC**, but for a big class of logics.

However, conservative translations do not exist in all cases: Scheer 2002, and Scheer/D'Ottaviano 2005 initiated the study of conservative translations involving cumulative non-monotonic logics and proved that there is no conservative translation from a cumulative non-monotonic logic into a Tarskian logic, and that there is no surjective conservative translation from a Tarskian logic into a non-monotonic cumulative logic.

4 Transfers and Contextual Translations

Motivated by the fact that conservative translations do not preserve the triviality of the source logic, Coniglio/Carnielli 2002 reintroduce the concept of translation and conservative translation from their concept of *transfer between two abstract logics* (previously called *meta-translation* by Coniglio 2005) over two-sorted languages.

However, in spite of transfers being pretty general, they require that every connective of the source logic must be translated into another connective, and this is too restrictive.

In 2009, Carnielli, Coniglio and D'Ottaviano introduce the concept of contextual translation, expecting to obtain an intermediate concept between translation and conservative translation. Contextual translations are mappings between languages preserving certain meta-properties of the source logics, that are defined in a formal first-order meta-language. In the following, for a given language, we formalize assertions and meta-rules, in order to introduce the definition of contextual translation.

Note 4.1. Observations and Notations:
(i) $X = \{X_i : i \in \mathbb{N}\}$ is the set of variables;
(ii) $\Sigma = \{\sigma_i : i \in \mathbb{N}\}$ is the set of schema variables;
(iii) $V = \{p_i : i \in \mathbb{N}\}$ is the set of propositional variables;
(iv) A propositional signature is a set $C = \{C_i : i \in \mathbb{N}\}$;
(v) $V \subseteq C_0$;
(vi) The elements of C_n are connectives of arity n.

Definition 4.2. If $L(C, \Sigma)$ and $L(C)$ are the C-algebras freely generated by $C_0 \cup \Sigma$ and C_0, respectively, then an assertion over C is a pair (Y, φ), written as $Y \vdash \varphi$, such that Y is a finite subset of $X \cup L(C, \Sigma)$ and $\varphi \in L(C, \Sigma)$.

Definition 4.3. A meta-rule (P) over C is a pair $(\{S_1, \ldots, S_n\}, S)$, written as $S_1, \ldots, S_n/S$, such that S_i, for $1 \leq i \leq n$, and S are assertions over C.

Definition 4.4. The C-algebra L has the meta-property (P) if every $\sigma, \pi \in L$ satisfies $(\sigma, \pi)(P)$.

The following definitions and results can be seen in detail in Carnielli et al. 2009.

Definition 4.5. (i) If $h : L(C_1, \Sigma_1) \to L(C_2, \Sigma_2)$ is such that $h(\sigma) = \sigma$, for every $\sigma \in \Sigma$ and $S = (Y, \varphi)$, then $\hat{h}(S) = (\hat{h}[Y], \hat{h}(\varphi))$ is such that $\hat{h}(\psi) = h(\psi)$, if $\psi \in L(C_1, \Sigma_1)$; and $\hat{h}(x) = x$, if $x \in X$.
(ii) if $(P) = (\{S_1, \ldots, S_n\}, S)$ is a meta-rule over C_1, then $\hat{h}(P)$ is the meta-rule $(\{\hat{h}(S1), \ldots, \hat{h}(Sn)\}, \hat{h}(S))$ over C_2. If (P) is structural, then $\hat{h}(P) = (P)$.

Definition 4.6. A contextual translation [c-translation] $h : L_1 \to L_2$ is a mapping $h : (C_1, \Sigma_1) \to (C_2, \Sigma_2)$ such that, if L_1 satisfies the meta-property (P) then L_2 satisfies the meta-property $\hat{h}(P)$.

Theorem 4.7. Any c-translation is a translation.

Proposition 4.8. For contextual translations:
(i) If L_1 satisfies a structural meta-property which is not satisfied by L_2, then L_1 is not c-translatable into L_2;
(ii) A trivial logic is not c-translatable into a non-trivial logic;
(iii) A monotonic logic is not c-translatable into a non-monotonic logic.

Carnielli, Coniglio and D'Ottaviano presented several examples, showing that contextual translations and conservative translations are essentially independent concepts (see Figure 4).

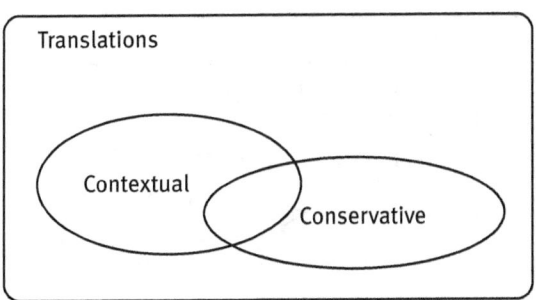

5 Abstract Contextual Translations

Moreira 2016 introduces the concept of abstract contextual translation, in order to avoid specific questions relative to logics generated from languages that are free algebras.

Abstract contextual translations also solve our previous question of obtaining an intermediate concept of translation between the concepts of translation, and conservative and contextual translation.

Definition 5.1. *Let $A = (A, C_A)$ and $B = (B, C_B)$ be two logics. An abstract contextual translation $t : A \to B$ is a function $t : A \to B$ such that, for every set $X_i \cup \{x_i\} \subseteq A$, with $i \in \{1, 2, ..., n\}$:*
if
$x_1 \in C_A(X_1), x_2 \in C_A(X_2), ..., x_{n-1} \in C_A(X_{n-1}) \Rightarrow x_n \in C_A(X_n),$
then
$t(x_1) \in C_B(t(X_1)), t(x_2) \in C_B(t(X_2)), ..., t(x_{n-1}) \in C_B(t(X_{n-1})) \Rightarrow t(x_n) \in C_B(t(X_n)).$

Proposition 5.2. *(i) If t is a contextual translation between Tarskian logics, then t is an abstract contextual translation.*
(ii) If t is a conservative translation, then t is an abstract contextual translation.

We observe that every contextual translation and every conservative translation are abstract contextual translations. And abstract contextual translations are cases of our general translations between logics (cf. Figure 5).

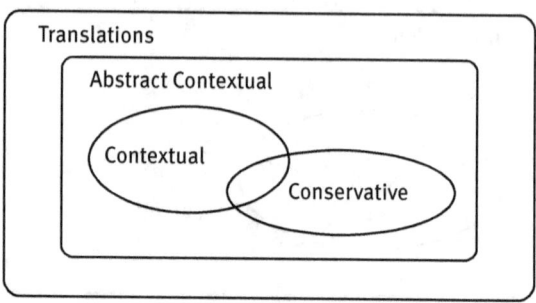

Moreira 2016 presents several examples involving these types of translations between logics; and obtains a necessary and sufficient condition for the existence of abstract contextual translations, very similar to the result previously obtained by Feitosa and D'Ottaviano.

Theorem 5.3. *Let A and B be logics, with the domain of B denumerable; and let $A|_{\sim A}$ and $B|_{\sim B}$ be the logics co-induced by A, Q_A and B, Q_B, respectively. Then there is an abstract contextual translation $t : A \to B$ if, and only if, there is an abstract contextual translation $t^* : A|_{\sim A} \to B|_{\sim B}$, such that the next diagram commutes:*

$$\begin{array}{ccc} A & \xrightarrow{t} & B \\ Q_A \downarrow & & \downarrow Q_B \\ A|_{\sim A} & \xrightarrow{t^*} & B|_{\sim B} \end{array}$$

Observe that it is not necessary that the function t^* be injective.

6 Some Other Concepts of Translations

Russo 2013 presents the concepts of *(weak) interpretation* and *(weak) representation* between propositional deductive systems [according to a specific algebraic approach], which are particular cases of our concepts of translation and conservative translation, respectively (see Moreira 2016).

Figallo 2013, and Coniglio/Figallo 2015 propose the concept of hypertranslation, according to which meta-properties of the domain logic, with more than one assertion in the conclusion (they call them *hypersequents*), are preserved in the target logic. Coniglio and Figallo (2013) introduce the categories **CHC** and **GHC** of calculi of commutative hypersequents and calculi of general hypersequents, respectively; the morphisms of such categories are called *hypertranslations*. These hypertranslations preserve general meta-properties.

Moreira 2016, by introducing the concept of hypercontextual translation, simplifies the concept of hypertranslation.

Definition 6.1. *Let $\mathbf{L_1}$ and $\mathbf{L_2}$ be propositional logics defined over the signatures C_1 and C_2, respectively. A hypercontextual translation (or contextual hypertranslation) is a function $h : L_1(C_1) \to L_2(C_2)$ such that, if $\mathbf{L_1}$ has a meta-property (P), then $\mathbf{L_2}$ has the meta-property $\hat{h}(P)$.*

It was shown that the hypercontextual translations are particular cases of contextual translations, and two new other concepts of translations were also introduced.

Definition 6.2. *Let $\mathbf{L_1}$ and $\mathbf{L_2}$ be logics on the signatures C_1 and C_2, respectively. A conservative-contextual translation (c-c translation) is a function $h : L_1(C_1) \to L_2(C_2)$, such that $\mathbf{L_1}$ has the meta-property (P) if, and only if, $\mathbf{L_2}$ has the meta-property $\hat{h}(P)$.*

Definition 6.3. *Let $\mathbf{L_1}$ and $\mathbf{L_2}$ be logics on the signatures C_1 and C_2, respectively. A strict conservative-contextual translation $h : L_1(C_1) \to L_2(C_2)$, is such that, if $\mathbf{L_1}$ has the meta-property (P) then $\mathbf{L_2}$ has the meta-property $\hat{h}(P)$, and, if $\mathbf{L_2}$ has the meta-property (P) then $\mathbf{L_1}$ has the meta-property $\hat{h}^{-1}(P)$.*

The definition of isomorphism between two logics is simple.

Definition 6.4. *Let $\mathbf{L_1} = (L_1, C_1)$ and $\mathbf{L_2} = (L_2, C_2)$ be two Tarskian logics. An isomorphism $t : \mathbf{L_1} \to \mathbf{L_2}$ is a bijective function t such that, for every set $X \cup \{x\} \subseteq L_1$, we have:*

$$x \in C_1(X) \Leftrightarrow t(x) \in C_2(t(X)).$$

7 Translations and Galois Pairs

Recently it was shown that the concept of translation between logics allows us to generate Galois adjunctions between Tarskian spaces (see Feitosa et al. 2017).

Definition 7.1. *If $(A, \leq_A)$ and $(P, \leq_P)$ are partial ordered sets, $a \in A$, $p \in P$ and $f : A \to P$ and $g : P \to A$ are functions, then:*

(i) (f, g) is a Galois connection if, and only if, $a \leq_A g(p) \Leftrightarrow p \leq_P f(a)$;
(ii) $(f, g)^d$ is a dual Galois connection if, and only if, $g(p) \leq_A a \Leftrightarrow f(a) \leq_P p$;
(iii) the pair $[f, g]$ is an adjunction if, and only if, $a \leq_A g(p) \Leftrightarrow f(a) \leq_P p$;
(iv) the pair $[f, g]^d$ is a dual adjunction if, and only if, $g(p) \leq_A a \Leftrightarrow p \leq_P f(a)$.

The name adjunction comes from the theory of categories. In some texts the pair [f, g] also is named residuated (see Dunn/Hardgree 2001, Ore 1944 and Orlowska/Rewitzky 2010).

Considering two logics **A** and **B** and a translation $t : \mathbf{A} \to \mathbf{B}$, we denote the sets of all theories or closed sets of **A** and **B**, respectively, by $\mathcal{F}(\mathbf{A})$ and $\mathcal{F}(\mathbf{B})$.

Now we take the functions $f : \mathcal{F}(\mathbf{A}) \to \mathcal{F}(\mathbf{B})$ defined by $f(D) = C_2(t(D))$, and $g : \mathcal{F}(\mathbf{B}) \to \mathcal{F}(\mathbf{A})$ defined by $g(E) = t^{-1}(E)$.

Theorem 7.2. *The pair $[f, g]$, above defined, is an adjunction for $(\mathcal{F}(\mathbf{A}), \subseteq)$ and $(\mathcal{F}(\mathbf{B}), \subseteq)$.*

In mathematics, Galois' pairs are used to take one problem from one context into another more adequate context, and vice-versa. Our translations between logics work similarly.

8 General Considerations and Relations among the Distinct Concepts of Translations

As discussed in this paper we have some general results:

- Our concept of translation is very general and wide.
- Conservative translations and contextual translations are independent concepts.
- There are abstract contextual translations that are neither conservative translations, nor contextual translations.
- There are translations that are not abstract contextual translations.
- Conservative translations preserve the non-triviality of the domain logic into the target logic, but, in general, they do not preserve the triviality of the domain logic.
- Contextual translations preserve the triviality of the domain logic into the target logic, but every non-trivial logic can be contextually translatable into any trivial logic.
- Translations and abstract contextual translations do not preserve either the triviality or the non-triviality of the domain logic.

- A non-trivial logic cannot be c-c translatable into a trivial logic.
- A trivial logic cannot be c-c translatable into a non-trivial logic.

We also have some results concerning the classes of logics and translations:

- The class of logics with translations between them is a bi-complete category (**Tr**).
- The class of logics with conservative translations is a co-complete sub-category (**TrCon**) of **Tr** (see Feitosa/D'Ottaviano 2001).
- The category of logics and abstract-conservative translations (**TrCx**) is a bi-complete sub-category of **Tr** (see Moreira 2016).
- The category **Tr** is a full sub-category of the category of the cumulative non-monotonic logics and translations (**TrNM**) (see Scheer/D'Ottaviano 2005).
- The category **TrTp**, whose objects are topological spaces and whose morphisms are the continuous functions between them, is a full sub-category of the bi-complete category **Tr** of logics and translations (see Feitosa/D'Ottaviano 2001).

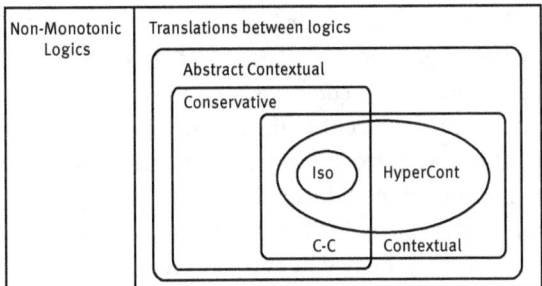

9 Final Considerations and the Ubiquity of Translations

Other developments of the wider notion of translation sprung forth.

Carnielli 2000 proposed a new approach to formal semantics for non-classical logics, initially called *non-deterministic semantics*. The underlying purposes of this semantical framework are to offer alternative semantic interpretations to a given logic in terms of a family of other simpler logics, and to combine logics so as to obtain other logics with a richer structure.

A previous study developed in Carnielli/Lima-M. 1999, *society semantics*, and Marcos 1999 proved that Carnielli's interpretation functions are in fact translations (in the sense of Da Silva et al. 1999) from the main logic into its components – the semantics became then known as *possible-translations semantics*.

Motivated by known problems concerning the non-algebraizability of some of da Costa's paraconsistent systems (see D'Ottaviano 1990) and inspired by Carnielli's possible translations semantics, Bueno-S. 2004 introduces possible-translations algebraic semantics, that lead to an unexpected relation between da Costa's paraconsistent logic C_1 and the three-valued MV-algebras (see Cignoli et al. 2000). It is also shown that the possible-translations semantics [completeness provided in categorical terms] appear as important conservative translations, into a categorical product of logics.

Carnielli/Veloso 1997 studies a new *ultra-filter logic* aiming at axiomatizing a purely quantitative logic of "most", which can also serve as an alternative logic for default reasoning. It is shown that this logic can be conservatively translated into a first-order theory, such translation being very helpful to understanding the model theoretical properties of this new logic.

Fernández 2005 uses translations in order to investigate *combinations of logics*, more particularly *fibring* of logics (see Carnielli et al. 2008).

We have recently obtained a conservative translation concerning a three valued paraconsistent logic subjacent to the da Costa's quasi-truth (see D'Ottaviano/Hifume 2007) and the logic of Tarski consequence operator for the notion of deductibility, an interpretation of the quasi-truth in the logic of deductibility.

The abundance of the categories **TrNM**, **Tr**, **TrCx** and **TrCon** call our attention to the richness of the tools provided by conservative and contextual translations, and the relevance of the study of interrelations between logics by the analysis of translations between them.

Of course, other notions and developments on interpretations, translations between logics, reductions and similar functions are welcome. All of them can be used for better understanding of general aspects of logics.

Bibliography

Béziau, J. Y. (1995): *Recherches sur la logique universelle (excessivité, négation, séquents)* [Investigations About the Universal Logic (Excessivity, Negations, Sequents)]. In French. Ph. D. Thesis. Paris: Université Paris 7.

Boolos, G. (1979a): "Solovay's Completeness Theorems." In: *The Unprovability of Consistency: An Essay in Modal Logic*. Cambridge: Cambridge University Press, 151 – 158.

Boolos, G. (1979b): "An S4-Preserving Proof-Theoretical Treatment of Modality." In: *The Unprovability of Consistency: An Essay in Modal Logic*. Cambridge: Cambridge University Press, 159 – 167.

Brown, D. J. and Suszko, R. (1973): "Abstract Logics." In: *Dissertationes Mathematicae*. Vol. 102, 9 – 41.

Bueno-Soler, J. (2004): *Semântica Algébrica de Traduções Possíveis* [Algebraic Semantics of Possible Translations]. In Portuguese. Master Dissertation. Campinas: Institute of Philosophy and Human Sciences, University of Campinas.

Carnielli, W. A. (2000): "Possible-Translations Semantics for Paraconsistent Logics." In: *Frontiers in Paraconsistent Logic*: Proceedings of the I World Congress on Paraconsistency. Batens D. et al. (eds.), London: Kings College Publications, 159 – 172.

Carnielli, W. A., Coniglio, M. E., Gabbay, D. M., Gouveia, P. and Sernadas, C. (2008) *Analysis and Synthesis of Logics: How to Cut and Paste Reasoning Systems*. Amsterdam, Springer.

Carnielli, W. A., Coniglio, M. E. and D'Ottaviano, I. M. L. (2009): "New Dimensions on Translations Between Logic." In: *Logica Universalis*. Vol. 3, 1 – 19.

Carnielli, W. A. and Lima-Marques, M. (1999): "Society Semantics and Multiple-Valued Logics." In: *Contemporary Mathematics*. Vol. 235, 435 – 448.

Carnielli, W. A. and Sette, A. M. A. (1995): "Maximal Weakly-Intuitionistic Logics." In: *Studia Logica*. Vol. 55, 181 – 203.

Carnielli, W. A. and Veloso, P. A. S. (1997): "Ultrafilter Logic and Generic Reasoning." In: *Lecture Notes in Computer Science*. Vol. 1289, 34 – 53.

Cignoli, R. L. O., D'Ottaviano, I. M. L. and Mundici, D. (2000) *Algebraic Foundations of Many-Valued Reasoning*. Trends in Logic, Vol. 2, Dordrecht: Kluwer.

Coniglio, M. E. (2005): "A Stronger Notion of Translation Between Logics." In: *Manuscrito – International Journal of Philosophy*. Vol. 28, No. 2, 231 – 262.

Coniglio, M. E. (2007): "Recovering a Logic from Its Fragments by Metafibring." In: *Logica Universalis*. Vol. 1, No. 2, 377 – 416.

Coniglio, M. E. and Carnielli, W. A. (2002): "Transfers Between Logics and Their Applications." In: *Studia Logica*. Vol. 72, No. 3, 367 – 400.

Coniglio, M. E. and Figallo, M. (2015): "A Formal Framework for Hypersequent Calculi and their Fibring." In: *The Road to Universal Logic: Festschrift for 50th Birthday of Jean-Yves Béziau*. Koslow, A.; Buchsbaum, A. (eds.).

da Silva, J. J., D'Ottaviano, I. M. L. and Sette, A. M. (1999): "Translations Between Logics." In: *Models, Algebras and Proofs*. Caicedo, X.; Montenegro, C.H. (eds.). New York: Marcel Dekker, 435 – 448 (Lectures Notes in Pure and Applied Mathematics, Vol. 203).

D'Ottaviano, I. M. L. (1973): *Fechos Caracterizados por Interpretações* [Closures Characterized by Interpretations], in Portuguese. Master Dissertation. Campinas: Institute of Mathematics, Statistics and Scientific Computation, University of Campinas.

D'Ottaviano, I. M. L. (1990): "On the Development of Paraconsistent Logic and da Costa's Work." In: *The Journal of Non-Classical Logic*. Vol. 7, No. 1–2, 9 – 72.

D'Ottaviano, I. M. L. (2013): "Translations as Representations Between Logics." In: *Representation and Explanation in the Sciences*. Evandro Agazzi. (ed.). Milano: Franco Angeli, 228 – 243.

D'Ottaviano, I. M. L. and Feitosa, H. A. (1999a): "Conservative Translations and Model-Theoretic Translations." In: *Manuscrito - International Journal of Philosophy*. Vol. 22, No. 2, 117 – 132.

D'Ottaviano I. M. L. and Feitosa, H. A. (1999b): "Many-Valued Logics and Translations." In: *Journal of Applied Non-Classical Logics*. Vol. 9, No. 1, 121 – 140.

D'Ottaviano I. M. L. and Feitosa, H. A. (2000): "Paraconsistent Logics and Translations." In: *Synthèse*. Vol. 125, 77 – 95.

D'Ottaviano, I. M. L. and Feitosa, H. A. (2006): "Translations from Łukasiewicz' Logics Into Classical Logic: Is It Possible?" In: *Essays in Logic and Ontology*. Malinowski J.; Pietrusczak A. (eds.). Poznan Studies in the Philosophy of the Sciences and the Humanities, Vol. 91, 157 – 168.

D'Ottaviano, I. M. L. and Feitosa, H. A. (2007): "Deductive Systems and Translations." In: *Perspectives on Universal Logic*. Béziau J.-Y.; Costa-Leite A. (eds.). Monza: Polimetrica International Scientific Publisher, 125 – 157.

D'Ottaviano, I. M. L. and Feitosa, H. A. (2011): "On Gödel's Modal Interpretation of Intuitionistic Logic." In: *Anthology of Universal Logic: from Paul Hertz to Dov Gabbay*. Basel: Springer. Studies in Universal Logic.

D'Ottaviano, I. M. L. and Hifume, C. (2007): "Peircean Pragmatic Truth and da Costa's Quasi-Truth." In: *Studies in Computational Intelligence*. Vol. 61, 383 – 398.

Dunn, J. M.; Hardegree, G. M. (2001): *Algebraic Methods in Philosophical Logic*. Oxford: Oxford University Press.

Epstein, R. L. (1990): *The Semantic Foundations of Logic*. Volume 1: Propositional Logics (with the collaboration of Carnielli, W. A., D'Ottaviano, I. M. L. and Krajewski, S.). Dordrecht: Kluwer Academic Publishers.

Feitosa, H. A. (1997): *Traduções Conservativas* [Conservative Translations]. PhD Thesis. Campinas: Institute of Philosophy and Human Sciences, University of Campinas.

Feitosa, H. A.; D'Ottaviano, I. M. L. (2001): "Conservative Translations." In: *Annals of Pure and Applied Logic*. Vol. 108, 205 – 227.

Feitosa, H. A., Lázaro, C. A. and Nascimento, M. C. (2017): "Pares de Galois e uma Adjunção Motivada pelas Traduções entre Lógicas." Bauru. *ERMAC 2017: Caderno de trabalhos completos e resumos*. 290 – 297.

Fernández V. L. (2005) *Fibrilação de Logicas na Hierarquia de Leibniz* [Logics Fibring in the Leibniz Hierarchy]. In Portuguese. PhD Thesis. Campinas: Institute of Philosophy and Human Sciences, University of Campinas.

Figallo, M. (2013) *Hipersecuentes y la Lógica Tetravalente Modal TML* [Hypersequents and the Four-Valued Modal Logic TML]. In Spanish. PhD Thesis. Bahía Blanca: mathematics Department, Universidad Nacional del Sur.

Gentzen, G. (1936): "Die Widerspruchsfreiheit der reinen Zahlentheorie." In: *Mathematische Annalen*. Translation into English in Gentzen 1969, Vol. 112, 493 – 565.

Gentzen, G. (1969): "On the Relation Between Intuitionist and Classical Arithmetic." In: *The Collected Papers of Gerhard Gentzen*. Szabo, M. E. (ed.). Amsterdam: North-Holland, 53 – 67.

Glivenko, V. (1929): "Sur quelques points de la logique de M. Brouwer." In: *Bulletins de la Classe de Sciences* (5e série). Vol. 15, 183 – 188.

Gödel, K. (1933): "On Intuitionistic Arithmetic and Number Theory." In: *Collected Works*. Feferman, S. et al. (eds.). Oxford: Oxford University Press, 1986, 287 – 295.

Gödel, K. (1933): "An Interpretation of the Intuitionistic Propositional Calculus." In: *Collected Works*. Feferman, S. et al. (ed.). Oxford: Oxford University Press, 1986, 301 – 303.

Goguen, J. A. and Burstall R. M. (1984): "Introducing Institutions." In: *Logics of Programs*. (Carnegie-Mellon University, June 1983), Vol. 164 of *Lecture Notes in Computer Science*, Springer, 221 – 256.

Goguen, J. A. and Burstall R. M. (1992): "Institutions: Abstract Model Theory for Specification and Programming." In: *Journal of the ACM*. Vol. 39, No. 1, 95 – 146.

Goldblatt, R. (1978): "Arithmetical Necessity, Provability and Intuitionistic Logic." In: *Theoria*. Vol. 44, 36 – 38.

Goodmann, N. D. (1984): "Epistemic Arithmetic Is a Conservative Extension of Intuitionistic Arithmetic." In: *The Journal of Symbolic Logic*. Vol. 49, 192 – 203.

Hilbert, D. (1923): "Die logischen Grundlagen der Mathematik." In: *Mathematische Annalen*. Vol. 88, 151 – 165.

Hoppmann, A. G. (1973): *Fecho e Imersão* [Closure and Embedding]. In Portuguese. Ph. D. Thesis. Rio Claro: Faculty of Philosophy, Sciences and Letters, São Paulo State University.

Jeřábek, E. (2012): "The Ubiquity of Conservative Translations." In: *The Review of Symbolic Logic*. Vol. 5, 666 – 678.

Kolmogorov, A. N. (1925): "On the Principle of Excluded Middle." In: *From Frege to Gödel: A Source Book in Mathematical Logic 1879–1931*. Hejenoort, J. (ed.). Cambridge (MA): Harvard University Press, 1977, 414 – 437.

Lewis, C. I. and Langford, C. H. (1932) *Symbolic Logic*. New York/ London: The Century Company.

Marcos J. (1999). *Semânticas de Traduções Possíveis* [Possible-Translations Semantics]. In Portuguese. Master Dissertation. Campinas: Institute of Philosophy and Human Sciences, University of Campinas.

McKinsey, J. C. C. and Tarski, A. (1948): "Some Theorems About the Sentential Calculi of Lewis and Heyting." In: *The Journal of Symbolic Logic*. Vol. 13, No. 1, 1 – 15.

Moreira, A. P. R. (2016) *Sobre Traduções Entre Lógicas: Relações Entre Traduções Conservativas e Traduções Contextuais Abstratas* [On Translations Between Logics: Relations Between Conservative Translations and Abstract Contextual Translations]. Ph. D. Thesis. Campinas: Institute of Philosophy and Human Sciences, University of Campinas.

Mossakowski, T., Diaconescu, R. and Tarlecki, A. (2005): "What Is a Logic?" In: *Logica Universalis: Towards a General Theory of Logic*. J-Y. Béziau, (ed.), Basel: Birkhäuser, 111 – 134.

Ore, O. (1944): "Galois Connections." In: *Transactions of the American Mathematical Society*. Vol. 55, 493 – 513.

Orlowska, E. and Rewitzky, I. (2010): "Algebras for Galois-Style Connections and Their Discrete Duality." In: *Fuzzy Sets and Systems*. Vol. 161, 1325 – 1342.

Prawitz, D. and Malmäs, P. E. (1968): "A Survey of Some Connections Between Classical, Intuitionistic and Minimal Logic." In: *Contributions to Mathematical Logic*. Schmidt, H. et al. (eds.). Amsterdam: North-Holland, 215 – 229.

Queiroz G. S. (1997) *Sobre a Dualidade Entre Intuicionismo e Paraconsistência* [On the Duality Between Intuitionism and Paraconsistency]. In Portuguese. PhD Thesis. Campinas: Institute of Philosophy and Human Sciences, University of Campinas.

Rasiowa, H. and Sikorski, R. (1953): "Algebraic Treatment of the Notion of Satisfiability." In: *Fundamenta Mathematicae*. Vol. 40, 62 – 95.

Russo, C. (2013): "An Order-Theoretic Analysis of Interpretations Among Propositional Deductive Systems." In: *Annals of Pure and Applied Logic*. Vol. 164, 112 – 130.

Scheer, M. C. (2002): *Para uma Teoria de Traduções Entre Lógicas Cumulativas* [Towards a Theory of Translations Between Cumulative Logics]. In Portuguese. Master Dissertation. Campinas: Institute of Philosophy and Human Sciences. University of Campinas.

Scheer, M. C. and D'Ottaviano, I. M. L. (2005): "Operadores de Consequência Cumulativos e Traduções Entre Lógicas Cumulativas." In: *Revista de Informação e Cognição*. Vol. 4, 47 – 60.

Solovay, R. M. (1976): "Provability Interpretations of Modal Logic." In: *Israel Journal of Mathematics*. Vol. 25, 287 – 304.

Szczerba, L. (1977): "Interpretability of Elementary Theories." In: *Logic, Foundations of Mathematics and Computability Theory*. Butts, H.; Hintikka, J. (eds.). Dordrecht: Reidel, 129 – 145.

Wójcicki, R. (1988) *Theory of Logical Calculi: Basic Theory of Consequence Operations*. Dordrecht: Kluwer (Synthèse Library, 199).

Jan Woleński
On the Relation of Logic to Metalogic

Abstract: This paper discusses the concept of metalogic, its methods and the role in logic. The problem of how logic is used in metalogic is also a topic of the author's considerations. Metalogic is understood here as that part of metamathematics which is restricted to logical systems, and refers to studies of logical systems by mathematical methods. The scope of metalogic depends on the range of the concept of logic. The paper focuses on classical first-order logic, but some remarks concern other systems as well. Metalogic exhibits various properties of logics via definitions and metatheorems, concerning completeness, consistency, etc. It is argued that attributes of logical systems without formal use in metalogic would be very vaguely understood. These methods cover an amount of mathematics as well as of logic itself. Hence the question of how strong metalogic should be in order to prove metalogical theorems. Any answer has to delimit the repertoire of logic used in metalogic. This leads to the problem of circularity of metalogic and the question concerning the source of meaning for logical constants.

According to the contemporary view, metalogic studies properties of logical systems and logical concepts. The qualification 'contemporary' is essential due to historical circumstances (see Rentsch 1980). The word 'metalogic' became popular in the 19th century, although its roots go back to the Middle Ages (*Metalogicus* of John of Salisbury). Philosophers, mainly Neokantians, understood metalogic as concerned with general considerations about logic, its nature, scope, relations to other fields, etc. Some curious uses appeared as well, for example, Ernest Troeltsch, an eminent German historian, referred to methods of concrete historical investigations as metalogical, and Walter Harburger, a German composer and musicologist, the author of the book (*Die Metalogik*, 1919) employed the word 'metalogic' as equivalent to 'logic of music'. The contemporary meaning crystallized around 1930 in Germany under the influence of Hilbert and Carnap, and Poland, particularly in Tarski's works. On the other hand, logical empiricists frequently qualified their definitions of meaningfulness (more generally, investigations about science and its language) as metalogical, because employing logical concepts and methods – it is a wider meaning of the word 'metalogic'.

The characterization of metalogic as investigations of logical systems and logical concepts is still very vague. Assume that we deal with propositional calculus

Jan Woleński, University of Information, Technology and Management, Rzeszów, Poland, jan.wolenski@uj.edu.pl

(**PC**). It is a logical system formulated in a language $\mathbf{L}^{PC}$ considered as a logical object-language. Since **PC**-metalogic is about this propositional calculus, it should use a metalanguage (**ML**). Typical metalogical problems related to **PC** include its axiomatization, completeness (in various meanings), decidability, etc. This picture can be easily generalized. Let **S** be a logical system. Thus, the symbol **MS** refers to a body of assertions about S, expressed in the related **ML**. Here is an example. A logical system **S** can be identified with the pair $\langle \mathbf{L}^S, Cn \rangle$, where $\mathbf{L}^S$ is a language of **S** and Cn – a consequence operation. This description applies to axiomatically generated systems as well as organized by a collection of deductive rules. If a logic has the deduction theorem, both mentioned codifications are equivalent via the deduction theorem to take the simplest case, if the formula A ⇒ B is a logical theorem, the rule A ⊢ B is logically valid (truth preserving). The above description of logic is metalogical, because it says something about logic as an object and is expressed in $\mathbf{ML}^S$. The only restriction limiting the above account consists in the assumption that **S** has the deduction theorem. This system can be classical or not, first-order or higher-order, modal or not, etc.

A simple temptation arises in order to propose the next generalization stating that if **S** is an arbitrary logical system, **MS** is its metalogic (or metatheory). Unfortunately, such a proposal raises several serious doubts – in fact, they can be addressed to earlier, less general, characterizations. For instance, we explain that the disjunction is a binary propositional connective symbolized by ∨ and translated by 'or' into the ordinary language. Logical for functioning as more or less informal remarks or just metalogical as formulated in the metalanguage? The first eventuality is commonly accepted. This means that the distinction of **L** and **ML** does not provide a sufficient criterion of distinguishing logic and metalogic. If we inspect books serving as manuals of logic (see Kleene 1967, Beth 1968 and Enderton 1971 as classical as some classical books, and Schechter 2005 and de Swaart 2018 as recent examples; Beth's monograph contains a very extensive treatment of relations between the intuitive and formalized metamathematics) or having in the title 'Metalogic' (see Hunter 1971 or Yaqūb 2015), we see that they mix metalogic and logic. Sometimes some topics are qualified as metalogical, sometimes not – in de Swaart's book, chapters 2.4 and 2.7 (on **PC**) are titled "Semantics: Meta-Logical Considerations" and "Syntax: Metalogical Considerations" (interesting that similar titles do not appear in chapters devoted to predicate calculus, **QC**, henceforth), respectively, but Yaqūb's book is not divided in logical and metalogical chapters (except Chapter Two "Resources of the Metatheory" having the metalogical character). Surely, metalogic is important for presenting several aspects of logic, but any exposition of the former without providing the elements of the latter appears as very problematic. One can even adapt Kant's famous dictum and say "Logic without metalogic is blind, metalogic without logic is empty". Thus, the interplay between both, logic and metalogic, must be taken as arguably indispensable from the practical as well as theoretical point of view.

Another reason to regard the explanation "if **S** is an arbitrary logical system, **MS** is its metalogic (or metatheory)" as too weak (or schematic) points out that the notion of logical systems requires further additions. I will take classical **PC** and first-order **QC** with identity as points of reference. One proposal consists in saying that logic is to be identified with the sum **PC** + **QC**. This claim restricts logic to so-called elementary logic (propositional calculus and first-order predicate (quantification) calculus with identity. Return to the triple the pair $\langle \mathbf{L}^S, Cn \rangle$. The language of **S** can be first-order, higher-order, have all connectives, infinitely denumerable or not, have finite formulas or not, etc.). Every choice depends on some metalogical settings associated with necessary resources to investigate particular cases. Similarly, Cn for classical logic is monotonic, finite and idempotent (see also below), but every mentioned condition can be modified in the case of other logics, for instance, non-monotonic one. Also the properties of Cn associated with non-classical logic are different than those generated by settings related to **QC**. These observations pertain to syntax, but semantic constraints related to logical are still more complicated in the case of semantics (for example, see rubrics: models for classical logic, models for intuitionistic logic, models for many-valued logic, models for fuzzy logic, models for modal logics). An elaboration of such differences should not be reduced to observing that they result from different philosophical presuppositions (it is true, of course), because there remain a lot of formal issues. In general, the claim concerning that such and such logic, for instance classical or intuitionistic, is very controversial (see also below). Anyway, any choice of a logical system as *the* logic has very serious consequences.

How is metalogic related to metamathematics? According to Łukasiewicz and Tarski (page-reference to the reprint mentioned in the bibliography at the end of the present paper; the same concerns other quotations):

> In the course of the years 1920–30 investigations were carried out in Warsaw belonging to that part of metamathematics – or better metalogic – which has as its field of study the simplest deductive discipline, namely the sentential calculus. (Łukasiewicz/Tarski 1930: 38)

Although this view concurs with the explanation of metalogic as studying properties of logical systems and logical concepts, it also alludes to the view that metalogic is a part of metamathematics. The latter, as understood presently, covers proof theory, model theory and recursion theory (computability), and sometimes is identified with the mathematical (as contrasted with philosophical) foundations of mathematics. The idea of metamathematics as exact study of formal systems arose in the Hilbert school, but the word 'metamathematics' was also used before Hilbert, but with a different meaning (see Schütte 1980). In the early 19th century, mathematicians, for instance, Gauss, spoke about metamathematics in an explicitly pejorative manner and considered it as a speculative way of looking at mathematics – something like the metaphysics of mathematics. The only one serious use of 'metamathematics' was restricted to so-called metageometry. This was due to the fact that the invention of various geometries in the 19th century stimulated investigations were undertaken of particular axiomatizations, their mutual relations, models of various geometrical systems, and attempts to prove their consistency. In this context, the word 'metageometry' referred to a well-established domain of formal studies. Hilbert extended this understanding of studying geometrical systems to arbitrary mathematical systems and added that it should be done by finitary methods. Nevertheless, this restriction was rejected in the due course and contemporary metamathematics employs arbitrary mathematical methods, finitary or not.

On the above characterization of metamathematics, it includes logic and metalogic. In particular, logic as such belongs to proof theory – the same concerns various, mostly syntactic investigations about various proof (logical) systems. Model theory constitutes (formal) semantics of logical systems. Finally, recursion theory deals with the problem of decidability of logic. Metaphorically speaking, limiting formal mathematical systems to logical ones immediately results in cutting metamathematics to metalogic. Once again, it is is a very rough picture. In particular, the latter uses various mathematical methods. Hence, it is important to describe the resources of metalogic. They cover (see Yaqūb 2015: 87 – 123): linguistic and logical items (English, modified mathematical English, logical constants, rules of inferences, definitions of various concepts, like completeness, consistency, etc.), arithmetical rules (for instance, the principle of mathematical induction) an amount of set theory (for instance, the axiom of choice) and topology (the closure operator), concepts of universal algebra (lattice, algebraic system), elements of category theory (for instance, notions of functor and transformation). Metalogic formulates various metatheorems about logical systems. Although they are expressed by using metalogical concepts, for instance, completeness or consistency, their proofs employ typical strategies of ordinary mathematics. Hence, metalogic can also be considered as a part of mathematics in the fairly normal sense. Once again, we see at the point a close interplay of logic, metalogic and, additionally, mathematics. However, there appears a specific point, namely the question, which mathematical theorems are necessary for proving metalogical (the same concerns metamathematical) facts. Has the continuum hypothesis or its negation any relevance for logic? Can axiom of choice be replaced by weaker similar principles, for instance, the Lindenbaum lemma? The second possibility looks as more appropriate, because the axiom of choice concerns all (non-empty) sets, but the Lindenbaum lemma – sets of sentences; in a sense, the latter is 'more' logical than the former. Although such questions are interesting from the general mathematical point of view, their role in metalogic (as well as in metamathematics) appears as basic, even disregarding hot philosophical controversies around the axiom of choice. I will exemplify specific relations between logic and metalogic by the classical case, eventually with some short digressions on non-classical systems.

To complete the general picture of metalogic, let me address to three additional points – they are rather motivated by philosophy than by formal issues. Firstly and historically speaking, we have the distinction between logic *sensu largo* (logic in a wide sense) and logic *sensu stricto* (logic in a narrow sense). Whereas the latter can be identified with formal logic, the former includes semantics (semiotics) and methodology of science. Logic and metalogic in the outlined sense, covers only a part of *logica sensu largo*, namely formal logic (*logica sensu stricto*) and semantics, provided that the latter is limited to formal semantics (model theory). Eventually, one can add that metamathematics functions as methodology of deductive sciences (this understanding was popular in Poland in the 1930s). However, the contemporary view on logic *sensu stricto* excludes from its scope various semantic problems, even if they are analyzed via formal methods, for instance, descriptions, *Sinn* and *Bedeutung* (in Frege's understanding), direct reference, compositionality (as applied to ordinary language), as well as metascientific issues, like the structure of empirical theories, the concept of scientific law or the logic of induction.

Secondly, a logical systems can be viewed from the strict Platonic view or other one. The former sees **S** as a fixed object with completely determined properties. Thus, **S** is either complete or incomplete, decidable or undecidable, etc. independently of our knowledge. Metalogic rather discloses these attributes than constitutes them. The opposite view has many different incarnations. The intuitonist would say that logical systems are our (mental) creations and depend of the nature of our minds as experiencing time. Hence, correct (observe that, it is passing from a description to an evaluation) systems must be constructive. The formalist's view points out that the ideal mathematics is, so to speak, pre-given, but it must be justified by finitary methods as used by mathematicians. My working assumption claims that, realistically speaking, logical systems are always shaped as an actual (in the sense, here and now) object, dependent of our accessible information. Usually, the later versions of logical systems are more perfect than earlier ones. More importantly, the aim of metalogic just consists in exhibiting various metaproperties of logical systems. On the Platonic view, **QC** is eternally undecidable, so to speak ontologically, and proving the related metatheorem by Church and Turing must be viewed only as an epistemological event. The working view of the present paper is different. I consider metatheorems about logical systems as relative to the actual amount of resources accessible for proving metalogical facts. These resources sometimes exist in advance (for instance, the principle of mathematical induction), but sometimes are intentionally created (for instance, the technique of arithmetization). On this view, a property of a logical system is proved or not, and all deliberations whether it is eternal or not, appear as redundant. In particular, metalogical activity is highly creative, similarly as what is done in other fields of mathematics. I do not claim that it is a new proposal for the philosophy of mathematics or logic, but, to say it once again, it functions as a working standpoint. As Errett Bishop expressed this attitude (quoted after Schechter 2005: 56): "Mathematics belongs to man, not to God. If god has mathematics of his own that to be done, let him do it itself". This is a constructivist manifesto, but, disregarding the philosophy of mathematics present in Bishop's words, we can paraphrase the above quotation as saying that logic and metalogic belong to man, not to the Platonic heaven and "if the Logician lining in this realm has logic of his own, let him do it himself". In particular, metalogic is not required in such an environment, because all properties of logic are pre-given in advance and should be (or even are) known to the ideal logician.

The third point concerns the distinction of logic into *logica docens* (theoretical logic) and *logica utens* (applied logic). Constructing (or discovering) logical systems and investigating their properties constitute the merit of *logica docens*. On the other hand, (almost) everybody agrees that we use logic in various ways and should do that (I do not enter into details concerning this claim). Petrus Hispanus a medieval logician said that *dialectica* (that is, logic) *est art artium et scientia scientiarum ad omnium aliarum scientiarum methodorum principia viam habent* (logic is the art of all arts and science of sciences, which provides methods for all other sciences). This immediately suggests that theoretical logic is universal and universally applied. The essence of Petrus Hispanus' quoted *dictum* about logic as *scientia scientiarum* contains the most fundamental intuition about logic. It was shared by Gödel 1944: 125: "[… logic] is a science prior to all others, which contains the ideas and principles underlying all sciences", Tarski 1994: xii:

> […] the word "logic" is used […] in the present book […] as the name of the discipline which analyses the meaning of the concepts common to all sciences, and establishes general laws governing these concepts

and Quine 1970: 102: "The lexicon is what caters distinctively to special tastes and interests. Grammar and logic are the central facilities, serving all comers". Thus, one can think about the view of Petrus Hispanus-Gödel-Tarski-Quine as suggesting that logic is universally applicable. One of the philosophical tasks of metalogic can be viewed as an attempt to explain why logic (to be more precisely, *logica sensu stricto*, that is, formal logic) finds its application in various and divergent subject-matters.

As I noted earlier, a logical system **S** can be constructed axiomatically or by natural deduction (including sequents in Gentzen's sense). Each concrete construction requires some metalogical decisions and leads to various consequences in resulting shapes of systems. In particular, the articulation of **S** assumes, definitions of alphabet and well-formulas, introducing symbolism (for instance, with brackets or free of them, like Łukasiewicz's notation), listing axioms and/or rules of inference, using schemata or concrete formulas, etc. Such metalogical decisions generate, for instance, whether rule of substitution occurs among principles of inferences, *Cn* is structural or a system is finitely axiomatizable or how many connectives suffice for constructing **PC**. Yet the results are equivalent, that is, a given logical system, codes the same rules of deduction (having the same deductive power). We can say that dependence of logic on metalogic is definitional, that is, metalogic provides definitions for concepts used in formulating a logical system. On the other hand, metatheoretical decisions exhibit various details of equivalent logical systems, unnoticeable without detailed metalogical studies. The Church thesis (it is rather a general metamathematical result) says that calculable = recursive nicely illustrated this issue. Since we agree that **S** is decidable means that **S** is calculable, we conclude, via the thesis in question, that logic is decidable if and only if the set of its theorems is recursive.

Let me take a concrete example. Łukasiewicz insisted that **PC** should be based on the minimal list of primitive connectives, axioms and rules of inferences. He constructed **PC** with one primitive (bi-negation, the negation of conjunction, symbolized by the letter *D*), one axiom (I will not quote it), the counterpart of the modus ponens for *D*) and the rule of substitution. Łukasiewicz's motivation was to build the simplest propositional logic. Not all logicians share such preferences and claim that logic should be simple, but this attribute cannot collide with practical usefulness. One polish philosopher, painter and writer (Stanisław Ignacy Witkiewicz) ironically presented a genial logician, presumably Leon Chwistek, who constructed the simplest logical system: one primitive – point, one axiom – point is point, one rule of inference – do nothing more. Łukasiewicz himself was very proud of his results. Father Bocheński told me the following story: "I visited Łukasiewicz in the middle of 1930s. He told me that he just discovered a new single axiom of propositional calculus, based on *C* (implication) and *N* (negation), presented his result, the formula having about 50 signs, beginning with something like *CCCNCNCC*... and said that it is so simple and so evident." Leaving, more or less colourful anecdotes aside, we frequently have to do with conflicts between theoretical preferences fully justified by metalogical results and practical claims (also based on metatheoretical assumptions).

How to define logic? Working with **QC** as a paradigmatic case (as I already noted) we have (I use the symbol **LOG** for logic):

(1) $A \in \mathbf{LOG} \Leftrightarrow A \in Cn\emptyset$, or, equivalently $\mathbf{LOG} = Cn\emptyset$.

This has two other equivalents (I follow Tarski's analysis given in his metamathematical papers in Tarski 1956, Chapters III, V, XII; see also Woleński 2004 for further remarks):

(2) $A \in \mathbf{LOG} \Leftrightarrow \neg A$ is inconsistent;
(3) **LOG** is the only non-empty product of all deductive systems (theories).

The proposal (1) looks artificial, but (2) and (3) – not. In fact, deducing something from nothing seems impossible. On the other hand we expect that negations of logical theorems are inconsistent as well as logic is contained in all deductive systems ((3) says that). However, axioms for Cn cover the entire logical deductive machinery. In the case of **PC**, we have (X, Y – arbitrary sets of formulas):

(4) (a) $\emptyset \leq \mathbf{L} \leq \aleph_0$;
 (b) $X \subseteq CnX$;
 (c) $X \subseteq Y \Rightarrow CnX \subseteq CnY$;
 (d) $CnCnX = CnX$;
 (e) $A \in CnX \Rightarrow \exists Y \subseteq X \wedge Y \in \mathbf{FIN} \wedge (A \in CnY)$;
 (f) $(A \Rightarrow B) \in CnX \Rightarrow B \in Cn(X \cup \{A\})$;
 (g) $B \in Cn(X \cup \{A\}) \Rightarrow (A \Rightarrow B) \in CnX$;
 (h) $Cn\{A, \neg A\} = \mathbf{L}$;
 (i) $Cn\{A\} \cap Cn\{\neg A\} = Cn\emptyset$.

The meaning of the items in (4) is as follows: (a) – the cardinality of **L** is not greater than the set of natural numbers; (b) – closure of X by Cn; (c) – monotocinity; (d) – idempotence; (e) finiteness; (f) and (g) – the deduction theorem; (h) – the theorem $A \wedge \neg A \Rightarrow B$; (i) – the product of consequences of mutually inconsistent sets is equal to the set $Cn\emptyset$. (4)(a)–(e) are so-called general axioms – they do not generate any logic. The deduction theorem explicate the meaning of implication, (h) and (i) – the meaning of conjunction and negation.

The equivalence of (1)–(3) immediately follows from (4). As I noted earlier, the universality of logic is one of its fundamental properties. Metalogic helps in its precise account. Take the definition (1) as the starting point for further analysis. We can read as the assertion that the derivation of logical theorems logic does not require any special premises belonging to **L** (the object language). What about the points recorded in (4)? They belong to **ML**. Thus, metalogic offers a set of conditions for understanding logic as $Cn\emptyset$ (note that an amount of set theory is assumed). The weak (because applied to logical systems only) completeness theorem

(5) $A \in Cn\emptyset$ if and only if A is true in any model **M**,

asserting that A is a logical theorem, provided that it is true in all models. The completeness theorem gives the exact sense of the assertion that logic is universal. In particular, this theorem explains why no premise is required to deduce logical theorems. In fact, the definition (3) displays the same attribute of logic, because it locates logical tautologies as a part of every deductive system – **X** is a deductive system or theory if and only if **X** = Cn**X**. Now, it is clear why logic as such does not distinguish any extralogical content – it is due to the fact that logic is universally valid (true in all models). The Löwenheim-Skolem-Tarski (**LST**) theorem says that **QC** (in general, all first order theories) has infinite (of arbitrary cardinality) models, provided that it has denumerable infinite models. This result is frequently considered as non-intuitive, because, for instance, arithmetic of real numbers has the model of the cardinality $\aleph_0$. However, in the case of **QC**, **LTS** is not a surprise, because if this logic is universally valid, it should not distinguish cardinalities of its models. The Lindström theorem (**LT**) says that any logic with **L** = $\aleph_0$ and classical connectives, quantifiers and identity, which is compact (in the sense, that X is consistent if and only if its every finite subset is consistent) and satisfies **LST** is equivalent to **QC**. In other words, **QC** is the only logic which satisfies **LT**. Interestingly, **LT** defines the universality of logic by compactness and **LTS** without appealing to (5).

My earlier remarks considered **QC** without dividing it into parts. However, there are important differences between **PC** and the full **QC**. The former is Post-complete (if a non-theorem is added to **PC**, it becomes inconsistent) and decidable, but **QC** is neither Post-complete nor decidable. There are also problems with the concept of identity. Having identity, we can define numerical quantifiers of the type 'there are n objects', where n is an arbitrary natural number. Consequently, we can characterize finite domains, although first-order logic is too weak in order to define the concept of finiteness. Now, if we add the sentence 'there are n objects' to first-order logic, its theorems are valid not universally, but in domains that have exactly n elements. Hence, it seems that identity brings some extralogical content to pure logic, contrary to the view (it can be expressed by a suitable metalogical theorem) that logic does not distinguish any extralogical content. Perhaps this is a very reason that the label 'the identity-predicate' is used, although logicians simultaneously remark that this is a very special predicate. Anyway, a qualification of identity as logical or extra-logical is conventional to some extent. Other reason to see identity as an extralogical notion stems from so-called inflation and deflation theorems (see Massey 1970: 249 – 255), both closely related to the definability of finite domains in **QC**. The former says that if a formula, let say A, is satisfied in a non-empty model **M** with n elements, it is also satisfied in any model **M'** with at least n elements. The deflation theorem asserts that if A is satisfied in **M**, it is also satisfied in any **M'** with at most n elements. Although these theorems hold for **QC** without identity, they fail for first-order logic with identity. The formula $\forall xy(x = y)$ provides a counterexample to the inflation theorem, because it is true in the one-element domain and no other, but the formula $\exists xy(x \neq y)$ is false in the domain with one element. Perhaps we could speak about the degrees of logicality, represented by **PC**, **QC** without identity and **QC** with identity. In a sense, properties of **PC** are stronger (Post-completeness, decidability) than **QC** but the strength of **QC** without identity is greater than the logic with the identity-predicate. All comparisons and decisions concerning qualification of a given system as logic, discussed in this paragraph, require an explanation of what are logical concepts on the level of metalogic. I will return to this question in the remarks at the end of the paper.

Logics (in the customary sense, that the variety of all systems customarily called logical) can be compared. For instance, look at first- and second order logic (**SO**). The first-order thesis (see Woleński 2004 for a discussion) says that **QC** is the only logic. This metalogical decision excludes **SO** from the family of logics. The main argument is that the latter is not universal. The counterargument in the favor of **SO** points out that its expressive power is much greater than **QC**. Thus, we have a metalogical dilemma: the universality property or the property a considerably powerful expressive power. Clearly, possible solutions depend on attributing to the disccussed properties some well-defined metalogical concepts and well-established metatheorems. A similar discussion concerns the question: is modal logic a logic in the proper sense? Consider the variety of modal logics associated with Kripke frames defined by the set of possible worlds and the accessibility relation (**AR**). Various modal systems are distinguished by properties of **AR**. For instance, **AR** for **K** has no specific properties, for **T** is reflexive, for **D** (deontic logic) is irreflexive, for **S4** is reflexive and transitive, for **S5** is reflexive, symmetric and transitive. The completeness theorem for a modal logic **ML** states that A is a theorem of **ML** if and only if A is true in all frames (models) associated with **AR** related to a given system. Now, only theorems of **K** are valid independently of any specific conditions, other obey additional constraints. In other words, the university of **K** is unconditional, and conditional in other cases. In fact, necessity and possibility as modal operators behave in **K**, like quantifiers (universal and existential, respectively), but their content in other systems is richer than in **K**. Similarly as at the end of the previous paragraph, we can conclude that understanding of modal concepts as logical depends on their metalogical properties.

Hitherto I discussed **QC** with some remarks concerning second-order logic and modal logic, particularly with respect to the problem of universality. What about classical and non-classical logic? Let us look at the case of intuitionistic logic, considered as the most important rival of the classical system. Assume (I limit further considerations to **PC** and omit rules of inference) that we characterize classical logic by the following axioms (see Pogorzelski 1994: 145):

(PCA 1) $A \Rightarrow (B \Rightarrow A)$;
(PCA 2) $(A \Rightarrow (A \Rightarrow B)) \Rightarrow (A \Rightarrow B)$;
(PCA 3) $(A \Rightarrow B) \Rightarrow ((B \Rightarrow C) \Rightarrow (A \Rightarrow C))$;
(PCA 4) $A \wedge B \Rightarrow A$;
(PCA 5) $A \wedge B \Rightarrow B$;
(PCA 6) $(A \Rightarrow B) \Rightarrow ((A \Rightarrow C) \Rightarrow (A \Rightarrow B \wedge C))$;
(PCA 7) $A \Rightarrow A \vee B$;
(PCA 8) $B \Rightarrow A \vee B$;
(PCA 9) $(A \Rightarrow C) \Rightarrow ((B \Rightarrow C) \Rightarrow (A \vee B \Rightarrow C))$;
(PCA 10) $(A \Leftrightarrow B) \Rightarrow (A \Rightarrow B)$;
(PCA 11) $(A \Leftrightarrow B) \Rightarrow (B \Rightarrow A)$;
(PCA 12) $(A \Rightarrow B) \Rightarrow ((B \Rightarrow A) \Rightarrow (A \Leftrightarrow B))$;
(PCA 13) $(A \Rightarrow B) \Rightarrow (\neg B \Rightarrow \neg A)$;
(PCA 14) $A \Rightarrow \neg\neg A$;
(PCA 15) $\neg\neg A \Rightarrow A$.

This set of axioms has a very nice feature. We can stratify (PCA 1)–(PCA 15) into subsets related to the properties of particular connectives that occur in PC. In particular, (PCA 1)–(PCA 3) characterize implication, (PCA 4)–(PCA 6) regulates the behaviour of conjunction, (PCA 7)–(PCA 9) show how disjunction behaves, (PCA 10)–(PCA 12) define equivalence, and, (PCA 13)–(PCA 15) explain negation. These subsets are useful for extracting some weaker logics being parts of **PC**, for instance the implicational logic defined by (PCA 1)–(PCA 3) (this system is also determined by the deduction theorem), and the positive logic (without negation) by (PCA 1)–(PCA 12).

If we drop axiom (PCA 15), we obtain intuitionistic logic. However, there is a problem. According to the presented formalization, intuitionistic logic is a subsystem of classical logic. On the other hand, the intuitionist will not say, that his logic is a part of **QC**. Clearly, the entire controversy is rooted in metalogic. The intuitionist will say to the classicist "Your error consists in accepting the classical metalogic – I do not agree that it is admissible, I claim that my metalogic is not classical". Yet we encounter a new problem, namely concerning proofs of metalogical properties, especially the completeness theorem. Now the intuitionistic counterpart of (5) has no intutionistically acceptable proof (see de Swaart 2018: 397 – 402 for a discussion). It is known that so-called Markov principle ($\forall n(Pn \lor \neg Pn) \land \neg\forall n \neg Pm) \Rightarrow \exists n Pn$ (n, m – natural numbers) suffices for the proof (it is not accepted by intuitionists, but recognized by Russian constructivists). Similar remarks concern other non-classical systems, for instance, whether metalogic of many-valued logic is (can be) many-valued or whether metalogic of paraconsistent logic is (can be) paraconsistent. Thus, logical strength of metalogic is relevant for presenting (proving) properties of logics.

The completeness theorem can be formulated as (the Gödel-Malcev theorem):

(6) A set of sentences if consistent if and only if it has a model.

Both versions have no intuitionistic (constructive, effective – I disregard various subtleties related to the meaning of these adjectives and consider them as equivalent) proofs, but the equivalence of (5) and (6) can be constructively demonstrated. Another example (from metamathematics) is that Gödel's incompleteness theorems are effectively (constructively) provable, but Tarski's result on the undefinability of truth has no such proof. Since metalogic contains an amount of logic, it is important how strong logical machinery is used in metalogic. Here, we have an example of dependence of metalogic on logic. Generalizing the issue, how to define the minimal metatheory sufficient for provability of important metalogical theorems. Truth-theory (Tarski's definition) and intuitionism suggest that metalogic must be generally stronger than logic, but no general rule seems available – for example, classical monadic predicate calculus is decidable, but intuitionistic – not. As far as the issue concerns truth-theory, a weak second order arithmetic with the comprehension axiom is enough for development of truth-theory for a theory **T**, provided that this theory is formalized in the first-order language (in principle, all mathematical theories admit such a formalization).

To repeat once again, logic and metalogic cannot be sharply divided. In other words, it is sometimes difficult to decide whether a given metatheorem or concept is logical or just metalogical. However, one thing is rather obvious. In general, metalogic must be less formal than logic. Assume that **QC** (or any other logical system) is perfectly (whatever it means) formalized. However, even if we use formal methods in metalogic, they must be accompanied by various more or less informal explanations. The word 'must be' should be understood literally. In other words, that metalogic is less formal than logic is a necessary consequence of the interplay of the former and the latter. The situation is to some extent similar as in the case of mathematics and metamathematics – if a mathematical theory is completely (whatever it means) formalized, its metatheory must be partially informal. Even if metalogic (the same concerns metamathematics) is formalized, informal elements must appear on the higher (let say, the third) level. Although a part of metamathematics can be embedded into the first level formalism via arithmetization the informal residuum always remains. Thus, here appears the danger of petitio principii and/or regressus ad infinitum in justification of mathematical concepts and theorems. Hilbert was conscious of this situation. His proposal to justify mathematics via finitary (that is, unquestionable) methods was an attempt to defend mathematical reasoning against a version of scepticism pointing out that deductive methods cannot avoid objections to be inaccurate due to suffering from petitio principii or regressus ad infinitium.

The situation in metalogic seems even worse than in general metamathematics, because we use logical operators and deductive defined in logic. Since logic is prior to any other field, sceptical objections seem to be more justified than in other cases. The answer that it is nothing special, because the situation is similar to that in the standard mathematics (see above) or physics (quantum theory is fundamental, but in order to establish it, we use macrophysics) is philosophically not convincing. We have two horns, namely either epoche in Husserl's sense or admitting that meaning of metatheoretical concepts is prior to their formal expositions. The second alterative seems coherent with the actual development of logic and metalogic. If we read Aristotle (a fairly advanced logician), his presentation of logic contains logical as metalogical factors. This means that he relied on ordinary meaning of logical terms. Perhaps we could use Ryle's (see Ryle 1953) observation that 'ordinary' is very frequently equivalent to 'standard'. Using this idea, one might say that the standard meaning of logical terms functioned earlier than logic became more or less formalized. If so, a typical picture that *logica docens* is prior to *logica utens*, because the former validates the latter, should be reversed. In fact, it seems that the humans earlier used logical principles intuitively and later discovered logical theories.

Bibliography

Beth, Evert (1968): *The Foundations of Mathematics*. Amsterdam: North-Holland Publishing Company.
Enderton, Herbert. B. (1971): *A Mathematical Introduction to Logic*. San Diego: Harcourt Academic Press.
Gödel, Kurt (1944): "Russell's Mathematical Logic." In: *The Philosophy of Bertrand Russell*. Schilpp, Paul (ed). LaSalle: Open Court Company. Reprinted in: *Collected Papers*. Vol. 2. New York: Oxford University Press 1989, 119 – 141.
Hunter, Geoffrey (1971): *Metalogic. An Introduction to the Metatheory of Standard First Order Logic*. Berkeley, Los Angeles: University of California Press.
Kleene, Stephen (1967): *Mathematical logic*. New York: John Wiley and Sons.

Łukasiewicz, Jan and Tarski, Alfred (1930): "Untersuchungen über den Aussagenkalkül'." In: *Comptes Rendus des séances de la Société des Sciences et des Lettres de Varsovie*. Classe III, 23, 30 – 50; Eng. tr. in: Tarski, Alfred (1959): *Logic, Semantics, Metamathematics. Papers from 1923 to 1938*. Oxford: Clarendon Press, Indianapolis: Hackett Publishing Company, 38 – 59.

Massey, Gerald (1970): *Understanding Symbolic Logic*. New York: Harper & Row.

Pogorzelski, Witold A. (1994): *Notions and Theorems of Elementary Logic*. Białystok: Warsaw University, Białystok Branch.

Quine, Willard van Orman (1970): *Philosophy of Logic*. Englewood, Clifts: Prentice-Hall.

Rentsch, Thomas (1980): "Metalogik." In: *Historisches Wörterbuch der Philosophie*. Ritter, Joachim; Gründer, Karlheinz (eds.), Vol. 5: L–Mo. Darmstadt: Wissenschaftliche Buchgesellschaft, 1171 – 1174.

Ryle, Gilbert (1953): "Ordinary Language." In: *The Philosophical Review*. Vol. 42, 301 – 319.

Schechter, Eric (2005): 2005: *Classical and Non-Classical Logic. An Introduction to the Mathematics of Propositions*. Princeton: Princeton University Press.

Schütte, Kurt (1980): "Metamathematik." In: *Historisches Wörterbuch der Philosophie*. Ritter, Joachim; Gründer, Karlheinz (eds.), Vol. 5: L–Mo. Darmstadt: Wissenschaftliche Buchgesselshaft, 1175 – 1177.

de Swaart, Harrie (2018): *Philosophical and Mathematical Logic*. Berlin: Springer.

Tarski, Aldred (1956): *Logic, Semantics, Metamathematics. Papers from 1923 to 1938*. Oxford: Clarendon Press. Reprinted, Indianapolis: Hackett Publishn Company 1983.

Tarski, Alfred (1994): *Introduction to Logic and to the Methodology of Deductive Sciences*. 4th edition, New York, Oxford: Oxford University Press.

Woleński, Jan (2004): "First-Order Logic: (Philosophical) Pro and Contra." In: *First-Order Logic Revisited*. Hendricks, Vincent; Neuhaus, Fabian; Pedersen, Stig A.; Scheffler, Uwe; Wansing, Heinrich (eds.), Berlin: λογος, 369 – 399. Reprinted in: *Essays on Logic and Its Applications in Philosophy*. Frankfurt am Main: Peter Lang, 61 – 80.

Yaqūb, Aladdin M. (2015): *An Introduction to Metalogic*. Peterborough: Broadview Press.

Edi Pavlović and Norbert Gratzl
Free Logic and the Quantified Argument Calculus

Abstract: The Quantified Argument Calculus (or Quarc for short) is a novel and peculiar system of quantified logic, particularly in its treatment of non-emptiness of unary predicates, as in Quarc unary predicates are never empty, and singular terms denote. Moreover, and as a consequence of this, the universally quantified formulas entail their corresponding particular ones, similar to existential import. But at the same time, Quarc eschews talk of existence entirely by having a particular quantifier instead of an existential one. To bring it back into consideration, we explicitly introduce the existence predicate, and modify the rules to make the existence assumption obvious. This, along with some modifications, leads to a version of negative free logic. A question that arises at this point, given that we are interested in free logic, is what happens when we remove the existence assumption on singular terms; here we can quite naturally choose the negative free logic framework as well. In this paper we shall therefore investigate interrelations between Quarc and free logic (especially with its negative variant), and approach these interrelations with proof-theoretic methods.

1 Introduction

Classical Logic (CL) is the most well-known approach to formal reasoning; it has its own quirks and features. Recently, a new alternative to CL has been developed. One of the reasons for developing a new alternative to CL is to provide for formal reasoning and formalization processes that are (arguably) closer to natural language. This system (or rather family of system) is called Quantified argument calculus (Quarc for short), (c.f. e.g. Ben-Yami 2014, Lanzet/Ben-Yami 2006, Pavlović/Gratzl 2019).

In the 1960ies CL has been investigated with respect to its specific existence assumptions and the outcome has been (again a family of) free logics, where free logic is short for first order logic free of existence assumptions. Existence assumptions vary but the central claims are: (1) the domain of an interpretation is not empty (this is respected in the CL theorem $\exists x(x = t)$), (2) every name denotes exactly one object in the domain, and (3) the quantifiers have existential import (expressed by $\forall x E!x$).

Edi Pavlović, University of Helsinki, Finland, edi.pavlovic@helsinki.fi
Norbert Gratzl, Ludwig Maximilian University of Munich, Germany, N.Gratzl@lmu.de
DOI 10.1515/9783110657883-7

Neither CL nor Free Logic, however, have existence assumptions on unary predicates. This is not so in Quarc – in Quarc unary predicates are never empty, and singular terms denote. But at the same time, Quarc eschews talk of existence entirely by having a particular quantifier instead of an existential one. A particular quantifier tells us that there is an instance of the unary predicate, so in this sense the said predicate is not empty. Furthermore, of this instance (at least) something can be truthfully predicated. So, in this sense it expresses that "there are" things, but is stops short of identifying this construction with the existence statement about it, and therefore remains agnostic on the claim of existence (it's possible we could say true things about non-existents, such as, for example, that there are some). To bring it back into the discussion, we explicitly introduce the existence predicate, modify the rules to make the existence assumption obvious, and introduce the rule for the new predicate. This leads to a version of negative free logic, and we investigate the versions both with and without identity.

In this paper we shall therefore investigate interrelations between Quarc and free logic (especially with its negative variant). Furthermore, this paper approaches these interrelations with proof-theoretic methods and results of Pavlović/Gratzl 2019. In it the authors claim that the rules of quantification in (the family of logics) Quarc resemble those of free logic. The results of this paper substantiate that claim.

1.1 Quarc

Quarc is a system of quantified logic which does away with variables and unrestricted predicates, but nonetheless achieves results similar to the Predicate Calculus by employing quantifiers, applied directly to unary predicates, which then appear as arguments of other predicates (hence the name Quantified Argument Calculus), along with operators that attach to predicates and subsequently modify the mode of predication, as well as anaphors. It is in these respects, as mentioned previously, closer to natural language.

Let us note that the quantifiers in Quarc do have particular import, a fact that is expressed semantically by the condition of non-emptiness of (unary) predicates. This is in contrast to first-order predicate logic, where, as it is well known, (unary) predicates can be empty. For the purpose of investigating logics free of existence assumptions, we will in the proceeding also eliminate this condition[1], and focus on the resulting systems, labelled $Quarc_B$ for version without identity, and $Quarc_2$

---

[1] Lanzet 2017 investigates a three-valued version of Quarc that also omits this assumption. Here, however, we will remain within the confines of a two-valued system.

for version with it, as well as their respective sequent-calculus representations, LK-Quarc$_B$ and LK-Quarc$_2$.

2 Free Logic

Let us start with admittedly very broad, but nonetheless instructive, explanation of what free logics are:

> A free logic is a formal system of quantification theory, with or without identity, which allows for some singular terms in some circumstances to be thought of as denoting no existing object, and in which quantifiers are invariably thought of as having existential import. (Bencivenga 2002: 148 – 149)

In this quote one might glimpse a connection between Quarc and free logics – the quantifiers having existential import. Of course, given that Quarc doesn't talk of existence, the connection will have to be refined in the proceeding, but this will serve as an initial point of contact – both Quarc and free logic challenge the standard commitments to existence. A more formal way to characterize (negative, as it among the many free logics will be the sole focus of this paper) free logic would be to describe it via axioms:

1. $\forall x(A \to B) \to (\forall xA \to \forall xB)$
2. $A \to \forall xA$, if x is not free in A
3. $\forall xA \to (E!t \to At)$
4. $\forall xE!x$
5. $R(t_1, \ldots, t_n) \to (E!t_1 \land \ldots \land E!t_n)$
6. $\forall x(x = x)$
7. $s = t \to (As \to At)$

On the other hand, in a sequent calculus the rules for quantifiers in free logic can be formulated as follows, following Bencivenga 2002 and slightly simplified:

$$\frac{A[t/x], \Gamma \Rightarrow \Delta}{\forall xA, \Gamma \Rightarrow \Delta} \text{L}\forall \qquad \frac{\Gamma \Rightarrow \Delta, E!t}{} \qquad \frac{E!t, \Gamma \Rightarrow \Delta, A[t/x]}{\Gamma \Rightarrow \Delta, \forall xA} \text{R}\forall^*$$

$$\frac{E!t, A[t/x], \Gamma \Rightarrow \Delta}{\exists xA, \Gamma \Rightarrow \Delta} \text{L}\exists^* \qquad \frac{\Gamma \Rightarrow \Delta, E!t \qquad \Gamma \Rightarrow \Delta, A[t/x]}{\Gamma \Rightarrow \Delta, \exists xA} \text{R}\exists$$

* – t does not occur below the inference line.

One can see that in all of these cases an extra requirement has been added – that of $E!t$. In the following section we will employ the same principle to transform Quarc into its free version.

3 Free Quarc

To produce the free versions of the systems LK-Quarc$_B$ and LK-Quarc$_2$ we add the new rule for the existence predicate $E!$, replace the rules for the quantifiers, and also replace one of the identity rules in the latter of the two systems. In the interest of brevity, the full systems will not be laid out here, but the reader can find both those, and a thorough discussion of their metatheoretical properties, in Pavlović/Gratzl 2019.

3.1 The Base System – FQ$_B$

To transform the system LK-Quarc$_B$ (which does not contain identity) into a system of free logic FQ$_B$, we modify the quantifier rules by an explicit condition on the existence of the singular term, in the same vein as above:

$$\frac{A[a/\forall M], \Gamma \Rightarrow \Delta}{A[\forall M], \Gamma \Rightarrow \Delta} L\forall \qquad \frac{\Gamma \Rightarrow \Delta, aM \quad \Gamma \Rightarrow \Delta, aE!}{} \qquad \frac{aM, aE!, \Gamma \Rightarrow \Delta, A[a/\forall M]}{\Gamma \Rightarrow \Delta, A[\forall M]} R\forall^*$$

$$\frac{aM, aE!, A[a/\exists M], \Gamma \Rightarrow \Delta}{A[\exists M], \Gamma \Rightarrow \Delta} L\exists^* \qquad \frac{\Gamma \Rightarrow \Delta, aM \quad \Gamma \Rightarrow \Delta, aE! \quad \Gamma \Rightarrow \Delta, A[a/\exists M]}{\Gamma \Rightarrow \Delta, A[\exists M]} R\exists$$

* – a does not occur below the inference line.

In addition to the rules for quantifiers, we also supply the rule for the (negative free logic) existence predicate:

$$\frac{aE!, A[a], \Gamma \Rightarrow \Delta}{A[a], \Gamma \Rightarrow \Delta} NE!^*$$

* – A is basic[2].

---

[2] In the terminology of Quarc, a basic formula corresponds to an atomic formula of PC.

With these in place, we can demonstrate the following axioms of negative free logic. Given that the system we were expanding did not contain identity, the resulting system will likewise not contain it.

Theorem 1. *All of the following axioms of negative free logic are derivable in FQ_B:*
1. $(\forall M_\alpha A \to \alpha B) \to (\forall MA \to \forall MB)$[3]
2. $A \to (\forall MM \land A)$
3. $\forall MA \to ((aE! \land aM) \to aA)$
4. $\forall ME!$
5. $(a_1, \ldots, a_n)A \to (a_1 E! \land \ldots \land a_n E!)$

Note that while these axioms characterize negative free logic, only axioms 3 – 5 are specific to it (i.e. 1 – 2 are likewise derivable in LK-Quarc$_B$).

3.2 Metatheoretic Properties of FQ$_B$

We now turn towards establishing some metatheoretic properties of FQ_B, first and foremost being the Cut elimination theorem. As everywhere in this paper, the proof is omitted, but it is a straightforward adaptation of the one found in Pavlović/Gratzl 2019.

Theorem 2. *The Cut elimination property holds of FQ_B. Namely, any sequent derivable in FQ_B is derivable without using the Cut rule.*

To consider some consequences of this theorem, we first define the Subformula property, specifically its weak version (which will suffice to establish the results required in this paper).

Definition 3 (Weak subformula property). *A sequent calculus system possesses the Weak subformula property just in case any formula occurring anywhere in a derivation of an endsequent is either a subformula of some formula occurring in that endsequent, or a basic formula.*

It follows straightforwardly from Theorem 2 that

Corollary 4. *FQ_B possesses the Weak subformula property.*

From this Corollary we can further demonstrate that

Corollary 5. *FQ_B is consistent.*

---

[3] This simplified formulation assumes the formula $\forall MB$ is governed by the Quantified Argument $\forall M$. Otherwise, a more involved form, namely $\forall M_\alpha M \land \alpha B$, must be used, like in the next axiom.

This Corollary represents a desirable property of a logical system, but will not be of particular note going forward. Quite the opposite holds of the following one, however.

Corollary 6. *FQ_B is a conservative extension of $LK\text{-}Quarc_B$. Namely, any derivation $\Gamma \Rightarrow \Delta$ derivable in FQ_B and such that Γ, Δ do not contain E! is likewise derivable in $LK\text{-}Quarc_B$.*

We will discuss these properties of FQ_B at some length in Section 4, but for the moment we will examine adding the identity predicate to the system at hand.

3.3 Adding Identity – FQ_2

We add the rule for identity into the mix. Given the close connection of identity and the existence predicate in negative free logic, it should come as no surprise that the rules for the two look the same. We add this rule (it replaces the rule $=_1$ of $LK\text{-}Quarc_2$) and the rule $=_2$ to FQ_B to produce the system FQ_2.

$$\frac{a = a, A[a], \Gamma \Rightarrow \Delta}{A[a], \Gamma \Rightarrow \Delta} \; N{=}^*$$

* – A is basic.

With the addition of the identity rules, we can now derive the remaining axioms:

Theorem 7. *In addition to those axioms mentioned in Theorem 1, the following are derivable in FQ_2:*

6. $\forall M a = a$, for any M
7. $s = a \rightarrow (sA \rightarrow aA)$

In addition to Theorem 7, several other results characteristic of a negative free logic are now derivable, namely

Theorem 8. *Equivalence of existence and self-identity, $aE! \leftrightarrow a = a$, and indiscernibility of non-existents, $(a\neg E! \wedge b\neg E!) \rightarrow (A[a] \rightarrow A[b/a])$, are both derivable in FQ_2.*

3.4 Metatheoretical Properties of FQ_2

Not much needs to be added here, as the results of this section closely resemble their counterparts in Section 3.2. It is straightforward to show that

Theorem 9. *Cut elimination property holds for FQ_2.*

And from this it again follows that

Corollary 10. *FQ_2 possesses the Weak subformula property.*

And again,

Corollary 11. *FQ_2 is consistent.*

An interesting consequence of Corollary 10 is

Corollary 12. *$a = a$ is not derivable in FQ_2.*

This corollary is of course, combined with Axiom 6 ($\forall x(x = x)$), characteristic of the way negative free logic treats the truth of self-identity sentences.

4 Comparing Quarc and Free Quarc

We have already seen that FQ_B is a conservative extension of LK-Quarc$_B$. Now we move on to compare their respective versions containing identity, FQ_2 and LK-Quarc$_2$. Given the equivalence of existence and self-identity in FQ_2, (Theorem 8), it will suffice that we observe the $E!$-free segment of FQ_2, FQ_2^*. The result we obtain in this case is that

Theorem 13. *FQ_2^* is a proper subset of LK-Quarc$_2$, $FQ_2^* \subset$ LK-Quarc$_2$. Namely, every rule of FQ_2^* is admissible in LK-Quarc$_2$, but (Corollary 12), $a = a$ is not derivable in FQ_2^*.*

This result should not come as a great surprise – in general, free logic is a restriction on classical logic. In this particular case, if we compare the differing identity rules in LK-Quarc$_2$ and FQ_2, respectively:

$$\frac{a = a, \Gamma \Rightarrow \Delta}{\Gamma \Rightarrow \Delta} =_1 \qquad \frac{a = a, A[a], \Gamma' \Rightarrow \Delta}{A[a], \Gamma' \Rightarrow \Delta} N=^*$$

* – A is basic.

We can see that the latter is really just a special case of the former – specifically, when Γ stands for the list of formulas $A[a], \Gamma'$. By placing a limitation on the rules of the system, we likewise limit the output of the said system.

But now it should strike us as most peculiar that the same situation did not occur in the case of LK-Quarc$_B$ and FQ_B. Much like the rules for identity above, the quantifier rules of FQ_B impose a limitation on the corresponding rules of LK-Quarc$_B$. And yet, we experience no loss of power (Corollary 6) – in fact, the only change has to do with the change in vocabulary that results from adding the existence predicate $E!$. This anomaly bears closer scrutiny.

4.1 Comparison of FQ$_B$ and LK-Quarc$_B$

As mentioned when we first introduced a free version of Quarc, we produce it by means of an additional restriction on the rules. This, however, as we have just seen, does not result in a loss of expressive power. Normally, the most notable formula that becomes underivable in free logic is unrestricted specification, $\forall x A x \to A t$. Instead, we have the weaker, restricted specification (Axiom 3). The corresponding version of unrestricted specification in Quarc would be:

Definition 14 (Unrestricted specification). $\forall M A \to (aM \to aA)$

This formula can be obtained even in the free version LK-Quarc$_B$:

Theorem 15. $\forall MA \to (aM \to aA)$ is derivable in FQ$_B$.

The formula follows from Axiom 3 (restricted specification, $\forall MA \to ((aE! \wedge aM) \to aA)$) and an instance of Axiom 5, $aM \to aE!$. The latter is what explains this anomaly – in negative free logic, there is a close connection between the truth of atomic sentences and existence, expressed by Axiom 5 (and the absence of the formula $a = a$). But in FQ$_B$, we only added the condition, in the appropriate place, depending on the rule, that $aE!$ (thus allowing for the derivation of some formulas containing the new predicate), but the aM requirement was already present in the non-free rule, i.e. the quantification was already restricted, and precisely in a manner that precludes the derivation of that formula which the free logic avoids. This demonstrates the point raised in Pavlović/Gratzl 2019,

Observation 16. *The quantification rules of Quarc, even on the non-free version, have a "free flavor", or a structural resemblance to those of free logic.*

This point is further strengthened in the following section, when we discuss free logic in relation to quantified arguments.

5 Empty Predicates

Given that quantified arguments containing predicates feature in the same syntactic roles as names in Quarc, it has two different sets of existence assumptions – those of non-emptiness of names, and also of predicates. As noted in the introduction, we will be dropping both of those in this paper.

In this section, we restate the axioms to talk not of individuals, referred to by constants (or singular arguments in the terminology of Quarc), but of "some M's", captured by unary predicates. In what Pavlović/Gratzl 2019 refers to as *full Quarc*, these are required to be non-empty, but both systems under consideration here, LK-Quarc$_B$ and LK-Quarc$_2$, omit that requirement.

It should be noted that restricted specification as applied to predicates instead of names, $\forall MA \to \exists MA$, is not valid in either of those systems (Pavlović 2017), and therefore neither is it so in FQ$_B$ (by Corollary 6), nor in FQ2 (by Theorem 13). This checks off the first requirement on being able to describe a system as a (negative) free logic. As importantly, all the axioms must likewise hold, and this is in fact the case, restated for "some M's":

Theorem 17. *All of the following axioms are derivable in FQ$_B$ [FQ$_2$]:*
1. $(\forall M_\alpha A \to \alpha B) \to (\forall MA \to \forall MB)$
2. $A \to (\forall MM \land B)$
3. $\forall MA \to (\exists ME! \to \exists MA)$
4. $\forall ME!$
5. $\exists MP \to \exists ME!$
6. $[\forall M_\alpha = \alpha, \text{for any } M]$
7. $[\exists M = \exists P \to (\exists MA \to \exists PA)]$

So, both FQ$_B$ and FQ$_2$ are free not just with respect to non-empty names, but also non-empty predicates. That this feature transfers back to LK-Quarc$_B$ and LK-Quarc$_2$ can be demonstrated by restating the axioms without the existence predicate $E!$. We are able to do this, when talking about some M's, since (given the close connection between unary predicates and existence predicate), for some M to exist is for it to be some unary predicate, namely M, $\exists MM$:

Lemma 18. $\exists ME! \leftrightarrow \exists MM$

The left-to-right direction is obtained by a simple use of R∃ and then L∃, and the right-to-left direction is an instance of the Axiom 5 from the Theorem 17.

It follows from Theorem 17 and Lemma 18, again using Corollary 6 and Theorem 13 that

Theorem 19. *All of the following axioms are derivable in LK-Quarc$_B$ [LK-Quarc$_2$]:*
1. $(\forall M_\alpha A \to \alpha B) \to (\forall MA \to \forall MB)$
2. $A \to (\forall MM \land B)$
3. $\forall MA \to (\exists MM \to \exists MA)$
4. $(\forall M_\beta M \land (\forall M_\alpha M \land A[\alpha, \beta])) \to (\forall M_\alpha M \land (\forall M_\beta M \land A[\alpha, \beta]))$ $[\forall M = \exists M]$
5. $\exists MP \to \exists MM$
6. $[\forall M_\alpha = \alpha, \text{for any } M]$
7. $[\exists M = \exists P \to (\exists MA \to \exists PA)]$

A note on Axiom 4 – for LK-Quarc$_2$, a very elegant axiom is available (as elsewhere, written within square brackets). However, since it requires the identity predicate, we cannot use it for LK-Quarc$_B$. Instead, what we do here is emulate the Permutation Principle (Fine 1983) in Quarc.

We can now strengthen the Observation 16 and conclude that

Observation 20. *The quantification rules of Quarc, even on the non-free version and with respect to both emptiness of names, as well as that of unary predicates, bear a structural resemblance to those of free logic, specifically negative free logic.*[4]

Bibliography

Baaz, Matthias; Leitsch, Alexander (2011): *Methods of Cut-Elimination*. Dordrecht: Springer.
Bencivenga, Ermanno (2002): "Free Logics." In: *Handbook of Philosophical Logic*. Gabbay, Dov; Guenthner, Franz (eds.), 2nd Edition, Vol. 5. Dordrecht: Springer, 147 – 196.
Ben-Yami, Hanoch (2004): *Logic and Natural Language: on plural reference and its semantic and logical significance*. Aldershot: Ashgate.
Ben-Yami, Hanoch (2014): "The Quantified Argument Calculus." In: *The Review of Symbolic Logic*. Vol. 7 No. 1, 120 – 146.
Buss, Samuel (1998): "An Introduction to Proof Theory." In: *Handbook of Proof Theory*. Vol. 137. Amsterdam: Elsevier, 1 – 78.
Fine, Kit (1983): "The Permutation Principle in Quantificational Logic." In: *Journal of Philosophical Logic*. Vol. 12, No. 1, 33 – 37.
Gentzen, Gerhard (1969): *The Collected Papers of Gerhard Gentzen*. Szabo, M.E. (ed.). Amsterdam: North-Holland Pub. Co.
Gratzl, Norbert (2010): "A Sequent Calculus for a Negative Free Logic." In: *Studia Logica*. Vol. 96, 331 – 348.
Kleene, Stephen Cole (2000): *Introduction to Metamathematics*. Amsterdam: Elsevier.
Lambert, Karel (1997): *Free Logics: Their Foundations, Character, and Some Applications Thereof*. Sankt Augustin: Academia Verlag.
Lambert, Karel (2001): "Free Logics". In:*The Blackwell Guide to Philosophical Logic*. Goble, Lou (ed.). Oxford: Blackwell Publishers, 258 – 279.
Lanzet, Ran; Ben-Yami, Hanoch (2006): "Logical Inquiries into a New Formal System with Plural Reference." In: *First Order Logic Revisited*. Hendricks, Vincent F. (ed.). Berlin: Logos Verlag, 173 – 223.
Lanzet, Ran (2017): "A Three-Valued Quantified Argument Calculus: domain-free model theory, completeness, and embedding of FOL." In: *The Review of Symbolic Logic*. Vol. 10, No. 3, 549 – 582.
Negri, Sara; von Plato, Jan (2001): *Structural Proof Theory*. Cambridge: Cambridge University Press.
Pavlović, Edi; Gratzl, Norbert (forthcoming 2019): "Proof-Theoretic Analysis of the Quantified Argument Calculus." In: *The Review of Symbolic Logic*.

---

4 This work was partially supported by the Academy of Finland, research project no. 1308664.

Pavlović, Edi (2017): *The Quantified Argument Calculus: an inquiry into its logical properties and applications*. PhD Thesis, Central European University, Budapest.
Schütte, Kurt (1960): *Beweistheorie*. Berlin-Göttingen-Heidelberg: Springer.
Takeuti, Gaisi (1987): *Proof Theory*. 2nd edition. Amsterdam-Lausanne-New York-Oxford-Tokyo: North Holland.

Gabriel Sandu
Dependencies Between Quantifiers Vs. Dependencies Between Variables

Abstract: I will argue that the most significant role of the logic of first-order quantifiers lies in its power to express functional dependencies and independencies between variables. The dependence of a variable x on another variable y has been standardly expressed by the formal dependence of a quantifier Qx on another quantifier Qy, which, in turn, is expressed by the former being in the syntactical scope of the latter. First-order logic, where scopes are required to be nested, cannot express all the possible patterns of dependence and independence between variables. To overcome this problem, two solutions have been proposed: to allow for more patterns of dependence and independence between quantifiers (Independence-Friendly (IF) logic); to express explicitly dependencies and independencies of variables (Dependence logic, Independence logic, etc). In both approaches the truth of a sentence amounts to the existence of appropriate "witness individuals" (Skolem functions). We have here a connection between the truth-conditions of quantified sentences and the existence of all the functions which produce these witness individuals. Hintikka has repeatedly argued that these functions codify winning strategies in certain (semantical) games and emphasized their connection to Wittgenstein's language games. In my contribution I will look at the interesting perspective that language games open for the discussion of logic in general. Some of these points have been discussed in Hintikka/Sandu 2007.

One of the revolutionary aspects of modern logic consists in considering valid inferences that involve multiple quantification. In this case one needs to consider quantifiers that appear in the scope of other quantifiers. In this paper I consider two kinds of dependencies: scopal dependencies between quantifiers and material dependencies between (the values of) variables. Some focus will be put on the discussion of mutual dependencies of both kinds.

1 The Other Function of Quantifiers

Many of the statements one encounters in mathematical practice involve multiple quantification. In this case one needs to consider quantifiers that appear in the

---

Gabriel Sandu, University of Helsinki, Gabriel.sandu@helsinki.fi
DOI 10.1515/9783110657883-8

scope of other quantifiers. When the sequence of quantifiers in a formula is linearly ordered, one indicates the scopal dependency of a quantifier on other quantifiers in a syntactic way by writing the former after the latter. The formal, scopal dependencies between quantifiers are indications of material dependencies between the values of the quantified variables in an underlying universe of discourse. The way these material dependencies are specified depends on the semantic representation. Each such representation has to solve the challenge that comes from the need to combine a semantic mechanism which corresponds to the "ranging over" semantic job of a quantifier and thereby considers one quantifier at a time, with a distinct mechanism which "glues" the successive steps together.

As pointed out in Sher 1990, in the Frege-Tarski tradition this challenge is solved in a relatively straightforward way. For any formula in which the quantifiers are linearly ordered one can consider only one singly quantified formula at a time and still account for the dependency of a quantifier on the previous one in the sequence. To illustrate, consider the scheme

1. $Q_1 x Q_2 y R(x, y)$

where $Q_1 x$ and $Q_2 y$ are any two (generalized) quantifiers. The interpretation of this sentence is specified in the following steps (again we follow Sher 1990): (1) is true if and only if there are q_1 a's for each one of which there are q_2 b's such that $R(a, b)$. Here q_1 and q_2 are the semantic conditions associated with the generalized quantifers Q_1, and Q_2, respectively. In this case the material dependence of the values of the variables, the "gluing together" is very weak being achieved by a relative expression synonymous with "for each one of which." One of Sher's examples is

2. Three frightened elephants were chased by a dozen hunters

represented by

3. $3x 12y\, (E(x) \wedge H(x) \wedge C(x, y))$

where $3x$ is the generalized quantifier interpreted in a universe of discourse M by the set of all subsets of M with three elements, and the generalized quantifier $12y$ is interpreted analogously. Thus (3) says that there are three frightened elephants, for each one of which there are 12 hunters such that every hunter chases it.

In the case in which Q_1 and Q_2 are the standard quantifiers $\forall$ and $\exists$, respectively, as in the sentence

4. $\forall x \exists y R(x, y)$

the truth-conditions state the existence of appropriate sets which are introduced sequentially: there exists a set X consisting of the whole universe such that for each of its elements a, there is a non-empty set Y such that a stands in the R-relation with each element of Y. The two sets can actually be composed into one binary relation S, making the truth-condition equivalent to: there is a set S which is a subset of R such that for each a there is at least a b such that $S(a, b)$. Under some weak set-theoretical assumptions, the last condition can be shown to be equivalent to the more familiar: for each a there is at least one b such that $R(a, b)$. As we see, the material dependence between the values of the variables x and y is rather weak, in the sense that any value of x does not constrain in any way the corresponding value of y except "externally" through the relation R (we assume that this relation is not an equation between the two variables). In other words, the material dependencies of the values of the variables take the form of a tree which is such that each arrow starting from the root points to an individual which represents a possible value of x, which in turn is further connected by arrows to all the individuals (leaves) with which it stands in the R-relation.

In this paper we are interested in arbitrary dependencies between standard first-order quantifiers; in particular we are concerned with the scopal dependencies of an existential quantifier on a sequence of other standard quantifiers. When these dependencies can be linearly ordered, as in (4), the semantic interpretation may follow the same pattern as that given above. An increase in the number of quantifiers may lead, however, to scopal dependence patterns which cannot be so linearised. One of these patterns, discovered long time ago by Henkin (Henkin 1961), involves four quantifiers:

- For every x and x', there exists y depending only on x and y' depending only on x' such that $Q(x, x', y, y')$ is true (here $Q(x, x', y, y')$ is a quantifier-free formula).

Henkin represented the four quantifier's prefix in a branching form:

$$\begin{pmatrix} \forall x & \exists y \\ \forall x' & \exists y' \end{pmatrix} Q(x, x', y, y')$$

to emphasize that $\exists y$ is only in the scope of $\forall x$ and $\exists y'$ is only in the scope of $\forall x'$. Finding a semantic interpretation for the branching prefix is not trivial. Barwise 1979 proposed a general scheme (for monotone quantifiers) which respects the spirit of the interpretation given by (4), except that now, to account for the partial ordering of scopes, the relevant sets are not introduced sequentially but right from the beginning. When these sets are combined into corresponding relations, the result may be expressed by the second-order sentence

$\exists R \exists S \left(\forall x \exists y R(x,y) \wedge \forall x' \exists y' S(x',y') \wedge \forall x \forall x' \forall y \forall y' \left(R(x,y) \wedge S(x',y') \to Q(x,x',y,y') \right) \right).$

This is not the semantic interpretation chosen by Henkin for his branching quantifier. Henkin's interpretation is based on a stronger dependence between the values of the variables, i.e., functional dependence, which goes back to Skolem. Perhaps the best way to introduce it is with respect to our earlier example (4). In this case the scopal dependence of $\exists y$ on $\forall x$ induces a functional correlation between the values of y on the values of x: for each a, which is a value of x there is exactly one b, which is a value of y such that $R(a,b)$. This correlation is assumed to be given by an unspecified function f so that the truth-conditions of (4) may be now expressed by the second-order sentence:

5. $\exists f \forall x R(x, f(x)).$

The function f is called a *Skolem function*.

Generalizing the Skolem functions approach to branching quantifiers yields Henkin's initial interpretation:

$$\begin{pmatrix} \forall x & \exists y \\ \forall x' & \exists y' \end{pmatrix} Q(x, x', y, y') \Leftrightarrow \exists f \exists g \forall x \forall x' Q(x, x', f(x), g(x')).$$

2 Game-Theoretical Semantics (GTS)

Henkin's interpretation of branching quantifiers based on the generalisation of the Skolem functions approach motivated Hintikka's game-theoretical interpretation of first-order quantifiers and connectives. A (semantic) game is associated with any first-order sentence and underlying model which interprets the non-logical constants of the sentence. The game is played by two players, the Verifier (whose moves corresponds to existential quantifiers and disjunctions) and the Falsifier (universal quantifiers and conjunctions). As an example we consider the game associated with the sentence φ

$$\forall x (\exists y L(x,y) \wedge \exists z H(x,z))$$

and a model M. The Falsifier chooses an individual from the universe of M, say a, to be the value of x after which he has the choice between the left and the right conjunct. If the former, the Verifier chooses an individual, say b, to be the value of y and the play stops. The Verifier wins the play if $L^M(a,b)$; otherwise the Falsifier wins. If the right conjunct is chosen, the Verifier chooses an individual, say c, to be the value of z and this play stops here, with similar conditions for winning and loosing.

Now as we see from the example, in the game-theoretical setting, a quantifier or a connective being in the scope of another quantifier or connective amounts to the *informational dependence* of the move prompted by the former on the move prompted by the latter. It is codified by the notion of *information set* associated with a given move, an epistemic notion which indicates which other moves the player making that move is aware of. Thus the scopal dependencies of quantifiers (and connectives) map directly into the knowledge of the players as codified by information sets.

On the other side, the truth (falsity) of a sentence S in a model M is defined as the existence of a winning strategy for the Verifier (Falsifier) in the appropriate semantic game. When the strategies are defined determistically (functionally), a winning strategy for the Verifier (when it exists) decomposes into Skolem functions (here we assume that each sentence is in prenex normal form, and we disregard the strategies associated with connectives). They express the *material dependencies* of the appropriate quantified variables. Referring to Skolem functions, say f, whenever $b = f(a)$ Hintikka thought of b as a *witness individual* which depends ontologically on the individual a. Thus the game-theoretical framework has two levels, each one coming with its own notion of dependence: the epistemic level of information sets which map isomorphically the scopal dependencies of quantifiers and connectives; and the ontological level of strategic functions which create a network of material dependencies between the individuals (values of the quantified variables) of the underlying universe.

3 Game-Theoretical Semantics As a Basis for General Logic

In this section I will draw on some ideas discussed in Hintikka/Symons unpubl..

The question asked in Hintikka/Sandu 1989 was: How can one extend scopal quantifier dependencies in order to express more material dependencies between the values of the quantified variables? Given the game-theoretical setting in the background, this question may be rephrased as: What are all the possible patterns of information flow (quantifier dependencies) compatible with the game rules for quantifiers and connectives? Recalling that quantifier dependencies map into dependencies between moves in the relevant semantic game, two minimal conditions suggested themselves quite naturally:

- A player's move can be informationally dependent only on an earlier move in the game
- The moves have to take place in linear time (linear time playability condition)

(The connection between quantifier dependencies and the linear time playability condition is also discussed in Hintikka/Symons unpubl..) The two conditions are seen to be easily satisfied by the semantic games associated with first-order sentences where a move is informationally dependent on all the earlier moves in the game. Given that we wanted to liberalise the patterns of dependencies among quantifiers beyond the first-order ones, this is a condition we did not want any longer to assume, so we gave it up and together with it we also gave up the transitivity of the dependence relation. All this together with the fact that in the conjunct $\varphi \wedge \psi$ the logical constants occurring in φ are not in the scope of those occurring in ψ, led naturally to the idea that the dependency relation governing scopes is an antisymmetric, partial, intransitive, and discrete ordering. The problem to be solved was to find a way to faithfully map this partial ordering onto a linear order which could be thought of as the temporal order of the moves in a semantical game. In this way the linear time playability condition would have been ensured. Notice that this condition is not fulfilled by Henkin's branching quantifiers prefix.

The way we chose to solve this problem in Hintikka/Sandu 1989 was to indicate separately the dependence (and independence) between moves. That is, we assumed that all the quantifiers and connectives depend *ceteris paribus* on all quantifiers and connectives before them in the ordering. The exceptions are indicated by the slash as in

$$\forall x \forall x' (\exists y/\{x'\})(\exists y'/\{x, y\}) Q(x, x', y, y')$$

which is our representation of the branching quantifier. Here $(\exists y/\{x'\})$ indicates that $\exists y$ is not in the scope (is independent) of $\forall x'$ and thus, given the assumption, it is only in the scope of $\forall x$. Similarly $(\exists y'/\{x, y\})$ indicates that $\exists y$ is independent of $\forall x$ and $\exists y$ and thus it is only in the scope of $\forall x'$. Game-theoretically the first condition means that the Verifier does not know the value chosen by the Falsifier for x; and the second condition means that the Verifier does not know the value chosen by the Falsifier for x' neither the one chosen by herself for y. There are some subtleties here concerning the Verifier "forgetting" her own earlier moves but they will not be my concern in this paper.[1] Notice, however, the time linearity of the 4 moves in the semantic game.

There would have been another way to implement the linear time playability condition, namely to assume that *ceteris paribus* all quantifiers and connectives are

---

[1] These questions are discussed in Barbero 2013.

scopally independent of each other. (This matter is also discussed in Hintikka 2002.) The *ceteris paribus* condition is now represented by

$$Q_2 y \parallel Q_1 x$$

which indicates the scopal dependence of $Q_2 y$ on $Q_1 x$ which occurs at its right, and likewise for connectives. The two representations are equivalent. We chose the first one, and called the result IF first-order logic.

4 Dependencies Among Quantifiers: Independence-Friendly Logic

The IF sentence φ_{inf}

6. $\forall x \exists y (\exists z/\{x\})(x = z \land c \neq y)$

defines (Dedekind) infinity, a property which cannot be expressed in ordinary first-order logic. The (scopal) dependence relation between quantifiers is antisymmetric and intransitive. The last claim follows from the fact that $\exists y$ depends on $\forall x$ and $\exists z$ depends on $\exists y$ but $\exists z$ does not depend on $\forall x$ (in other words, in her move correspoding to $\exists y$ the Verifier knows the value chosen for x by the Falsifier, and in the move corresponding to $\exists z$ she knows her own earlier move but does not know the choice made by the Falsifier). Instead, the material dependence between (the values of) variables expressed by the Skolem functions $y = f(x)$ and $z = g(y)$ is transitive: one can define a new function which expresses the dependence of the values of z on the values of x: $h(z) = g(f(x))$.

Now returning to our earlier question "What are all the possible patterns of information flow (quantifier dependencies) compatible with the game rules for quantifiers and connectives?", we shall rephrase it, following Hintikka/Symons unpubl. as:

> What patterns of scopal dependencies between quantifiers and correspondingly, what patterns of material dependencies between variables does the linear time playability condition exclude?

The answer given in Hintikka/Symons unpubl. is: mutually dependent variables. The challenge is now to understand what such variables are. One way to understand them is as in Hintikka 2002, where Hintikka talks about "strongly correlated variables which are mutually dependent so that they cannot be represented separably as functions of some third variable (non-commuting variables in quantum theory)". One of the examples Hintikka gives is the following IF-sentence

7. $\forall t \forall x \forall y \, (\exists z/\{x\})\,(\exists u/\{y, z\})\,[x = z \land y = u \land S(t, x, y)]$

which expresses the scopal dependence of $\exists z$ on $\forall t$ and $\forall y$ and that of $\exists u$ on $\forall t$ and $\forall x$. (Actually Hintikka has only u in the second slash-set, but this is due to the fact that he used the convention that existential quantifiers are always independent of each other.) Here the quantifier $\forall t$ ranges over moments of time. Making explicit the dependencies of the variables induced by the quantifier dependencies yield the following second-order sentence which expresses the truth-conditions of (7):

8. $\exists f \exists g \forall t \forall x \forall y \, [x = f(t, y) \land y = g(t, x) \land S(t, x, y)]$

(8) shows that x is a function of time and of the variable y whereas y is a function of time and of the other variable x.

Unfortunately Hintikka's example can be shown to be flawed, that is, (8) is true only in a model (set) with one element. To see this, suppose, for a contradiction, that (8) is true in a model which has two distinct elements, say a and b. It follows that there are no functions f and g as described in (8). For let $t = a$, $x = a$ and $y = a$. Then we must have $a = f(a, a)$. If on the other side we let $t = a$, $x = b$ and $y = a$, we should have $b = f(a, a)$, which is impossible.

The problem here is with the scheme

$$\forall t \forall x \forall y \, (\exists z/\{x\}) \ldots [\ldots x = z \ldots],$$

which can be true only in a model with one element. By analogy the same holds of the scheme

$$\forall t \forall y \, (\exists u/\{y\}) \ldots [\ldots y = u \ldots].$$

We may try to ignore the time variable t. Then (8) becomes

9. $\exists f \exists g \forall x \forall y \, [x = f(y) \land y = g(x) \land S(x, y)]$.

But now (9) may be shown to lead to the same problem as before, that is, it is true only in models with one element (by choosing $x = a$ and $y = b$ we get $a = f(b)$ and $b = g(a)$; on the other side, by choosing $x = b$ and $y = b$ we get $b = f(b)$ which is impossible). The problem is, as in the preceding example, with the scheme

$$\forall x \forall y \, (\exists z/\{x\}) \ldots [x = z \ldots]$$

which can be true only in a singleton set.

Perhaps considerations of this sort determined Hintikka to abandon later on (Hintikka/Symons unpubl.) the claim that mutual dependencies of two variables are expressible in IF logic and to conclude that mutual dependence illustrates a pattern of variable dependence which cannot be analysed game-theoretically "along the lines typically followed by logicians". In that paper Hintikka and Symons endorsed explicitly the linear time playability condition.

If I am allowed to speculate, I think that Hintikka wanted to get a mutual correlation between the values of two variables, x and y so that the values of y depend on those of x in one way, and conversely, the values of x depend on those of y in (possibly) another way. The problem now springs from the fact that he wanted to get both correlations using quantifiers. In this case, in order to get the first correlation, Hintikka needed something like

(i) $\exists y$ depends only on $\forall x$

and in order to get the second correlation he needed

(ii) $\forall x$ depends only on $\exists y$.

But taken jointly (i) and (ii) violate the linear time playability condition, which may have led Hintikka and Symons in Hintikka/Symons unpubl. to the conclusion of their paper mentioned earlier. Independently of Hintikka and Symons' motivation, I think that the question of the logical representation of mutually dependent variables is of independent interest. Hintikka tried to represent such correlations in IF logic, that is, using scopal dependencies among quantifiers. It might help to approach the same question from a different but related angle.

5 Dependencies Among Variables: Dependence Logic, Väänänen 2007

As already pointed out, the formal dependencies of quantifiers on each others induce a material, functional dependence between the values of the corresponding variables which is encoded in the semantics using (generalised) Skolem functions. For an illustration, we recall the equivalence between (7) and (8). But one can try to represent the functional dependencies of the values of variables directly in the syntax, disentangled from the scopal dependencies of the quantifiers which bind them. To this effect, Väänänen 2007 extends the syntax of first-order logic with *dependence atoms*

$$= (x_1, \ldots, x_n; y)$$

which have the intended interpretation: the value of y is (functionally) determined by the values of $x_1, \ldots, x_n$. The semantic unit of evaluation of a dependence atom is a team X, that is, a set of partial assignments (i.e., assignments defined on a finite set of variables) sharing a common domain of variables with values in the universe of an underlying model. The semantic clause for a dependence atom is then expressed by

(i) $M \models_X = (x_1, \ldots, x_n; y)$ if and only if for all $s, s' \in X$, if s, s' agree on $x_1, \ldots, x_n$, then they also agree on y.

For an example, let $X = \{s_1, s_2, s_3\}$ be a team where the three assignments share the same domain $\{x, y, z\}$ as shown below:

	x	y	z
s_1	1	1	0
s_2	2	1	1
s_3	1	1	1

It is easy to check that $M \models_X = (x; y)$ but $M \not\models_X = (x, y; z)$. The clauses for complex formulas generalise the semantic clauses for first-order logic. We give here only the clauses for quantifiers:

(ii) $M \models_X \exists x \psi$ if and only if there is a function $f : X \to M$ such that $M \models_{X[x,f]} \psi$, where $X[x, f]$ is the team formed by extending each assignment s in X with $(x, f(s))$.

(iii) $M \models_X \forall x \psi$ iff $M \models_{X[x,M]} \psi$, where $X[x, M]$ is the team formed from X by extending each assignment s in X with (x, a) for each a in M.

Notice that this interpretation induces, in the same way as the game-theoretical interpretation, a functional dependency between the values of the existentially quantified variable x and the values of the other variables bound by the quantifiers in the scope of which $\exists x$ occurs.

Finally we define:

(iv) A sentence (formula with no free variables) φ is true in the model M if $M \models_{\{\emptyset\}} \varphi$, where $\emptyset$ is the empty assignment.

It should not be too difficult to see, based on the definitions, that the truth of $\forall x \exists y R(x, y)$ is equivalent to the truth of the second-order sentence $\exists f \forall x R(x, f(x))$. But now the functional dependency between the values of 'y' and those of 'x' may be explicitly asserted in the object language:

$$\forall x \exists y R(x, y) \Leftrightarrow \forall x \exists y (= (x; y) \land R(x, y)) \Leftrightarrow \exists f \forall x R(x, f(x)).$$

The dependence atom $= (x; y)$ may be read off from the scopal dependency of the two quantifiers. Given the semantic interpretation of the quantifiers as described by the two clauses above, this atom is redundant and it may be omitted. The interesting cases (i.e. those which lead to an increase expressive power) are the ones in which the dependencies between variables differ from those induced by the scopal dependencies of quantifiers. For an example consider the sentence

10. $\forall x \exists y \exists z (= (x; y) \land = (y; z) \land x = z \land c \neq y)$

which is equivalent with

11. $\forall x \exists y \exists z (= (x; y) \land = (y; z) \land = (x; z) \land x = z \land c \neq y)$

and with the second-order sentence

12. $\exists f \exists g \forall x (x = g(f(x)) \land c \neq f(x))$.

The truth of (12) asserts the existence of an injective function f whose range is not the whole universe. In other words, (12) is true in a model M if and only if the universe of M is (Dedekind) infinite. In (11) we could have added another conjunct, $= (x, y; z)$, but we did not do it because it is redundant. Its redundancy follows from $= (y; z)$. That is, as a general principle, if z depends on y, then z depends on any larger sequence of variables which contains y.

6 Mutual Dependency Between Variables

The last example should make it clear that the dependence among variables is transitive, i.e., if y depends on x and z depends on y then z depends on x, reflexive, x depends on x, but not symmetric. That is, there are cases (e.g. many-one correlations) in which y depends on x but x does not depend on y. The mutual dependence of two variables can now be expressed in a straightforward way:

$$= (x; y) \land = (y; x).$$

It should be clear that any team X which verifies

$$= (x; y) \wedge\; = (y; x)$$

establishes a one-to-one correlation between the individuals which are the values of x and those which are the values of y. To see this, we observe that the functional correlation f which associates the values of x with the values of y cannot be such that it sends two distinct individuals, say a and b to one and the same individual, say c, because the truth of the dependence atom $= (y; x)$ forces a and b to be identical. Using this fact, it can be shown (Kontinen et al. 2014) that (Dedekind) infinity can be defined in Dependence Logic using only two quantifiers and one dependence atom:

13. $\forall x \exists y (= (y; x) \wedge y \neq c)$

(recalling that the formula $= (x; y)$ is redundant.)

It is not difficult to see that all the scopal dependencies of quantifiers in IF logic may be expressed using just standard quantifiers and dependence atoms. It turns out that the converse is also true, i.e. dependence atoms can be contextually eliminated using scopal dependencies of IF quantifiers. Thus an occurrence of

$$\left(= (\vec{X}; y) \wedge \ldots\right)$$

in a sentence may be replaced by

$$(\exists z/W)\,(z = y \wedge \ldots)$$

where W is the set of variables dominating z minus $\vec{X}$.

Here is an example. The Dependence logic sentence

$$\forall x \exists y (= (y; x) \wedge y \neq c)$$

has the same truth-conditions as the IF logic sentence

$$\forall x \exists y\, (\exists z/\{x\})\,(z = x \wedge y \neq c)$$

which is our earlier sentence (6) whose truth-conditions are expressed by (12). We take note that the IF counterpart obeys the "linear time playability condition" in the underlying game. But we also observe that one can induce a mutual dependence between the values of x and the values of y in IF logic, by introducing the extra quantifier $\exists z$ which is not in the scope of $\forall x$ but only in the scope of $\exists y$. We recall in this context Hintikka's endeavour to express mutual dependence by the IF sentence (we ignore the time variable)

14. $\forall x \forall y\, (\exists z/\{x\})\, (\exists u/\{y\})\, [x = z \wedge y = u \wedge S(x,y)]$.

As we pointed out earlier, the problem with Hintikka's proposal lies with the pattern $\forall x \forall y\, (\exists z/\{x\})\, (\ldots x = z \ldots)$. On the other side, one way to describe what we are trying to do is that in order to express a mutual correlation, we can start with $\forall x \exists y$ which gives one side of the correlation, and then to add either a dependence atom $=(y;x)$, or, equivalently, an IF quantifier $(\exists z/\{x\})$ together with the conjunct $x = z$, to obtain the other side.[2]

To conclude this section, I hope to have shown that if we take variable dependence as a basic feature of our general logic, then we obtain a substantial increase in expressive power when we combine it with the linear dependence of first-order quantifiers. The increasing growth in expressive power over ordinary first-order logic is due, as our last example has suggested, to the mismatch between the two kinds of dependencies, one induced by quantifiers and the other one displayed by variables. All these developments show that the distinction between first-order and second-order logic is blurred as we argued in Hintikka/Sandu 2007, given that Dependence Logic, as well as IF logic capture concepts that were thought to be expressed only by means of second-order logic (at the level of formulas Dependence logic is known to have greater expressive power than IF logic, but I will not enter into these matters here).

7 Kit Fine: Dependence Between Arbitrary Objects

Finally let us shortly describe another framework which deals with the same problems we have tried to tackle in this paper. It is the framework of arbitrary objects introduced in Fine/Tennant 1983. In this framework there are also two kinds of dependence, a (material) dependence at the level of individual objects which is sustained, not by a relation of dependence among quantifiers, as in IF logic, but by a relation of dependence between arbitrary objects. Arbitrary objects are introduced by quantifiers (and are named by constants in the object language) and the dependence relation among them is represented in the object language. Thus when b is an arbitrary object that depends only on the arbitrary object a, then the values assigned to b are determined by the values assigned to a. Arbitrary objects are divided into independent and dependent ones. At this stage an example might help. Consider the natural language discourse fragment:

---

[2] I am indebted to Fausto Barbero for the material of the last two sections.

15. Every farmer owns a donkey. He beats it; he feeds it rarely.

Here 'Every farmer' introduces an independent arbitrary object, and 'a donkey' introduces another arbitrary object which is dependent on the first. The first arbitrary object is associated with a set of individual farmers, and so is the second. Every pair (a, b) of such individual objects stand in the relation of ownership, a owns b. The anaphoric pronoun 'He' is a place holder for the name of the first arbitrary object, and 'it' is a placeholder for the name of the second arbitrary object. And so on.

Fine formulates an objection against the use of Skolem functions to encode the material dependencies of individuals which are associated with arbitrary objects. According to it, the problem with Skolem functions is that they cannot handle multi-dependencies. Fine illustrates his objection with an example involving 3 arbitrary objects, a, b, and c, such that c depends on b in a particular way, and b depends on a in another way. Recalling that arbitrary objects are introduced by quantifiers, the independent ones by universal quantifiers, and the dependent ones by existential quantifiers, I take Fine's worry (if I correctly understood him) to be about the impossibility to sustain the dependencies encoded by the two Skolem functions by appropriate scopal dependencies of three quantifiers, one universal and two existential ones. Fine's worry is justified, as can be seen from our discussion in the previous sections: there is no way to arrange the three first-order quantifiers to match the dependencies encoded by the two Skolem functions. That was one of the motivations for introducing IF logic, i.e. to liberate the patterns of dependencies between quantifiers and thereby generalise the notion of Skolem function. To get the two dependencies Fine wants, we need

$$\forall x \exists y \, (\exists z / \{x\})$$

where $\exists z$ corresponds to Fine's arbitrary object c and is only in the scope of $\exists y$ (Fine's arbitrary object b), which, in turn, depends only on $\forall x$ (Fine's arbitrary object a). We can also achieve the same result using Dependence logic as a framework (see example (10)), adding to the linear sequence of quantifiers $\forall x \exists y \exists z$ the dependence atoms $= (x; y)$ and $= (y; z)$. Notice that the dependencies between the three variables asserted by the two atoms match Fine's dependencies between the three arbitrary objects, a, b and c. It is true that in this case one cannot, for instance, represent (15) in such a way that 'He' is a place holder of "Every farmer", but the gain in ontological parsimony is considerable. I have tackled some of these issues in Sandu, forthcoming.

Bibliography

Barbero, F. (2013): "On Existential Declarations of Independence in IF Logic." In: *The Review of Symbolic Logic*. Vol. 6, No. 2, 254 – 280.
Barwise, J. (1979): "On Branching Quantifiers in English." In: *Journal of Philosophical Logic*. Vol. 8, No. 1, 47 – 80.
Kontinen, J., Kuusisto, A., Lohmann, P. and Virtema, J. (2014): "Complexity of Two-Variable Dependence Logic and IF-Logic." In: *Information and Computation*. Vol. 239, 237 – 253.
Fine, K. and Tennant, N. (1983): "A Defence of Arbitrary Objects." In: *Proceedings of the Aristotelian Society*. Supplementary Volumes, Vol. 57, 55 – 77, 79 – 89.
Henkin, L. (1961): "Some Remarks on Infinitely Long Formulas." In: *Infinitistic Methods: Proceedings of the Symposium on Foundations of Mathematics, Warsaw, 2 – 9 September 1959*. Bernays, P. (ed.), Oxford: Pergamon Press, 167 – 183.
Hintikka, J. (2002): "Quantum Logic As a Fragment of Independence-Friendly Logic." In: *Journal of Philosophical Logic*. Vol. 31, 197 – 209.
Hintikka, J. and Sandu, G. (1989): "Informational Independence As a Semantical Phenomenon." In: *Logic, Methodology and Philosophy of Science*. Fenstad, J. E. et al. (eds.), Vol. 8, Amsterdam: Elsevier, 571 – 589.
Hintikka, J. and Sandu, G. (2006): "What Is Logic?" In: *Philosophy of Logic*. Jacquette, Dale (ed.), Amsterdam: Elsevier, 13 – 39.
Hintikka, J. and Symons, J.: "Game-Theoretical Semantics As a Basis of a General Logic." Unpublished manuscript.
Sandu, G. (forthcoming): "Functional Anaphora and Arbitrary Objects." In: *Metaphysics, Meaning, and Modalities: Themes from Kit Fine*. Dumitru, Mircea (ed.). Oxford University Press.
Sher, G. (1990): "Ways of Branching Quantifiers." In: *Linguistics and Philosophy*. Vol. 13, 393 – 422.
Väänänen, J. (2007): *Dependence Logic*. Cambridge (UK): Cambridge University Press.

Wolfgang Kienzler
Three Types and Traditions of Logic: Syllogistic, Calculus and Predicate Logic

Abstract: Modern logic grew out of the work of Frege and of the tradition which Boole initiated. However, as the Quine-Putnam exchange illustrates, the relations between the respective camps are far from being well understood. We can get some clues from the way Frege critically discusses the Boole-Schröder tradition. Furthermore Michael Wolff has suggested that there is a close and internal relatedness of all three major types of logic, even declaring syllogistic logic to be the one and only "strictly formal" type of logic. A closer look at the Euler diagrams and their influence on the understanding of logic in the 19th century can highlight something of a silent revolution under way, preparing logicians to accept the non-exclusive alternative as basic, to accept tautologies as the paradigm of truth, and to introduce truth-functionality. The second half of this contribution offers an overview of the three traditions, in giving brief answers to the same series of questions. In addition, Wittgenstein's *Tractatus* is included in the questionnaire. All of this will may help to view the history of logic as the interaction of the three distinct, yet intrinsically related paradigms of Syllogistic, Calculus and Predicate logic.

1 The Quine-Putnam Muddle[1]

In describing the history of logic it is customary to distinguish traditional from modern logic. In doing this we soon run into a difficulty which we can call the "Quine-Putnam Muddle": For one thing, it seems natural to date modern logic from the publication of Frege's *Begriffsschrift* in 1879, simply because this little book contains the first full and precise system of propositional and predicate logic as we know and practise it today. We can call this the "Quine view" (or the "textbook view", as it is inspired by the contents of modern logic textbooks), expressed in his well-known opening: "Logic is an old subject, and since 1879 it has been a great one." (Quine 1959: vii, Preface, first sentence)

---

[1] In this paper I distinguish *Syllogistic logic*, *Calculus logic* and *Predicate logic*, respectively. These labels can in some cases be misleading, but there is no fixed and established usage concerning the distinctions I want to highlight. In German, my favorite labels would be "disjunktive, adjunktive und prädikative Logik", but the term "adjunctive" is not well established in English usage.

Wolfgang Kienzler, University of Jena, wolfgang.kienzler@uni-jena.de

On the other hand it seems every bit as natural to date the beginning of modern logic with Boole's 1847 and 1854 books. In this spirit Putnam writes in protest against Quine:

> It seemed inconceivable to me that anyone could date the continuous effective development of modern mathematical logic from any point other than the appearance of Boole's two major logical works, the *Mathematical Analysis* and the *Laws of Thought*. (Putnam 1990: 255)

Towards the end of his article, Putnam claims to have found "a fair-minded statement of the historical importance of the different schools of work, a statement that does justice to each without slighting the others" (Putnam 1990: 255). He finds it on the first page of Hilbert and Ackermann's *Logic*:

> The first clear idea of a mathematical logic was formulated by Leibniz. The first results were obtained by A. de Morgan (1806–1876) and G. Boole (1815–1864). The entire later development goes back to Boole. Among his successors, W. S. Jevons (1835–1882) and especially C. S. Peirce (1839–1914) enriched the young science. Ernst Schröder systematically organized and supplemented the various results of his predecessors in his *Vorlesungen über die Algebra der Logik* (1890–1895), which represents a certain completion of the series of developments proceeding from Boole.
>
> In part independently of the development of the Boole-Schröder algebra, symbolic logic received a new impetus from the need of mathematics for an exact foundation and strict axiomatic treatment. G. Frege published his *Begriffsschrift* in 1879 and his *Grundgesetze der Arithmetik* in 1893–1903. G. Peano and his co-workers began in 1894 the publication of the *Formulaire des Mathematiques*, in which all the mathematical disciplines were to be presented in terms of the logical calculus. A high point of this development is the appearance of the *Principia Mathematica* (1910–1913) by A. N. Whitehead and B. Russell. (Hilbert/Ackermann 1950: 1f.)

This account may do justice to the two main schools, or origins, of modern logic – but it does not tell us much about their mutual relation: Putnam seems to presuppose that both schools arrive at more or less the same thing, called "modern mathematical logic". This view (we might call it the "Putnam view" or the "logical theory view") is further encouraged by the reference to Leibniz. It makes it seem that mathematical logic is something first envisioned by Leibniz, and then furthered by the respective efforts of the Booleans and of Frege, leading up to

*Principia Mathematica*² and the mainstream of logical research of the 20ᵗʰ and 21ˢᵗ centuries.³

From a perspective of the 1950s or later it seems natural to view both traditions as really being parts of one and the same development of "modern symbolic, or mathematical logic".⁴ However, this approach has blurred the perception of the elementary and very basic conceptual differences between both approaches.

The main effect of Putnam's article seems to have been that Quine silently dropped the sentence about the importance of the year 1879 in later editions of his book – without offering any replacement.⁵

So we are left without a clear answer to our paradox that in one sense it seems obvious that modern logic derives from Frege's *Begriffsschrift*, while in another sense today's logical outlook seems more closely related to the ways the Boolean tradition introduces logical calculi and then discusses possible interpretations of the signs and rules involved.

While the unravelling of these intricate interrelations is too tall an order for one article, it may be worthwhile to shed some light on the origin of the two traditions. This may help to gain a better understanding of some of the 19ᵗʰ century developments.

2 Frege about the Basic Differences between Calculus Logic and Predicate Logic

Most early proponents of the development had a strong inclination to speak of one more or less coherent tradition of modern logic. In his review of Frege's *Be-*

---

2 The overwhelming success of *Principia* (the back cover of the 1962 paperback reprint carries this recommendation from Quine: "This is the book that has meant most to me") which stands firmly in the Fregean tradition (or tries to do so) has obscured the fact that the spirit of most later logical work is very different from the style of that monumental work.
3 In his comprehensive 'A Bibliography of Symbolic Logic' (1936) Church encourages such a view: Starting with Leibniz, all varieties of logic making extensive use of "symbols" are listed and arranged into one seemingly continuous stream. This is one way of *avoiding* the question about the relation between the different traditions.
4 However, the mere use of symbols or a close affinity to mathematics are rather vague and superficial ways to characterize the nature of modern logic.
5 Incidentally, the edition in question appeared in 1982, the same year as Putnam's article. Künne voices this suspicion (which in Boston may be common knowledge) – while adding 1837, the year of publication of Bolzano's *Wissenschaftslehre*, to the list of potential beginnings of modern logic (Künne 2010: 17, n. 9). Below we will suggest 1761 as another answer.

griffsschrift Schröder claimed that Frege's system was a mere notational variant of Boole's logic, and a cumbersome one at that (Schröder 1880). Thus, from the beginning, Frege was commonly regarded as part of the movement towards creating "mathematical, symbolic logic", and he was usually lumped together with Schröder into the mathematical camp.

Frege himself viewed matters quite differently. While he acknowledged some common ground, he refused to be placed in the same tradition with Boole and Schröder. He would even go on to exclude the Boolean tradition from the domain of logic altogether. However, it took him quite some time to do so effectively. In a series of answers to Schröder Frege tried to explain the essential differences between his *Begriffsschrift* logic and the diverse versions of Boolean Algebra of logic, and in doing so he stressed especially the points where his logic was technically more powerful than Boolean logic, mainly the use of quantifiers and of concepts of different levels. Frege showed that his kind of logic could do everything Syllogistic logic could do, as well as everything Boolean logic could do – while the reverse was definitely not the case.[6]

It was not until 1895 that Frege succeeded in giving a principled account of the basic conceptual differences involved. In his review (Frege 1895) of Schröder (Schröder 1890) Frege elaborated that Schröder basically practised a "calculus of areas" (*Gebietekalkül*) – where areas are compounded, divided and so on. He also pointed out that circles in a plane, the Euler diagrams, are a natural and quite perfect instrument to illustrate the moves in this calculus of areas. All areas are located on the same plane, and the lines compound these areas into larger areas, or they demarcate smaller areas of intersection. These moves of logical addition and multiplication can be performed with any number of areas.

Frege further remarked that for the same enterprise the interpretation as a "calculus of classes" was also possible – but he pointed out that the notion of a class invites moves which are impossible to perform consistently in a calculus of areas – moves which had infested Schröder's system with confusion, the most striking symptom of which was a version of the paradox of classes. The fundamental relation in the calculus of areas is the relation of part and whole – but there is nothing that could express the relation of an element to the class it belongs to.[7] If

---

[6] Frege called Boolean logic, in Leibnizian terms, a mere calculus (*calculus ratiocinator*) while his own version was to be both a calculus and a language able to express contents (*lingua characteristica*). This, however, still made the differences seem merely technical and thus "too slight" (as Cora Diamond, following Wittgenstein, might say).

[7] One might thus call this a logic of *coordination*, while logic in the basic sense of the term is and should be a logic of *subordination* (this would be Kant's objection to the use of Euler diagrams). Frege would respond to Kant by pointing out that concentric circles can illustrate the subordination

we have no extension on the plane there will be no area at all, so there is nothing in the calculus of areas equivalent to an empty class: there is no "zero area" – a zero area would be no area at all. So we have to sharply keep apart a calculus of classes (or areas) without any class-element relation on one hand, from a calculus of classes which includes the relation of an element to a class. Only this second type of calculus deserves to be called "logic" in the sense Frege wants to use this term. Frege thus would not accept anything as being logic which did not have a concept-object distinction.

To sum up: We should keep apart a tradition of logic where we have type distinctions (between elements and classes, or objects and concepts) from traditions which have no such distinction. And, of course, only if we have type distinctions we can have quantifiers ranging over objects.

While Frege suggested that we exclude the calculus tradition from the domain of logic, the reverse could also be argued for. After all, Syllogistic logic, the paradigm for logic for 2000 years, also lacks type distinctions – and it is fairly easy to reproduce syllogistic inferences in the symbols of calculus logic. Actually Boole's first book contained precisely such a "mathematical analysis of logic" (Boole 1847) – it was not designed to introduce a new version of logic. It was only due to some by-products of this first attempt that Boole decided to develop his system on its own terms in his 1854 *Laws of Thought*. Now, according to Frege's criterion, Syllogistic logic and Calculus logic are of the same stripe, and neither one can count as being "logic". Frege turns out to be the Bolshevik who is destroying the old system. But we could also argue the other way around that Frege is the odd man out and that his system should not be called (formal) logic.

3 Michael Wolff on the Basic Differences between Syllogistic and Modern Logic

Michael Wolff has developed a provocative view about the true nature of logic, and about the relation between Syllogistic and Modern logic (both of the Calculus and Predicate version). Building on his studies especially of Frege and Kant, Wolff discusses the notion of what it means to be "formal" in a strict sense. He then argues that only syllogistic logic, if interpreted in an adequate way, deserves to be called purely formal. Now, if "purely formal" means that logic is not concerned with

---

of concepts, and that the real difference lies in the notion of *subsumtion*, the falling of an object under a concept.

objects at all, this seems quite straightforward: the most elementary expressions of predicate logic have argument places that are to be filled with expressions signifying objects. The concept-object distinction lies at the heart of Frege's logic – and predicate logic simply makes no sense if we don't want to speak about objects. But to Wolff, this is only one of three features where modern logic is not purely formal.

Wolff suggests that syllogistic theory can be transformed into Fregean predicate logic by simply adding three extra axioms, and in Wolff's sense of "formal" these axioms are not purely formal. These are the three extra axioms or postulates (Wolff 2005 §82: 365):

1. The *tertium non datur* for propositions, stating that the negation of each proposition has the opposite truth value to the original proposition. This principle entails the *duplex negatio affirmat* which (as Wolff points out) is not part of syllogistic logic, and it makes the truth-functional use of negation possible. This introduces truth-functionality.
2. The axiom of the "arbitrary sufficient condition": this expresses the idea that tautologies (i.e. propositions of the form "if p, then p") are considered to be logically true. This introduces the tautological nature of logic (which Wolff takes to be very different from the formal nature of logic). It introduces the idea that every proposition follows from itself.
3. The axiom of the non-empty domain of individuals. This is the postulate that at least one object exists. This allows the use of predicate logic.

Without discussing these suggestions in detail I want to point out that according to these principles we have two basic kinds of doing logic:

Syllogistic logic is not truth-functional, and in it neither the *ex falso quodlibet*, nor the *duplex negatio affirmat* are valid. Negation only means that a proposition is "not true", but it does not entail that there is a true ("negated") proposition. False propositions and nonsensical or defective propositions come out very similar: both are defined as being "not true", and neither type can be made true by negating it.

In this kind of logic we also have no use for tautologies, because we only use propositions which are substantially different from the start. Propositions that are "equipollent" are formally counted as one and the same proposition. We also insist that a logical deduction only takes place if the original proposition, or propositions, are different from the deduced proposition. "Deducing something from itself" simply is not counted as a logical deduction (although it will be "materially" true). And finally, no objects are mentioned within the entire scope of this tradition of logic.

Now, while some of these details need further discussion, "Wolff's Proposal", as I want to call it, states that we have exactly one logical world. This one world is based on purely formal, syllogistic logic, and this world can be expanded through the successive addition of extra principles or axioms.[8]

I would like to take up Wolff's suggestion that syllogistic and modern logic are related much more closely than is usually acknowledged. We might put it this way: While the main lines in the history of logic can be described as three paradigms, each of which has its own basic images and moves, opening up a world of its own – these paradigms are in various ways internally related to each other. We can understand each tradition much better if we direct our attention to these interrelations.[9]

4 Three Traditions

If we take Frege's suggestions and consider what Boole is doing, it seems natural to put it this way: Boole created a version of logic that can be called "Calculus logic", but his logic does not know any type distinctions, and thus the Boolean tradition should conceptually be kept apart from the tradition beginning with Frege. In many respects Boole gives a "mathematized" version of traditional logic – and this is exactly what he claims to do in his *Mathematical Analysis of Logic*. Frege, on the other hand, builds up a new kind of logic from scratch, with hardly any reference to traditional logic. We can thus distinguish these three types of doing logic:

A Syllogistic logic deals with concepts and their relations. It is based on the exclusive disjunction of concepts into mutually exclusive subordinated concepts. We might call it "disjunctive" logic. The Tree of Porphyry illustrates this version: each concept is either on one or on the other branch of the tree – but no concept can be on both branches ("at the same time").

B Calculus logic: Leibniz first suggested the introduction of a calculus of concepts or classes. It is based on the non-exclusive relationship between

---

8 Wolff discusses "Stages in the history of logic" (Wolff 2005 §84: 371–375) but there he does not offer the threefold division suggested here. According to his considerations, the first two extra postulates will establish Calculus logic while adding the third should establish Predicate logic. Wolff has also explained the results of adding each one of the three extra axioms separately (Wolff 2013).

9 The writings of Michael Wolff first introduced me to the world and workings of non-truth-functional, Syllogistic logic. This also included novel ways to understand Kant's notions of general and transcendental logic.

classes which may or may not overlap. There is no talk of elements of classes. The traditional alternative of "disjunction" is now re-interpreted in a non-exclusive way. We could call this "adjunction" and speak of "adjunctive" logic. Euler circles (or other diagrams) go well with it, and Boole and Schröder (and some others) revived and developed this tradition.

C Predicate logic, as invented by Frege, is based on the distinction between concepts and objects. This introduces levels into logic, and also a two-way division of logic: first we have propositional logic without any type distinctions, and here Frege sticks to the idea of a calculus, and also to the idea of the non-exclusive alternative. Secondly, we have predicate logic, and here levels are introduced, and we meet quantifiers – and this is the type of logic which allows us to analyze language and capture the distinction between terms standing for objects and terms expressing properties of objects.

One striking feature in this tangled tale is the way the calculus tradition was established – without being regarded as a revolutionary new version of logic. The main reason for this seems to be that the Calculus tradition was taken to be something else, something not replacing traditional Syllogistic logic, but an enterprise of a different character and aim. Calculus logic was quite naturally regarded as a branch of mathematics, using mainly mathematical techniques. When Leibniz declared arithmetic to be a branch of logic, he had already re-interpreted logic in terms of his notion of a calculus. From the point of view of syllogistics, such a claim would be simply incoherent. This has to do with the very notion of a calculus: syllogistic logic plainly is no calculus at all, and thus it could not be replaced by a better calculus.

Thus one real revolution that took place in logic is the transformation of something which has no calculus, no truth functions, no operations, and thus no operators, no logical connectives serving as operational signs, into the Calculus version of logic we take for granted today. But before the notion of a calculus could seem natural, the notion of truth-functionality would have to be introduced firmly. This was effected through the way Euler diagrams superseded the Tree of Porphyry as the main image for logic.

5 The Silent Euler Revolution

When Leibniz first developed the idea of a logical calculus, he started no tradition. In 1847 most of his ideas were re-invented by Boole who knew nothing about

Leibniz as a logician. However, Boole's suggestions almost immediately led to the development of a Boolean tradition of mathematicians and logicians, further discussing and expanding his ideas. What had happened between Leibniz' death in 1716 and 1847 that had made the calculus ideas so much more attractive?

I want to suggest that the paradigm shift which brought about the calculus tradition in logic was prepared by the quite inconspicuous introduction of the Euler diagrams.[10]

Euler used his diagrams to illustrate the inferences of traditional, syllogistic logic, in his popular book *Letters to a German Princess* (the first logic installment, the 102nd letter, is dated 14 February 1761). He had no intentions whatsoever of changing anything in the way logic was done – except to make some things more obvious.

Right in his first installment Euler introduced three ways that two circles can be situated towards each other, in order to illustrate the four forms of syllogistic judgments:

1. One circle A may lie entirely inside another circle B (with a part of B not taken up by A). This can illustrate "All A are B": $((B(A)B))$[11]
2. A and B may lie entirely apart, without touching each other. This can illustrate "No A is B": $(A)(B)$
3. A and B may intersect. This can be used to illustrate "Some A are B" as well as "Some A are not B": $(\ (\ )\ )$

Euler uses this third diagram to illustrate two different logical cases which therefore need to be distinguished in some way. In order to get two different diagrams Euler inserts letters in different places into the same figure:

The "Some A are B" diagram places an A in the intersecting area and B in the non-intersecting part of B: $(\ (A)\ B)$

The "Some A are not B" diagram places both A and B in their respective non-intersecting parts: $(A\ (\ )\ B)$.

In the second case the intersecting part remains empty.

---

10 There are hints that diagrams to illustrate logical relations were used even in antiquity, and the history of logic abounds with suggestions to introduce various symbolic and diagrammatic tools (many of these are discussed in Venn 1881, ch. 20: "Historic Notes"). It was, however, only with the advent of Euler circles that a standard way of visualization came into widespread use.
11 Corresponding double parentheses indicate a large circle: $((\ ))$, while two intersecting circles will be written thus: $(\ (\ )\ )$.

These two cases can accommodate traditional cases like: "some animals are human, and some animals are not human" – where particular judgments are created through superordination.

In letter 105 (dated 24 February 1761) Euler introduces a new technique, using a star for the purpose: ((∗))[12]

This star indicates "there is a part of the concept A which overlaps with concept B". Strictly speaking, we still have no existential statements but only the comparison of spheres of concepts – but now we find that spheres can have a nonempty common part which exhausts neither concept entirely. While in syllogistic logic particular judgments were simply defined as the negation of general ones ("some" would mean simply "not all"), now we meet a positive notion of particular judgments with "some" meaning "not none". Euler explained his example thus: "Some scholars are stingy" – in place of the technically more correct: "The sphere of the concept *scholar* and the sphere of the concept *stingy* overlap". Now the two concepts were on the same level, neither one subordinated to the other. Such a case had been outside the scope of Syllogistic logic. Now it seemed natural that we should put things thus: "there exist beings, or humans, which have the property of being scholars and also the property of being stingy".

The importance of this shift can hardly be overestimated as it introduced, or at least prepared, the notion of particular judgments as being existential judgments – and of logic as being concerned with existential questions. To put it bluntly: This may have been the point where logic was reversed from a top-down approach (in being about concepts and their spheres only) to a bottom-up approach (in being about matters of existence).[13]

Arguably, this is also the main source of the notorious discussions about the "existential import" of the judgments in the Square of Opposition.[14]

The use of symmetrical circles also suggested that concepts can be compared in a symmetrical way, and that this might be done using equations. This had not been possible with a tree diagram.

---

12 Quine uses the same technique of inserting a small cross into such a diagram which he calls "Venn's Diagrams" (Quine 1959: 69), and he uses Venn's technique of shading to indicate emptiness in order to uniformly express Euler's first two cases.

13 In Hume's terms: logic was no longer being regarded as being about relations of ideas but rather about matters of fact (preferably general facts, but still facts).

14 Pre-Euler discussions of the square knew nothing of "existential import", but the discussions after Euler made it seem that there was something wrong about the square – as cases of judgments with empty subject positions would invalidate most inferences in the square. Changing interpretations of the square can show elementary changes in the logical landscape, but studies of these connections will have to wait for another occasion.

This situation also opened up new possibilities for forming judgments. Syllogistc theory knows only the four forms SaP, SeP, SiP and SoP (A, E, I and O). These forms can all be generated by introducing negations into "All A are B". This results in:

    A "All A are B"
    E "All A are not B" [= "No A is B"]
    O "Not: All A are B" [= "Some A are not B"]
    I "Not: All A are not B" [= "Some A are B"]

Seemingly these old forms are generated by negation and by what could be called "quantifying the subject" as being "general" or "particular": *All A are B, No A are B, Some A are B, Some A are not B*.

The diagrams made it seem natural that A and B could be symmetrically exchanged, and thus four extra forms could be generated through "quantifying the predicate" – and this made it seem that syllogistic logic really should discuss all eight elementary forms of judgment:

    A All A *are* all B; All A are *some B*
    E Any A is *not any B*; Any A is *not some B*
    I Some A is *all B*; Some A is *some B*
    O Some A is *not any B*; Some A is *not some B* (Venn 1881: 9 and 31)

In this way Hamilton introduced his notion of the much debated "quantification of the predicate". The diagrams thus had opened up the (seeming) necessity to introduce some major innovations into Syllogistic logic.

Some of these new forms of judgment were of a quite different character than the traditional forms: "All A are *all B*" describes the situation that the circles A and B are coextensive, and thus that the concepts A and B were identical. In this way tautological situations had been introduced into what still seemed to be Syllogistic logic. However, these questions were not regarded as stemming from a *new* conception of logic, but rather they were seen as problems not seen before, which needed to be solved within the framework of traditional logic.

These tensions can be seen in an especially striking way in John Venn's *Logic* (1881). Venn compares all the different versions in one table. He gives four columns: (i) Diagrammatic, (ii) Common Logic, (iii) Quantified, (iv) Symbolic.

He takes the diagrammatic version as basic, distinguishing five diagrams:[15]

1. A and B are coextensive: $((AB))$
2. A lies entirely within B: $((B(A)B))$
3. B lies entirely within A: $((A(B)A))$
4. A and B overlap: $(A(\ )B)$
5. A and B lie separate: $(A)(B)$

Syllogistic or "Common Logic" can capture all these five possibilities, but in all but one case more than one judgment must be used. In "Quantified Logic" each diagram has exactly one corresponding judgment:

1. All A is all B
2. All A is some B
3. Some A is all B
4. Some A is some B
5. No A is B

The only trouble with Hamilton's "(redundant) eight-fold scheme" (Venn 1881: 31) is that we have three idle forms of judgment which seem to have no function at all.

In Symbolic (i.e. Boolean) notation, there is again more than one equation to all but one of the diagrams.

Venn treats the different approaches as mainly notational variants. His entire approach thus stands between the old Syllogistic logic and new versions of logic – especially when he introduces his own diagrams. His own diagrams are all symmetrical while the type of logic he prefers is not.[16]

While the Euler diagrams suggested the form of an equation as natural in logic, this innovation was not to last. Against Boole and Schröder (1877), Peirce re-introduced an asymmetrical relation as logically basic, which Schröder (1890) adopted. In this way, the calculus tradition moved away from merely imitating mathematics.[17]

---

15 Euler did not use the case of co-extension and he did not distinguish the cases (2) and (3). The five-fold scheme had been introduced by Gergonne (Venn 1881: 6). This quite natural development introduced the notion of "good" tautologies into logic.

16 It would be necessary to go deeper into the way Venn moves between Syllogistic and Calculus logic. (I want to thank Gabriele Mras for discussions about this topic – which are to be continued.)

17 Actually all the basic ideas of a Boolean calculus can be illustrated by a simple diagram of two intersecting circles A and B on a plane, dividing the plane into exactly four areas (Schröder 1877). In reference to De Morgan's Laws this diagram might be called a "De Morgan Diagram". While the use of diagrams is uniform in using divisions in a plane, the extra use of devices like shading and

6 Three Types of Logic Introduced and Contrasted

In conclusion I want to give a brief, if somewhat schematic survey of the three logical traditions discussed here. The form of a questionnare, where the same questions are answered by each of the traditions may help to highlight the most important analogies as well as the disanalogies. As it may not be obvious that the same questions can meaningfully be directed at the three traditions, the mere fact of the triplicate questionnaire may help to introduce some new questions into this area. As noted above, the first and foremost field of application of these ideas may be to the study of the development of 19th century logic.

6.1 Syllogistic Logic

Persons: Aristotle invented Syllogistc logic but it was Kant who offered its purest version.

Basic Image: The basic image of this tradition is the *Arbor Porphyrii*, the Tree of Porphyry, sometimes called "pyramid of concepts".

Elements: All concepts used should be generated on the Tree of Porphyry, and concepts are defined by their sphere. There is no talk of anything falling under a concept.

Proposition: Every proposition, or judgment, puts two concepts in relation.

Basic move: Basic is the subordination of one concept under another concept (as in "all *humans* are *mortal*") – and the second move is a syllogism like Modus Barbara (consisting of two moves of the same nature). We also have disjunction, where two concepts have no part of their sphere in common ("no humans are immortal"), and superordination, which is the reverse of subordination ("some mortals are humans").

Connectives: There are no connectives; there are four elementary types of judgments, A, E, I, and O. There is nothing to connect except two concepts S and P into a judgment: the copula is regarded as the single connective (of sorts).

Other relations: We have a hypothetical relation and (exclusive) disjunction between concepts: None of these are truth-functional.

Type of relations: All relations should be essential relations between concepts on the tree.

---

placing letters in the diagrams calls for more investigations. The differences between the versions of Euler, Gergonne, De Morgan, Venn and Peirce deserve to be studied in more detail.

Structure of relations: Almost all structures, except disjunction, are asymmetrical. All inferences are asymmetrical.

Generality: Logic is by its very nature general because concepts are general. Degrees of generality are due to the concepts involved.

Identical propositions: "Equipollent" propositions are excluded from discussion, as they are counted as one and the same proposition, only in a different guise. There is no room for any logical relation between a proposition and itself.

Technical form: The form of logic consists of syllogisms. These are ways to derive new true propositions from old true propositions. The "system" consists of an enumeration of all valid moods of inference. There is no "system" of logic, only some mnemotechnical devices.

Metaphysical form: The judgments taken into consideration should really all be pure a priori truths about concepts.

Related science: Logic has some relation to biology, taken as as science of levels of classification according to genus and species.

Particular judgments: The negation of generality generates particular judgments. These do not stand on their own feet. Superordination always includes both *Some* and *Some Not* judgments. If all humans are mortals, then it will be true that "some mortals are human" and also that "some mortals are not human".

Relation to mathematics: Syllogistic logic has no connection to mathematics. Mathematics starts from arbitrary elements which can be put repeatedly – logic is about our (everyday) concepts. Logic does not calculate.

"All humans are mortal": This is a priori true because it expresses the essential (analytic) relation between the concepts "mortal", "animal" and "rational" on the Tree. Logically analyzed this comes out as: "All animals which are both rational and mortal are animals that are mortal." However, even in Antiquity, this was sometimes considered to be a factual statement.

Taken to be: Logic is supposed to be the form of "all thought whatsoever", the "laws of thought". Even Moritz Schlick, in his 1918 book *Erkenntnislehre*, declared that Modus Barbara was the logical form of all scientific thought (Schlick 1918 §14).

An aim not reached: Syllogistic logic cannot successfully analyze the simple sentence "Socrates is wise" because simple predication cannot be expressed in it. We only have concepts, and individuals are not concepts. An ad-hoc solution to the problem was given: individuals (i.e. objects and persons) are taken to be "individual concepts". Once they are "conceptified" they can be treated logically.

An aim reached: Concepts can be ordered according to levels of increasing generality.

6.2 Calculus Logic

Persons: Leibniz, Boole, Peirce, Schröder.
Basic Image: The basic images are Euler Circles or other diagrams. Diagrams are seen as an important, or at least interesting, research and illustrative tool.
Elements: We deal with classes but we can interpret everything just as well in terms of concepts – and there are also interpretations as propositions.
Proposition: Strictly speaking there are at bottom no propositions but just classes put into mutual relations – this results in a new class. "Propositions" are basically complex names of classes compounded from other classes, which are taken to be simple. It is a fairly long way of interpretation to start with circles on a plane and to come up with judgments or propositions. "Primary propositions" concern classes, and only "secondary propositions" are concerned with sentences.
Basic move: The basic situation is the overlap of classes. This favors the use of non-exclusive disjunction as basic. It would have been apt to use a new name like "adjunction" – but the old word "disjunction" was simply used in a slightly different sense.
Connectives: There are two basic connectives. Their names are taken from mathematics (by ways of analogy): logical multiplication (overlap area only: $A.B$) and logical addition (total area covered: $A + B$).
Other relations: Using a De Morgan diagram we can naturally derive 16 ways how A and B (and their respective complements) can be in- or excluded. Fairly early Peirce showed that all 16 can be expressed using just one connective.
Type of relations: The relations favored are the factual statements of natural science. These relations can be treated in a truth-functional way.
Structure of relations: Basically everything is symmetrical. If A comprises part of B, then B comprises part of A: they "overlap" (mutually). Asymmetrical relations are much harder to introduce.
Generality: Everything is general because entire areas ("all of area A") are considered. If we work with symbols, it seems that there is no generality. We do get levels of generality according to the sizes of the areas considered. Particular judgments are difficult to treat.
Identical propositions: Identity, i.e. co-extensionality, is no longer excluded, but it becomes the paradigm of truth: $A = A$, and $A + A = A$ become axioms.
Technical form: We build a calculus to express more complex cases of overlap, combination and exclusion; a system of operations (see Schröder 1877).
Metaphysical form: Logic now is a tool for calculating the relations holding between empirical data as they are expressed by classes. Logic helps with sorting out complex systems of factual relations.

Related science: The paradigm science is chemistry, dealing with co-existing substances. This is exemplified in Boole's prime example about co-existing elements (Boole 1854: 146 and Schröder 1890: 522 – 528).

Particular judgments: Calculus logic has no precise expression for individuals, but talk of "classes" suggests that there should be a way to handle individual cases and thus questions of existence. Boole and Schröder made some unsuccessful attempts to introduce logical machinery to do this. Peirce invented "quantifiers" (of a kind).[18]

Relation to mathematics: Calculus logic was invented by mathematicians who applied mathematical methods to logical problems. Consequently it seemed a mere matter of convention whether Calculus logic should be classified as a branch of mathematics or not.

"All humans are mortal": This proposition is about two classes. Calculus logic really has no way to deal with the individuals concerned, and thus with the issue of truth.

Taken to be: Calculus logic was mainly seen as a tool to be used in natural science, and also as a new branch of mathematics.

An aim not reached: Simple predication as in "Socrates is wise", cannot be expressed, because there are just classes and still no individuals.

An aim reached: Truth-functional propositional logic can be expressed naturally: two intersecting circles can visualize 16 truth-combinations.

6.3 Predicate Logic

Person: Frege.

Basic Image: There is no basic image but substitution can be regarded as the basic tool, and the empty argument place would then be the basic image: $f(\ )$

Elements: Predicate logic deals with objects falling under concepts of various order and arity.

---

[18] From the point of view introduced here, Calculus logic has no resources to consistently introduce quantifiers, as there is no notion of an individual a quantifier could range over. Peirce introduced two separate signs for "some" and for "all", but these are not inter-definable, except in that they are interpreted as infinite collections of logical sums and products. Thus it is not merely a notational variant to Frege's notation, as Putnam 1990 claims. The details of these differences seem not to have been worked out on a principled basis. In a similar way, Peano introduces two different signs for "all" and for "there exists". It is only with Frege that one uniform "generality notation" (his term) is introduced. Incidentally, it seems that it was only A. N. Whitehead who figured this out entirely in 1904, and then taught Russell how to use the "quantifiers". The ways of set theory (also introducing levels of objects and classes) further complicated these connections.

Proposition: Entire propositions (judgments, thoughts) are the basic items. Concept-expressions are generated by creating argument places through taking a proper name away from a proposition.

Basic move: The fundamental logical relation is subsumtion: an object falling under a concept: $f(a)$.

Connectives: All 16 connectives are equally basic. Frege takes implication and negation as basic. In his work he uses no ther connectives.

Other relations: While earlier logics could only treat subordination, Frege has four basic uses of "is", three of them involving objects: subordination, predication, identity between objects and existence.

Type of relations: The main aim of logic is the expression of general laws through the use of quantifiers.

Structure of relations: Most relations are asymmetrical.

Generality: Quantifiers can express and limit generality.

Identical propositions: Mere identities are of little interest. Frege wants to give proofs that have content. His fundamental axioms are, however, all (what we would call) tautologous. He also needs to explain why the equations of arithmetic are not "merely identical propositions". He tried to solve this problem with the distinction between sense and reference.

Technical form: Frege provides gapless proofs of theorems, conducted exclusively in logical signs (symbols).

Metaphysical form: There are no modalities and therefore no necessities in Frege's logic. His propositions express generalities about all objects.

Related science: Predicate logic has a special affinity to physics, as the science expressing general laws, but also to the infinitesimal calculus.

Particular judgments: They express existence claims: "there exist some f's". Frege is the first one who can express them precisely, and who can explain how they work.

Relation to mathematics: Frege's logic is motivated by the desire to give a logical foundation to arithmetic, as a part of mathematics. In order to achieve this, he organizes his logic in a mathematical fashion.

"All humans are mortal": This proposition is a general statement about all individuals which are human. Strictly speaking we cannot claim to know that it is true.

An aim not reached: Frege's "logicism" was to supply a logical foundation for arithmetic but Russell's paradox destroyed all hope of attaining this aim.

An aim reached: Predicate logic is very successful in the logical analysis of language, as $f(a)$ comes out as the basic form of an ordinary sentence, saying that some object a has a certain property f.

6.4 Appendix: *Tractatus* Logic

Note

This section is *not* part of the article. It is offered as an attempt to open another, extra angle on the matters discussed – just trying to give some suggestions for further inquiry.

In his *Tractatus*, Wittgenstein tried to clarify the nature of logic once and for all. While doing this, he pointed out several inconsistencies and defects in the conceptions of Frege and Russell. While he had a very limited knowledge of Syllogistic logic, he did express sympathies with what he took to be older conceptions of logic.

Person: (Early) Wittgenstein.
Basic Image: There is no basic image. But in the end, almost everything in logic is "shown" and thus in some sense pictorial.
Elements: Names (standing for objects) are the material to be concatenated in logical forms.
Proposition: Propositions are concatenation of names.
Basic move: "We make to ourselves pictures of facts."
Connectives: The usual 16 possibilities.
Other relations: Outside logic, everything is chance.
Type of relations: We must distinguish between expressions of facts (all of them logically unrelated) and a priori internal relations between logical forms (these can only be shown).
Generality: We need to keep apart empirical generality (expressed by quantifiers) and essential generality (expressed through internal relations between logical forms).
Identical propositions: Identity cannot even be stated, thus there simply are no identical propositions. What we want to express through "identity" can only be expressed by use of the same symbol.
Technical form: Wittgenstein gives no system. He only gives examples which intend to show the mistakes still inherent in Frege's and Russell's logical systems.
Metaphysical form: Logic shows the structure of the world. It is thus transcendental. (Without a world there would be no point in there being logic.)
Related science: Logic is widely different from any science, with the partial exception of probability theory.
Particular judgments: Particular judgments don't really exist, essential are singular judgments which are replaced by elementary propositions.
Relation to mathematics: Logic and mathematics are parallel enterprises, each of them "transcendental". Both show the structure of the world, either in

tautologies, or in equations. Mathematics can be seen as a "method of logic". Conversely number can be regarded as the fundamental idea of logic.

"All humans are mortal": This is not a meaningful proposition, as it cannot be verified. (Unless it is regarded as expressing an internal relation between concepts.)

An aim not reached: To make things so clear that everybody will understand that all philosophical problems are solved.

An aim reached: One of Wittgenstein's main logical achievements is the clarification of the tautologous nature of logic.[19]

Bibliography

Boole, George (1847): *The Mathematical Analysis of Logic*. Cambridge: Macmillan.
Boole, George (1854): *Laws of Thought*. Cambridge: Macmillan.
Church, Alonzo (1936): "A Bibliography of Symbolic Logic". In: *Journal of Symbolic Logic* 1, pp. 121–218.
Euler, Leonard (1802): *Letters of Euler on Different Subjects in Physics and Philosophy addressed to a German Princess*. London: Murray and Highley. [First published in French in 1768]
Frege, Gottlob (1879): *Begriffsschrift – eine der arithmetischen nachgebildeten reinen Formelsprache des Denkens*. Halle: Nebert.
Frege, Gottlob (1895): "A Critical Elucidation of Some Points in E. Schröder, *Vorlesungen über die Algebra der Logik*". In: Frege, Gottlob (1984), *Collected Papers*. Oxford: Blackwell, pp. 210–228.
Hilbert, David and Ackermann, Wilhelm (1950): *Principles of Mathematical Logic*. New York: Chelsea. [Translation of the 1937 German edition.]
Künne, Wolfgang (2010): *Die Philosophische Logik Gottlob Freges. Ein Kommentar*. Frankfurt am Main: Klostermann.
Putnam, Hilary (1990): "Peirce the Logician." In: *Realism with a Human Face*. Cambridge, MA: Harvard University Press, pp. 252–260. [First publication 1982]
Quine, Willard V. O. (1959): *Methods of Logic*. New York: Holt, Rinehart and Winston. [Fourth edition (1982): Cambridge, MA: Harvard University Press]
Schlick, Moritz (1918): *Erkenntnislehre*. Berlin: Springer.
Schröder, Ernst (1877): *Operationskreis des Logikkalkuls*. Leipzig: Teubner.
Schröder, Ernst (1880): Review of: Frege, *Begriffsschrift*. Tr. and repr. in: Bynum, T.W. (1972): *Conceptual Notation and related articles*. Oxford: Oxford University Press, pp. 218–232.
Schröder, Ernst (1890): *Vorlesungen über die Algebra der Logik*. Vol. 1. Leipzig: Teubner.
Venn, John (1881): *Symbolic Logic*. London: Macmillan.
Whitehead, Alfred North and Russell, Bertrand (1962): *Principia Mathematica to *56*. Cambridge: Cambridge University Press.

---

19 I wish to thank the editors for their patience and support. Hanno Birken-Bertsch, Tabea Rohr and Astrid Schleinitz saved me from some errors and mistakes.

Wittgenstein, Ludwig (1922): *Tractatus Logico-Philosophicus*. London: Kegan Paul, Trench, Trubner.
Wolff, Michael (2005): *Prinzipien der Logik*. Frankfurt am Main: Klostermann.
Wolff, Michael (2013): "Viele Logiken – Eine Vernunft. Warum der logische Pluralismus ein Irrtum ist". In: *Methodus* 7: pp. 79–134.

Günther Eder
Truth, Paradox, and the Procedural Conception of Fregean Sense

Abstract: In his seminal article *On Sense and Reference*, Frege introduced his famous distinction between sense and reference. While Frege's notion of reference is relatively clear, the notion of sense was viewed with suspicion right from the beginning. The aims of this article are two-fold. First, I will motivate and discuss what I will call the *procedural conception of Fregean sense*, according to which senses are understood as procedures to determine referents. Senses of sentences, in particular, are identified with procedures to determine truth-values. Based on a formal explication of the procedural conception of sense proposed by John Horty, I will, secondly, give an outline of a theory of the semantic paradoxes and related semantic anomalies, drawing on the idea that paradoxical sentences correspond to sense-procedures that, because of their internal structure, fail to determine a truth value.[1]

1 Introduction

One of the foremost contributions of Gottlob Frege to the philosophy of logic and language is his celebrated distinction between sense and reference. While Frege's notion of *reference* (*Bedeutung*) is relatively clear and eventually found its way into mainstream semantic theorizing, his notion of *sense* (*Sinn*) proved to be harder to pin down exactly. Following up on some of Frege's remarks, in this article I take up a suggestion to identify senses with *procedures*, specifically, procedures to determine referents. Given the close connection between sense and reference that is postulated by such an account, this conception may also be expected to have a bearing on the semantic paradoxes. Indeed, a central goal of this article is to make plausible that a procedural understanding of sense provides a natural way to think

---

This article is a shortened version of a more detailed article I am currently preparing. The extended version will also include formal details that can only be indicated here. Research on this article was funded by the Austrian Science Fund (FWF, Project number P 30448-G24). An earlier draft was presented at a work-in-progress seminar at the Department of Philosophy in Salzburg. I would like to thank the audience as well as Lorenzo Rossi and Alexander Hieke for helpful discussions and valuable feedback.

Günther Eder, University of Salzburg, guenther.eder@univie.ac.at

DOI 10.1515/9783110657883-10

about the semantic paradoxes and related phenomena, the basic idea being that some sentences correspond to sense-procedures whose internal structure preclude them from determining a truth value. In order to show that this idea can be made precise, I will indicate how a formal model of sense-procedures proposed by John Horty can be extended and applied to an arithmetical language that contains a self-applicable truth predicate.

The article is organized as follows: in the next section I will briefly discuss the basics of Frege's theory of sense and reference and the general notion of a procedure as it will figure in the procedural account of sense. In section 3, I will introduce a specific procedural framework for a simple, arithmetical language. In section 4, this framework will be extended and put to use to develop an account of paradoxicality and other semantic anomalies for a language that contains a self-applicable truth predicate. I conclude with some tentative remarks relating to the scope of the suggested proposal and further topics to be studied.

2 Senses, Procedures, and Sense-Procedures

One of the most influential innovations of Frege in the philosophy of logic and language is his famous distinction between sense and reference, officially introduced in his *On Sense and Reference* in 1892 (Frege 1984: 157 – 177). It is well-known that Frege originally introduced the distinction in order to solve a puzzle about identity statements. Roughly, Frege asks himself how come that a true identity statement like

(1) The morning star is identical with the evening star.

is informative, whereas the statement

(2) The morning star is identical with the morning star.

is not, in spite of the fact that one sentence arises from the other by substituting a singular term for another one that denotes the same object, namely, the planet Venus. Frege's proposed solution to the puzzle crucially involves distinguishing between two semantic features of a singular term, its sense and its reference. His solution is based on two central traits of these notions: first, two singular terms may have the same referent, but express different senses, and, second, knowledge of the sense of a singular term does not in general entail knowledge of its referent. Because the singular terms flanking the identity sign in (1) express different senses, grasping their respective senses is not enough to determine whether their referents are the same. Since substantial empirical investigations are required to determine whether their respective referents are identical, (1) therefore contains a "valuable extension of our knowledge", whereas (2) does not.[2]

Frege later extends his distinction to other kinds of expression, including sentences and predicates. Sentences, according to Frege, refer to one of two truth values, the True and the False, and predicates refer to functions from objects to truth values. Also, both sentences and predicates express senses, which, in the case of sentences, Frege also calls "thoughts".

Whereas Frege's notion of reference is relatively clear and eventually found its way into mainstream semantic theorizing, his notion of sense proved to be harder to pin down exactly, the million-dollar question being: what *are* senses? Frege himself was aware of the problem. In a letter to Husserl he explicitly notes that "[i]t seems to me that an objective criterion is necessary for recognizing a thought again as the same, for without it logical analysis is impossible" (Frege 1980: 70). And yet, Frege makes no decisive attempt to tackle this question head-on. Instead, we are left with a variety of hints and metaphors which sometimes point to conceptions of sense that are directly opposed to each other. My aim, however, is not to discuss all of these conceptions and to develop an account that is maximally consistent with the scattered remarks on the concept of sense that can be found in Frege's writings. Rather, my focus will be on a particular kind of metaphor, the conception of sense they suggest, and how this conception can be made precise.

The metaphors I am thinking of are ones where Frege describes senses as "ways of determining" or "arriving at" a referent. For example, in an unpublished manuscript from 1902 Frege writes that "different signs for the same thing [...] indicate the different ways in which it is possible for us to arrive at the same thing" (PW: 85). In a letter to Russell, Frege remarks in a similar spirit that different signs for the same object might not be interchangeable because "they determine the same object in different ways" (Frege 1980: 152). To be sure, quotes like these certainly do not force a particular conception of sense. But a natural way to conceptualize the idea that senses represent "ways of determining" a referent is by thinking of such a "way" as a particular *method* or *procedure*. Accordingly, understanding an expression, grasping its sense, consists in knowledge of such a procedure. This is what I will call the *procedural conception of Fregean sense*.

The procedural conception of sense is not unheard-of in the literature and several scholars have developed procedural accounts of structured meanings based

---

[2] See (Frege 1892: 157). A detailed discussion of Frege's distinction from a scholarly point of view can be found, for instance, in Textor 2011.

on some of Frege's ideas.³ The aim of what follows, however, is not to present a worked-out procedural account of Fregean senses that could claim to provide a full-fledged account of structured meanings. Instead, the goal will be to make use of the basic idea underlying the procedural conception in order to provide the basis for an account of the *semantic paradoxes*. Indeed, following this conception, there is a natural way to think about paradoxical and otherwise semantically deficient sentences. We merely have to accept the notion that a procedure may, for a variety of reasons, fail to determine an output. The idea, then, is that paradoxical sentences simply correspond to sense-procedures whose internal structure makes it impossible for them to determine a truth value.⁴

Before we can see what such an approach might look like more exactly, let us try to get a clearer picture of the general notion of a procedure that is at issue. A procedure, as it will be understood here, may be characterized informally as follows:

(PR) *A procedure consists in a list of instructions that determines a set of actions to achieve a certain goal in an orderly manner.*

To be sure, this characterization should not be taken as a definition. It should merely indicate the direction in which we are heading. For example, a cooking recipe may be thought of as a specification of a procedure. The recipe provides a set of instructions that prescribe a sequence of actions to process certain ingredients in a certain way to obtain a certain result. So we have a list of instructions, a set of actions, a goal, and order. Other obvious examples of procedures are the kind of algorithmic procedures known from mathematics and formal logic, including procedures like the Euclidean algorithm, Newton's method, or the truth-table method. In each of these cases, we have a set of instructions that prescribe a series of calculations to compute a certain result.

But the kind of procedures we are interested in are not exhausted by procedures like the ones just mentioned, and certainly not by "effective procedures" in the sense of computability theory. For example, nothing in the informal characterization of a procedure above places any restrictions on the number of instructions.

---

3 See, for instance, Duží et al. 2010, Jespersen 2010, Moschovakis 1994, Moschovakis 2006, Muskens 2005, and Tichy 1988. Some of Michael Dummett's views can be traced to a particular procedural understanding of Frege's notion of sense. See, for instance, chapter 5 in Dummett 1973. A more recent discussion of the procedural conception of sense can be found in Penco 2009.
4 This idea is not entirely new. Aaron Sloman, in his Sloman 1971 already anticipates the outlines of such an approach from a broadly Fregean point of view. Attempts to spell it out in a precise setting have been made in Moschovakis 1994 and Muskens 2005.

Hence, a procedure may contain infinitely many instructions. Also, in contrast to the actions involved in a cooking recipe or Newton's method, an action need not be *effectively executable* in the sense that it could actually be performed by an agent. Furthermore, it is crucial to distinguish between a procedure and a *linguistic description* of that procedure. Procedures are abstract, language-independent entities. A (written or spoken) description of a procedure is just that, a description of it. Finally, it is important to see that, even though the specification of a potential goal is required for a procedure, nothing in our informal conception of a procedure requires that this goal is actually *achieved*, or even *achievable*. Accordingly, a procedure may be *non-terminating* in the sense that the actions prescribed by it may never lead to an output.

Now, according to the procedural conception of sense, senses are procedures, namely procedures to determine referents. Following Frege's theory of reference sketched earlier, the sense-procedure associated with a singular term will therefore consist in a procedure to determine a certain object, the sense-procedure associated with a sentence will be a procedure to determine a truth value, and the sense-procedure associated with a predicate will be a procedure to determine a truth value, given certain inputs. A few remarks will be helpful to put this into perspective. (I will limit myself to the discussion of sense-procedures corresponding to sentences. Similar remarks apply to singular terms and predicates.)

One of the basic intuitions that inform the procedural conception of sense is the idea that grasping the sense of a sentence consists in knowledge of what, in some idealized sense, would have to be done to determine its truth value, given the way the world is or might have been. In order to develop a procedural account of sense that captures this idea, two distinctions are crucial. First, we have to distinguish between a procedure and its potential *output*. Clearly, we do not want our account of sense to have the effect that grasping the sense of a sentence amounts to (explicit or implicit) knowledge of its truth value. Secondly, we also have to distinguish between a sentence's actual truth value and its truth *conditions*. Indeed, properly speaking, we shouldn't identify sense-procedures with procedures to determine truth values at all, but rather truth conditions. Once again, knowing what a sentence means should in general entail neither explicit nor implicit knowledge of its actual truth value. But if sense-procedures were identified with procedures to determine *actual* truth values, then understanding an empirical sentence would amount to implicit knowledge of its actual truth value, which means that we would have to attribute empirical omniscience to a person merely in virtue of her being a competent language user, which is absurd. So if a procedural conception of sense is to have any plausibility, the senses associated with sentences ought to be procedures to determine *truth conditions* (*intensions*), rather than *actual* truth values. Having said that, in the case of *mathematical* sentences (which will be our main focus in what follows) we might restrict ourselves to sense-procedures to determine actual truth values after all, simply because a mathematical statement is true if true in all possible worlds, and false if false in all possible worlds.[5]

Next, as mentioned earlier, procedures in general and sense-procedures in particular are not linguistic in nature, they are language-independent. Bearing this in mind is especially important in the case of sense-procedures. If understanding a sentence consists in knowledge of a certain sense-procedure, and knowledge of that procedure were to consist in understanding a certain set of linguistic instructions, then grasping the sense of a sentence S_1 would require grasping a (set of) sentence(s) S_2, which would require grasping a (set of) sentence(s) S_3, and so on. Thus, it seems that we would be either facing a vicious circle or an infinite regress. Understanding would be impossible.

Finally, if senses are identified with procedures to determine referents, then, assuming (as we do) that the general notion of a procedure permits non-terminating procedures, the procedural conception of sense permits that some sentences may fail to determine a truth value. Indeed, paradoxical sentences will turn out to correspond to sense-procedures of that kind. The general approach pursued here therefore falls in the broad category of "gappy approaches" to the paradoxes. It is worth emphasizing though that this is a consequence of our general conception of a procedure, and not some ad hoc maneuver designed to deal with the paradoxes.

So far, we've looked at some of the general heuristics underlying the procedural conception of sense. The question, then, is: how is all of this made precise? In the next section I will introduce a specific formal model of sense-procedures that is particularly suitable for the purpose at hand. Based on this model, we will then be able to (indicate how to) define precise notions of paradoxicality and related notions.

3 A Simple Procedural Model of Fregean Senses

The model of sense-procedures I want to adopt is due to John Horty, who introduced it in his (Horty 2007) in order to deal with problems relating to the semantics of defined expressions.[6] Horty develops his model for a simple, quantifier-free,

---

[5] The problem of logical and empirical omniscience in connection with a procedural understanding of meaning is discussed e.g. in Duží et al. 2010: 12 – 14.
[6] In order to make use of Horty's model to study the semantic paradoxes, several modifications will be made. Readers who are interested in Horty's original model are referred to chapter 5 of Horty 2007.

arithmetical language. In what follows, we assume that this toy language contains individual constants for each natural number, function symbols + and × for the addition and multiplication function, relation symbols = and > for the identity and the greater-than-relation, and propositional connectives ¬ and ∧ for negation and conjunction. (Other connectives may be introduced by definition.)

The semantics of this language follows the familiar pattern. First, primitive expressions are assigned suitable referents. The referents of complex expressions are then determined compositionally by the referents of their sub-expressions. More precisely, each constant symbol *n* refers to the natural number n, and the function symbols + and × refer to the addition function add and the multiplication function mult respectively. Following Frege's view that each sentence refers to one of the two truth values, TRUE or FALSE, the relation symbols = and > refer to the identity function id and the greater-than function gr from pairs of natural numbers to truth values, and the connectives ¬ and ∧ refer to classical truth-functions not and and that assign truth values to (pairs of) truth values in the familiar way. Finally, in accordance with the principle of compositionality for reference, the referent of a complex expression is functionally determined by the referents of its immediate sub-expressions. So, for example, the referent of the expression "2 + 2 = 5" is id(add(2, 2), 5), which is simply the value FALSE.

In order to assign senses to expressions, a similar strategy is followed. First, each simple expression is assigned a basic sense-procedure. The sense-procedures associated with complex expressions are then compositionally determined by the sense-procedures associated with their sub-expressions. Specifically, we assume that to each individual constant *n* there corresponds a sense-procedure **n** that computes the natural number n; to the functions symbols + and × there correspond sense-procedures **add** and **mult** that compute the functions add and mult; to the relation symbols = and > there correspond sense-procedures **id** and **gr** that compute the functions id and gr; and to the connectives ¬ and ∧ there correspond sense-procedures **not** and **and** that compute the classical truth-functions not and and.

Given this assignment of basic procedures to primitive expressions, the sense-procedure associated with a complex expression can then be identified with a tree which represents the fundamental structure of the complex procedure that determines the referent of that expression.[7] Of course, for such a tree to represent a procedure to determine the expression's referent, further specifications have to be made. In particular, given our informal conception of a procedure from earlier,

---

[7] Instead of identifying complex procedures with their associated procedure trees, Horty uses a linear bracket-notation to represent sense-procedures. We will follow the tree-approach, however. Under suitable provisions, both versions are equivalent.

we have to specify how exactly such a tree canonically determines a sequence of actions to determine the referent of an expression. For this, the main task is to describe how coordination among basic procedures is effected. In Horty's model, this is achieved by making use of *registers*. The idea is that registers will serve as pick-up and drop-off points for values (in our case, *numbers* and *truth values*). They provide the memory space, as it were, on which basic procedures can operate. Basic procedures will use the values stored in some register as input, then do whatever they do, and then deposit the output in some other register, where it can be used by another procedure, and so on.

Since a procedure is supposed to determine a sequence of *actions*, Horty next introduces a notation to represent such actions, which he calls *procedure executions*. Procedure executions are actions of the sort "apply the procedure **P** to the value stored in registers $s_1, s_2, \ldots$ and store the output in register s". With this, he then goes on to show how the procedure tree associated with an expression canonically determines a sequence of basic procedure executions (executions of basic procedures) that provides a detailed specification of the actions to be performed in order to determine the referent of that expression.

This is roughly the model proposed by Horty. In order to make this a full-blown account of sense that could claim to elucidate the concept of linguistic meaning, several amendments would have to be made. But it should be clear from this brief outline that sense-procedures have a number of features that one would expect from an account of Fregean senses. For one thing, sense-procedures can be put together to form more complex procedures. On a procedural account, senses are therefore *compositional* in nature, and the structure of an expression by and large mirrors the structure of the sense expressed. Furthermore, since a procedure is not determined by its input-output behavior, but by the particular *way* this input-output behavior is effected, sense-procedures are *finely individuated*. Thus, while the sentences "2+2 = 4" and "5 is a prime number" agree in their truth-conditional content in virtue of expressing necessary truths, they differ in *sense* because they correspond to different procedures to determine their respective truth values. Again, more would have to be said in defence of the proposed account. But instead of improving on Horty's model to make it a more plausible account of Fregean sense (or of meaning in general), in what follows, I will try to indicate how it may be used to provide the basis of an account of the semantic paradoxes and related semantic deficiencies.

4 Sense-Procedures and Semantic Paradox

So, first, let us be more specific about our setting and what we are trying to achieve. As is common in the modern study of the semantic paradoxes, we will be working in a sufficiently expressive, arithmetical language. In this context, "sufficiently expressive" means that the language is capable of referring to its own syntax via arithmetization. In what follows, we will work with a standard first-order language $\mathcal{L}_T$ that contains the connectives $\neg$ and $\wedge$, the universal quantifier $\forall$ (including infinitely many individual variables), as well as symbols for certain recursive functions in addition to the arithmetical vocabulary mentioned in the previous section. This will ensure that our language has the resources to achieve self-reference so that for each formula $A(x)$ of $\mathcal{L}_T$ there will be a term t of $\mathcal{L}_T$ such that t denotes the (Gödel number of the) sentence $A(t)$.[8] Finally, we assume that $\mathcal{L}_T$ contains a unary predicate T which is supposed to represent a self-applicable truth predicate for $\mathcal{L}_T$, that is, a predicate that may be applied to sentences that contain the truth predicate T. Since $\mathcal{L}_T$ allows for self-reference, it follows that there will be a term λ that denotes the sentence $\neg T\lambda$, which is just the standard liar-sentence.

The overall goal, then, is to develop an account of the liar and similar semantically deficient sentences, based on the notion that the sense of a sentence is a procedure to determine its (potential) truth value. Since the internal structure of atomic sentences will be largely irrelevant for what follows, to simplify exposition, we will assume that we are given a non-repeating enumeration $A_1, A_2, \ldots$ of the atomic sentences of $\mathcal{L}_T$. Also, sentences of the form Tt will be counted among the A_i's if t does not denote a sentence.

In order to achieve our main goal, the first and foremost task is to associate with each sentence of $\mathcal{L}_T$ a sense-procedure that determines its potential referent. For this, we first have to specify what the referents of the basic expressions of $\mathcal{L}_T$ are and how the referents of complex expressions compositionally depend on the referents of their parts. Given our simplified setting, we assume that to each atomic sentence A_i there corresponds a constant truth-function A_i such that A_i gives the value TRUE if A_i is true in the standard model of arithmetic and the value FALSE if A_i is false in the standard model of arithmetic or A_i is of the form Tt where t does not denote a sentence. Furthermore, the logical symbols $\neg$, $\wedge$, and $\forall$ refer to certain partial truth-functions. Here, different choices are possible. To be specific, we will assume that $\neg$, $\wedge$, and $\forall$ refer to the functions not, and and all that are defined as follows: the function not maps the value TRUE to FALSE and FALSE to TRUE, and is undefined whenever it does not get an input. The two-place function and maps

---

[8] As is common, I will usually identify sentences with their Gödel numbers in what follows.

the pair (TRUE, TRUE) to the value TRUE, pairs of truth values to the value FALSE if at least one of them is FALSE, and is undefined in every other case. Similarly, the infinitary truth function all maps the infinite sequence all of whose components are TRUE to the value TRUE, each sequence that contains at least one value FALSE, to the value FALSE, and does not assign a truth value to any other sequence.[9] Finally, we assume that the truth predicate refers to the function tr that maps the value TRUE to TRUE, FALSE to FALSE, and is undefined otherwise. Hence, the truth predicate refers to the identity function on truth values. By definition, then, a truth-predication Tφ will refer to the value TRUE if φ refers to the value TRUE, and it will refer to the value FALSE if φ refers to the FALSE.[10]

Given these stipulations, we now want to associate sense-procedures with each sentence of $\mathcal{L}_T$. Following Horty's model outlined earlier, we assume that to each primitive expression of $\mathcal{L}_T$ there corresponds a basic procedure. In particular, to each A$_i$ there corresponds a basic procedure **A**$_i$ that computes A$_i$, and to the truth functions not, and, all and tr, there correspond basic procedures **not**, **and**, **all** and **tr** that compute the respective truth functions.

To associate sense-procedures with complex sentences, we first introduce what I call the *semantic tree associated with a sentence*, which is a labelled, ordered tree where each node is labelled by some sentence of $\mathcal{L}_T$. More precisely, we assume that the children of a node are labelled by the immediate sub-sentences of the label of their common parent, where we count the instances of a quantified sentence as immediate sub-sentences, and φ as an immediate sub-sentence of the truth predication Tφ. The *procedure tree associated with a sentence* is then defined as the isomorphic copy of this semantic tree where each label in the semantic tree is replaced by the basic procedure corresponding to its main operator (see Figure 2).

Note that the procedure tree associated with a sentence of $\mathcal{L}_T$ will generally be neither finite in breadth nor in height. In the presence of quantifiers and the truth predicate, procedure trees may be complex structures that may even contain infinite branches, that is, branches that contain infinitely many nodes. The procedure trees corresponding to the standard liar and the standard truth-teller are cases in point (see Figure 1).

---

9 Note that, even though we are allowing that a sequence of truth values may be "gappy" in that one or more components may be neither TRUE nor FALSE, we do not assume that there is a "third" truth value. Also, a sentence may refer to a truth value even though one of its immediate sub-sentences does not. The evaluation scheme that is used here is called the *strong Kleene evaluation scheme*. As indicated, other evaluation schemes may be used instead.
10 That the truth predicate is supposed to satisfy some version of the T-scheme is already indicated by various remarks made by Frege throughout his writings, for instance, in (PW: 233) or (Frege 1984: 354–355).

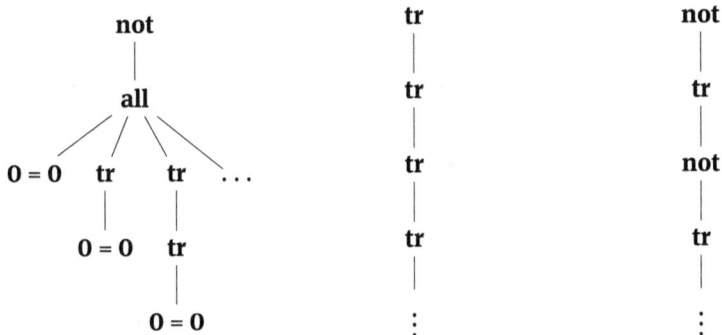

Fig. 1: Procedure trees associated with the sentence $\neg \forall n T^n(0 = 0)$, the truth-teller sentence $\tau = T\tau$ (middle), and the standard liar sentence $\lambda = \neg T\lambda$ (right). In the first example, T^n represents the n-th iteration of the truth predicate. That is, $T^0\varphi = \varphi$, $T^1\varphi = T\varphi$, $T^2\varphi = TT\varphi$, etc.

Following Horty, we will make use of *registers* to give substance to the idea that these trees actually represent sense-procedures, that is, procedures to determine referents. To this end, we assume that we are given a *tree of registers*. The tree of registers is again a labelled tree whose nodes are the registers, and whose labels are sequences of ordinal numbers, which we call *addresses*. (We will identify a register with its address, even though, strictly speaking, these are different entities.) The root node of the tree is labelled by the empty sequence and the children of the register with address s are registers with addresses s⌢α, where s is an address, α an ordinal, and ⌢ represents concatenation of sequences of ordinals. Given some sufficiently large tree of registers, we can then canonically associate with each sentence φ a unique sub-tree of the tree of registers, namely the leftmost-topmost copy of the procedure tree within the tree of registers, which I will call the *register tree associated with φ* (see Figure 2).

Next, we have to explain how the procedure tree associated with a sentence φ canonically determines a sequence of actions to determine the truth value of φ. For this, we introduce two further notions. The first is the notion of a *partial filling*, which will be needed to keep track of which truth values are stored in which register. So, a partial filling V for the register tree associated with φ is simply an assignment of truth values to registers in the register tree associated with φ. This assignment need not be total, but we assume that, first, it should be compatible with the truth values assigned to the atomic sentences by our arithmetic ground model and, second, that the same truth value is assigned to registers that correspond to nodes in the semantic tree with the same label.

The second notion that will be central in what follows is the notion of an *evaluation trial from a register s*, which is a certain sequence of basic procedure

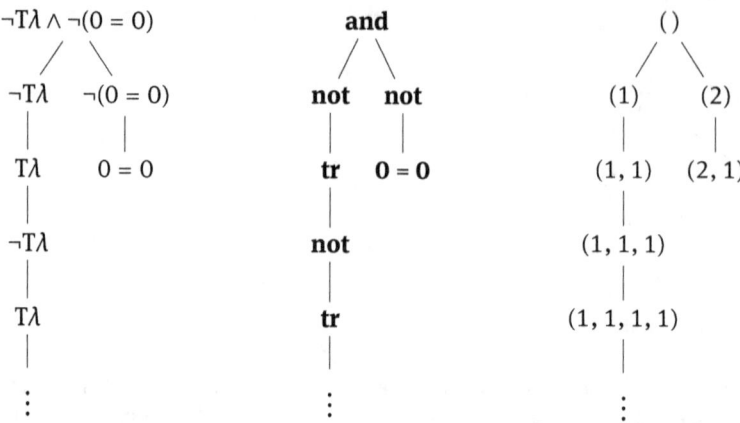

Fig. 2: Semantic tree (left), procedure tree (middle), and register tree (right) associated with $\neg T\lambda \wedge \neg(0 = 0)$, where λ is the liar sentence $\lambda = \neg T\lambda$.

executions. The sequence starts with the execution of the procedure **P** in the procedure tree that corresponds to the register s in the register tree. That is, the procedure **P** is applied to the truth value(s) in the register(s) immediately below s and the result is stored in register s. If **P** happens to be an atomic procedure $\mathbf{A}_i$, no input is required and the truth value computed by $\mathbf{A}_i$ is immediately stored in s. We then successively move upwards in the register tree, at each node applying the corresponding procedure in the procedure tree to the value(s) stored in the register(s) immediately below the current register s_i and storing the result in s_i (perhaps overwriting or deleting default-values that might be stored in s_i), until we finally reach the top node in the register tree.

Note that each evaluation trial can be understood as an operator that maps a given partial filling to another partial filling, namely, the partial filling that results after executing the evaluation trial. The idea, then, is that we first impose a suitable order on the register tree associated with φ, and then try to determine the truth value in the top register by moving through the register tree in the stipulated order, successively performing evaluation trials at each register.[11] Once we have visited all registers, we simply start over and do the same again, and again, and again. Given some initial partial filling $V = V_0$, the process just outlined gives rise to an ordinal-indexed sequence of partial fillings

---

[11] There are several alternatives of how to fix a suitable well-order through the register tree. For definiteness, one may think of the registers as being ordered left-to-right and top-to-bottom. The details won't matter though in what follows.

$$V_0, V_1, \ldots V_\alpha, V_{\alpha+1}, \ldots$$

and the following notions can be defined:[12]

Definition. *For each sentence φ and each initial partial filling V:*
i) *φ is eventually stably true$_V$ iff there is an ordinal β such that V_α assigns* TRUE *to the top register of the register tree associated with φ for all α ≥ β.*
ii) *φ is eventually stably false$_V$ iff there is an ordinal β such that V_α assigns* FALSE *to the top register of the register tree associated with φ for all α ≥ β.*
iii) *φ is eventually stably undefined$_V$ iff there is an ordinal β such that V_α assigns no truth value to the top register of the register tree associated with φ for all α ≥ β.*

Thus, a sentence is eventually stably true (false, undefined) with respect to some initial filling V if the top register of its associated register tree is at some point filled with the value TRUE (FALSE, undefined), and remains filled with TRUE (FALSE, undefined) forever after. We say that a sentence has a stable truth value$_V$ if it is either eventually stably true$_V$ or eventually stably false$_V$. Finally, we can then define precise notions of paradoxicality and hypodoxicality as follows:

Definition. *For each sentence φ*
i) *φ is paradoxical iff there is no initial partial filling V such that φ has a stable truth value$_V$.*
ii) *φ is hypodoxical iff there are initial partial fillings V_1 and V_2 such that φ is eventually stably true$_{V_1}$ and eventually stably false$_{V_2}$.*

It is easy to see that the standard truth-teller is classified as hypodoxical and that the standard liar is classified as paradoxical by these definitions. The same is true for related paradoxical sentences such as versions of the strengthened liar that are formulated by means of a determinateness operator.[13]

---

[12] Making this construction precise is not a trivial matter and involves several substantive decisions, in particular about 1) what happens at limit stages in the ordinal-indexed sequence of partial fillings and 2) how the periodic nature of the process is captured. Concerning 1), the idea is that, for limit ordinals α, $V_\alpha(s)$ will be TRUE (FALSE) if there is an ordinal β less than α such that $V_y(s)$ = TRUE (FALSE) for all ordinals y between β and α and $V_y(s)$ will be undefined if there is no such ordinal.

[13] D is a determinateness operator if Dφ receives the value TRUE whenever φ receives the value TRUE, and Dφ receives the value FALSE in every other case. It is easy to see that the "strengthened liar" $\lambda_S = \neg DT\lambda_S$ is paradoxical in the sense of our definition. A determinateness operator is definable e.g. in a language that contains a Łukasiewicz conditional. See e.g. chapter 4 of Field 2008.

It seems to me that these definitions (once unpacked) capture the intuitive idea that paradoxical (hypodoxical) sentences correspond to sense-procedures that fail to determine a truth value in a fairly straightforward way. It is also worth emphasizing that, on the current conception, sentences like the liar are perfectly meaningful. They are put together from meaningful parts according to admissible principles of composition. It is just that, due to their specific composition, their "procedural contents" fail to determine a truth value.

In the proposed framework, we can also give a precise definition of the intuitive concept *semantic groundedness* (see Kripke 1975). Intuitively, a sentence is semantically grounded if its truth value is determined by non-semantic facts alone. In the current framework, a sentence can be defined to be grounded just in case it has a stable truth value with respect to the unique initial filling V that assigns truth values *only* to atomic, arithmetic sentences. Once again, unpacking the definitions, this can be seen to be a direct implementation of the idea that a sentence is grounded if its truth value is procedurally determined by non-semantic facts.[14]

5 Conclusion

The proposed account has several similarities with other approaches to the semantic paradoxes. Generally speaking, the approach followed here falls into the category of "gappy approaches" like, for instance, Kripkean fixed point semantics. Looking at the details, it also becomes clear that there are strong commonalities with the revision theory of truth. Indeed, the basic construction in terms of a traversal through the register tree and the performance of evaluation trials essentially amounts to a process in which partial fillings are constantly being revised. Furthermore, at limit stages, the truth value in a register is determined in a way that closely resembles the standard limit rule in a revision sequence. The same is true of the definitions of eventual stable truth and falsity. Finally, the approach suggested here also has considerable overlap with various recent studies of the semantic

---

[14] Hence, no detour via fixed point models or the like is required, as in Kripke's approach. Also, the current account arguably captures two aspects of groundedness that have been distinguished by Steve Yablo in his Yablo 1982, the *dependence* and the *inheritance* aspect. Roughly, the idea is that the dependence aspect is captured by a "top-down" tree search, while the inheritance aspect is captured by a "bottom-up" evaluation trial.

paradoxes that make use of graph-theoretic methods. So there is plenty to be said on relations between the current approach and that of others in the literature.[15]

But there are other themes which are worth further exploration. One direction in which the current conception may be further developed is connected to the phenomenon of contingent liars which has been drawn attention to, again, by Kripke in his Kripke 1975. The basic idea here is that, under unfavorable circumstances, perfectly normal sentences lead to paradox. A full-blown procedural account of sense, where sense-procedures corresponding to sentences are identified with procedures to determine truth conditions (relative to some context) may be able to provide a systematic account of these phenomena.

Another direction in which the general approach suggested here might be expanded comes from the observation that, following a broadly Fregean view of semantics, "reference" is understood as an umbrella-term that covers several relations between expressions and their interpretations. A Fregean perspective therefore suggests that "paradoxical" singular terms like

(3) the man who came in after the last person to enter

or "ungrounded" ones like

(4) the father of the person referred to by this expression

are studied within a framework that is suitable to study the behaviour of semantically defective expressions more generally, not just sentences (see Sloman 1971).

People working on the paradoxes are typically not overly interested in questions relating to fuzzy notions like *meaning*. Formal semanticists, on the other hand, usually do not worry about the paradoxes. This is an unfortunate state of affairs, since clearly there is a close connection between meaning and truth. The aim of this article was to make plausible that a suitable procedural reading of Frege's theory of sense and reference may have the potential to connect these areas and to provide an overarching perspective on meaning, truth, and paradox.

---

15 See Gupta 1993 for an exposition of the revision theory of truth. Recent graph-theoretic investigations of the semantic paradoxes can be found e.g. in Cook 2004, Rabern 2013, Beringer/Schindler 2017, and Rossi forth. Some of the relations to other approaches will be dealt with in a more comprehensive article that is currently in preparation.

Bibliography

Beringer, Timo and Schindler, Thomas (2017): "A Graph-Theoretical Analysis of the Semantic Paradoxes." In: *Bulletin of Symbolic Logic*. Vol. 23, No. 4, 442 – 492.

Cook, Roy T. (2004): "Patterns of Paradox." In: *Journal of Symbolic Logic*. Vol. 69, No. 3, 767 – 774.

Dummett, Michael (1973): *Frege: Philosophy of Language*. New York/ Evanston/San Francisco/London: Harper & Row.

Duží, Marie, Jespersen, Bjorn and Materna, Pavel (2010): *Procedural Semantics for Hyperintensional Logic. Foundations and Applications of Transparent Intensional Logic*. Heidelberg/London/New York: Springer Dordrecht.

Field, Hartry (2008): *Saving Truth from Paradox*. Oxford: Oxford University Press.

Frege, Gottlob (1979): *Posthumous Writings*. Oxford: Basil Blackwell.

Frege, Gottlob (1980): *Philosophical and Mathematical Correspondence*. Oxford: Blackwell Publishers.

Frege, Gottlob (1984): *Collected Papers on Mathematics, Logic and Philosophy*. Oxford: Basil Blackwell.

Gupta, Anil and Belnap, Nuel (1993): *The Revision Theory of Truth*. Cambridge, MA: MIT Press.

Horty, John (2007): *Frege on Definitions. A Case Study of Semantic Content*. Oxford: Oxford University Press.

Jespersen, Bjorn (2010): "How Hyper are Hyperpropositions." In: *Language and Linguistics Compass*. Vol. 4, 96 – 106.

Kripke, Saul (1975): "Outline of a Theory of Truth." In: *The Journal of Philosophy*. Vol. 72, No. 19, 690 – 716.

Moschovakis, Yiannis (1994): "Sense and Denotation as Algorithm and Value." In: *Lecture Notes in Logic*. Väänänen, J.; Oikkonen, J. (eds.), Vol. 2. Berlin: Springer, 210 – 249.

Moschovakis, Yiannis (2006): "A Logical Calculus of Meaning and Synonymy." In: *Linguistics and Philosophy*. Vol. 29, 27 – 89.

Muskens, Reinhard (2005): "Sense and the Computation of Reference." In: *Linguistics and Philosophy*. Vol. 28, 473 – 504.

Penco, Carlo (2009): "Rational Procedures: A Neo-Fregean Perspective on Thought and Judgement." In: *Yearbook of Philosophical Hermeneutics: The Dialogue*. Vol. 4, No. 1, 137 – 153.

Rabern, Landon, Rabern, Brandon and Macauley, Matthew (2013): "Dangerous Reference Graphs and Semantic Paradoxes." In: *Journal of Philosophical Logic*. Vol. 42, No. 5, 727 – 765.

Rossi, Lorenzo (forthcoming): "A Unified Theory of Truth and Paradox."

Sloman, Aaron (1971): "Tarski, Frege, and the Liar Paradox." In: *Philosophy*. Vol. 46, No. 176, 133 – 147.

Textor, Mark (2011): *Frege on Sense and Reference*. London/New York: Routledge.

Tichý, Pavel (1988): *The Foundations of Frege's Logic*. Berlin: DeGruyter.

Yablo, Steve (1982): "Grounding, Dependence, and Paradox." In: *Journal of Philosophical Logic*. Vol. 11, No. 1, 117 – 137.

Christoph C. Pfisterer
Wittgenstein and Frege on Assertion

Abstract: In the *Philosophical Investigations*, Wittgenstein famously criticizes Frege's conception of assertion. "Frege's opinion that every assertion contains an assumption", says Wittgenstein, rests on the possibility of parsing every assertoric sentence into two components: one expressing the assumption that is put forward for assertion, the other expressing that it is asserted. But this possibility does not entail that the "assertion consists of two acts, entertaining and asserting" – any more than the possibility of rendering assertions as pairs of questions and affirmative answers entails that they consist of questions. Frege scholars protest that such criticism is inappropriate, not only because Frege doesn't speak about assumptions, but also – and crucially – because Wittgenstein fails to address the logical nature of assertion as reflected in Frege's use of the judgment stroke. They seem to read Wittgenstein's argument in the light of a remark in the *Tractatus* saying that the judgment stroke is "logically meaningless" because it simply indicates that the author holds the propositions marked with this sign to be true. In this paper, I argue that Wittgenstein's criticism of Frege is not that the latter's conception of judgment and assertion contains a corrupting psychological element. Rather, the criticism is that for Frege judgment and assertion are composed of two separate acts, i.e. an act of *referring* to a truth value and an act of *determining* which of the two it is. Through a detailed examination of the "black-spot analogy" in the *Tractatus*, I want to show that Wittgenstein presents a serious objection to Frege's conception of judgment and assertion.

1 Introduction

The notion of *judgment* is essential to Frege's conception of logic. He considers judgment to be a "logically primitive activity" (Frege 1979: 15) and introduces a special symbol for it in order to make judgments recognizable in logical derivations. Indeed, the "judgment stroke" represents such an important discovery for Frege that he wishes to have cited it in first place when responding to the question "*What may I regard as the Result of my Work?*" (Frege 1979: 184).

For all his admiration for Frege's work, Wittgenstein doesn't share Frege's enthusiasm, and both in early and late periods he expresses reservations about Frege's conception of judgment. Ignoring certain subtleties such as that Frege

---

Christoph C. Pfisterer, University of Zurich, pfisterer@philos.uzh.ch

distinguishes between judgment and assertion, or that he only calls the vertical part of the complex symbol "judgment stroke", Wittgenstein disapproves of the logical significance of assertion and its representation in logical symbolism. He says that "assertion is merely psychological" (Wittgenstein 1913: 95) and that "Frege's 'judgment-stroke' '⊢—' is logically quite meaningless" because it simply indicates that the author holds the propositions marked with this sign to be true (TLP 4.442).[1]

Wittgenstein's negative verdict seems to be repeated in §22 of *Philosophical Investigations*, where he discusses "Frege's opinion that every assertion contains an assumption". The assertion sign turns out to be "superfluous" if its function is to indicate assertion, since the specification of what is asserted characteristically takes the form of an assertion already. Wittgenstein grants that the "assertion sign" can be used to distinguish assertions from questions, fictions or assumptions, but it is a mistake to think of assertion as composed of two separate acts, one of which is represented by the judgment stroke.

Dummett made no bones about his take on these considerations: "The confused objection of *Philosophical Investigations*, §22, is not to the point" (Dummett 1991: 247). And many Frege scholars seem to share this negative assessment. For one thing, Wittgenstein still seems to slide over the complexity of the assertion sign as well as the fact that for Frege it is not "assumptions" but *thoughts* that are put forward as true in judgments and assertions.[2] For another, it has been objected that Wittgenstein's criticism misfires because it completely ignores the *normative* dimension judgment. Frege's notion of judgment is essentially normative, since the judgments he is dealing with are made on the grounds of other judgments in accordance with logical laws as the "guiding principles for thought" (Frege 1964: xv). Against this background, the judgment stroke is anything but superfluous, since far from indicating what anyone holds to be true, it indicates what everyone *should* acknowledge as true.[3]

Can Wittgenstein's objection be defused by highlighting the link between Frege's notion of judgment and norms for logical inference? Does Wittgenstein

---

[1] Wittgenstein's hostility echoes in a letter from Philip Jourdain, who asks Frege for permission to publish some passages from *Grundgesetze* in *The Monist*, assuring him that Wittgenstein has agreed to check the translation: "Also, will you tell me, […] whether you now regard assertion (⊢—) as merely psychological" (Frege 1980: 78).

[2] This has given rise to the suspicion that Wittgenstein targets Russell's Frege, rather than Gottlob Frege, since Russell portrays Frege as thinking that judgments consist of assumptions (Russell 1903 §477); see Anscombe 1959: 105.

[3] Among the interpreters who try to dissolve the tension between the logical role of Frege's judgment stroke and the threat of psychologism in valuable ways are Smith 2000, Greimann 2000, Taschek 2008, Textor 2010, Pedrali 2017 and van der Schaar 2018.

really complain about a corrupting psychological element in Frege's conception of logic? The fact that he doesn't sound particularly hostile when he claims that assertion is merely psychological suggests that the core of his objection doesn't concern psychologism at all. Admittedly, one easily gets this impression when reading §22 of the *Investigations* in the light of Wittgenstein's negative remark in the *Tractatus*. However, what Wittgenstein, early and late, reject is Frege's idea that "the assertion consists of two acts, entertaining and asserting" (PI §22). Defending Frege against this charge is more difficult than showing that the judgment stroke is not logically superfluous. In order to avoid barking up the wrong tree, therefore, I suggest getting a proper understanding of Wittgenstein's criticism by reading his objection in the *Tractatus* in the light of §22 of the *Investigations*. Before that, however, we need to define clearly the proper target of Wittgenstein's criticism, viz. Frege's view of assertion.

2 A Problem for Frege's Conception of Assertion

For Frege, the distinction between thought, judgment and assertion is crucial. He famously distinguishes between "the grasp of a thought – thinking, the acknowledgement of the truth of a thought – the act of judgment, the manifestation of this judgment – assertion" (Frege 1918a: 62). The difference between the content of a judgment and the acknowledgment of its truth is logically relevant and thus has to be expressed in *Begriffsschrift*, in which "everything that is necessary for a valid inference is fully expressed" (Frege 1879 §3). Frege uses a horizontal line to express that a content is "judgeable" (—— A), and he draws a vertical line at the left end of the horizontal line to express the recognition of its truth (⊢—— A). In order to paraphrase the difference in natural language, Frege suggests reading the former as "the circumstance that A" and the latter as "the circumstance that A is a fact". The complex symbol "⊢——" is something like "the common predicate for all judgments" (Frege 1879 §3).[4]

In order to explain the difference between content and judgment, Frege makes essential use of *nominalizations*. For example, the content of the judgment that Archimedes was killed at the capture of Syracuse is expressed as "the violent death of Archimedes at the capture of Syracuse", or some other nominalization that goes

---

[4] The comparison must be taken with a pinch of salt, since Frege dispenses with the grammatical distinction between subject and predicate, distinguishing instead between function and argument. Taken at his word, the "single predicate for all judgments, namely 'is a fact'" (Frege 1879 §3) would turn out to be logically irrelevant.

with "...is a fact", and makes it clear that we are not yet dealing with a judgment. However, this early conception of judgment is exposed to two objections. First, the distinction between assertables and unassertables seems to be *ad hoc*; e.g., nominalizations such as "the death of Cesar" can be asserted, but nouns such as "house", "the number 2" as well as propositions involving vague concepts can't, as they don't express assertable contents. As we will see later, this problem does not arise on Frege's mature conception of the content of a judgment.

Second, and more importantly, Frege faces a grammatical *dilemma*: If the content of a judgment is to be expressed by the use of a nominalization, then the content is nothing that can strictly speaking be judged. Grammatically, nominalizations introduced with "the circumstance that..." function like names or other noun-phrases whose function is referential rather than expressive. If, on the other hand, the content of a judgment is not nominalized but paraphrased with a sentence in the indicative mood, then the content is assertoric and there is no point in adding the judgment stroke. Therefore, the content of a judgment is either nominalized and hence unfit for being the subject of a possible judgment, or it is in the indicative mood and thereby steals the judgment stroke's thunder. To be clear, the cause of this predicament resides in Frege's proposal on how to paraphrase the difference between content and judgment. Frege's development of his semantic theory of sentences and names in the early 1890s should allow him to cope with this problem too.

According to Wittgenstein, however, the dilemma just outlined is not merely a problem of the paraphrase, but rather a problem of what is so paraphrased. Immediately after his misleading remark about Frege's opinion that every assertion contains an assumption, he seems to allude to an argument that runs parallel to Frege's dilemma:

> But "that such-and-such is the case" is not a sentence in our language – it is not yet a move in the language game. And if I write, not "It is asserted that...", but "It is asserted: such-and-such is the case", the words "It is asserted" simply become superfluous. (PI §22)

Although Wittgenstein is not strictly targeting Frege's nominalizing device "the circumstance that...", the structure of his argument corresponds to what I called the grammatical dilemma above. If the content of a judgment or of an assertion is to be isolatable, such as an assumption that we agree or disagree with, then the content needs to be expressed in a sentence that is either true or false. But neither "that p" nor "the circumstance that p" meets this requirement, as with these phrases one makes no move in the language game; i.e., they can be used as parts of utterances, but one doesn't perform a speech act when expressing them in isolation. Remedying this deficiency means understanding the content assertorically, which ipso facto makes the judgment stroke redundant.[5]

According to Baker and Hacker, this predicament already shows that Frege's conception of judgment and assertion is incoherent:

> Any attempt to represent Frege's claim that every assertion contains an assumption by transformations permissible in language is thus subject to contradictory demands. For the linguistic expression of the contained assumption must both be, and not be, a sentence. (Baker/Hacker 2005: 80)

However, such general verdict might be premature, as it is essentially Frege's early conception of a judgment that causes all the trouble, and so the question naturally arises whether Frege's mature conception avoids the dilemma.

3 Does Semantics Come to the Rescue?

With the discovery of the sense-reference distinction and the prior extension of the notion of a *function*, Frege modifies his *Begriffsschrift* in a way that also affects his conception of judgment and assertion. The "horizontal" now represents a truth function, the value of which is the True if the argument is true, and the False in all other cases (see Frege 1891: 21; Frege 1964 §5). Thanks to this function, Frege no longer needs to stipulate that the content of a judgment is assertable: "—— x" already represents something that can be judged for any meaningful instance of x. Consequently, "⊢ 2" expresses a judgment just as "⊢ 3 > 2" does.

As is well known, a function whose value is always a truth value is for Frege a *concept*. Since the horizontal stands for such a function, the question arises as to *which* concept it represents? There is no agreement among Frege scholars on this point. Some have suggested reading "—— x" in the sense of "x is identical to the True".[6] The merits of this reading are obvious: as a relational concept under which nothing but the True falls, the concept represented by the horizontal applies equally to assertables and unassertables. However, the proposal faces the problem

---

[5] Just to be clear, Frege can't possibly agree to paraphrasing "⊢ p" with "it is asserted: p", as one is entitled to apply the judgment stroke only to *true* propositions. Regardless of whether truth is a norm for assertions, contrary to "⊢ p", "it is asserted: p" does not entail the truth of p (see Künne 2009b: 337). Yet, as for the distinction between content and judgment, the tension is real, since Frege makes demands that are difficult to reconcile. The plausibility of Frege's distinction depends crucially on what is "contained" in judgment and assertion.

[6] See Walker 1965: 132 and Noonan 2001: 150 for explicit statements of this suggestion; for a critical but nevertheless approving discussion of the proposal see Greimann 2000: 232.

that "3 > 2 is identical to the True" is both ungrammatical and assertoric. In terms of grammar, it won't help to opt for a metalinguistic alternative, such as "'3 > 2' is identical to the True", since the expression "3 > 2" in this reading does not refer to the True but to itself, and hence the value of the horizontal would be the False, not the True. Moreover, the assertoric mode of the metalinguistic alternative preempts the role of the judgment stroke.[7]

In order to overcome this second problem, David Bell has suggested reading "—— x" as corresponding to the complex noun phrase "x's being identical to the True" (Bell 1979: 23). Thus nominalized, the horizontal can be equally applied to names and sentences without assertoric import, but in the latter case the resulting expression still sounds grammatically odd. For how are we to understand the judgment that 3 is greater than 2, or that Caesar is dead? According to Bell's proposal, "Caesar's death's being identical to the True is a fact" expresses a true judgment stating identity between Caesar's death and the True. But since truth values are abstract objects, Caesar's death should also be an abstract object, and perhaps even more controversially, it should be the *same* abstract object as the True. To stay within Frege's framework, one can say that "Caesar's death" *refers* to the True, just as any other true sentence does (via its sense), but reference and identity are not one and the same.

Apart from this difficulty, Bell's proposal departs from Frege's language use. In a letter to Husserl, he explicitly says how the horizontal of the modified *Begriffsschrift* is to be read: "Instead of speaking of a 'circumstance', one should speak of a 'truth-value'" (Frege 1980: 64). Unfortunately, the key formulation is lost in the English translation, since Frege gives precise instructions on *how* to speak of a truth value: "Wahrheitswert davon, dass" (Frege 1976: 98). The definite description "the truth value of (that) x" is applicable to assertables as well as unassertables without assertoric import and without infringing on grammar. Hence, Frege's mature semantic theory seems to provide the resources necessary to avoid the quandaries related to his early conception of judgment and assertion.

Wittgenstein's charge, however, is not completely settled, as will be shown in the next section. But before examining whether Frege's semantic conception of judgment and assertion is coherent, let me just highlight an immediate consequence of his modified account. If the content of a judgment is *referential*, insofar as it is referring to a truth value, then a judgment is by its very nature something linguistic, since the reference relation essentially holds between a linguistic object and something else. Bluntly put, therefore, a judgment for Frege is always

---

[7] Heck/Lycan 1979 conclude from this that it is impossible to determine which concept is represented by the horizontal.

about a name's reference to the True. This may be acceptable for the judgments made within a framework such as the *Grundgesetze*, but how does this referential conception work for judgments that are not put into writing?[8]

4 Wittgenstein's Criticism in the *Tractatus*

On Frege's mature conception, assertion essentially involves reference to a truth value. When introducing the horizontal and the judgment stroke in *Grundgesetze*, he explicitly states that the part of a "proposition of Begriffsschrift" (*Begriffsschriftsatz*) that determines the judgment's content "simply designates a truth-value, without saying which of the two it is" (Frege 1964 §5). He continues to say that "we therefore need a special sign to be able to assert something as true". Thus, it falls to the judgment stroke to say which truth value is denoted by the rest of a Begriffsschriftsatz. As I read him, Wittgenstein opposes this division of labor both in the *Investigations* and in the *Tractatus* (as well as in *Notes on Logic*). For a proper understanding of his objection, however, one has to look at the earlier of these writings too, since in *Investigations* he only presents the diagnostics of the mistake:

> Of course, one has the right to use an assertion sign in contrast with a question mark [...] It is a mistake only if one thinks that the assertion consists of two acts, entertaining and asserting (assigning a truth-value, or something of the kind), and that in performing these acts we follow the sentence sign by sign roughly as we sing from sheet music. (PI §22)

Wittgenstein seems to grant the use of an assertion sign (⊢—) for the purpose of contrasting assertions with other speech acts such as questions and demands. However, the mistake to which he wants to draw our attention distinctively in connection with Frege is to think that this contrastive sign would represent the performance of *two separate* acts. Since Fregean assertion and judgment essentially involve the act of referring to a truth value, he seems to be guilty of making this mistake.

Wittgenstein does not, however, say in this passage *why* the two-stage model of assertion is mistaken. The model could be rejected simply on the intuitive ground

---

[8] It has been argued that "judging that *p* is attempting to refer to the True, by thinking that *p*" (Heck/May 2007: 19). It seems to me that one can maintain the Fregean spirit of this general proposal only if one is prepared to accept the controversial claim that thought presupposes language, since Fregean reference is essentially a relation between linguistic signs and their denotation.

that in making a judgment or an assertion one is not doing *two* things in a row, as one reads note after note when singing from a score.⁹ But Wittgenstein has a stronger objection to Frege's model of judgment and assertion, although it occurs elsewhere and Wittgenstein obviously feels no need to repeat it. The decisive argument can be found in the *Tractatus*, where Wittgenstein criticizes Frege's notion of truth with a comparison:

> Imagine a black spot on white paper: you can describe the spot by saying for each point on the sheet, whether it is black or white. To the fact that a point is black there corresponds a positive fact, and to the facts that a point is white (not black), a negative fact. If I designate a point on the sheet (a truth-value according to Frege), then this corresponds to the supposition that is put forward for judgment, etc. etc. (TLP 4.063; see also NL B10)

The thought is that a random black stain on white paper representing a totality of facts can be completely described by, for example, indicating whether each spot is black or white by means of Cartesian coordinates. In this analogy, each point on the sheet corresponds to a Fregean truth value, and pointing to a particular spot corresponds to a Fregean supposition (*Annahme*). Just as one can point to the color of, say, J9, so one can refer to the truth value of p. Moreover, just as the reference to J9 is a substantial component of the judgment that this particular spot is black, so reference to the truth value of p is a substantial component (on Frege's terms) of the judgment that p is true. So according to this comparison, "— p" refers to a truth value without telling whether it is the True or the False, just as "the color of x" refers to a color without telling whether it is black or white.¹⁰

However, Wittgenstein thinks that the analogy breaks down because referring to a truth value is not relevantly similar to pointing at color stains:

> But in order to be able to say that a point is black or white, I must first know when a point is called black, and when white: in order to be able to say, "'p' is true (or false)", I must have

---

9 The two-stage model of assertion is not a strawman's position and surfaces in many of Frege's characterizations. He sometimes describes judgment in terms of "taking steps" (Frege 1892: 34), "advances from a thought to a truth-value" (Frege 1892: 35), or making a "choice between opposite thoughts" (Frege 1979: 198). The literal interpretation of these characterizations is critically discussed in Stepanians 1998.

10 Note that "the truth value of p is the True" is an identity statement with definite descriptions on both sides; accordingly, the corresponding sentence in Wittgenstein's analogy would have to be "the color of spot x is the color black". Otherwise, the comparison would not make sense, since truth values are *objects* and colors are *properties*. Whether it is plausible to use "the color of x" as a referring device that parallels the horizontal's reference to a truth value seems to be more problematic and will be discussed below.

determined in what circumstances I call "*p*" true, and in so doing I determine the sense of the proposition.
Now the point where the simile breaks down is this: we can indicate a point on the paper even if we do not know what black and white are, but if a proposition has no sense, nothing corresponds to it, since it does not designate a thing (a truth-value) which might have properties called "false" or "true". The verb of a proposition is not "is true" or "is false", as Frege thought: rather, that which "is true" must already contain the verb. (TLP 4.063; see also NL B10)

This passage is rich and notoriously difficult to understand, partly because it contains some elements that do not fit the Fregean picture at all.[11] Although Frege's notion of truth is the declared target of the whole section, it remains unclear, for example, whether Wittgenstein's argument builds upon Fregean or Tractarian *sense*. Furthermore, according to the conclusion of Wittgenstein's argument, Frege allegedly took "is true" (and "is false") to be the verb of a proposition, thus imposing a *predicational* conception of judgment and assertion that Frege couldn't possibly accept.[12] Regardless of these incongruities, I am going to suggest a reading of this section according to which Wittgenstein raises a serious objection to the two-stage model of judgment and assertion. In my interpretation, the middle section of 4.063 is a straightforward continuation of the analogy because it makes explicit the similarity between the statement that a particular thought is true and the statement that a particular color patch is black; i.e., one has to know the conditions of application for expressions such as "true" and "black".[13]

Having emphasized the similarity between the color case and the semantic case at the level of judgment and assertion, Wittgenstein goes on to explain why the analogy breaks down. One can point to a particular spot on the paper – either

---

[11] Unfortunately, many commentators end up rephrasing this passage instead of elucidating it; noteable exceptions are Proops 1997: 129ff., Ricketts 2002: 239ff., and Potter 2009: 89ff.
[12] For the difference between Fregean and Tractarian sense, see Künne 2009a: 45ff. and Hacker 2001: 206f.; for the second supposition, see Proops 1997: 131. I will give reasons below why the predicational conception of judgment is not acceptable for Frege.
[13] In this respect, I deviate from Proops, who argues that this paragraph "is not a continuation of the analogy", but "presents Wittgenstein's own views about what it is to have a grasp of the notion of truth" (Proops 1997: 131). According to Proops, "to sustain the analogy, truth and falsity would have to be applicable to truth-values, not propositions" (Proops 1997: 143). Yet, this is not a result of sustaining the analogy but a result of misconceiving judgment and assertion as *predicating* truth. In my reading, *at the level of judgment and assertion* – and that is what the middle section is about – the color case *is* similar to the semantic case, regardless of how truth attaches to thought. Frege can confidently accept what Wittgenstein says in the middle section: one cannot *judge* a proposition to be true without determining its sense. It is *at the level of thought* that the analogy breaks down, because designating a truth value is in relevant respects not like pointing to a color patch.

ostensively or by using Cartesian coordinates – without knowing the application conditions for such expressions as "black" and "white". But one cannot designate a truth value by the use of a sentence without knowing the application conditions for such expressions as "true" and "false". The pointing device in the semantic case may be a sentence or a definite description such as "the truth value of x"; either way, one cannot make use of the device without a prior understanding of what the device is supposed to refer to. This is in stark contrast to the use of a pointing device in the color case, because one can make use of a finger or of coordinates without knowing anything of the colors of the point thus indicated. Wittgenstein explains the dissimilarity by alluding to *some* notion of *sense*: the target in the semantic case (truth) surfaces in the requirements for semantic pointing, insofar as the use of a sentence, for whatever purpose, cannot be detached from grasping the thought expressed, and grasping the thought is grasping the truth conditions of the sentence. As I read him, Wittgenstein is not saying that grasping the truth conditions of a sentence is knowing *whether* the sentence is true, for this would obviously forestall the point of judgment and assertion. He seems to make the more subtle observation that by referring to a truth value with a sentence one has to make use of the notion of *truth* as it occurs in judgment and assertion, because one has to know that the sentence is either true or false. This is where the semantic case differs from the color case, as pointing to a specific spot of a stain can be done without knowing that it is either black or white.[14]

5 Drawing the Right Conclusion

In this last section, I want to discuss the conclusion to be drawn from this argument. On the one hand, as Künne (Künne 2009a: 57) and others have pointed out, Wittgenstein's official conclusion is indeed bewildering, since Frege never said that "is true" is the verb of the proposition. On the other hand, if the analogy only

---

[14] Thus, the argument does not necessarily presuppose Wittgenstein's notion of *sense*, as it seems to be equally valid for logical tautologies, which characteristically lack Tractarian sense. When grasping the truth conditions of, say, "—— $(c \to (b \to a)) \to ((c \to b) \to (c \to a))$", one is computing a large number of conditionals (including a case differentiation) regarding the truth and falsity of the whole sentence with respect to the truth or falsity of its parts. For these computations one has to make tentative use of the notion of *truth* as it occurs in judgment and assertion. However, the argument does presuppose that by simply writing down a well-formed formula preceded by the horizontal one is not referring to a truth value. But this assumption is compatible with Frege's demand that it must be possible to express a thought without acknowledging its truth, for by merely writing down a sentence one has not yet grasped the thought it expresses.

stresses the fact that judgment and assertion involve grasping truth conditions, then it is hard to see why this is an objection to Frege. In short, Wittgenstein's criticism is either unjustified or inconsequential, or so it seems.

Let me begin with the first half of the lesson that Wittgenstein officially wants us to draw: "The verb of a proposition is not 'is true' or 'is false', as Frege thought[.]" This conclusion will strike Fregeans as puzzling, since Frege almost always insists that judgment and assertion are *not* predications of truth (the only exception is in *Begriffsschrift* §3; see my footnote 4 above). He famously observes that "the thought that 5 is a prime number is true" contains the same thought as "5 is a prime number", and that the relation of the thought to the True may therefore not be compared with that of subject to predicate (Frege 1892: 34). So he would oppose at least the first part of Wittgenstein's conclusion by pointing to the *redundancy* of the truth predicate. Moreover, Frege not only thinks that the truth predicate so used contributes nothing to a thought; he also offers a compelling argument against the mistaken conception of judgment and assertion as *predications* of truth:

> By combining subject and predicate, one reaches only a thought, never passes from a sense to its *Bedeutung*, never from a thought to its truth-value. One moves at the same level but never advances from one level to the next. (Frege 1892: 35)

This rules out the predicational view of judgment and assertion that Wittgenstein allegedly ascribes to Frege. Judgment and assertion cannot consist in predicating "is true" of a thought, for the result of combining thought and truth in terms of predication yields just another, and more complex, thought. As an account of judgment and assertion, the predicational view rather amounts to an *infinite regress* at the level of thought than of judgments and assertions, which are at another level to stay with Frege's picture (cf. Textor 2010: 637f.).

Wittgenstein's misrepresentation of Frege's views harbors the danger of concealing the second part of the lesson to be drawn from the analogy: "that which 'is true' must already contain the verb". The claim is not that the truth predicate is the verb of the proposition, but rather that that to which such a predicate applies, whether redundant or not, must already contain *some* verb. If this is the conclusion that follows from the analogy, then it must stand independently of the failed prelude. Whatever "is true" contributes, it cannot make its contribution to something that doesn't already contain a verb and thereby is assertable. Regarding the dilemma that arises from paraphrasing Frege's judgment stroke (see section 2 above), Wittgenstein seems to be willing to accept one of the alternatives, namely that the content of an assertion is assertoric because of the verb. However, Frege cannot agree to this conclusion, as it confuses predicating with judging (cf. Frege 1979: 185). Therefore, if it follows from the analogy that the content of the assertion is assertoric, Wittgenstein seems to have a point which seriously threatens Frege's conception of judgment and assertion, and which is valid regardless of Wittgenstein's unfortunate portrayal.

To reiterate, Frege wants to drive a wedge between merely grasping a thought on the one hand and acknowledging its truth on the other. What is sometimes characterized informally as temporally distinct acts (cf. Frege 1918b: 151; Frege 1979: 7, 138) is formally represented by symbols depicting the logical relation between them. The representation of an overt speech act of assertion and of its silent counterpart, a judgment, incorporates the representation of an act whose performance is logically independent of the first type of act. Just as "—— p" is a graphical component of "⊢—— p" that has an independent meaning, so the act of grasping a thought, of referring to a truth value, is a component of judgment and assertion that can be performed independently of these latter acts. It is precisely this 'logical anatomy' that Wittgenstein's analogy addresses, since it questions the logical autonomy of truth value reference in Frege's two-stage model of judgment and assertion. As a separate act one should be able to perform it without performing the other.

For Frege, judgment and assertion are composed of two separate acts, represented by the horizontal and the judgment strokes. This makes it comparable to the two-stage process of pointing to a particular spot on a piece of paper and telling what color it is. But whereas the color has no bearing on the autonomous act of pointing at the spot, the act of designating a truth value doesn't have this autonomy. For one can only refer to a truth value by means of a proposition that is either true or false, that is, by the use of a vehicle containing a verb. According to Wittgenstein then, the two-stage model of judgment and assertion is mistaken because it conceives of reference to a truth value as a separate act on which judgment and assertion have no bearing. If truth value designation is a component of making a judgment, but cannot be described independently from judgment and assertion, then Frege's attempt to drive a wedge between designating a truth value (—— p) on the one hand, and the judgment that the thing so designated is identical with the True (⊢—— p), will not succeed. Therefore, it is a mistake to think of judgment and assertion as containing a separate act of reference to a truth value.

By way of conclusion, I shall briefly respond to an objection that has been raised by one of Frege's most insightful scholars. In his discussion of Wittgenstein's analogy, Wolfgang Künne (Künne 2009a: 55ff.) suggests reading *Begriffsschriftsätze* as pairings of sentence questions and affirmative answers. The idea is taken from Frege's remarks about questions as a form of words that can be used to express a truth without asserting it (Frege 1918a: 62, Frege 1918b: 143 – 147), and it is launched against Hacker's negative verdict that there is no such form of words (Hacker 2001: 211). Künne realizes that for grammatical reasons we cannot simply paraphrase the judgment that the Earth moves as "Is the Earth moves identical with the true? Yes", and that some kind of nominalization is needed. His proposal is to parse the judgment as follows: "Is the truth-value of the thought that the Earth moves identical with the True? Yes." I consider it to the merit of this proposal that the paraphrase for the judgment stroke (Yes) applies to something containing a verb (moves) without rendering the content of the judgment assertoric, as it is embedded in the wordy nominalization "the truth-value of the thought that …".

However, I see no way of reconciling this proposal with Frege's function-theoretic interpretation of the horizontal, according to which simple nouns can also be used as arguments (see section 3). Does it make sense to ask whether the truth value of the thought that 2 is identical with the True, if there is no such thing as the thought that 2? The hesitation at this point could be an indication that Wittgenstein is not wrong in claiming that assertables should contain a verb – the grammatical dilemma is hard to overcome. Apart from technical sophistry, Künne seems to be glossing over Frege's remark that "a judgement is often preceded by questions" (Frege 1976: 7), because according to Künne's own proposal, Frege should rather have made the general claim that judgments are *always* preceded by questions. However, if my interpretation of Wittgenstein's criticism is conclusive, then Künne's proposal seems to be grist to Wittgenstein's mill. Not only is Künne's analysis of judgments as pairings of questions and affirmative answers a clear manifestation of the two-stage model in terms of two separate speech acts; it also demonstrates how the speech act of asking a question already draws on the notion of truth.

Bibliography

Anscombe, G.E.M. (1959): *An Introduction to Wittgenstein's Tractatus*. 2nd edition. New York: Harper & Row.
Baker, Gordon and Hacker, P.M.S. (2005): *Wittgenstein: Understanding and Meaning II: Exegesis §§1 – 184*. Oxford: Wiley-Blackwell.
Beaney, Michael (ed.) (1997): *The Frege Reader*. Oxford: Blackwell.
Bell, David (1979): *Frege's Theory of Judgement*. Oxford: Oxford University Press.
Dummett, Michael (1991): *Frege and Other Philosophers*. Oxford: Oxford University Press.
Frege, Gottlob (1879): *Begriffsschrift: Eine der Arithmetischen nachgebildete Formelsprache des reinen Denkens*. Halle/S.: Louis Nebert. In: Beaney 1997.
Frege, Gottlob (1891): "Function and Concept." In: Beaney 1997.
Frege, Gottlob (1892): "On *Sinn* and *Bedeutung*." In: Beaney 1997.
Frege, Gottlob (1918a): "Thought." In: Beaney 1997.
Frege, Gottlob (1918b): "Negation." In: Beaney 1997.
Frege, Gottlob (1964): *The Basic Laws of Arithmetic*. Furth, Montgomery (ed.), Berkeley: University of California Press.

Frege, Gottlob (1976): *Wissenschaftlicher Briefwechsel*. Gabriel, Gottfried; Hermes, Hans; Kambartel, Friedrich; Thiel, Christian; Veraart, Albert (eds.). Hamburg: Meiner.

Frege, Gottlob (1979): *Posthumous Writings*. Hermes, Hans; Kambartel, Friedrich; Kaulbach, Friedrich (eds.), Long, Peter; White, Roger (trans.). Oxford: Blackwell.

Frege, Gottlob (1980): *Philosophical and Mathematical Correspondence*. Gabriel, Gottfried; Hermes, Hans; Kambartel, Friedrich; Thiel, Christian; Veraart, Albert; McGuinness, Brian F. (eds.), Kaal, Hans (trans.). Oxford: Blackwell.

Greimann, Dirk (2000): "The Judgment-Stroke as a Truth-Operator: A New Interpretation of the Logical Form of Sentences in Frege's Scientific Language." In: *Erkenntnis*. Vol. 52, 213 – 238.

Hacker, P.M.S. (2001): *Wittgenstein: Connections and Controversies*. Oxford: Oxford University Press.

Heck, William C. and Lycan, William G. (1979): "Frege's Horizontal." In: *Canadian Journal of Philosophy*. Vol. 9, 479 – 492.

Heck, Richard G. and May, Robert (2007): "Frege's Contribution to Philosophy of Language." In: *The Oxford Handbook of Philosophy of Language*. Lepore, Ernest; Smith, Barry (eds.). Oxford: Oxford University Press, 3 – 39.

Künne, Wolfgang (2009a): "Wittgenstein and Frege's *Logical Investigations*." In: *Wittgenstein and Analytic Philosophy: Essays for P.M.S. Hacker*. Glock, Hans-Johann; Hyman, John (eds.). Oxford: Oxford University Press, 26 – 62.

Künne, Wolfgang (2009b): *Die Philosophische Logik Gottlob Freges: Ein Kommentar*. Frankfurt: Klostermann.

Noonan, Harold W. (2001): *Frege: A Critical Introduction*. Oxford: Polity Press.

Pedriali, Walter B. (2017): "The Logical Significance of Assertion." In: *Journal for the History of Analytical Philosophy*. Vol. 6, No. 8, 1 – 22.

Potter, Michael (2009): *Wittgenstein's Notes on Logic*. Oxford: Oxford University Press.

Proops, Ian (1997): "The Early Wittgenstein on Logical Assertion." In: *Philosophical Topics*. Vol. 25, No. 2, 121 – 144.

Ricketts, Thomas (2002): "Wittgenstein Against Frege and Russell." In: *From Frege to Wittgenstein*. Reck, Erich (ed.). Oxford: Oxford University Press, 227 – 251.

Russell, Bertrand (1903), *Principles of Mathematics*. 2nd edition. London: Allen and Unwin.

Smith, Nicholas, J.J. (2000): "Frege's Judgement Stroke." In: *Australasian Journal of Philosophy*. Vol. 78, No. 2, 153 – 175.

Stepanians, Markus S. (1998): *Frege und Husserl über Urteilen und Denken*. Paderborn: Schöningh.

Taschek, William W. (2008): "Truth, Assertion, and the Horizontal: Frege on the 'Essence of Logic'." In: *Mind*. Vol. 117, No. 466, 375 – 401.

Textor, Mark (2010): "Frege on Judging as Acknowledging the Truth." In: *Mind*. Vol. 119, No. 475, 615 – 655.

van der Schaar, Maria (2018): "Frege on Judgement and the Judging Agent." In: *Mind*. Vol. 127, No. 505, 225 – 250.

Walker, Jeremy D.B. (1965): *A Study of Frege*. London: Oxford University Press.

Wittgenstein, Ludwig (1913): *Notes on Logic*. In: Potter 2009.

Wittgenstein, Ludwig (1961): *Tractatus Logico-Philosophicus*. Pears, David F.; McGuinness, Brian F. (eds.). London: Routledge & Kegan Paul.

Wittgenstein, Ludwig (2009): *Philosophical Investigations*. Hacker, P.M.S.; Schulte, Joachim (eds.), 4th edition. Oxford: Wiley-Blackwell.

Maria van der Schaar
Assertions and Their Justification: Demonstration and Self-Evidence

Abstract: In Frege's epistemic account of logic, the notions assertion, justification and being evident play a central role. Although the notion of judging agent plays an important role in the explanation of these notions, this does not mean that Frege's logic is committed to a form of psychologism. How can we use Frege's account of these notions to illuminate the notions of demonstration and being evident in Constructive Type Theory (CTT)? As the judging agent also plays a role in CTT, how can it prevent a form of psychologism? Although the notion of demonstration cannot be understood without invoking a judging agent, such a judging agent is a first person, which is not to be understood as an empirical subject. And similarly for being evident. The latter notion is often taken to imply a form of psychologism. Although the appeal to the notion of being evident involves a form of fallibilism, the notion is normative, and therefore not psychological. It can thus be used to account for a justification of the inference rules.

1 Introduction

In Constructive Type Theory (CTT), epistemic notions like judgement and demonstration, play a central role. Is it possible to give an epistemic foundation of logic, without being committed to a form of psychologism? Frege's epistemic account of logic (section 2) is taken as a starting-point to see whether he is able to prevent a form of psychologism, given the central role of the judgement stroke, the judging agent, justification and being evident in his logic. In section 3, the tradition notion of judgement is related to modern accounts of assertion, and it turns out that both judgement and assertion can be understood in epistemic terms. Both in Frege's writings and in CTT, the act of demonstration, a special kind of act of judgement, plays a central role. Can we give a non-psychological account of such acts? Finally, in section 4, we have to appeal to the idea that some of our judgements and inference rules are self-evident. Although most readers of Frege think of self-evidence as a property of Thoughts, it is rather the *being evident* of the truth of a thought *to* a certain agent that is relevant to Frege's writings and to CTT. I will therefore speak of *being (self-)evident* rather than of *self-evidence*. As claims to being evident

---

Maria van der Schaar, Institute for Philosophy, Leiden University, The Netherlands, e-mail: m.v.d.schaar@phil.leidenuniv.nl

DOI 10.1515/9783110218091-12

are fallible, Wittgenstein seems to be right in his criticism of Frege: "If, from a proposition being evident to us, it does not follow that it is true, then its being evident is also not a justification for our belief in its truth." (TLP 5.1363, translation by R.H. Schmitt). Is it possible to deal with the notion of being evident in a way that is not psychologistic?

2 Frege on Judgement, Demonstration and Being Self-Evident

In Frege's logic, the threat of psychologism is not only introduced by the notion of justification (*Rechtfertigung*), but also by the mere presence of the judgement stroke. It seems that Wittgenstein is right when he says that the judgement stroke in Frege's ideography only shows that Frege holds the relevant proposition to be true (TLP 4.442). Frege himself seems to confirm this reading: "With this judgement stroke I close a sentence, [...] and the content of the sentence thus closed I assert as being true by the same sign." (Frege 1896: 232, orig. 377; cf. Frege 1891: orig. 22). The role of the judgement stroke is thus related to the first person in Frege's ideography, but this does not mean that the judgement is about Frege, or a process in the asserter's mind. For, that the content of the judgement is true, is independent of the knowing agent.[1] Within the ideography, the use of the judgement stroke indicates that the content is acknowledged to be true. As truth is independent of the judging agent, the empirical fact that I make the judgement is irrelevant to the truth or falsity of the content. Furthermore, one can fully understand the assertion without knowing who made the assertion. Fully understanding the assertion is rather to make the judgement oneself, as a first person.[2] Essential to logic is the truth claimed in the act of judgement, truth as we use it in our practice of judgements and inferences, for logic aims at the laws of truth. Frege, though, explicitly claims in "My Basic Logical Insights [1915]" that the essence of logic cannot be found in the word "true", but lies in the assertive force (NS: 272; PW: 252). Apart from using the judgement stroke in front of a sentence expressing what we take to be a law, within the ideography, there is no other way for us to say what the laws of truth are. A truth-predicate is not able to do this, for a sentence of the form "so and so is true" need not be

---
[1] "Gewiss ist das Urteilen (das als wahr anerkennen) ein innerer seelischer Vorgang; aber dass etwas wahr ist, ist unabhängig vom Erkennenden, ist objektiv. Wenn ich etwas als wahr behaupte, will ich nicht von mir sprechen, von einem Vorgang in meiner Seele." (WB: 126f.)

[2] The notion of first person is essential for understanding the role of the judgement stroke in Frege's ideography, see (Schaar 2018).

asserted. When Frege has determined that a Thought is a logical law, he writes down a *Begriffsschriftsatz*, an ideographical theorem, which is not a name, but an assertion, and he needs the judgement stroke to indicate that it is.

For his logicist project Frege has not only to define the arithmetical concepts in logical terms. He also has to show that these definitions render the arithmetical theorems demonstrable by logical means alone. Frege's logicist project is avowedly epistemic, for the central question is: What is the epistemic nature of the demonstrated laws of arithmetic (GG: vii)? This question can only be answered when we have made everything in our demonstrations explicit, and, in particular, for good overview have determined a small number of basic laws (*Urgesetze*, GG: vi). In *Die Grundlagen der Arithmetik*, Frege formulates his logicist thesis in terms of a notion of analyticity that is explained in epistemic terms. Whether the judgement is analytic depends on the "justification" of the judgement (*die Berechtigung der Urteilsfällung*, GLA §3). A truth is analytic, if and only if the justification or demonstration (*Beweis*) of our judgement can be given by logical laws, definitions and their known presuppositions, and logical inference alone. The notion of analytic truth in these passages is an epistemic one, because Frege is speaking here of theorems, justification, and judgement.

Does Frege's use of his notion of justification of a judgement constitute a form of psychologism? In the Preface to the *Begriffsschrift*, Frege writes that the question how we arrived at a certain judgement is to be sharply separated from the question how in the end our judgement is most securely to be grounded.[3] Whereas the way we arrive at a certain truth may differ for different people, a grounding justification is in principle the same for every agent making the judgement. With a reference to Leibniz in section 17 of the *Grundlagen*, Frege presupposes a natural order of truths. The justifications of our judgements thus have to correspond to a metaphysical proof-structures in the independent "Third" realm.[4] This forms a contrast with the earlier passage in the *Grundlagen* on analyticity, where Frege gave a purely epistemic account of judgement and justification. In fact, this purely epistemic account accords better with Frege's view that metaphysics follows logic, or as the master puts it himself: "Ich halte es für ein sicheres Anzeichen eines Fehlers wenn die Logik Metaphysik und Psychologie nöthig hat, Wissenschaften, die selber der

---

[3] "Es kann daher einerseits nach dem Wege gefragt werden, auf dem ein Satz allmählich errungen wurde, andrerseits nach der Weise, wie er nun schliesslich am festesten zu begründen ist." (Bs, Preface; cf. GLA §3).

[4] I thank Göran Sundholm for pointing out this tension to me. See also (Sundholm 2011 and Shapiro 2009). Burge 1998 does not see a tension here in Frege's writings, and rather explains the epistemic notions of judgement and justification in terms of objective proof-structures, which are independent of the judging agent, at least if conceived as human agent.

logischen Grundsätze bedürfen." (GG: xix). Neglecting the metaphysical presupposition, one may say that if a demonstration is given for a judgement, each judging agent should be able to follow the inference steps, insofar as he has acknowledged the rules of inference and definitions used. In this sense, these demonstrations are not accidental ways in which *others* may have accidentally come to these truths (GLA §3). These demonstrations do not only have the aim to secure the truth of these conclusions, but also give us an insight into the dependency of truths among each other (GLA §2; cf. NS: 171, 220). As Frege puts it in the *Begriffsschrift*, the way truths depend on each other becomes clear if one demonstrates complex truths from more simple ones (Bs §13). These dependency relations are commonly read in terms of a total ordering of propositions by means of their postulated metaphysical proof-structures, but they can also be read in an epistemic way (cf. An. Post.: 72a and ÜG §1). To see the dependency relations between truths, one has to demonstrate the less general truths by means of more general laws (GLA §4). In the *Grundlagen*, new demonstrations become possible, when the content of the judgement is analysed in a new way, showing itself in new definitions. The more we are able to demonstrate, the fewer truths we have to take to be undemonstrable.

If one would thus have demonstrated the arithmetical laws from a certain group of *basic* laws, the important question remains: What is the epistemic nature of these basic laws? What makes one entitled to use the judgement stroke in front of sentences expressing these basic laws? These most general laws neither can nor need be demonstrated, as Frege repeats the traditional phrase going back at least to Leibniz (GLA §3). But, if they cannot be demonstrated, how can they be known? Such a truth is known insofar as it is immediately evident (*unmittelbar klar, unmittelbar einleuchtend*, GLA §5). Traditionally, one makes a distinction between mediately evident judgements, which can be seen to be true by a demonstration, and immediately evident judgements, which can be seen to be true without the mediation of any further judgement. In the latter case, I will simply call the judgement evident, or (self-)evident. Here, the judgement is made on the basis of a single act, or deed, of knowing (*Erkenntnisthat*, GG: vii). "Immediacy" should not be understood as immediate in time; being obvious can be characterised as being immediate in time, but obviousness is a psychological notion, as we will see below, and cannot be used to elucidate the notion of being (self-)evident. What is the nature of this act of seeing or insight (*das Einleuchten*): is it logical or of an intuitive nature (logisch oder anschaulich, GLA §90)? The crucial question for logicism is whether this act of knowing involves a spatial or time intuition (*Anschauung*), or whether the act is based on a purely conceptual insight. In order to vindicate logicism, the basic laws from which the arithmetical laws are demonstrated have to be logical laws, that is, they have to be knowable by conceptual means alone. And the inference-rules and definitions have to be purely logical, as well.

In the *Begriffsschrift* Frege uses the term "einleuchtend". Axiom I, ⊢ $a \to (b \to a)$: "besagt 'der Fall, wo a verneint, b bejaht und a bejaht wird, ist ausgeschlossen." Dies leuchtet ein, da a nicht zugleich verneint und bejaht werden kann." (Bs §14). Because it can be difficult to grasp its content, Frege explains, first, the function of implication (GG §12), and then shows that on this understanding of implication the first basic logical law can be known to be true, purely on the basis of grasping its content (GG §18). Frege's arguments here can be understood as an unfolding of what implication means. The reader may thus be convinced that the sentence expressing the basic law has to be true, but the understanding part he has to do for himself. And this last epistemic part is essential for an entitlement to use the judgement stroke. We may generalise these remarks on the first axiom by saying that a basic law is a logical law, if and only if one is entitled to judge it purely on the basis of grasping its content.[5]

At the time of writing the *Grundgesetze*, Frege held Basic Law V to be a logical law (GG: vii), and later confesses that it was not as evident (*einleuchtend*, GG: 253, *Nachwort*) as the others. This must mean that he never fully grasped its content, and that he realizes this, now that he knows about Russell's paradox. Writing about his judgement in the past, he is describing a psychological fact about the degrees of perceived clarity. Frege understands that he is not entitled to use the judgement stroke in front of the sentence expressing Basic Law V, and withdraws his former assertion.

Wittgenstein thus seems to be right: "Wenn daraus, dass ein Satz uns einleuchtet, nicht *folgt*, dass er wahr ist, so ist das Einleuchten auch keine Rechtfertigung für unseren Glauben an seine Wahrheit." (TLP 5.1363; the translation can be found in section 1). Is being evident then a psychological notion?

For Frege, being evident insofar as it plays a role in our knowing the basic laws is not a psychological notion. It is neither to be identified with (1) the mental state of conviction; nor with (2) obviousness; and it does not allow for (3) degrees. Concerning (1), Frege writes: "In mathematics one cannot be satisfied with the fact that something is evident (*dass etwas einleuchte*), that one is convinced of something." (NS: 221; compare GLA §90).[6] We are strongly convinced of our prejudices, but they are not known: they are not self-evident. The epistemic value of

---

5 As Frege puts it in his very late paper *Gedankengefüge*, the truth of a basic law of logic is immediately evident on the basis of the sense of its expression: "(weil) die Wahrheit eines logischen [Grund] Gesetzes unmittelbar aus ihm selbst, aus dem Sinne seines Ausdrucks einleuchtet." (Frege 1923: 50)

6 As this passage shows, Frege does not have any technical terminology relating to the notion of being evident; we always have to invoke the context in order to determine the meaning of terms such as "einleuchtend" and "selbstverständlich".

being convinced is derived from our judgement being justified or evident. (2) A truth is obvious if it is patent: we do not need time to make the judgement. The truth that 2 + 3 = 5 is obvious, but not self-evident, for we can give a demonstration for it. Self-evident truths may be obvious, but they may also not be obvious at all. When writing the *Grundgesetze*, Frege did not take Basic Law V to be obvious, but he thought it would be possible to make it evident by conceptual means alone. Finally, (3) as Frege puts it, "Whether the falsity of a Thought is easy or difficult to recognize is not relevant to logic." (Frege 1923: 42). Such degrees may differ for different people, and should be irrelevant when considering the notion of being evident. Just as we cannot speak of degrees of demonstration or knowledge, so we cannot speak of degrees of being evident. Being evident, for Frege, has to be epistemic, and epistemology is to be sharply separated from psychology (Frege 1885: orig. 329). Since being evident is an epistemic notion, it is normative. It is precisely for this reason that we can make mistakes regarding the question whether the truth of a judgeable content (or a Thought) is evident.

The question whether being evident is a psychological notion may be illuminated by relating it to the question how numbers are given to us. The latter is not a question that can be investigated by empirical means, whether of a psychological or other nature. The objects of arithmetic are not given to us as objects external to us (*als etwas Fremdes*) that can only be known by means of the senses; these objects are independent of the individual judger, but at the same time are immediately given to us as reasoning or judging agents (*unmittelbar der Vernunft gegeben*, GLA §105). It is precisely for this reason that the arithmetical laws are not subjective.

For Frege, to say that a Thought is true, or that its truth is demonstrated or self-evident is not predicating a property of that Thought. Its being true is not an empirical phenomenon to be described by means of a predicate. These normative notions rather show themselves in our first-person acts of judgement, inference and demonstration, and in the use of the ideographical judgement stroke. The fact that these acts are fallible does not make them subjective or psychological. For, when we use a judgement stroke in front of a sentence expressing a basic law, we implicitly claim that any judging agent can, in principle, make the truth of the Thought known to himself. Although these notions cannot be understood without invoking the notion of judging agent, this does not make them subjective or psychological.

3 Assertion and Demonstration

The idea that logic is an epistemic project is an old one (Sundholm 2009), although since Wittgenstein's *Tractatus* and the meta-mathematical turn, epistemological questions seem to have been banished from the realm of logic. Swedish Proof-Theory and the Dutch tradition of mathematical intuitionism, though, have retained some elements of the old conception of logic. Today an epistemic view of logic is advocated by Per Martin-Löf in his Constructive Type Theory (CTT), where logic is to comprise the theory of assertion and (epistemic) inference (Martin-Löf 2015).

In a constructivist account of logic, as indeed for Frege, each inference step needs to be knowledge preserving, that is, the act of inference needs to be epistemic. If the premises are actually known, the act of inference is an act of demonstration, that is, an explicitly epistemic notion. Before explaining this point let me first show what the role of judgement is in logic in general, and in CTT in particular, and to what extent judgement is an epistemic notion. As inferences bring us from judgements made to new judgements, the notion of judgement is essential to logic. As there is, from a logical point of view, no distinction between judgement and assertion, I will speak of assertion, instead, because this makes it possible to relate the idea that judgement is relevant for logic to current discussions on assertion.

At first sight, it seems that one is entitled to make an assertion, only if the asserted content is true. This seems to be confirmed by Frege's thesis that judgement is the acknowledgement of the truth of a Thought. A disadvantage of this view is that the asserter is often not able to determine whether the content is true, at least, if truth is understood in realistic terms. It is therefore sometimes proposed that one is entitled to make an assertion if and only if one is sincerely convinced of the truth of its content, if one feels it to be true. The asserter is thus able to determine whether he is entitled to make the assertion. This demand on assertion seems not to be strict enough, for an interlocutor is entitled to ask "How do you know that?", when an assertion is made.[7] When George W. Bush asserted that there are weapons of mass destruction in Iraq, he was, *perhaps*, sincerely convinced of the truth of what he asserted, but this made him not entitled to make the assertion, for he did not know that there were such weapons. Such an account of assertion would make assertion, and thereby judgement, a psychological notion. The logical or epistemic notion of judgement and the psychological notion of belief must be kept apart (see Schaar 2018). Are we then to defend a knowledge account of assertion, as Timothy Williamson 2000 has done? If knowledge entails truth, the same argument that holds against the truth-account of assertion can be raised here. Instead, one may

---

7 See Sundholm 1988.

argue that the "How do you know that?' question can be answered by giving a justification for one's assertion. One may thus adopt a justification account of assertion, which constitutes an epistemic account of assertion and judgement. One is entitled to make an assertion only if one is able to give a justification for it.[8] Within Frege's ideography, it seems that one is entitled to use the judgement stroke only if what is asserted is known or justified, either through an act of demonstration or through an act of immediate insight. The judgement stroke would thus present an implicit knowledge-claim. All assertions can be understood as preceded by "I know that". Such a claim to knowledge is not a part of what is asserted; it is not part of the descriptive content. It has rather a performative function.

What would *count* as a justification for an assertion can be determined by grasping the meaning of the declarative sentence that is standardly used to make the assertion. Understanding a sentence S is knowing its meaning, that is, knowing what justification one must possess in order to be entitled to use an utterance of S with assertive force. Meaning thus determines what counts as a justification.

There are two ways to justify a logical or mathematical thesis: One may justify it by means of an act of demonstration, based on known premises. But in order to have any epistemic value such demonstrations need to have finite length, whence they have to end in first principles. And these first principles have to be justified, too, for the demonstrations to result in knowledge. The question how the first principles are to be justified I address in the next section.

Within Constructive Type Theory, in each act of demonstration a proof-object is constructed that makes the proposition true. A judgement of the form *A is true* is demonstrated if one has constructed a proof-object for the proposition A, and thus has made a judgement of the form $a : Proof(A)$. That is, the proof-object a is a proof for proposition A, where proof-objects are understood as mathematical objects, or *constructions* (see Sundholm 1994). A judgement of the elliptical form *A is true* thus suppresses the constructed proof-object, and is essentially of the form *there exists a proof-object of A*. A proposition is identified with the set of proof-objects that make it true, and is explained by what counts as a canonical proof-object (and by what it is for two such objects to be equal). A judgement of the form $a : A$ can thus be read as proof-object a is element of the set *Proof (A)*. There is an internal relation between the proposition and its proof-objects, as they define the proposition as set. If one has constructed a non-canonical proof-object for the proposition whose truth is to be demonstrated, one thereby possesses a method, or program, for obtaining a canonical proof-object. Thus, when one has constructed a proof-object, be it canonical or non-canonical, in both cases one exhibits a reason

---

[8] I have developed such an account of assertion in Schaar 2011.

why the proposition is true, at least, if one displays the proof-object via a judgement in fully explicit form *a: Proof (A)*.

The distinction between *demonstratio propter quid*, sometimes called "knowledge of the reason why", and *demonstratio quia*, sometimes called "knowledge of the fact", is a distinction at the level of demonstrations, not at the level of ontological proof-objects, as spelled out in Sundholm 2011: 69. The distinction goes back to Aristotle's *Posterior Analytics*, and it translates into CTT as the distinction between the full judgemental form *a is a proof-object for A* and the elliptic form *A is true*. If someone makes an assertion without giving the proof-object, this is a judgement of the form *A is true*. When asked "How do you know that?" he might be giving the proof-object, thereby making the judgement *a is a proof-object for A*, and thus showing that he knows the reason why. If the asserter is not able to give such a ground, he has to withdraw his assertion, for his judgement has turned out to be blind. The judgemental form *A is true* thus derives its epistemic value from the fully explicit form. We can call it knowledge, though only in a derived sense.

The distinction between the ontological notion of proof-object and the epistemic notion of act of demonstration (proof-act) is crucial for understanding that the judging agent plays an important role in Constructive Type Theory. One is entitled to assert a judgement of the form *A is true*, only when a proof-object for *A* is constructed in an act of demonstration. As we cannot construct infinite proof-objects in an act of demonstration, there is nothing that may entitle us to assert that a proposition for which such proof-objects would be needed is true.

As one cannot have a demonstration without someone making the demonstration, the notion of judging agent is presupposed in the notion of demonstration. Does this mean that Constructive Type Theory is committed to a form of psychologism? When I have demonstrated a thesis, that is, have justified my assertion, and have left traces on paper, or on a blackboard, my justification lies open to others. These traces or tracks can be found in mathematical texts, and today we are used to call them proofs, though in another sense then it is used above. They are traces of the act of demonstration given by the writer (Sundholm 2004: 455). By means of the traces left, the reader is able to carry out the demonstration for himself. He has to take care that he knows the premises, and that each step in the demonstration preserves knowledge. In principle, each reader is thus able to demonstrate the theorem in the same way. While there is thus no demonstration without a judging agent, each judging agent can carry out the same demonstration, as soon as he has acknowledged the first principles, definitions and rules of inference.

4 Judgement and Self-Evidence

As we have seen in Frege, the basic laws are known by means of grasping their content alone; these laws are self-evident on logical, that is, conceptual grounds. Besides, on Frege's account, *Modus Ponens* has to be evident to the judging agent in order for him to be entitled to use the rule. The reader of Frege's ideography has to make the truth of the basic laws, and the validity of at least one inference rule evident to himself, for without an epistemic account of the inference rule(s), there will be no demonstration. If we allow for first judgements, we may endorse Frege's account of being evident. But, how can inference rules be justified within Constructive Type Theory? When we want to understand the conjunction introduction rule, we first have to know how a proposition of the form $A \& B$ is explained in terms of its canonical proof(-object).

When a is a proof of A and b is a proof of B, then the ordered pair $< a, b >$ is a canonical proof of $A \& B$ (Martin-Löf 1984: 12).

The rule of &-introduction is then:

$$\frac{a:A \quad b:B}{< a, b >\, :\, A \& B}$$

On the assumption that the premises are known (while assuming that A and B are propositions), and on the basis of the above explanation of conjunction, and with it our grasp of the proposition $A \& B$ in terms of what counts as a canonical proof for it, we see that we are entitled to make the judgement $< a, b >: A \& B$.[9] One has thus made the introduction-rule for conjunction evident to oneself. For someone who does not know yet what conjunction means, we may say, with Gentzen, that the introduction rule gives the meaning of conjunction. In this sense introduction rules can be understood as meaning giving. If someone is not willing to assert the conclusion under the assumption that the premises are known, one will have to say that he has not understood what conjunction is, and that there will be no way to communicate with him. There is no further guarantee that our inference rule is correct, as it is ultimately based on what we take to be self-evident, and thereby on our understanding of the terms involved. Does this mean that, after all, our justification of the inference rules, founded upon the notion of being (self-)evident as they are, implies a form of psychologism?

Robin Jeshion has proposed in her paper "On the Obvious" a distinction between a psychological notion of *obviousness* and an objective property of self-

---

[9] I thank Göran Sundholm for giving me his notes relating to the subject. See also Martin-Löf 1983.

evidence in order to account for the fallibility of our judgements of self-evidence: "Self-evidence is here taken to be the objective correlate of the subjective notion of obviousness" (Jeshion 2000: 354). Her account of self-evidence relates to truths, not to inference rules, but it can be easily extended to the justification of the inference rules proposed here.

According to Jeshion, "A proposition p is obvious to an agent A at time t [...] if and only if at t A finds p true on the basis of her [occurrent] conceptual understanding alone." (Jeshion 2000: 345). Because of the term "A finds p true at t', obviousness is a psychological notion, for finding something true at t is not normative. Even if I now understand that p is not true, I can without contradiction admit that I found p true at a former time t. I am just describing a psychological fact about my past. Jeshion's notion of self-evidence is not explained in psychological terms: "a proposition p is self-evident if and only if understanding the concepts in p provides sufficient and compelling basis for recognition of p's truth." (Jeshion 2000: 354). My criticism of her definition of self-evidence concerns two points. First, self-evidence does not pertain to the proposition as such: it is the truth of a proposition that is evident, as Frege put it. Or, as I prefer to put it, being evident pertains to judgements rather than to propositions. Second, there is a problem with the way Jeshion relates obviousness and self-evidence. According to Jeshion, obviousness can make us "*a priori* justified" (Jeshion 2000: 334). But, a psychological notion can never give us such a justification. In questions of being evident, the A for the agent and the t for the time of judging are irrelevant. These aspects of obviousness show that it is an empirical, psychological notion. And, if obviousness is a psychological notion, how can it provide an entitlement to take the proposition to be self-evident in any objective sense? The notion of obviousness is a *third-person notion*, an empirical factual phenomenon. Obviousness may be relevant when we ask a psychological question, for example: "For how many people is this obvious?". If being evident is at stake, our question is rather of the form: "Is it evident *to me*?" We thus see that being evident is a non-empirical, first person notion. We have to rely on *an epistemic notion of being evident* right from the start.

A good feature of Jeshion's account is that she allows for a role of the judging agent, but such a judging agent should not be understood in any psychological sense. When a judgement or an inference rule is evident to me, I take it that each judging agent is able to make it evident to himself. Being evident is in this sense objective. This notion of objectivity is neither independent of the notion of judging agent, nor does it provide a rock-bottom, for the judging agent can be mistaken. Being evident is essentially related to a first person. For, the fact that the judgement is evident to me gives no one else an entitlement to judge. Here, each agent has to see for himself; the most one can do is to give elucidations of the primitive notions to the reader. As Per Martin-Löf once put it: "there are also certain limits to what verbal explanations can do when it comes to justifying axioms and rules of inference. In the end, everybody must understand for himself." (Martin-Löf 1979: 166).

5 Conclusion

Essential to Frege's logic is the judgement stroke, and thereby the notion of judgement, and Frege is thus able to give an epistemic account of logic. Crucial notions in Frege's ideography, such as demonstration or justification and being evident, cannot be understood without invoking the notion of judgement stroke, and thereby that of the judging agent. As the judgement stroke is a sign of assertive force, it thus becomes clear that the notion of assertion has to play a central role in logic if one understands logic, like Frege, as an epistemic endeavour.

If one aims at an epistemic account of logic, as is done in Constructive Type Theory, one has to answer the question whether such epistemic notions as judgement, demonstration and being evident do not bring in a form of psychologism. This question is more easy to answer in the case of demonstration, because (traces of) demonstrations are found elsewhere in mathematical and logical texts, and each of us is thus able to make the demonstration, which gives demonstration the required objectivity, although it cannot be understood without invoking the notion of judging agent. The question is harder to answer in the case of being (self-)evident, but it needs to be answered because there will be no demonstration without our knowledge of the first principles and our knowledge of the inference rules.

Recently, philosophers like Robin Jeshion and Tyler Burge have proposed to make a distinction between a subjective counterpart to the notion of self-evidence, called *obviousness*, and an objective notion of self-evidence. The problem of this approach is that the psychological notion of obviousness does not give any entitlement to make an assertion, nor can it give us any entitlement to use the inference rules. Here it is argued that the psychological notion of obviousness is not to play a logical role at all. It is precisely because demonstration and self-evidence are normative notions that we can be mistaken: our claims to self-evidence are fallible. These notions cannot make sense without introducing the judging or knowing agent – each knowing agent has to make the first judgements and the inference rules evident to himself. This does not imply a form of psychologism, because the agent does not describe an empirical fact about himself when he claims a judgement or inference rule to be self-evident; it is rather that he thereby implicitly claims that any judging agent can make these judgements or inference rules evident to himself. In this sense, demonstration and self-evidence gain an independence from the empirical subject, although not any independence of the judging agent.[10]

Bibliography

Aristotle (1994): *Posterior Analytics*. Barnes, J. (ed.), second edition. Oxford: Clarendon Press, 1994.

Burge, Tyler (1998): "Frege on Knowing the Foundation." In: *Mind*. Vol. 107, 305 – 347.

Frege, Gottlob (Bs): *Begriffsschrift; eine der arithmetischen nachgebildete Formelsprache des reinen Denkens*. In: *Begriffsschrift und andere Aufsätze*. Angelelli, I. (ed.). Hildesheim: Olms, 1971^2 (1879).

Frege, Gottlob (GLA): *Die Grundlagen der Arithmetik*. Breslau: Koeber, 1884.

Frege, Gottlob (1885): *Das Prinzip der Infinitesimal-Methode und seine Geschichte*, Review of Cohen, H. In: KS, 99 – 102.

Frege, Gottlob (1891): "Funktion und Begriff." In: *Funktion, Begriff, Bedeutung*. Patzig, G. (ed.). Göttingen: Vandenhoeck & Ruprecht, 1980, 18 – 39.

Frege, Gottlob (GG): *Grundgesetze der Arithmetik*. Hildesheim: Georg Olms, 1998 (1893). Facsimile of Jena: Hermann Pohle.

Frege, Gottlob (1896): "Über die Begriffsschrift des Herrn Peano und meine Eigene." In: KS, 220 – 233.

Frege, Gottlob (1923): "Gedankengefüge." In: *Beiträge zur Philosophie des deutschen Idealismus*. Vol. 3, 36 – 51.

Frege, Gottlob (KS): *Kleine Schriften*. Angelelli, I. (ed.). Hildesheim: Olms, 1990^2 (1967), 281 – 323.

Frege, Gottlob (WB): *Wissenschaftliche Briefwechsel*. Gabriel, G. et al. (eds.). Hamburg: Felix Meiner, 1976.

Frege, Gottlob (PW): *Posthumous Writings*. Hermes, H. et al. (eds.), Long, P.; White, R. (trans.). Oxford: Blackwell, 1979.

Frege, Gottlob (NS): *Nachgelassene Schriften*. Hermes, H. et al. (eds.). Hamburg: Felix Meiner, 1983.

Jeshion, Robin (2000): "On the Obvious." In: *Philosophy and Phenomenological Research*. Vol. 40, 333 – 355.

Martin-Löf, Per (1979): "Constructive Mathematics and Computer Programming." In: *Logic, Methodology and Philosophy of Science VI*. Cohen, L.J.; Los, J.; Pfeiffer, H.; Podewski, K.-P. (eds.), Proceedings of the 1979 international congress at Hannover. Amsterdam: North-Holland Publishing Company, 1982, 153 – 175.

Martin-Löf, Per (1983): "On the Meaning of the Logical Constants and the Justifications of the Logical Laws.", lectures held at Siena. In: *Nordic Journal of Philosophical Logic*. Vol. 1, 1996, 11 – 60.

Martin-Löf, Per (1984): *Intuitionistic Type Theory*. Naples: Bibliopolis.

---

[10] I am indebted to Göran Sundholm for extensive comments on two former versions of this paper.

Martin-Löf, Per (2015): "Is Logic Part of Normative Ethics?" Lecture University of Utrecht, 16 April 2015 & Paris, CNRS, 15 May 2015, manuscript.

Schaar, Maria van der (2011): "Assertion and Grounding: A Theory of Assertion for Constructive Type Theory." In: *Synthese*. Vol. 183, 187 – 210.

Schaar, Maria van der (2018): "Frege on Judgement and the Judging Agent." In: *Mind*. Vol. 127, 225 – 250.

Shapiro, Stewart (2009): "We Hold these Truths to be Self-Evident: But what Do We Mean by That?" In: *The Review of Symbolic Logic*. Vol. 2, 175 – 207.

Sundholm, Göran (1988): *Oordeel en Gevolgtrekking. Bedreigde Species?* Inaugural lecture, Leiden.

Sundholm, Göran (1994): "Existence, Proof and Truth-Making: A Perspective on the Intuitionistic Conception of Truth." In: *Topoi*. Vol. 13, 117–126.

Sundholm, Göran (2004): "Antirealism and the Roles of Truth." In:*Handbook of Epistemology*. Niiniluoto, I.; Sintonen, M.; Wolenski, J. (eds.). Dordrecht: Kluwer, 437 – 466.

Sundholm, Göran (2009): "A Century of Judgment and Inference, 1837-1936: Some Strands in the Development of Logic." In: *The Development of Modern Logic*. Haaparanta, L. (ed.). Oxford: Oxford University Press, 263 – 317.

Sundholm, Göran (2011): "A Garden of Grounding Trees." In: *Logic and Knowledge*. Cellucci, C.; Grosholz, E.; Ippoliti, E. (eds.). Newcastle upon Tyne: Cambridge Scholars Publishing, 57 – 74.

Williamson, Timothy (2000): *Knowledge and its Limits*. Oxford: Oxford University Press.

Wittgenstein, Ludwig (TLP): *Tractatus*. In: *Werkausgabe, Band 1*. Frankfurt am Main: Suhrkamp, 1984 (1921).

Wittgenstein, Ludwig (ÜG): *Über Gewissheit*. In: *Werkausgabe, Band 8*. Frankfurt am Main: Suhrkamp, 1984 (1951).

Elena Dragalina-Chernaya
Surprises in Logic: When Dynamic Formality Meets Interactive Compositionality

Abstract: This paper addresses Ludwig Wittgenstein's claim that "there can never be surprises in logic" (TLP 6.1251) from a perspective of the distinction between substantial and dynamic models of formality. It attempts to provide an interpretation of this claim as stressing the dynamic formality of logic. Focusing on interactive interpretation of compositionality as dynamic formality, it argues for the advantages of dynamic, i.e., game-theoretical approach to some binary semantical phenomena. Firstly, model-theoretical and game-theoretical interpretations of binary quantifiers are compared. Secondly, the paper offers an analysis of Wittgenstein's idea that mixed colours (e.g., bluish green, reddish yellow, etc.) possess logical structures. To answer some experimental challenges, it provides a game-theoretical interpretation of the colours opponency in Payoff Independence (PI) logic. Comparing Nikolay Vasiliev's logical principles and Wittgenstein's internal properties and relations, Wittgenstein's approach is argued for as an attempt of modelling a balance between logic and the empirical.

1 Introduction

In TLP 6.1251, Wittgenstein pointed out that "there can never be surprises in logic". A way of understanding this claim is to view him as stressing the dynamic formality of logic. Given the distinction between substantial and dynamic formality, logic may be considered either as formal ontology, i.e., the domain of higher order formal objects, or as formal deontology, i.e., the domain of rules-governed and goals-directed activity (Dragalina-Chernaya 2016). This distinction is based on the Edmund Husserl's dichotomy. For Husserl, logic is two-sided: apophantic logic belongs to the sphere of assertive statements (judgments), while logic as formal ontology is the domain of abstract higher-level categorical objects (Husserl 1906/07).

Model-theoretically, the substantial formality of logic is specified in terms of being invariant under permutations of objects in the domain (Tarski 1986) or under isomorphisms (Sher 1991), homomorphisms (Feferman 1999), partial isomorphisms (Bonnay 2008) of structures. In his seminal lecture *What are Logical Notions?*, Alfred Tarski proposed to call a notion logical if and only if "it is

Elena Dragalina-Chernaya, National Research University Higher School of Economics, edragalina@hse.ru
DOI 10.1515/9783110657883-13

invariant under all possible one-one transformations of the world onto itself" (Tarski 1986: 149). This definition extends Klein's Erlangen Program to the domain of logic. Felix Klein classified various geometries according to invariance under suitable groups of transformations. For Tarski, logic deals with our most general notions, i.e., notions invariant under permutations. I suggest considering classes of permutations as model-theoretic analogues of Husserl's abstract categorical objects of higher order. Thus, logic as formal ontology does not distinguish between individual objects, but deals with individuals of higher order, i.e. classes of permutations, hypostases of structurally invariant properties of models.

In contrast, logic as formal deontology addresses effective formal agency rather than higher level formal objects. In the substantial model of formality, formal objects are given as structures. As Gila Sher pointed out, "Speaking in terms of objects we can say that formal objects are not just elements of formal structures, they are themselves formal structures" (Sher 1996: 678). In TLP 2.033, however, form is not a structure but the possibility of structure. In Wittgenstein's view,

> The structures of propositions stand in internal relations to one another. (TLP 5.2)
>
> In order to give prominence to these internal relations we can adopt the following mode of expression: we can represent a proposition as the result of an operation that produces it out of other propositions (which are the bases of the operation). (TLP 5.21)
>
> An operation is the expression of a relation between the structures of its result and of its bases. (TLP 5.22)

If we understand a language as a system of complex actions, its structural aspects should be considered from the perspective of how they make different actions possible. Given that, the dynamic formality of a theory means its interactive compositionality, i.e. its ability to demonstrate how simple structured actions make more complex structured actions possible. From a dynamic perspective, as Ahti-Veikko Pietarinen points out, "[m]eaning is that form of interactive processes that gives rise to the sum total of all actions, possible or actual, that arise, or may, will or would arise, as a consequence of playing the game across different contexts and in varying environments" (Pietarinen 2007: 232).

For Wittgenstein, internal relation is equivalent to the operation (TLP 5.232). However, considering internal relations as formal operations switches attention from substantial to dynamic model of formality. This shifting in focus offers some important insights into the logical modelling of contextual factors in the compositional analysis of binary semantical phenomena, e.g., binary quantifiers and mixed colours.

2 Binary Quantifiers

According to Tarski, "our logic is logic of cardinality" (Tarski 1986: 151). As he stresses, "it turns out that our logic is even less than a logic of extension, it is a logic of number, of numerical relations" (Tarski 1986: 151). However, not only second-order properties (e.g., Mostowski quantifiers) but also second-order relations between first-order relations on the universe (e.g., Lindström binary quantifiers) may be considered as logical. In contrast to the Tarski's claim, as binary quantifiers iterated quantifier prefixes distinguish between equicardinal relations (Mikeladze 1979: 289). As a result, the theory of binary quantification is not "a logic of numerical relations". It deals not only with cardinalities, but also with patterns of ordering of the universe. The dynamic formality addresses the procedures that constitute the possibility of this patterns of ordering.

Quantifiers may be viewed in two different ways, i.e., as higher-order predicates and as embodying choice functions (Hintikka/Sandu 1994). The second way of looking at quantifiers is normally explained in game theoretical semantics which can be used to justify both classical and constructivist deductive practices. However, only a constructive semantical game may be viewed as a viable way to build up compositionally connected acts of assertions. From a dynamic perspective, compositional analysis must show how the competence to perform simple acts of assertions allows for the competence to perform structurally more complex acts. In turn, to have the competence to assert a proposition is to be able to assert it with evidence, while the evidence for an assertion of a proposition is essentially an act of proving it. In constructive game theoretical semantics, the meaning of a quantifier proposition is determined by specifying its canonical proof, i.e., an effective winning strategy of the Verifier in a constructive semantic game with the proposition. As Jaakko Hintikka argued,

> The demand of playability might seem to imply that the set of the initial verifier's strategies must be restricted. For it does not seem to make any sense to think of any actual player as following a nonconstructive (nonrecursive) strategy... This playability of our "language games" is one of the most characteristic features of the thought of both Wittgenstein and Dummett. (Hintikka 1996: 214 – 215)

The key notions of constructive game theoretical semantic are as follows: a tree T of semantical game for a formula A with respect to atomic base B (an atomic base B is a pair $< L, R >$ where L is a set of descriptive constants and R is a set of inference rules from atomic formulas to atomic one); a semantic pay-off function Φ; players set functions of the Verifier and the Falsifier on T. The tree of a semantical game is a binary structure such that every node n_i is a pair (μ_1, μ_2), where $\mu 1$ is a formula from L and μ_2 is an informational characteristic of n_i. A formula A is valid with respect to atomic base B if and only if there is an effective strategy of the Verifier in the tree T of semantical game for a formula A with respect to B. Assertion of A with respect to atomic base B leads to the absurd, if such a strategy is available to the Falsifier. In constructive semantical game for a formula A with respect to atomic base B, effective winning strategy of the Verifier is both a subtree of T and a closed canonical argument for A with respect to B.

Game-theoretical technique is powerful enough to deal with different patterns of ordering of quantifiers. For example, a branching quantifier proposition

$$\begin{matrix}(\forall x \in A)(\exists y \in B)\\ (\forall z \in C)(\exists u \in D)\end{matrix} F(x, y, z, u) \qquad (1)$$

may be defined as a type

$$(\exists f \in A \to B)(\exists g \in C \to D)(\forall x \in A)(\forall z \in C)F(x, ap(f, x), z, ap(g, z)), \qquad (2)$$

where $ap(f, x) = h(x)$, if $f = (\lambda x)h(x)$ (Ranta 1988: 394). As a result, the meaning of the formula (1) is determined by specifying its canonical proof, i.e., the winning strategy of the Verifier in a constructive semantical game with (1). Yet, branching quantifier prefixes are known as a standard argument against the principle of compositionality[1]. The trouble is that the meaning of a branching quantifier expression seems to be undefined in terms of its constituent parts by explaining one quantifier in a time. I suggest considering a formula ϕ as a part of a formula ψ in the context of a complex formula A if and only if the informational conditions of the assertion of ϕ are included into the informational conditions of the assertion of ψ in a constructive semantical game with A. Thus, scope relations of quantifiers are viewed as informational relations of the nodes of semantical games. This approach presupposes the post-Fregean notation which explicitly indicates the domains of quantification, since all substitutions are treated as parts of a quantified proposition. Moreover, meanings of quantified propositions may vary from context to context. Literally, this fact is not in conflict with compositionality, as context of evaluation may be considered as an extra factor, contributing to the complex meaning.

---

[1] Gabriel Sandu and Jaakko Hintikka famously generalized branching quantifiers by the formalism of Independence-Friendly logic. (Non)-Compositionality of IF-logic is discussed in a series of papers (see, for instance, Hodges 1997, Hodges 2001, Hintikka 2001, Janssen 2002, Abramsky/Väänänen 2009).

3 Mixed Colours

Wittgenstein's logic of colours addresses challenging binary phenomena, i.e., mixed colours. His claim that "there can be a bluish green but not a reddish green" is puzzling, since he considered reddish green not only as an empirically, but as a formally forbidden colour.

Logic as formal ontology does not discriminate between the colours. According to Husserl, the ontological region of coloured individuals is an extension of material concepts of colour. He considered the necessity of propositions about colours as synthetic, as they are true by virtue of essential relations among the material concepts involved. For Husserl, the pure logical study of the concepts of colours is impossible.

Contrary to this, Wittgenstein claims that logic deals with internal relations.

> A property is internal if it is unthinkable that its object should not possess it. (This shade of blue and that one stand, *eo ipso*, in the internal relation of lighter to darker. It is unthinkable that these two objects should not stand in this relation.) (Here the shifting use of the word "object" corresponds to the shifting use of the words "property" and "relation".) (TLP 4.123)

However, in *Tractatus*, the logic of colours faces the colour exclusion problem. On the one hand, colour-ascriptions should be elementary and logically independent, on the other hand, they cannot be elementary and logically independent:

> the simultaneous presence of two colours at the same place in the visual field is impossible, in fact logically impossible, since it is ruled out by the logical structure of colour (It is clear that the logical product of two elementary propositions can neither be a tautology nor a contradiction. The statement that a point in the visual field has two different colours at the same time is a contradiction.) (TLP 6.3751)

Wittgenstein tries to solve this problem during his middle period. In *Some Remarks on Logical Form*, he is interested in examining the logical structure of the "phenomena themselves":

> we can only arrive at a correct analysis by, what might be called, the logical investigation of the phenomena themselves, *i.e.*, in a certain sense *a posteriori*, and not by conjecturing about *a priori* possibilities. (SRLF: 163)

Colour-incompatibility claims are *a posteriori*, but they are not empirical generalization, as their necessity is based on the geometrical organization of colour space. In *Philosophical Remarks*, Wittgenstein adapts Alois Höfler's colour octahedron for the representation of this geometrical organization. Following the middle Wittgenstein's path, it seems reasonable to consider the logic of colour as a domain ontology for a colour space, i.e., as a "geometry" in Klein's sense. The invariance criterion generalized in this way would presuppose not only one type of invariance (e.g., permutation invariance) but also a variety of invariances respecting different types of ordering of the universe (e.g., relations of colours). As a result, however, the colour octahedron imposes *ad hoc* restrictions on logical space. Logical space breaks down into different regions which are governed by physiology, psychology, and physics rather than by logic.

On the contrary, Wittgenstein viewed the colour octahedron as a part of grammar (RPP II, §8). Yet, could colour octahedron provide criteria for the grammatical use of colour terms? Considering context-free monochromatic colours as the base of our ordinary use of colour terms turns out to be problematic. Contrary to the classical Brent Berlin and Paul Kay hypothesis of basic colour terms (Berlin/Paul 1969), the data from national language corpora do not support the idea that the determinate system of basic colour samples makes us consider other colours as derived from them (see, for instance, Rakhilina 2011). Our inclination to think of the ordinary use of colour terms as reduced to monochrome patches of colour is analogous to the bias of supposing that Euclidean geometry determines the grammar for describing visual space. In both cases, we presuppose that a unique description of the properties and relations involved lies behind our ordinary grammar.

Gradually, Wittgenstein himself turns away from the idea of using a colour octahedron in the logical representation of colour terms. In *Philosophical Remarks*, he calls an octahedron a rough representation and prefers to consider the colour scales as more or less convenient measuring instruments. In *Remarks on Colour*, he argues that colour terms are not labels for pure colour concepts.

> There is no such thing as *the* pure colour concept. (RC III, §73)

> Where does the illusion come from then? Aren't we dealing here with a premature simplification of logic like any other? (RC III, §74)

> I.e., the various colour concepts are certainly closely related to one another, the various "colour words" have a related use, but there are, on the other hand, all kinds of differences. (RC III, §75)

Claims about colour do not reflect the essence of colour. On the contrary, they are necessary as parts of different language games. Since the criteria for the sameness of colours are variable, it is impossible to possess unique logic or geometry of colours.

> The difficulties which we encounter when we reflect about the nature of colours (those difficulties which Goethe wanted to deal with through his theory of colour) are contained in

the fact that we have not one but several related concepts of the sameness of colours. (RC III, §251)

Remarks on Colour casts doubt on the idea of a geometry of colour. As Gabriele Mras pointed out,

> Wittgenstein's treatment of the question about the relation amongst colours in his *Remarks on Colour* is meant to bring that out – to bring out the idea that what is thought of as a geometry of colour is really an illusion. (Mras 2017: 48)

It seems reasonable to shift focus from static geometrical to dynamic game theoretical models of colour which are more flexible in accounting for different contexts of mixed colours, e.g., forbidden mixed colours. Some mixed colours, e.g., reddish green, are normally forbidden for human observers. According to the Ewald Hering's opponent-processing model of colours, red and green are spectrally opposing, i.e., the perceiving of both red and green would presuppose the simultaneous transmission of positive and negative signals in the same channel. Does it entail the analyticity of the statement that forbidden mixed colours are impossible?

For Moritz Schlick, the validity of the statements "If a thing is uniformly coloured of red, then it is not, at the same time and under the same respect, uniformly coloured of green" is not Husserlian synthetic *a priori* truth, as it rests on the meanings of colour terms (Schlick 1979). Similarly, Hilary Putnam considered the statement "Nothing is red (all over) and green (all over) at the same time" as analytically valid. As a result, in defining second-level predicates $Red(F)$ (for "F is a shade of red") and $Grn(F)$ (for "F is a shade of green") we are restricted by the postulate: "Nothing can be classified as both a shade of red and a shade of green" (i.e., "that shade of red" and "that shade of green" must never be used as synonyms) (Putnam 1956: 215 – 216). Moreover, as Larry Hardin argues, "not being red is part of the concept of being green" (Hardin 1988: 122).

However, if forbidden mixed colours are impossible in the same way that a married bachelor is, then a report that these colours can be perceived may testify nothing but the ignorance of the rules of native language. Yet, the report that forbidden mixed colours can be perceived in a "stabilized-image" experiment is one of the most surprising statements in modern literature on colour vision (Crane/Piantanida 1983; Billock et al. 2001; Billock/Tsou 2010). [2]. Interestingly, Vincent Billock and Brian Tsou provide a game-theoretical interpretation of this experiment. In their view, stabilization of the retinal image between red and green

---

[2] Although in the very titles of their papers the authors reported "seeing reddish green and yellowish blue" or "seeing forbidden colours", this interpretation of the results of "stabilized-image" experiments has been criticized by neuropsychologists (see, e.g., Hsieh/Tse 2006).

equiluminant fields "turns off" the winner-takes-all competition between "red" and "green" antagonistic neurons.

> In our model, populations of neurons compete for the right to fire, just as two animal species compete for the same ecological niche – but with the losing neurons going silent, not extinct. A computer simulation of this competition reproduces classical color opponency well – at each wavelength, the "red" or "green" neurons may win, but not both (and similarly for yellow and blue). Yet if the competition is turned off by, say, inhibiting connections between the neural populations, the previously warring hues can coexist. (Billock/Tsou 2010: 75)

In contrast, I suggest modelling "stabilized-image" effects in the framework of Payoff Independence (PI) logic (Pietarinen 2006; Pietarinen 2009). As opposed to winner-takes-all games, in over-defined PI-games both "red" and "green" teams of neurons may have winning strategies. Thus, stabilization "turns off" the information exchange rather than the competition between the opponent neurons populations. PI-logic is powerful enough to simulate the variety of competitive and non-strictly competitive behavior of teams of "red" and "green" neurons.

Clearly, the role of neuropsychological models in the exegesis of Wittgenstein's logic of colour is ambivalent. In *Philosophical Investigations*, he famously points out: "It was true to say that our considerations could not be scientific ones." (PI §109) In *Remarks on Colour*, Wittgenstein stresses:

> We do not want to establish a theory of colour (neither a physiological one nor a psychological one), but rather the logic of colour concepts. And this accomplishes what people have often unjustly expected of a theory. (RC I, §22)

In his view, however, the border between logic and the empirical is determined by use and, thus, it is flexible and dynamic.

> Sentences are often used on the borderline between logic and the empirical, so that their meaning changes back and forth and they count now as expressions of norms, now as expressions of experience.
> (For it is certainly not an accompanying mental phenomenon – this is how we imagine "thoughts" – but the use, which distinguishes the logical proposition from the empirical one.). (RC I, §32)

Normatively, Wittgenstein's prohibition against reddish green casts doubt on the sensational interpretation of the results of a "stabilized-image" experiment (see Suarez/Nida-R. 2009, Ritter 2013, Lugg 2017). It is not surprising that subjects of this experiment "were tongue-tied in their descriptions of these colors, using terms like 'green with a red sheen' or 'red with green highlights' " (Billock et al. 2001: 2398). Furthermore, they reported on the dynamic effects of simultaneous or serial reorganizations of both visual and colour spaces.

> We discovered an entirely novel percept (4 out of 7 subjects) in which the red and green (or blue and yellow) bipartite fields abruptly exchange sides before fading or returning to the veridical percept; a digital-like switching phenomenon that may indicate a nonlinear dynamic process in operation...One subject – an expert psychophysical observer – saw a 90° reorganization of the bipartite field so that red and green were now over and under rather than side by side. (Billock et al. 2001: 2399)

Does it mean that subjects of a "stabilized-image" experiment reported on perception of a new colour? Wittgenstein seems to anticipate this puzzling question.

> But even if there were also people for whom it was natural to use the expressions 'reddish-green' or 'yellowish-blue' in a consistent manner and who perhaps also exhibit abilities which we lack, we would still not be forced to recognize that they see colours which we do not see. There is, after all, no commonly accepted criterion for what is a colour, unless it is one of our colours. (RC I, §14)

In Howard Lovecraft's horror short story, *The Colour Out of Space*, an unnamed narrator pointed out that the meteor's extra-cosmic colour was almost impossible to describe, thus, it was only by analogy that it may be called a colour. A colour-blind subject with a grapheme-colour synesthesia claimed to see numbers in "Martian colours", i.e., colours that he could never see in the real world (Ramachandran/Hubbard 2001: 26). Wittgenstein, however, does not associate our system of colours with "the nature of things" in our cosmic space.

> We have a colour system as we have a number system. Do the systems reside on our nature or in the nature of things? How are we to put it? – Not in the nature of numbers or colours. (RPP II, §426)

Language-games seem to eliminate the dilemma of "our nature" and "the nature of things" (see Schulte 2017). For Wittgenstein, our concepts do not reflect our life, but "stand in the middle of it" (RC III, §302). Language should be considered as a collection of instruments that makes it possible to achieve the desired outcome using available resources rather than as a game with arbitrary rules. Wittgenstein prefers to reckon the colour samples among the "instruments of the language" (PI §16). As he puts it, "'There is no greenish red' ... What would go wrong if we denied these laws? ... It would come to building a system which would be decidedly impractical." (OC §235)

However, neuropsychological experiments may contribute to our language games which, in turn, constitute what the colours are. Language games with colour concepts involve not only perception, but also memory, imagination, cultural codes. It is possible, for example, that the invention of cheap eyetracker will make the narrative on "red with green highlights" or "both red and green" a habitual part of our everyday discourse. In *Remarks on Colour*, Wittgenstein himself invites us to imagine different colour systems designed for hypothetical language games. Finally, he addresses the key question:

> "Can't we imagine certain people having a different geometry of colour than we do?" That, of course, means: Can't we imagine people having colour concepts other than ours? And that in turn means: Can't we imagine people who do not have our colour concepts but who have concepts which are related to ours in such a way that we would also call them "colour concepts"? (RC I, §66)

In what sense (if any) Wittgenstein aims to establish the logic rather than the anthropology or the linguistic typology of colour concepts? In my view, partially in the same sense in which Nicolay Vasiliev tried to develop his imaginary logic (Vasiliev 1993, Vasiliev 2003). According to Vasiliev, "The possibility of another geometry must convince us of the possibility of a logic other than ours" (Vasiliev 1993: 332). He distinguished between two levels of logical structures: metalogic, i.e., the level of necessary principles of logic which cannot be revised without distracting the logic itself, and of ontology, i.e., the level of logical laws depending on the objects under consideration. For him, formal principles are unchangeable and immune to factual evidence. However, logical laws are empirically based and, thus, revisable.

> Since the law of contradiction is an empirical and real law, we can reason without it as well, and then we will get an imaginary logic. In fact, on empirical grounds I can arbitrarily build whatever imaginary objects and imaginary disciplines. I can create centaurs, sirens, griffins and imaginary zoology. I can create utopias, an imaginary sociology, or an imaginary history ... Empirical and real laws are about reality, but their opposite is always conceivable. (Vasiliev 2003: 140)

Vasiliev considered the law of excluded contradiction as a reduced formula comprising various facts, e.g., that red is incompatible with blue, white, and black. The opposite to these facts is not unthinkable and, as a result, the law of excluded contradiction is revisable in imaginary logic.

> *The law of contradiction expresses the incompatibility between an assertion and [its] negation.* A cannot be non-A. No object contains a contradiction, [i.e.] allows us to at once make an affirmative and a negative proposition (about it). But if we ask ourselves what in fact negation is, we can define it only in one way: *negation is that which is incompatible with affirmation.* We call 'red' the negation of 'blue' and say that a red object is not a blue one, because red is incompatible with blue. Where there is no incompatibility, we are not allowed to speak about negation. (Vasiliev 2003: 132)

Vasiliev distinguished between formal and material aspects of negative propositions and, consequently, between properties of negation and grounds for negation.

> The formal aspect manifests [the fact] that the truth of a negative proposition implies the recognition of the falsehood of the affirmative one, but it leaves open the question on what grounds we can ascertain the truth of negative propositions. The material aspect gives an answer to this question. Therefore, the formal aspect manifests the properties of negation; the material aspect manifests the grounds for negation. While preserving the formal aspect, we can change the material one and then obtain a different kind of negation. (Vasiliev 2003: 135)

In *Tractatus*, unthinkability is a criterion of formality: a property is internal, i.e., formal, if it is unthinkable that its object should not possess it. As Wittgenstein stresses in *Remarks on Colour*, "When dealing with logic, 'One cannot imagine that' means: one doesn't know what one should imagine here" (RC I, §27). Yet, internal properties are not eternal entities which we try to conceive using our concepts[3]. On the contrary, they exist in virtue of their roles in our practice. Our language games may manifest the grounds of internal properties of colours, revealing new internal properties of colours.

> We must always bear in mind the question: How do people learn the meaning of colour names? (RC III, §61)

> What does, "Brown contains black," mean? There are more and less blackish browns. Is there one which isn't blackish at all? There certainly isn't one that isn't yellowish at all. (RC III, §62)

> If we continue to think along these lines, 'internal properties' of a colour gradually occur to us, which we hadn't thought of at the outset. And that can show us the course of a philosophical investigation. We must always be prepared to come across a new one, one that has not occurred to us earlier. (RC III, §63)

Since " 'internal properties' of a colour gradually occur to us", there can be surprises in logic. Surprises occur not only on the level of Vasiliev's "ontology". Contrary to the absolute principles of Vasiliev's metalogic, Wittgensteinian logical "mythology" is dynamic and revisable.

---

[3] When scholasticism distinguishes between formal and material consequences it often appeals to imagination. For example, Paul of Venice defines formal consequence as that in which "the opposite of the consequent is formally repugnant to the antecedent... 'Formally repugnant' means that these two sentences are not imaginable to stand simultaneously without a contradiction (non sunt imaginabilia stare simul sine contradictione)" (Paulus 1395/96: 167). In Scholastics, however, internal (transcendental) relations which justify the formality of a consequence "This thing is white thus it is not black" are rooted in eternal truths (see, for instance, Dragalina-Chernaya 2018: 31f.).

> The mythology may change back into a state of flux, the river-bed of thoughts may shift. But I distinguish between the movement of the waters on the river-bed and the shift of the bed itself; though there is not a sharp division of the one from the other. (OC §97)
>
> But if someone were to say "So logic too is an empirical science" he would be wrong. Yet this is right: the same proposition may get treated at one time as something to test by experience, at another as a rule of testing. (OC §98)

Logic is not an empirical, but a formal science. Dealing with internal properties and relations, it is about constructing concepts rather than about empirical objects.

"An internal relation is never a relation between two objects, but you might call it a relation between two concepts. And a sentence asserting an internal relation between two objects ... is not describing objects but constructing concepts." (LFM: 73)

Two concepts are internally related if the competence to construct the one presupposes the competence to construct the other. Logic intended to explain this structural complexity, i.e., interactive compositionality of language games, and its contribution to truth conditions and inferences. It is not surprising that Wittgenstein's answer to the question "Now to what extent is it a matter of logic rather than psychology that someone can or cannot learn a game?" (RC III, §114) is as follows: "I say: The person who cannot play this game does not have this concept" (RC III, §115).

4 Conclusion

Dynamic formality meets interactive compositionality in language games the logic plays. From the dynamic perspective, internally related concepts give rise to structural complexity, i.e., interactive compositionality of language games. The dynamic of the interaction provides a general framework for dealing with binary semantical phenomena, e.g., binary quantifiers and mixed colours.

However, it turns out to be dramatically hard to distinguish between normativity drawn from logic and non-logical features of the hierarchy of concepts in language games. Revision in logic often took the form of Proclus' porism, i.e., an unexpected prize for a researcher. A discovery of new internal properties and relations of concepts involved was not planned by a researcher although they are deductible within the framework, i.e. the "mythology" of her investigation, which, in turn, is also revisable.[4]

Bibliography

Abramsky, Samson and Väänänen, Jouko (2009): "From IF to BI. A tale of dependence and separation." In: *Synthese*. Vol. 167, 207 – 230.
Berlin, Brent and Kay, Paul (1969): *Basic Color Terms: Their Universality and Evolution*. Berkeley: U. of California.
Billock, Vincent, Gleason, Gerald and Tsou, Brian (2001): "Perception of forbidden colors in retinally stabilized equiluminant images: an indication of softwired cortical color opponency?" In: *Journal of the Optical Society of America*. Vol. 18, No. 10, 2398 – 2403.
Billock, Vincent and Tsou, Brian (2010): "Seeing Forbidden Colours." In: *Scientific American*. Vol. 302, No. 2, 72 – 77.
Boyer, Julien and Sandu, Gabriel (2012): "Between proof and truth." In: *Synthese*. Vol. 187, 821 – 832.
Bonnay, Denis (2008): "Logicality and invariance." In: *Bulletin of Symbolic Logic*. Vol. 14, 29 – 68.
Crane, Hewitt and Piantanida, Thomas (1983): "On Seeing Reddish Green and Yellowish Blue." In: *Science, New Series*. Vol. 221, No. 4615, 1078 – 1080.
Dragalina-Chernaya, Elena (2016): "The Roots of Logical Hylomorphism." In: *Logical Investigations*. Vol. 22, No. 2, 59 – 72.
Dragalina-Chernaya, Elena (2018): "Consequences and Design in General and Transcendental Logic." In: *Kantian Journal*. Vol. 37, No. 1, 25 – 39.
Feferman, Solomon (1999): "Logic, logics, and logicism." In: *Notre Dame Journal of Formal Logic*. Vol. 40, 31 – 54.
Hardin, Larry (1988): *Color for Philosophers*. Hackett, Indianapolis and Cambridge MA.
Hintikka, Jaakko and Sandu, Gabriel (1994): "What Is a Quantifier?" In: *Synthese*. Vol. 98. No. 1, 113 – 129.
Hintikka, Jaakko (1996): *The Principles of Mathematics Revisited*. Boston: Cambridge University Press.
Hintikka, Jaakko and Sandu, Gabriel (2001): "Aspects of Compositionality." In: *Journal of Logic, Language, and Information*. Vol. 10, No. 1, 49 – 61.
Hodges, Wilfrid (1997): "Compositional Semantics for a Language of Imperfect Information." In: *Logic Journal of the IPGL*. Vol. 5, 539 – 563.
Hodges, Wilfrid (2001): "Formal features of compositionality." In: *Journal of Logic, Language and Information*. Vol. 10, No.1, 7 – 28.
Hsieh, Po-Jang / Tse, Peter (2006): "Illusory color mixing upon perceptual fading and filling-in does not result in 'forbidden colors'." In: *Vision Research*. Vol. 46, 2251 – 2258.

---

[4] The article was prepared within the framework of the HSE University Basic Research Program and funded by the Russian Academic Excellence Project '5-100'

Husserl, Edmund (2008): "Introduction to Logic and Theory of Knowledge. Lectures 1906/07." In: *Husserliana: Edmund Husserl – Collected Works*. Vol. 13, Netherlands: Springer.
Janssen, Theo (2002): "Independent Choices and the Interpretation of IF Logic." In: *Journal of Logic, Language and Information*. Vol. 11, No. 3, 367 – 387.
Lugg, Andre (2017): "Impossible Colours: Wittgenstein and the Naturalist's Challenge." In: *How Colours Matter to Philosophy*. Silva, Marcos (ed.). Springer, 107 – 121.
Mikeladze, Zurab (1979): "On a class of logical concepts." In: *Logical inference*. Moscow, 287 – 299 (In Russian).
Mras, Gabriele (2017): "Propositions About Blue: Wittgenstein on the Concept of Colour." In: *Wittgenstein on Colour*. Gierlinger, Frederik; Riegelnik, Stefan (eds.). Berlin: De Gruyter, 45 – 55.
Paulus Venetus (1984): *Logica Parva*, Perreiah, Alan R. (transl., introd. and annotated). München/Wien: Philosophia.
Pietarinen, Ahti-Veikko (2006): "Independence-Friendly Logic and Games of Information." In: *The Age of Alternative Logics: Assessing Philosophy of Logic and Mathematics Today*. van Benthem, Johann et al. (eds.). Springer, 243 – 259.
Pietarinen, Ahti-Veikko (2007): "The Semantics / Pragmatics Distinction from the Game-Theoretic Point of View." In: *Game Theory and Linguistic Meaning*. Pietarinen, Ahti-Veikko (ed.). Amsterdam: Elsevier, 229 – 242.
Pietarinen, Ahti-Veikko and Sandu, Gabriel (2009): "IF Logic, Game-Theoretical Semantics, and the Philosophy of Science." In: *Logic, Epistemology and the Unity of Science*. Rahman, Shahid et al. (eds.). Springer, 105 – 138.
Putnam, Hilary (1956): "Reds, Greens, and Logical Analysis." In: *The Philosophical Review*. Vol. 65, No. 2, 206 – 217.
Rakhilina, Ekaterina and Paramei, Galina (2011): "Colour terms: Evolution via expansion of taxonomic constraints." In: *New Directions in Colour Studies*. Biggam, Carole et al. (eds.). Amsterdam: John Benjamins Publishing Company, 2011, 121 – 132.
Ramachandran, Vilayanur and Hubbard, Edward (2001): "Synaesthesia – A Window Into Perception, Thought and Language." In: *Journal of Consciousness Studies*. Vol. 8, No. 12, 3 – 34.
Ranta, Aarne (1988): "Propositions as Games as Types." In: *Synthese*. Vol. 76, 377 – 395.
Ritter, Bernhard (2013): " 'Reddish Green': Wittgenstein on Concepts and the Limits of the Empirical." In: *Conceptus: Zeitschrift für Philosophie*. Vol. 42, No. 101–102, 1 – 19.
Schlick, Moritz (1979): "Is there a Factual a Priori? [1930]" In: *Philosophical Papers*. Mulder, Henk et al. (eds.), Vol. II (1925–1936), Dordrecht: D. Reidel, 161 – 170.
Schulte, Joachim (2017): "We Have a Colour System as We Have a Number System." In: *Wittgenstein on Colour*. Gierlinger, Frederik; Riegelnik, Stefan (eds.). Berlin: De Gruyter, 21 – 32.
Sher, Gila (1991): *The Bounds of Logic: A Generalized Viewpoint*. Cambridge, MA: MIT Press.
Sher, Gila (1996): "Did Tarski Commit 'Tarski's Fallacy'?" In: *The Journal of Symbolic Logic*. Vol. 61, 653 – 686.
Suarez, Juan and Nida-Rümelin, Martine (2009): "Reddish Green: A Challenge for Modal Claims About Phenomenal Structure." In: *Philosophy and Phenomenological Research*. Vol. 78. No. 2, 346 – 391.
Tarski, Alfred (1986): "What are Logical Notions?" In: *History and Philosophy of Logic*. Vol. 7, 143 – 154.
Vasiliev, Nikolay (1993): "Logic and Metalogic." In: *Axiomathes*. Vol. 3, 329 – 351
Vasiliev, Nikolay (2003): "Imaginary (Non-Aristotelian) Logic." In: *Logique et Analyse*. Vol. 182, 127 – 163.

Wittgenstein, Ludwig (1929): "Some Remarks on Logical Form." In: *Proceedings of the Aristotelian Society*. Supplementary Volumes, Vol. 9, Knowledge, Experience and Realism, 162 – 171.
Wittgenstein, Ludwig (1961): *Tractatus Logico-Philosophicus*. Pears, D. F.; McGuinness, B. (trans.). London: Routledge.
Wittgenstein, Ludwig (1964): *Philosophical Remarks*. Oxford: Blackwell.
Wittgenstein, Ludwig (1975): *On Certainty*. Anscombe, G.E.M.; von Wright, G. H. (eds.). Oxford: Blackwell.
Wittgenstein, Ludwig (1976): *Lectures on the Foundations of Mathematics*. Cambridge 1939, Diamond, Cora (ed.). Ithaca: Cornell University Press.
Wittgenstein, Ludwig (1977): *Remarks on Colour*. Anscombe, G. E. M. (ed.), McAlister, L.; Schattle, M. (trans.). Oxford: Blackwell.
Wittgenstein, Ludwig (1980): *Remarks on the Philosophy of Psychology*. Vol. II, von Wright, G.H.; Nyman, Heikki (eds.). Oxford: Blackwell.
Wittgenstein, Ludwig (2009): *Philosophical Investigations*. P. M. S. Hacker; J. Schulte (eds.), Anscombe, G. E. M.; Hacker, P. M. S.; Schulte, J. (trans.), 4th edition. Oxford: Blackwell.

Part II: **Philosophy of Mathematics**

Paolo Mancosu and Benjamin Siskind
Neologicist Foundations: Inconsistent Abstraction Principles and Part-Whole

Abstract: Neologicism emerges in the contemporary debate in philosophy of mathematics with Wright's book *Frege's Conception of Numbers as Objects* (1983). Wright's project was to show the viability of a philosophy of mathematics that could preserve the key tenets of Frege's approach, namely the idea that arithmetical knowledge is analytic. The key result was the detailed reconstruction of how to derive, within second order logic, the basic axioms of second order arithmetic from Hume's Principle

$$\text{HP} \quad \forall X, Y (\#X = \#Y \leftrightarrow X \cong Y)$$

(and definitions). This has led to a detailed scrutiny of so-called abstraction principles, of which Basic Law V

$$\text{BLV} \quad \forall X, Y \big(\partial X = \partial Y \leftrightarrow \forall x \, (X(x) \leftrightarrow Y(x)) \big)$$

and HP are the two most famous instances. As is well known, Russell proved that BLV is inconsistent. BLV has been the only example of an abstraction principle from (monadic) concepts to objects giving rise to inconsistency, thereby making it appear as a sort of monster in an otherwise regular universe of abstraction principles free from this pathology. We show that BLV is part of a family of inconsistent abstractions. The main result is a theorem to the effect that second-order logic formally refutes the existence of any function F that sends concepts into objects and satisfies a 'part-whole' relation. In addition, we study other properties of abstraction principles that lead to formal refutability in second-order logic.

1 Neologicism As a Foundation for Mathematics

Neologicism emerges in the contemporary debate in philosophy of mathematics with Wright's book *Frege's Conception of Numbers as Objects* (Wright 1983). Wright's project was to show the viability of a philosophy of mathematics that could preserve the key tenets of Frege's approach. *Prima facie* this could have appeared as an impossible task. After all, Frege's foundational system, as presented first in *The Foundations of Arithmetic* (Frege 1884) and in final formulation in *Basic Laws of*

Paolo Mancosu, UC Berkeley, USA, mancosu@socrates.berkeley.edu
Benjamin Siskind, UC Berkeley, USA, bsiskind@math.berkeley.edu

DOI 10.1515/9783110657883-14

Arithmetic (Frege 1893 and Frege 1903), relied on the use of extensions codified by means of the notorious Basic Law V in the latter work, a law that was shown by Russell to lead to contradiction. Basic Law V postulates the existence of a function (or operator) ∂ that maps the concepts into the objects of the domain in such a way as to satisfy the following statement:

$$\text{BLV} \quad \forall X, Y \big(\partial X = \partial Y \leftrightarrow \forall x\, (X(x) \leftrightarrow Y(x)) \big).$$

It is easy to show that Basic Law V yields an unrestricted comprehension axiom which, in turn, yields the existence of Russell's paradoxical set. Frege himself, and a few others after him, tried to tweak Basic Law V but with no success and, until Wright's book, Frege's foundational project was no longer considered viable. Wright's approach emerged as a combination of a powerful technical result and of a host of philosophical arguments. The technical result consisted in the observation that in the *The Foundations of Arithmetic* (1884) Frege used extensions in order to prove what is now called Hume's Principle, namely

$$\text{HP} \quad \forall X, Y(\#X = \#Y \leftrightarrow X \approx Y).$$

Hume's Principle provides a criterion of identity for the operator # (to be interpreted as "number of"): it states that the number of X and the number of Y are identical if and only if there is a one-to-one correspondence between the objects falling under X and those falling under Y. It is important to remark that just as in the case of Basic Law V, the right-hand side of the biconditional can be expressed in purely logical terms. What Wright, and later Boolos and Heck, showed was that once Hume's Principle was established, Frege did not have any further use for extensions and proceeded, in the *Foundations of Arithmetic*, to provide an outline of how to prove from Hume's Principle the basic axioms for second-order arithmetic. The technical result, thus, was Wright's detailed reconstruction of how to derive, within second-order logic, the basic axioms of second-order arithmetic from Hume's Principle (and definitions). If we call FA (for Frege Arithmetic) the system of second-order logic with HP and with Z_2 the axiomatic theory for second-order arithmetic, Wright's result can be summarized as:

$$\text{FA} \vdash Z_2.[1]$$

It could still be objected that this is a far cry from the logicism championed by Frege. After all, HP does not look like a logical axiom. Wright agrees with this but

---

[1] More properly, the result should be stated as FA + FD ⊢ Z_2, where FD is a set of definitions for zero, successor, and natural number.

his Frege was interested in logicism in order to prove, contra Kant, the analyticity of arithmetic. And Wright's philosophical point was that this part of the Fregean project could be vindicated by taking HP on board and arguing that it was analytic (or definitional) of the "number of" operator or something akin to analytic. Obviously, this is where much of the philosophical debate has focused but that discussion is irrelevant for our goals.

Rather, what is important is that the discussion has led to a detailed scrutiny of abstraction principles, of which BLV and HP are the two most famous instances. Indeed, one of the first objections to accepting Hume's Principle was an argument, the first form of "bad company" objection, which questioned the legitimacy of using definitions such as Hume's Principle arguing that something deeply problematic must be going on here as shown by the fact that BLV has exactly the same definitional form as HP but gives rise to contradiction. This led to the first requirement, of a long series to come, for separating good from bad abstractions, namely consistency.

Thus, much of the philosophical work in neologicism is aimed at explaining what is special (epistemologically, ontologically, or otherwise) about good abstraction principles that allows them to provide a foundation for mathematics along the neologicist lines (see Hale/Wright 2001 or the recent collection Ebert/Rossberg 2016). Mathematical work in this area is aimed at identifying good abstraction principles with enough deductive or interpretive power to recover some part of mathematics (arithmetic, analysis, set theory), as well as identifying good features of abstraction principles so as to avoid bad company problems (see Cook 2012 or, again, Ebert/Rossberg 2016).

We conclude this section by remarking that, if we restrict our attention to operators from monadic concepts to objects[2], BLV has been the only natural example of an operator giving rise to inconsistency, thereby making it appear as a sort of monster in an otherwise regular universe of abstraction principles free from this pathology. One of the tasks of this paper is to show that BLV is no monster, or in any case that it is part of a family of inconsistent abstraction principles. But in order to explain the background for this result we need to talk about the part-whole relation among concepts and their extensions.

---

[2] We are restricting this claim to first-level concepts, as Boolos has given examples of monadic second-level concepts that entail a contradiction. See Boolos 1998: 173.

2 Part-Whole Principles and Abstraction Operators

Interest in the relation of part-whole is as old as the history of philosophy and mathematics. In *Two New Sciences*, Galileo presented a paradox of infinity one horn of which rested on a part-whole intuition. The collection of squares of natural numbers is properly included in the collection of natural numbers, thus the number (or numerosity) of the former must be strictly less than that of the latter. More formally, if A is strictly included in B then $Num(A) < Num(B)$ (which yields $Num(A) \neq Num(B)$). On the other hand, and this is the second horn of the dilemma, we can associate univocally the squares of natural numbers with the natural numbers and hence the two collections have the same number (or numerosity). More formally, if there is a one-to-one correspondence between A and B then $Num(A) = Num(B)$. The two intuitions lead to a contradiction, in that according to the first the squares are less numerous than the natural numbers (and thus have different numerosity) and according to the second they have the same numerosity.

In the history of philosophy and mathematics there have been defenders of the part-whole principle, most notably Bolzano.[3]

But it was the second principle, based on one-to-one correspondence, that was enshrined in Cantorian set theory, through the notion of cardinal number, as the proper way to generalize counting from the finite to the infinite. One reason for the dominance of the Cantorian solution was that no mathematical implementation of part-whole that extended to infinite sets seemed possible.

Recent work on the theory of numerosities by Benci, Di Nasso, and Forti (see Benci et al. 2003, Benci et al. 2006, Benci et al. 2007; see also Mancosu 2016, chapter 3) has shown that it is possible to implement an assignment of numerosities to sets that preserve the part-whole principle, namely the property that if $A \subsetneq B$, the numerosity of A is strictly less than the numerosity of B. The context in which this construction of numerosities takes place is that of ordinary set-theoretic mathematics. The system of numerosities can be seen as a refinement of Cantorian cardinalities: if two sets have the same numerosity, then they have the same Cantorian cardinality. However, the converse does not hold. For instance, the numerosities of the square numbers is different from the numerosity of the natural numbers (as it should be, given that we are implementing the part-whole principle) but their cardinality is the same. Since the relation of equivalence among sets induced by the

---

[3] For a survey of the history see Mancosu 2009; Mancosu 2016, chapter 3.

theory of numerosities is much finer than the one yielding Cantorian cardinalities, it turns out that one needs many more numerosities (and this many more is spelled out in terms of ordinary set theory) than the objects whose collections are being counted. For instance, a theory of numerosities for arbitrary collections of natural numbers requires at least uncountably many numerosities. By contrast, a theory of cardinality for arbitrary collections of natural numbers requires only countably many cardinalities. In this observation is contained the essence of an important result which will show the impossibility, in a Fregean setting, of implementing the part-whole principle for operators defined by abstraction.[4] We will cover the reasons, formally and informally, in the next section.

3 Basic Law V and Zermelo's Result

Abstraction principles in the Fregean context compress the concepts into the domain of objects according to an equivalence relation on concepts. When one looks at the situation set-theoretically this corresponds to the idea of mapping the power set of a set X into the set X itself according to an equivalence relation defined among the elements of the power set. In the case of BLV this relation is the extensional identity of the sets, in the case of HP it is the existence of a one-to-one mapping between the sets.

In a Fregean setting an abstraction operator is an operator on concepts ∂ such that $\partial X = \partial Y$ iff XEY, where E is some equivalence relation on concepts. What would it mean to say that ∂ must satisfy a part-whole constraint? We say that an abstraction operator satisfies part-whole if the following condition holds:

$$\text{PW} \qquad \forall X, Y (X \subsetneq Y \to \partial X \neq \partial Y),$$

where $X \subsetneq Y =_{def} \forall x\, (X(x) \to Y(x)) \land \exists y\, (Y(y) \land \neg X(y))$.[5]

But PW does not have a standard model in a Fregean setting, i.e. a model of the form $\langle X, P(X), f \rangle$ for $f : P(X) \to X$, where $P(X)$ is the full power set of X. As Mancosu has pointed out in Mancosu 2016, an old result by Zermelo (see Zermelo 1904 and Kanamori 1997, Kanamori 2004) shows that for any X and for any f, if $f : P(X) \to X$

---

4 By saying this we do not mean to subscribe to the widely held view that the "real cause" of the paradoxicality of BLV is that it contradicts Cantor's theorem. It is indeed the case that BLV contradicts Cantor's theorem, but some variations of BLV are inconsistent even though the inconsistency does not rest on cardinality issues. See Paseau 2015, whose considerations are also applicable to the formal inconsistency emerging from operators satisfying part-whole.
5 We will also say that an equivalence relation $E(X, Y)$ *satisfies part-whole* iff $X \subsetneq Y \Rightarrow \neg E(X, Y)$.

then there must be sets $A, B \subset X$ such that $A \subsetneq B$ and $f(A) = f(B)$.[6] Since Frege's systems are second-order systems, this unsatisfiability does not automatically lead to a formal refutation. In this paper, we will show that second-order logic can actually refute the existence of any such operator.

Among the consequences of this result are the following. First, the formal inconsistency of Basic Law V can be seen in the context of a wider family of formally inconsistent abstraction principles (i.e. abstraction principles whose main operator satisfies part-whole), and thus the inconsistency of BLV is not an isolated fact. Indeed, note that the extension operator characterized by BLV trivially satisfies the part-whole principle, for if $A \subsetneq B$ then A and B are extensionally different and thus $\partial A \neq \partial B$. Mancosu's principles BLV-F and FSP (Mancosu 2015 and Mancosu 2016, Appendix, section 4.10) also are shown inconsistent by this result since they both satisfy part-whole. BLV-F is the following abstraction principle.

BLV-F $\quad \forall X, Y (\partial X = \partial Y \leftrightarrow ((\text{Fin}(X) \wedge \text{Fin}(Y) \wedge X \cong Y) \vee (\text{Inf}(X) \wedge \text{Inf}(Y) \wedge X = Y)))$,

where $\text{Fin}(X)$ expresses that X is finite in second-order logic by saying there is no injection $f : X \to Y$ where $Y \subsetneq X$ and so $\text{Inf}(X) =_{def} \neg \text{Fin}(X)$ expresses that X is infinite.

We now define FSP. First we define a relation $\cong$ between concepts as follows. $A \cong B$ iff there is an injective function from A to B, every injective function from A to B is surjective, and every injective function from B to A is surjective. Only pairs of finite concepts can satisfy $\cong$. However, $\cong$ is not an equivalence relation over concepts since reflexivity fails for infinite sets. Now we define $\cong^*$ by means of $\cong: A \cong^* B$ iff there is a $C \subseteq A$ and a $D \subseteq B$ such that $A \setminus C = B \setminus D$, and $C \cong D$. The equivalence relation is due to Frank Sautter, hence the label FSP:

FSP $\quad \forall X, Y (\partial(X) = \partial(Y) \leftrightarrow X \cong^* Y)$

Second, neologicist foundations of mathematics will never be able to implement a foundational approach which will vindicate the part-whole intuition on all its concepts (contrary to what the theory of numerosities achieves in a standard set-theoretical foundational framework). Third, we will look at other intuitive properties of counting studied in the theory of numerosities (such as the Aristotelian principle) and show their refutability in second-order logic. In addition, as a consequence of the study of the Aristotelian principle, we will show that there are infinitely many abstraction principles that (provably) satisfy part-whole and

---

[6] Boolos (see Boolos 1997, Boolos 1998) had reached the same result from scratch by thinking about Cantorian counterexamples. As pointed out by Kanamori himself, Boolos's proof is essentially equivalent to that contained in the cited work by Zermelo and Kanamori.

which are thus all refutable in second-order logic. Finally, we will also mention some results in the foundational framework of reverse mathematics, showing that part-whole principles can be refuted there too.

But let us now say something more about the foundational situation with Basic Law V. The proof of its inconsistency relies on impredicative comprehension principles (for the expanded language). But BLV is consistent with predicative comprehension and one can pursue a version of logicism more faithful to Frege in this system (see Burgess 2005) In fact, the bad company problem generated by inconsistent abstraction principles disappears at the level of predicative abstraction.[7] Sean Walsh, improving on results of others, has shown that the theory of all purely second-order definable abstraction principles (whose right-hand side is an equivalence relation, provably in second-order logic) is consistent with Δ^1_1 comprehension in the expanded language (and, indeed, full comprehension for pure second-order formulas; see Walsh 2016 and the discussion below Theorem 3.5). Beyond this and some related results, inconsistent abstraction principles haven't been given much attention in the literature. But the above results already suggest that a formal refutation of the existence of abstraction operators satisfying the part-whole condition in pure second-order logic (in the precise sense specified below) must use at least Π^1_1 comprehension.

4 Background Logic

For a standard introduction to second-order logic, see Shapiro 1991. We work in the signature of second-order logic with the additional function symbol ∂ whose arguments are monadic predicates (i.e. concepts in Fregean jargon) and whose values are objects. To show that some abstraction principles are inconsistent, we show that the negations of certain sentences in this language are provable, keeping track of the amount of comprehension used in the proofs.

$\vdash$ denotes provability in second-order logic with just the usual natural deduction rules for quantifiers and connectives, where we treat ∂X as a term for an object. We also assume extensionality for predicates, though the results go through without this (with some care). In particular, we do not assume any amount of choice and flag any comprehension assumptions in our theorems.

We are interested in using syntactic complexity classes to give a stratification of comprehension principles. Since we don't have the existence of a pairing function

---

[7] In fact, the bad company problem for jointly inconsistent abstractions disappears, though this is not relevant for most of our results.

in our general context, we need to allow finite strings of quantifiers instead of single quantifiers in the definitions of our syntactic classes.

Definition 4.1.
- A term is either an object variable x or $\partial(X)$ for a unary predicate variable X.
- A formula is atomic iff it has the form $s = t$ or $R(\bar{t})$ for s, t terms and $\bar{t}$ an n-tuple of terms, where R is an n-ary predicate variable.
- FO is the smallest class of formulas which contains the atomic formulas and is closed under Boolean combinations and object quantification.
- Σ^1_1 is the class of all formulas of the form

$$\exists X_1 \ldots X_n\, \varphi$$

  for φ in FO and $n \in \omega$ (the case $n = 0$ just ensures FO formulas are Σ^1_1).
- Π^1_1 is the class of all formulas of the form

$$\forall X_1 \ldots X_n\, \varphi$$

  for φ in FO and $n \in \omega$
- Σ^1_2 is the smallest class of formulas of the form

$$\exists X_1 \ldots X_n \varphi$$

  for φ in Π^1_1 and $n \in \omega$.

Definition 4.2.
- For φ a formula with free variables including the n-tuple $\bar{v}$, we put

$$\mathrm{CA}(\varphi, \bar{v}) =_{def} \text{(the universal closure of)}\ \exists R\, \forall \bar{v}\, (R(\bar{v}) \leftrightarrow \varphi(\bar{v})),$$

  where R is an n-ary predicate variable.
- For Γ a set of formulas (e.g. one of the classes described above), we let

$$\Gamma\text{-CA} =_{def} \{\mathrm{CA}(\varphi, \bar{v}) \mid \varphi \in \Gamma \text{ and } \bar{v} \text{ free variables of } \varphi\}.$$

- We also define

$$\Delta^1_1\text{-CA} =_{def} \{(\varphi \leftrightarrow \psi) \to \mathrm{CA}(\varphi, \bar{v}) \mid \varphi \in \Sigma^1_1, \psi \in \Pi^1_1, \text{ and } \bar{v} \text{ free variables of } \varphi\}$$

  This says that we have comprehension for the provably Δ^1_1 predicates.

We also use some standard notational variations, e.g. "$x \in X$" instead of "$X(x)$".

It'll be useful to use the following closure properties of comprehension.

Proposition 4.3. *The following are provable in FO-CA, for all formulas φ, ψ with appropriate free variables.*

- If $CA(\varphi, x, \bar{v})$, then $CA(\neg\varphi, x, \bar{v})$, $CA(\exists x\, \varphi, \bar{v})$, and $CA(\forall x\, \varphi, \bar{v})$.
- If $CA(\varphi, \bar{v})$ and $CA(\psi, \bar{v})$, then $CA(\varphi \wedge \psi, \bar{v})$ and $CA(\varphi \vee \psi, \bar{v})$.

Proof. These are all fairly similar, so we just check one of them. Work in FO-CA and suppose $CA(\varphi, \bar{v})$. We show $CA(\neg\varphi, \bar{v})$. Let A witness $CA(\varphi, \bar{v})$, i.e. let A be such that for all $\bar{x}$,
$$A(\bar{x}) \Leftrightarrow \varphi(\bar{x}).$$
By comprehension for the FO formula $\neg X(\bar{v})$, we have (plugging in A in for the second-order variable X) that there is a B such that for all $\bar{x}$
$$B(\bar{x}) \Leftrightarrow \neg A(\bar{x}).$$
Recalling our choice of A, we have for all $\bar{x}$,
$$B(\bar{x}) \Leftrightarrow \neg\varphi(\bar{x}),$$
so that B witnesses $CA(\neg\varphi, \bar{v})$. □ Proposition 4.3

From this proposition we get the following.

Corollary 4.4. $\Pi^1_1\text{-CA} \equiv \Sigma^1_1\text{-CA}$.

Proof. This is immediate from the previous proposition since every Σ^1_1 formula is equivalent to the negation of a Π^1_1 formula, and vice-versa. □ Corollary 4.4

Proposition 4.3 tells us that Π^1_1-CA also implies comprehension for the syntactic class formed by closing the Π^1_1 formulas under Boolean operations and object quantifiers and that Σ^1_2-CA implies comprehension for the closure of Σ^1_2 under Boolean operations and object quantifiers. We shall often use these consequences of Π^1_1-CA and Σ^1_2-CA without additional explanation.

We also need the following observation about Π^1_1 formulas.

Proposition 4.5. *Finite conjunctions and disjunctions of Π^1_1 formulas are equivalent to Π^1_1 formulas.*

Proof. For conjunctions, this is just because universal quantifiers distribute over conjunctions, i.e. $\forall X\, (\varphi \wedge \psi)$ is equivalent to $\forall X\, \varphi \wedge \forall X\, \psi$. For disjunctions, we just use that we are allowing strings of universal quantifiers, plus some care with free variables. Suppose φ, ψ are formulas such that X is *not* free in ψ and Y is not free in φ. Then $\forall X\, \varphi \vee \forall Y\, \psi$ is equivalent to $\forall X, Y\, (\varphi \vee \psi)$. □ Proposition 4.5

5 Inconsistent Abstraction Principles

In this section we prove the main results of the paper.

5.1 Basic Law V

First note that under extensionality, BLV becomes $\forall X, Y\, (\partial X = \partial Y \leftrightarrow X = Y)$.

Theorem 5.1. Π_1^1-CA $\vdash \neg$BLV

We give a short proof which is really just the proof of Cantor's theorem that there is no injection $f : P(X) \to X$.

Proof. Work in Π_1^1-CA. Suppose BLV. By Π_1^1-CA, let A be a predicate such that
$$x \in A \Leftrightarrow \exists X\, (x = \partial X \wedge x \notin X).$$
Suppose $\partial A \in A$. Let X be such that $\partial X = \partial A$ and $\partial A \notin X$. By BLV, we have $A = X$. So also $\partial A \in X$, a contradiction. So $\partial A \notin A$. But then A is a witness to $\exists X\, (\partial A = \partial X \wedge \partial A \notin X)$, so that, by choice of A, $\partial A \in A$, a contradiction. Hence, $\neg$BLV. □ Theorem 5.1

The hypothesis of this theorem has no improvement among the natural comprehension hypotheses, since as mentioned in the introduction, BLV + Δ_1^1-CA is consistent.

5.2 Part-Whole Principles

We define the following sentence in the expanded language.

$$\text{PW} \quad \forall X, Y\, (X \subsetneq Y \to \partial X \neq \partial Y).$$

PW says that ∂ respects part-whole, i.e. (speaking set-theoretically for a moment) each set has a ∂-value which is distinct from all of the ∂-values of its proper subsets.

Of course, PW is not literally an abstraction principle, though there are abstraction principles which imply PW, for example BLV, and BLV-F and FSP. So, the following theorem gives a new proof of the inconsistency $\neg$BLV, and the first proof of the inconsistency of the other principles.[8] In §5.3 we will show that there are infinitely many abstraction principles which deductively imply PW.

---

[8] An unpublished proof of the refutability of BLV-F was communicated by Roy Cook to the first author on June 23, 2014 (see acknowledgment in Mancosu 2016: 200). The proof used a modification

Theorem 5.2. Σ_2^1-CA $\vdash \neg$PW.

The idea is to build witnesses $X \subsetneq Y$ to the failure of PW by transfinite induction. We first look at what object our abstraction operator ∂ assigns to the empty set, call this x_0, and put $x_0 \in Y$. We then look at what ∂ assigns to $\{x_0\}$. If this is again x_0, we're done since we can take $X = \emptyset$ and $Y = \{x_0\}$ to violate PW. Otherwise $\partial\{x_0\}$ is some $x_1 \notin \{x_0\}$, and we put $x_1 \in Y$. If $\partial\{x_0, x_1\} = x_0$ or $\partial\{x_0, x_1\} = x_1$, we get witnesses to the failure of PW by letting $Y = \{x_0, x_1\}$ and $X = \emptyset$ or $X = \{x_0\}$, respectively. Otherwise, $\partial\{x_0, x_1\} \notin \{x_0, x_1\}$, so we add it to Y. In the end of this process, we get some well-ordered set Y such that for all proper initial segments Z of Y, ∂Z is the next element in the well-order. But then $\partial Y \in Y$ since otherwise we could continue building Y by adding ∂Y. By construction, ∂Y got into Y by being the ∂ value of some proper initial segment X of Y, i.e. $\partial Y = \partial X$ and $X \subsetneq Y$, violating PW.

In order to make this work in our setting, we just need to see that we can describe this Y in a Σ_2^1 way: it is the field of the unique maximal well-order R such that for all proper initial segments S of R, the R-least upper bound of S is the ∂ value of the field of S.

Before we flesh out this argument, we introduce some abbreviations. Below, R, S are binary predicate variables.

- $x \in \mathsf{Field}(R) =_{def} \exists y\, (R(x, y) \vee R(y, x))$
- $\mathsf{LO}(R) =_{def} \forall x, y, z \in \mathsf{Field}(R)\, \big(R(x, x) \wedge (R(x, y) \vee R(y, x) \vee x = y) \wedge ((R(x, y) \wedge R(y, x)) \to x = y) \wedge ((R(x, y) \wedge R(y, z)) \to R(x, z))\big)$
- $\mathsf{WO}(R) =_{def} \mathsf{LO}(R) \wedge \forall X\, ((X \subseteq \mathsf{Field}(R) \wedge X \neq \emptyset) \to \exists x \in X\, \forall y \in X\, R(x, y))$
- $R \trianglelefteq S =_{def} \forall x, y\, (y \in \mathsf{Field}(R) \to (R(x, y) \leftrightarrow S(x, y)))$
- $R \trianglelefteq\!\!\!\!/\, S =_{def} R \trianglelefteq S \wedge \exists y\, (y \in \mathsf{Field}(S) \wedge y \notin \mathsf{Field}(R))$
- $R_x(y, z) =_{def} R(y, x) \wedge R(z, x) \wedge x \neq y \wedge x \neq z \wedge R(y, z)$
- $(R \oplus x)(y, z) =_{def} R(y, z) \vee (y \in \mathsf{Field}(R) \wedge z = x) \vee (y = x \wedge z = x)$.

So $\mathsf{Field}(R)$ defines the field of the relation R, $\mathsf{LO}(R)$ states that R is a non-strict linear order, $\mathsf{WO}(R)$ is the Π_1^1-formula asserting that R is a well-order of its field, $R \trianglelefteq S$ expresses that R is an initial segment of S, $R \trianglelefteq\!\!\!\!/\, S$ expresses that R is a *proper* initial segment of S, $R_x(y, z)$ defines the initial segment of R below some element, and $R \oplus x$ is the relation obtained from R by adjoining x as the new largest element (at least when $x \notin \mathsf{Field}(R)$).

---

of the standard proof of the refutability of BLV. While of independent interest, Cook's proof is not generalizable to the other abstraction principles satisfying part-whole that are being considered in this paper.

A useful observation is that $R_x \trianglelefteq R_x \oplus x \trianglelefteq R$ whenever $x \in \text{Field}(R)$ and if x is the R-largest element, then $R_x \oplus x = R$, granting these predicates exist (they do under FO-CA).

The following Lemma is the key step in the proof, outlined above. The set-theoretic analogue is due to Zermelo, isolated in Kanamori 1997 and Kanamori 2004.

Lemma 5.3. *The following is provable in Σ_2^1-CA. There is a unique binary predicate R such that*
(i) $\text{WO}(R)$;
(ii) *for all $x \in \text{Field}(R)$, $\partial \text{Field}(R_x) = x$; and*
(iii) $\partial \text{Field}(R) \in \text{Field}(R)$

Formally, by (ii) we mean
$$\forall X \forall x (x \in \text{Field}(R) \rightarrow (\forall y (y \in X \leftrightarrow \exists z (R_x(y,z) \vee R_x(z,y))) \rightarrow \partial X = x).$$

Under Σ_2^1-CA, for every R and $x \in \text{Field}(R)$, R_x and $\text{Field}(R_x)$ exist, so that the only X with the property $\forall y (y \in X \leftrightarrow \exists z (R_x(y,z) \vee R_x(z,y)))$, by extensionality, and this quantification is non-vacuous.

By (iii), we mean
$$\forall X (\forall x (x \in X \leftrightarrow x \in \text{Field}(R)) \rightarrow \partial X \in \text{Field}(R)).$$

Again, by extensionality, the only X which satisfies the antecedent is $\text{Field}(R)$, which exists by our comprehension hypothesis.

Notice that each of (i), (ii), (iii) is Π_1^1 ((ii) and (iii) are easily seen to be Δ_1^1, though we won't use this).

Proof. Let $\Phi(R)$ be the conjunction of (i) and (ii). By Proposition 4.5, $\Phi(R)$ is equivalent to a Π_1^1 formula.

We'll need the following observation about Φ relations.

Claim 5.4. *Suppose $\Phi(R)$ and $\Phi(S)$ and for all $x \in \text{Field}(R)$, $R_x \trianglelefteq S$. Then either $R \trianglelefteq S$ or $S \trianglelefteq R$.*

Proof. There are two cases, depending on whether R has a largest element or not.

Suppose first that R has no largest element. Then for every $y \in \text{Field}(R)$ there is a $z \in \text{Field}(R)$ such that $y \in \text{Field}(R_z)$ (since y is not the R largest element there is a z such that $y \neq z$ and $R(y,z)$, so that $y \in \text{Field}(R_z)$). So for all x, y such that $y \in \text{Field}(R)$,
$$R(x,y) \Leftrightarrow \exists z\, R_z(x,y) \Leftrightarrow S(x,y),$$
so $R \trianglelefteq S$ by definition (using here $R_z \trianglelefteq S$ for every z).

Now suppose R has a largest element, x. We have $R_x \trianglelefteq S$. If $S = R_x$, we're done, so suppose $R_x \not\trianglelefteq S$. Since S is a well-order, there is an S-least element $y \notin \text{Field}(R_x)$. It is easy to see that this y is the S-least upper bound of $\text{Field}(R_x)$, so that $S_y = R_x$. By clause (ii) of Φ we have

$$x = \partial(\text{Field}(R_x)) = \partial(\text{Field}(S_y)) = y.$$

So then $R = R_x \oplus x \trianglelefteq S$. □ Claim 5.4

From Claim 5.4 we actually get that the Φ relations are totally ordered by initial segment.

Claim 5.5. *For all R, S such that $\Phi(R)$ and $\Phi(S)$, either $R \trianglelefteq S$ or $S \trianglelefteq R$.*

Proof. First notice that if $\Phi(R)$ and $S \trianglelefteq R$, then $\Phi(S)$. In particular, if $\Phi(R)$ and $x \in \text{Field}(R)$, then $\Phi(R_x)$.

Now let R and S be such that $\Phi(R)$ and $\Phi(S)$.
Let X be such that for all $x \in \text{Field}(R)$

$$X(x) \Leftrightarrow R_x \not\trianglelefteq S.$$

If X is empty, we have for all x, $R_x \trianglelefteq S$, so that Claim 5.4 implies either $R \trianglelefteq S$ or $S \trianglelefteq R$, as desired.

So suppose X is nonempty. Since R is a well-order, let x be the R-least element of X. As remarked at the start of the proof, we have that $\Phi(R_x)$. Also, by our choice of x, for all $y \in \text{Field}(R_x)$, $R_y \trianglelefteq S$ (since $R(y, x)$ implies $y \notin X$). Now we apply Claim 5.4 to R_x and S. $R_x \not\trianglelefteq S$, by assumption, so we must have $S \trianglelefteq R_x$. But then $S \trianglelefteq R$, so we're done.

□ Claim 5.5

We now define the witness R to the Lemma. As in our informal proof above, R will just be the maximal Φ relation, which exists by our comprehension hypothesis and Claim 5.5.

$$R(x, y) \Leftrightarrow \exists S \, (\Phi(S) \wedge S(x, y)).$$

Set-theoretically, this amounts to taking the union over all the Φ relations, as in Kanamori's rendering of Zermelo's proof.

The formula on the right-hand side is equivalent to a Σ_2^1 formula since $\Phi(S) \wedge S(x, y)$ is equivalent to a Π_1^1 formula by Proposition 4.5, again. So by Σ_2^1-CA, R exists. Now we just need to check that (i)-(iii) hold of this R.

For (i), first notice that if $x, y \in \text{Field}(R)$, then there are S, T such that $\Phi(S), \Phi(T)$ and $x \in \text{Field}(S)$, $y \in \text{Field}(T)$. By Claim 5.5, $S \trianglelefteq T$ or $T \trianglelefteq S$, so that either both $x, y \in \text{Field}(S)$ or both $x, y \in \text{Field}(T)$. In either case, we have $R(x, x)$,

$R(x, y) \vee R(y, x) \vee x = y$, and $(R(x, y) \wedge R(y, x)) \rightarrow x = y$, since these hold of S and T. We can check transitivity similarly. So R is a linear order on its field. To show it is a well-order on its field, let $X \subseteq \text{Field}(R)$ such that X is nonempty. Then there is an $x \in X$ and some S such that $\Phi(S)$ and $x \in \text{Field}(S)$. Suppose y is such that $R(y, x)$. Then $T(y, x)$ for some T such that $\Phi(T)$. By Claim 5.5, we have either $T \trianglelefteq S$ or $S \trianglelefteq T$. In either case, it follows that $S(y, x)$. So we have for all y, $R(y, x) \Leftrightarrow S(y, x)$. Hence, letting $Y(y) \Leftrightarrow R(y, x)$ (this predicate exists by Σ_2^1-CA), Y is a nonempty subset of $\text{Field}(S)$, and so has an S-least element, which then is the R-least element of X. So R is a well-order.

(ii) is similar. For $x \in \text{Field}(R)$, $x \in \text{Field}(S)$ for some S such that $\Phi(S)$. By the proof of (i), $S_x = R_x$, so $\partial \text{Field}(R_x) = \partial \text{Field}(S_x) = x$, by $\Phi(S)$. So we have that $\Phi(R)$ (since Φ was just the conjunction of (i) and (ii)).

Finally, for (iii), suppose $\partial \text{Field}(R) \notin \text{Field}(R)$. Define $R' = R \oplus \partial(\text{Field}(R))$. We'll show that $\Phi(R')$ and then derive a contradiction. R' is a well-order since R is and we've only added a new largest element (every subset of $\text{Field}(R')$ is just a subset of $\text{Field}(R)$ possible with the extra element $\partial \text{Field}(R)$). Also, for every $x \in \text{Field}(R)$, $R'_x = R_x$, so the only new case of (ii) to check is that $\partial \text{Field}(R'_{\partial(\text{Field}(R))}) = \partial \text{Field}(R)$. But, by definition, $R'_{\partial(\text{Field}(R))} = R$, so indeed $\partial \text{Field}(R'_{\partial(\text{Field}(R))}) = \partial \text{Field}(R)$. So we have $\Phi(R')$ and $R'(\partial \text{Field}(R), \partial \text{Field}(R))$, so by definition of R, $R(\partial \text{Field}(R), \partial \text{Field}(R))$. So $\partial \text{Field}(R) \in \text{Field}(R)$ after all, a contradiction.

For uniqueness, notice that any S satisfying (i)-(iii) has no proper end extension which satisfies Φ, which we can see as follows. Suppose $S \triangleleft S'$ and $\Phi(S')$. Then since $\text{Field}(S) \subsetneq \text{Field}S'$ and is nonempty (by (iii)), there is some S'-least element x S'-greater than everything in $\text{Field}(S)$. So then $S = S'_x$ and $x = \partial(\text{Field}(S))$, by (ii). But then $\partial(\text{Field}(S)) \notin \text{Field}(S)$, contradicting (iii) for S.

So, any S satisfying (i)-(iii) is the unique maximal binary predicate satisfying Φ, by Claim 5.5. Hence R is the unique binary predicate satisfying (i)-(iii).

□ Lemma 5.3

We can now easily prove Theorem 5.2.

Proof of Theorem 5.2. Let R be as in Lemma 5.3. Let $Y = \text{Field}(R)$ and $X = \text{Field}(R_{\partial(Y)})$ (these exist by Σ_2^1-CA). By (iii), $\partial Y \in Y$. So, $R_{\partial(Y)}$ is a proper initial segment of R and $X \subsetneq Y$. On the other hand, (ii) implies $\partial(X) = \partial(Y)$, so ¬PW.

□ Theorem 5.2

Notice that Lemma 5.3 implies that R is actually Δ^1_2, since it can be equivalently defined by

$$R(x, y) \leftrightarrow \forall S \left([\Phi(S) \wedge \partial \text{Field}(S) \in \text{Field}(S)] \to S(x, y) \right).$$

Still, to get that the antecedent is non-vacuous, we need to prove there is such a relation, and so this argument seems to need Σ^1_2-CA.

Two natural questions arise from the above result. The first is whether Σ^1_2-CA is the optimal comprehension hypothesis for refuting PW. The second question was asked to us by Leo Harrington and is motivated by the fact that PW is clearly expressible in monadic second-order logic with the function symbol ∂ taking monadic concepts to objects. Our proof seems to use binary relations in an essential way, so Harrington asked whether PW can be refuted in a monadic setting, without appeal to binary relations. This second question was positively answered by Guillaume Massas, after the first version of this paper had been circulated. He gave a shorter refutation of PW which we include here with his permission. His proof also sheds some light on the first question, which we'll discuss after giving the proof. The following result is carried out in monadic second-order logic with an additional function symbol ∂ and Σ^1_3-CA for monadic fomulas with just one distinguished free variable. First we give a useful definition. For y an object variable and Y a predicate variable, we let $Y = \{y\}$ abbreviate the formula expressing that y is the unique element of Y, i.e.

$$Y = \{y\} =_{\text{def}} Y(y) \wedge \forall z \left(Y(z) \to z = y \right).$$

Theorem 5.6. Σ^1_3-CA $\vdash_{MON} \neg$PW

Proof. Work in Σ^1_3-CA[1] restricted to monadic formulas with one distinguished free variable. Let $\psi(X)$ be the formula

$$\forall y \exists Z \left(X(y) \to (Z \subsetneq X \wedge \partial Z = y) \right).$$

$\psi(X)$ asserts that every member of X is the ∂ value of some proper subset of X. Notice that $\psi(X)$ is provably equivalent to the formula

$$\forall Y \exists X \left(\exists y \left(Y = \{y\} \wedge X(y) \right) \to (Z \subsetneq X \wedge \partial Z \in Y) \right),$$

where we replace the quantification over objects by, basically, quantification over the singleton predicates.

As in our previous proof, we use an outer existential predicate quantifier to take a union of the X such that $\psi(X)$.

By our comprehension assumption, we let S be such that

$$S(x) \leftrightarrow \exists X \left(X(x) \wedge \psi(X) \right).$$

The formula on the right-hand side is not literally Σ_3^1, but it is equivalent to a Σ_3^1 formula by the observation above.

Notice that if $\psi(X)$, then $X \subseteq S$. It follows that $\psi(S)$ holds. $\psi(S)$ means that if $y \in S$, then $y = \partial Z$ for some $Z \subsetneq S$. So suppose $y \in S$. By definition of S, there is an X such that $y \in X$ and $\psi(X)$. This means that $y = \partial Z$ for $Z \subsetneq X \subseteq S$, as desired.

We now show that $\partial S \in S$. Suppose $\partial S \notin S$. It is enough to see that in that case $\psi(S \cup \{\partial S\})$ holds, since then $S \cup \partial S \subseteq S$, so that we have $\partial S \in S$, a contradiction. Again, we need to see that if $y \in S \cup \{\partial S\}$, then $y = \partial Z$ for $Z \subsetneq S$. If $y \in S$, then by $\psi(S)$ we have that $y = \partial Z$ for $Z \subsetneq S \subseteq S \cup \{S\}$. So consider $y = \partial S$. By assumption, $S \subsetneq S \cup \{\partial S\}$, so S witnesses that $y = \partial Z$ for some $Z \subsetneq S \cup \{\partial S\}$. This shows $\psi(S \cup \{\partial S\})$, so we reach a contradiction, as desired.

So $\partial S \in S$. By $\psi(S)$, we have that $\partial S = \partial Z$ for $Z \subsetneq S$, i.e. S, Z witness $\neg PW$.

□ Theorem 5.6

This proof is definitely simpler than our original proof, though we will need our proof (really Lemma 5.3) elsewhere in the paper. However, whether it is an improvement in terms of lowering the comprehension assumption is a delicate issue.

In order to discuss that issue we return to our original set up where we allow binary relations (and relations of any arity). In this setting we can express the choice schema Σ_1^1-AC.

$$\Sigma_1^1\text{-AC} \quad \forall x \exists Y \varphi(x, Y) \to \exists R \forall x \varphi(x, R_x),$$

$$\varphi \text{ is } \Sigma_1^1, \text{ possibly with parameters.}$$

In the above, we use the abbreviation $\varphi(x, R_x)$ for the translation of $\varphi(x, Y)$ obtained by replacing all instances of the formula $Y(t)$ by $R(x, t)$, where t is any term.

The formula we applied comprehension to was provably equivalent to one of the form $\exists X \forall y \exists Z \varphi$ for φ first-order. Under Σ_1^1-AC, we have that this is provably equivalent to a Σ_1^1 formula. So actually, the above proof shows the following.

Theorem 5.7. $\Pi_1^1\text{-CA} + \Sigma_1^1\text{-AC} \vdash \neg PW$.

As far as we know, the theories Σ_2^1-CA and Π_1^1-CA + Σ_1^1-AC are incomparable, i.e. neither implies the other.[9] Perhaps this is evidence that our comprehension hierarchy is not natural, in some way. Maybe we should build in closure under object

---

[9] In fact, Σ_2^1-CA and comprehension for formulas of the form $\exists X \forall y \exists Z \varphi$, where φ is first-order, are incomparable comprehension principles, as far we know. It should be clear to the reader that a detailed study of the possible alternatives to the Σ_n^1 and Π_n^1 hierarchies and their resulting comprehension principles in pure second order logic without choice (and with or without the extra function symbol ∂) is needed. To our surprise, this has been neglected in the literature.

quantifiers into the syntactic classes Π_1^1, Σ_2^1, etc, so that the previous theorem becomes a refutation of PW from Π_1^1-CA. We've decided not to do this for several reasons. First, the syntactic hierarchy we're using is equivalent to the one used in reverse mathematics, where issues like this come up, albeit higher up (at the level of Σ_3^1). Moreover, we think that the question of whether we can drop the assumption of Σ_1^1-AC in the previous theorem is an interesting question which would be obfuscated in a coarser comprehension hierarchy.

5.3 The Aristotelian Principle

In the context of the theory of numerosities, see Benci et al. 2012, one has studied equivalence relations that have special properties. One interesting property of an equivalence relations has been labeled the Aristotelian principle. This is something of a misnomer as the principle captures two of Euclid's common notions, namely that adding equals to equals, the results are equal and that subtracting equals from equals, the results are equal. We will keep the name Aristotelian principle in order not to create confusion with the literature. We first discuss some basic results about equivalence relations satisfying the Aristotelian principle and then apply these results to the context of abstraction principles.

Let $\cong$ be an equivalence relation. We say that $\cong$ satisfies the *Aristotelian principle* if and only if

$$\text{AP} \quad \forall A, B \, (A \cong B \leftrightarrow A \setminus B \cong B \setminus A).$$

This is a very natural principle for counting on finite sets (interpreting the equivalence as one-to-one correspondence) but the principle fails in general for Cantorian cardinalities. For instance, let A stand for $\omega \setminus \{1\}$ and B for ω. Then $|A| = |B|$ but $A \setminus B = \emptyset$ and $B \setminus A$ is $\{1\}$, so $|A \setminus B| \neq |B \setminus A|$. On the other hand, the trivial equivalence relation (on any domain) $X \cong Y \leftrightarrow \top$ satisfies AP.

In order to show that AP is a compact form to express the two aforementioned common notions from Euclid, we state, following Benci et al. 2012, two different principles, the sum principle (SP) and the difference principle (DP), and show that AP is equivalent to the conjunction of the two principles.

$$\text{SP} \quad \forall A, A', B, B' \, ((A \cap B = A' \cap B' = \emptyset \wedge A \cong A' \wedge B \cong B') \to A \cup B \cong A' \cup B')$$
$$\text{DP} \quad \forall A, A', B, B' \, ((A \cong A' \wedge A \cup B \cong A' \cup B') \to B \setminus A \cong B' \setminus A').$$

Proposition 5.8. *An equivalence relation $\cong$ satisfies AP iff it satisfies SP and DP.*

Claim 5.9. *If $\cong$ satisfies SP and DP, then $\cong$ satisfies AP.*

Proof. Assume $\cong$ satisfies SP and DP. We show first if $A \cong B$, then $A \setminus B \cong B \setminus A$.

Assume $A \cong B$. We obtain our result by letting in the statement of DP $B = A'$ and $A = B'$. This automatically gives $A \cup B = A' \cup B'$ and thus $A \cup B \cong A' \cup B'$. Thus, by DP we obtain
$$B \setminus A \cong B' \setminus A'.$$
For the other direction, assume $A \setminus B \cong B \setminus A$. Letting $B = B' = A \cap B$, $A = A \setminus B$, and $A' = B \setminus A$, by SP we obtain
$$(A \setminus B) \cup (A \cap B) \cong (B \setminus A) \cup (A \cap B).$$
The left-hand side is A and the right hand side is B, so $A \cong B$. □ Claim 5.9

Claim 5.10. *If $\cong$ satisfies AP, then $\cong$ satisfies SP and DP.*

See Appendix A.1 for the proof of Claim 5.10.

Equivalence relations which satisfy part-whole need not satisfy AP. We give an example of such an equivalence relation that is definable in pure second-order logic; this will be important later for our discussion about abstraction principles. Let $\text{Sing}(X)$ be the formula saying that X holds of exactly one object (this is our way of speaking about singletons). We define $\cong_1$ by

$$X \cong_1 Y \Leftrightarrow ((\text{Sing}(X) \wedge \text{Sing}(Y)) \vee (\neg\text{Sing}(X) \wedge \neg\text{Sing}(Y) \wedge X = Y));$$

i.e. $\cong_1$ identifies the singletons and works like Basic Law V on everything else.

It's easy to see that $\cong_1$ satisfies part-whole, since if $X \subsetneq Y$, then at most one of X and Y is a singleton, so $X \not\cong_1 Y$.

On the other hand, if there are at least three objects, $\cong_1$ does not satisfy AP and, in fact, doesn't satisfy either SP or DP. Let a, b, and c be distinct objects. Consider $\{a, b\}$ and $\{b, c\}$. Since these are distinct and not singletons, $\{a, b\} \not\cong_1 \{b, c\}$, but also $\{a, b\} \setminus \{b, c\} = \{a\} \cong_1 \{c\} = \{b, c\} \setminus \{a, b\}$, so $\{a, b\} \setminus \{b, c\} \cong_1 \{b, c\} \setminus \{a, b\}$, so AP fails. To see the failure of DP, let $A = \{b\}$, $B = \{a, c\}$, $A' = \{c\}$ and $B' = \{a, b\}$. DP fails since the conditions are satisfied but $\{a, c\} \not\cong_1 \{a, b\}$. It's easy to see that the assignment $A = \{b\}$, $A' = \{a\}$, $B = \{c\}$, $B' = \{d\}$ witnesses the failure of SP.

Indeed, the above can be suitably modified to yield infinitely many equivalence relations that satisfy part-whole but for which AP, SP and DP fail.

For each $n \geq 1$ we can define purely logically what it means for exactly n objects to fall under a concept. We express that relation by writing $|X| = n$ and henceforth we often treat concepts extensionally. Define now

$$X \cong_n Y \Leftrightarrow (|X| = n \wedge |Y| = n) \vee (|X| \neq n \vee |Y| \neq n \wedge X = Y).$$

It is easy to show that $\cong_n$ is an equivalence relation. If the domain of objects has cardinality $\leq n$, then the equivalence relation collapses into extensional identity (i.e. the equivalence relation used for BLV), so in particular satisfies part-whole. If there are at least $n + 1$ objects, then part-whole is satisfied on concepts X, Y on account of the following. If X and Y both have cardinality n, then they cannot be in the relation of part-whole. Moreover, if they have different cardinalities (with one of them perhaps of cardinality n) they are in the same equivalence class if and only if they are extensionally equivalent and thus part-whole is preserved. Since as we saw, extensional identity satisfies AP, SP, and DP, we will be able to falsify the three principles, while holding on to part-whole, by making, for each specific n, an assumption on the cardinality of the domain of objects.

Claim 5.11. *For each $n \in \omega$, if there are at least $n + 2$ objects, then $\approx_n$ doesn't satisfy AP, SP, or DP.*

Proof. Fix n. We show that SP and DP fail; the failure of either already implies AP fails by Proposition 5.8.

First we show DP fails. We actually only need $n + 1$ many objects to see this, so also only need $n + 1$ objects to have AP fail. Fix $n + 1$ distinct objects $a_1, \ldots, a_{n-2}, a, b, c$. Let

$$A = A' = \{a\},$$
$$B = \{a_1, \ldots, a_{n-2}, b\},$$
$$B' = \{a_1, \ldots, a_{n-2}, c\}.$$

We have $A \approx_n A'$ because of extensionality. Notice that

$$A \cup B = \{a, a_1, \ldots, a_{n-2}, b\},$$
$$A' \cup B' = \{a, a_1, \ldots, a_{n-2}, c\};$$

so $A \cup B \approx_n A' \cup B'$ since both sets have cardinality n. But

$$B \setminus A = \{a_1, \ldots, a_{n-2}, b\}$$
$$B' \setminus A' = \{a_1, \ldots, a_{n-2}, c\},$$

so $B \setminus A \not\approx_n B' \setminus A'$ since they have cardinality $n - 1$ but are not extensionally equivalent.

To show failure of SP, fix $n + 2$ distinct objects $a_1, \ldots, a_{n-2}, a, b, c, d$. Define

$$B = B' = \{a\},$$
$$A = \{a_1, \ldots, a_{n-2}, b, c\},$$
$$A' = \{a_1, \ldots, a_{n-2}, b, d\}.$$

Now $A \cap B = A' \cap B' = \emptyset$ and $A \cong_n A'$ because both have cardinality n. However,

$$A \cup B = \{a, a_1, \ldots, a_{n-2}, b, c\},$$
$$A' \cup B' = \{a, a_1, \ldots, a_{n-2}, b, d\}$$

So $A \cup B \not\cong_n A' \cup B'$, since they have cardinality $n + 1$ but are not extensionally equivalent.

□ Claim 5.11

As mentioned above, AP holds in the trivial case where we identify all subsets of our domain. If $\cong$ satisfies DP and additionally for all $X \neq \emptyset$, $X \not\cong \emptyset$, then actually $\cong$ satisfies part-whole, since if $Y \subsetneq X$, then $Y \cup X = X \cup \emptyset$, so if $Y \cong X$, DP implies $X \setminus Y \cong \emptyset$, but $X \setminus Y \neq \emptyset$, contradicting our assumption. The additional property for all $X \neq \emptyset$, $X \not\cong \emptyset$ is natural in the context of the theory of numerosities (it is a very special instance of the part-whole principle). We will return to this property below.

We now apply the above results in the context of abstraction principles, by considering the equivalence relation $\partial X = \partial Y$.

We define the following principles, which express in our augmented second-order language that the equivalence relation $\partial X = \partial Y$ has the various properties from the previous section. We omit the universal quantification over the (apparently) free concept variables.

APO $\partial X = \partial Y \leftrightarrow \partial(X \setminus Y) = \partial(Y \setminus X)$

SPO $(X \cap Y = \emptyset \wedge X' \cap Y' = \emptyset \wedge \partial X = \partial X' \wedge \partial Y = \partial Y') \rightarrow \partial(X \cup Y) = \partial(X' \cup Y')$

DPO $(\partial X = \partial X' \wedge \partial(X \cup Y) = \partial(X' \cup Y')) \rightarrow \partial(Y \setminus X) = \partial(Y' \setminus X')$.

The proof of Proposition 5.8 shows the following

Lemma 5.12. FO-CA ⊢ APO ↔ (SPO ∧ DPO)

Using our previous work on equivalence relations, we can now define for $n \geq 1$,

$$C_n \quad \forall X, Y \, (\partial(X) = \partial(Y) \leftrightarrow X \cong_n Y),$$

using here that $X \cong_n Y$ is expressible in pure second-order logic.

Proposition 5.13. *There are infinitely many abstraction principles that are refutable in second-order logic.*[10]

---

[10] There is a sense in which purely syntactic variants of BLV yield infinitely many abstraction principles which are refutable in second-order logic (for instance, one can modify the right hand side of BLV by disjoining it with a tautology; different tautologies will give rise to different abstrac-

Proof. Each C_n defines an abstraction operator that implies PW. Thus, by Theorem 5.2, Σ_2^1-CA $\vdash \neg C_n$. ☐ Proposition 5.13

Semantically, it is easy to find examples of abstraction principles that do not satisfy part-whole and that have no standard second-order models $\langle M, P(M), f \rangle$. All our examples of inconsistent abstraction principles provably imply part-whole. There are however infinitely many abstraction principles which defy part-whole and yet are formally refutable. This was shown by Richard Kimberly Heck in reply to a previous version of this paper. For each $n \geq 2$, let $E_n(X, Y)$ be the equivalence relation on concepts which puts the empty concept and all n-membered concepts in the same equivalence class and uses equi-extensionality on all the other concepts. Let P_n be the abstraction principle associated with $E_n(X, Y)$, namely

$$P_n \quad \forall X, Y (\partial X = \partial Y \Leftrightarrow E_n(X, Y)).$$

Then P_n is formally refutable. We sketch for the case $n = 2$. Define membership as usual:

$$a \in b \Leftrightarrow \exists X(\partial(X) = b \wedge X(a)).$$

Consider the first four Zermelo ordinals

$$0 = \{x \mid x \neq x\} = \emptyset,$$
$$1 = \{x \mid x = 0\} = \{0\},$$
$$2 = \{x \mid x = 1\} = \{\{0\}\},$$
$$3 = \{x \mid x = 2\} = \{\{\{0\}\}\}.$$

Using the definition of membership given above, we can talk about the Zermelo ordinals in our language by setting $0 = \partial(x \neq x)$, $1 = \partial(x = 0)$, $2 = \partial(x = 1)$, and $3 = \partial(x = 2)$.

With the exception of 0, the remaining three sets are all one-membered. Using equi-extensionality on one-membered concepts, which is decreed by P_2, one proves that 0, 1, 2, and 3 are all different. Moreover, $1 \notin 1$, $2 \notin 2$, and $3 \notin 3$. Hence, $x \notin x$ is neither empty nor two-membered. For refutability, we now use a slight modification of the Russell paradox. Let P_2 be given. Let $R = \partial(x \notin x)$. Is $R \in R$? Suppose so. Then

$$\exists X(\partial(X) = \partial(x \notin x) \wedge X(R)),$$

---

tion principles). Of course, we mean something stronger. We lack the space here to discuss how to specify criteria that equivalence relations have to satisfy in order to insure that two abstraction principles are different in more than a trivial syntactic sense. But for all the uses needed in this paper the following will do: it is consistent with Δ_1^1-CA that one abstraction principle holds, but the other doesn't.

i.e., for some F:
$$E_2(F, x \notin x) \wedge F(R).$$

We have already argued that $x \notin x$ is neither empty nor two-membered. So, by P_2, F must be co-extensive with $x \notin x$, whence $R \notin R$. But then
$$\partial(x \notin x) = \partial(x \notin x) \wedge R \notin R$$
and so
$$\exists X(\partial(X) = \partial(x \notin x) \wedge X(R)),$$
by comprehension and E_2, so $R \in R$ after all. Contradiction.

Heck has communicated to us further variations on refutable inconsistent abstractions defying part-whole.

Our examples of equivalence relations that satisfies part-whole, but not AP, SP, or DP can be used to show the following.

Proposition 5.14. Δ_1^1-CA + PW $\not\vdash$ APO, SPO, DPO

Proof. Consider, e.g. C_1. By the results in Walsh 2016, Δ_1^1-CA + C_1 has an infinite model, which, by the discussion above, satisfies PW but not APO, SPO, or DPO.
□ Proposition 5.14

The abstraction principle used in the proof of Proposition 5.14 was motivated by an example of Denis Hirschfeldt which showed that PW doesn't imply APO in the context of arithmetic. We give this example here. Consider a model of PW with domain the natural numbers. Let § be the abstraction operator satisfying PW. Define a new operator $ in the following way. If A is an infinite set let $(A) = 2§(A)$. If A is empty let $(A) = 1$. If A is a singleton then $(A) = 3$. Finally, one assigns odd values $(A) > 3$ to all other finite sets A in an injective way. This model still satisfies PW, but not APO, as witnessed by letting $A = \{1, 2\}$ and $B = \{0, 2\}$. Indeed $(A \setminus B) = (B \setminus A)$ (for $A \setminus B$ and $B \setminus A$ are singletons) but $(A) \neq (B)$.

Since the trivial equivalence relation satisfies AP, the sentence APO is not refutable in full second-order logic. We define
$$\text{NE} \quad \forall X(X \neq \emptyset \rightarrow \partial X \neq \partial \emptyset).$$

Lemma 5.15. FO-CA + DPO + NE $\vdash$ PW.

Proof. If $\neg$PW, there are $A, B, A \subsetneq B$ and $\partial(A) = \partial(B)$. Since also $\partial(A \cup B) = \partial(B \cup A)$, we get, by DPO (with $X = Y' = A$ and $X' = Y = B$), $\partial(B \setminus A) = \partial(A \setminus B)$. But $A \setminus B = \emptyset$ and $B \setminus A \neq \emptyset$, contradicting NE.
□ Lemma 5.15

Theorem 5.2 and Lemmas 5.15 and 5.12 immediately give

Theorem 5.16. Σ_2^1-CA $\vdash \neg$(DPO + NE), $\neg$(APO + NE).

Notice that BLV and FSP both imply NE and APO.

5.4 Relation-Type Principles

For this section ∂ will be a *binary* abstraction operator (i.e., it applies to binary predicates). The generalization to n-ary abstraction operators, although cumbersome, does not present essential problems. For reasons of space, we omit the details.

We define the following abstraction principles.

$$\text{RTP} \quad \forall R, S\,(\partial R = \partial S \leftrightarrow R \cong S),$$

$$\text{ORD} \quad \forall R, S\Big((\text{WO}(R) \wedge \text{WO}(S)) \to (\partial R = \partial S \leftrightarrow R \cong S)\Big),$$

where, again, WO(R) is the Π_1^1-formula expressing that R is a well-order on its field, as in §5.2, and $R \cong S$ is the Σ_1^1-formula expressing that there is an isomorphism from R to S, as binary relations.

Of course, RTP ⊢ ORD. The proof of ¬RTP produces well-orders witnessing its failure, so actually shows ¬ORD.

The proof of ¬PW from Σ_2^1-CA adapts to a proof of ¬ORD (so ¬RTP) and ¬PW$_2$ (where PW$_2$ is the natural generalization of PW to the binary relation context).

The refutability of ORD and RTP is well known, though it appears not to be written up anywhere. It is indicated in various places (for example, Burgess 2005: 116, Hazen 1985: 253f., and Hodes 1984: 138) that the proof is more or less through the Burali-Forti paradox. Our proof is basically along these lines. We've worked through a proof that is more faithful to this idea and it gives us refutability of ORD and RTP from the same hypotheses as below.

We just need the following version of Lemma 5.3 and a result on the comparability of well-orders.

Lemma 5.17. *The following is provable in Σ_2^1-CA. There is a unique binary predicate R such that*
  (i) WO(R);
  (ii) *for all* $x \in$ Field(R), $\partial R_x = x$; *and*
  (iii) $\partial R \in$ Field(R).

The only difference between Lemmas 5.3 and 5.17 is that in the former we apply our unary ∂ to the field of the well-orders, whereas in the latter we apply our binary ∂ to the well-order itself. Making this change to the proof gives us the proof of Lemma 5.17.

We also need the theorem that any pair of well-orders are *comparable*, i.e. that one is isomorphic to a unique initial segment of the other. Formally, this is the following principle.

CWO $\forall R, S\big((\text{WO}(R) \wedge \text{WO}(S)) \to ((R < S \vee S < R \vee R \cong S) \wedge (R < S \to R \not\cong S))\big),$

where $R < S$ is the Σ_1^1-formula expressing that R is isomorphic to a proper initial segment of S. CWO is provable from comprehension:

Lemma 5.18. Π_1^1-CA $\vdash$ CWO.

See Appendix A.2 for a proof of Lemma 5.18.

Theorem 5.19. Σ_2^1-CA $\vdash \neg$ORD

Proof. Let R be as in Lemma 5.17. Let $S \triangleleft R$ such that $\partial S = \partial R$ ($S = R_{\partial R}$ is such an S). Then by CWO, $S \not\cong R$ (since $S < R$, as witnessed by inclusion). So $\neg$ORD.
□ Theorem 5.19

For the R, S in the proof of Theorem 5.19, we have that $S \subsetneq R$ but $\partial S = \partial R$, so this also gives a direct proof of $\neg$PW$_2$.

We can weaken the comprehension hypothesis and still prove $\neg$RTP because we can code all well-orders by a single (first level) predicate and use this to lower the complexity of the R of Lemma 5.17.

Theorem 5.20. Π_1^1-CA $\vdash \neg$RTP.

Proof. We show there is a (first-level) predicate coding the second-level predicate WO via its image under ∂. We can't seem to do this under just ORD, since we could have $\partial R = \partial S$ when R is a well-order but S is not.

Claim 5.21. RTP $\vdash \exists Z \forall x (Z(x) \leftrightarrow \exists R (\text{WO}(R) \wedge \partial R = x))$.

Proof of Claim Work in Π_1^1-CA + RTP. Let $Y = \text{Range}(\partial)$ (this exists under our comprehension assumptions). Let Z be such that

$Z(x) \leftrightarrow Y(x) \wedge \forall R \forall X ((X \neq \emptyset \wedge X \subseteq \text{Field}(R) \wedge \forall x \in X \exists y \in X R(y, x)) \to \neg \partial(R) = x).$

The right-hand side is equivalent to a Π_1^1 formula by Proposition 4.5, so there is such a Z by our comprehension hypothesis (using Y as a parameter). But then Z is as desired since under RTP, if $\partial R = x$ for some well-order R and $\partial S = x$, then S is a well-order too.
□ Claim 5.21

This Z allows us to replace the Π_1^1 formula WO(R) with the atomic formula $\partial R \in Z$, reducing complexities of some definitions.

We need only observe that the R that witnesses Lemma 5.17 has a Σ_1^1 definition. Working in Π_1^1-CA, let Z as in Claim 5.21. R is defined by the following formula (this is just going through the necessary modification to the R from Lemma 5.3, as discussed above).

$$R(x, y) \Leftrightarrow \exists S \, (\text{WO}(S) \wedge \forall z \in \text{Field}(S) \, (\partial S_z = z) \wedge S(x, y))$$

As mentioned above, we can use Z to replace the formula WO(S) with the atomic formula $\partial(S) \in Z$:

$$R(x, y) \Leftrightarrow \exists S \, (\partial S \in Z \wedge \forall z \in \text{Field}(S) \, (\partial S_z = z) \wedge S(x, y)).$$

This is a Σ_1^1 definition, so R exists by Π_1^1-CA. We then proceed exactly as in the proof of Theorem 5.19.

□ Theorem 5.20

6 Further Results

6.1 Heck's Trick

In all results above, Σ_2^1-CA was sufficient to prove the negations of the inconsistent principles. It is easy to see that for any n, there is an abstraction principles consistent with Σ_n^1-CA, but inconsistent with Σ_{n+1}^1-CA. The technique we use is known as Heck's trick in the literature.[11]

Let φ be a sentence such that Σ_{n+1}^1-CA $\vdash \varphi$, but Σ_n^1-CA $\nvdash \varphi$. We define the abstraction principle A_φ as follows.

$$A_\varphi \quad \forall X, Y \left(\partial X = \partial Y \leftrightarrow (\neg \varphi \vee X = Y) \right)$$

Notice that under $\neg \varphi$, A_φ is equivalent to $\forall X, Y \, \partial X = \partial Y$ (which is, of course, consistent with Σ_n^1-CA). Since Σ_n^1-CA $\nvdash \neg \varphi$, Σ_n^1-CA + $\neg \varphi$ is consistent, hence Σ_n^1-CA + A_φ + $\neg \varphi$ is consistent. So Σ_n^1-CA $\nvdash \neg A_\varphi$.

On the other hand, under φ, A_φ is equivalent to BLV. Since Σ_{n+1}^1-CA $\vdash \varphi$ and Σ_{n+1}^1-CA $\vdash \neg$BLV (since $n + 1 \geq 1$), Σ_{n+1}^1-CA $\vdash \neg A_\varphi$. So A_φ is as desired.

---

[11] We thank Richard Kimberly Heck for pointing out to us the usefulness of their "trick" in this connection.

6.2 Refuting PW in Higher-Order Arithmetic

In this section we present a proof of the following result due to Ted Slaman, included here with his permission.

Theorem 6.1. $\mathrm{RCA}_0 + \Pi^1_1\text{-CA} \vdash \neg\mathrm{PW}$

The proof works by obtaining a witness to BLV recursively in a given witness to PW. This is done by showing that there is a perfect Σ^0_1 subset of $P(\omega)$ which is totally ordered by $\subseteq$. As far as we know, this interesting fact is new as well.

Proof. Let $\mathcal{M} = \langle M, S(M), +, \times, \leq \rangle \models \mathrm{RCA}_0$ and suppose $f : S(M) \to M$ is such that $\langle \mathcal{M}, f \rangle \models \mathrm{PW}$. We find $g : S(M) \to M$ which is $\Delta^0_1(f)$ such that $\langle \mathcal{M}, g \rangle \models \mathrm{BLV}$. Since every set $X \in P(M)$ that is $\Sigma^1_n(g)$ over $\mathcal{M}$ is $\Sigma^1_n(f)$ (since $g \, \Sigma^0_1(f)$ over $\mathcal{M}$), it follows that if $\langle \mathcal{M}, f \rangle \models \Sigma^1_n\text{-CA}$, then $\langle \mathcal{M}, g \rangle \models \Sigma^1_n\text{-CA}$. In particular, since $\Sigma^1_1\text{-CA} + \mathrm{BLV}$ is inconsistent, $\langle \mathcal{M}, f \rangle \not\models \Sigma^1_1\text{-CA}$.

Working in $\langle \mathcal{M}, f \rangle$, let T be the recursive binary tree, i.e. the set of (codes of) finite binary sequence ordered by end extension. For b a branch of T, let X_b be the set of nodes of T to the left of b, i.e. $x \in X_b$ iff $x \in b$ or for the unique n such that $b \upharpoonright n = x \upharpoonright n$ and $b(n) \neq x(n)$, $x(n) < b(n)$. Put also $b < c$ iff for the unique n such that $b \upharpoonright n = c \upharpoonright n$ but $b(n) \neq c(n)$, $b(n) < c(n)$. Then $X_b \subsetneq X_c$ whenever $b < c$ (in fact $X_c \setminus X_b$ is infinite since the order on branches is dense). Let j be the recursive function $b \mapsto X_b$. Let i be the recursive bijection from $P(\omega)$ to $[T]$. Then $i \circ j$ is a recursive function $P(\omega) \to P(\omega)$ such that for all $X, Y \in P(\omega)$,

$$X \neq Y \Leftrightarrow i \circ j(X) \subsetneq i \circ j(Y) \vee i \circ j(Y) \subsetneq i \circ j(X).$$

Let $g = f \circ i \circ j$. Then if $X \neq Y$, $i \circ j(X) \subsetneq i \circ j(Y)$ or $i \circ j(Y) \subsetneq i \circ j(X)$, so since PW holds for f, $g(X) = f(i \circ j(X)) \neq f(i \circ j(Y))$, in either case. So g is as desired.

□ Theorem 6.1

Recall that $\mathrm{BLV} \vdash \mathrm{PW}$, and the above proof worked by showing that there is a perfect set of predicates on which $\subseteq$ is a total order. Restricted to this perfect set, *any* function witnessing PW looks like a function witnessing BLV. This indicates that PW and BLV are equivalent in some reverse math-theoretic sense. We can make this precise as follows. In the theorem below, $\vdash$ denotes provability in *third-order logic*, the theory RCA^3_0 is as defined in Schweber 2015, and the formulas $\exists^2 f\, \mathrm{BLV}(f)$, $\exists^2 f\, \mathrm{PW}(f)$ are the natural formulas in pure third-order logic asserting that BLV and PW hold for f.

Theorem 6.2. $\mathrm{RCA}^3_0 \vdash \exists^2 f\, \mathrm{BLV}(f) \leftrightarrow \exists^2 f\, \mathrm{PW}(f)$.

Proof. (sketch) As mentioned above, we have that $\mathsf{RCA}_0^3 \vdash \exists^2 f\,\mathsf{BLV}(f) \to \exists^2 f\,\mathsf{PW}(f)$, so we just need to see that $\mathsf{RCA}_0^3 \vdash \exists^2 f\,\mathsf{PW}(f) \to \exists^2 f\,\mathsf{BLV}(f)$. This follows by the above proof of Theorem 6.1, using that the function g we defined there provably exists under RCA_0^3, since it was $\Sigma_1^0(f)$.

☐ Theorem 6.2

7 Conclusion

The preceding results more or less sum up what is known about inconsistent abstraction principles, save for the detailed investigations into consistency with weaker comprehension principles, as in Burgess 2005, Walsh 2016, and elsewhere. In particular, there is no general picture of the inconsistent abstraction principles, though our work on PW is a step in this direction.[12]

We saw that sufficiently strong comprehension principles (which are *true* principles of second-order logic) imply ¬BLV, ¬RTP, and ¬PW. Moreover, BLV ⊢ PW, so our proof of ¬PW gives us a new more general proof of ¬BLV.[13] Natural questions include the exact comprehension lower bounds for the proofs of ¬BLV, ¬PW, ¬(APO + NE), and ¬ORD. Additionally, in light of the fact that part-whole properties on the entire realm of concepts are not necessary for formal inconsistency, it remains to be understood the extent to which satisfying the part-whole property on a subclass of the entire realm of concepts is responsible for formal inconsistency. It would be especially interesting to find a formally inconsistent but natural abstraction principle that is not just a modification of Basic Law V and whose inconsistency does not depend on its satisfying part-whole on all concepts.

We also saw that BLV and PW are equivalent in the context of (higher-order) arithmetic, which also improves the comprehension lower bound for refuting PW to Π_1^1-CA in this context. Investigating similar implications between BLV and ORD and optimal lower bounds for their refutability in this setting is also natural and potentially more tractable. This would likely shed light on the problems in the general logical setting, as well.

---

[12] As pointed out above, some results by Richard Kimberly Heck written in reply to a first version of this paper, constitute interesting progress in this direction. After the final version of this paper was completed we became aware of the paper Ebels-Duggan 2018 which solves the long standing problem of the refutability in second-order logic of the conjunction of HP and the Nuisance principle. This also leads to further refutable abstraction principles defying part-whole but we cannot address this here.

[13] Our proof of ¬PW uses more comprehension in the background, though it is open whether this is necessary.

Acknowledgments

We would like to thank Vieri Benci, Marco Forti, Mauro Di Nasso, Theodore Slaman, Denis Hirschfeldt, Richard Kimberly Heck, Roy Cook, Leo Harrington, and Guillaume Massas for fruitful conversations on the topic of this paper.

Appendix

A.1 Proof of Claim 5.10

Proof. We first observe two useful consequences of AP.

Subclaim 7.1. *Suppose $A \cong B$ and $C \cap A = C \cap B = \emptyset$. Then $A \cup C \cong B \cup C$.*

Proof. Suppose $A \cup C \not\cong B \cup C$. By the contrapositive of AP, $A \cup C \setminus B \cup C \not\cong B \cup C \setminus A \cup C$. $A \cup C \setminus B \cup C = A \setminus B$ and $B \cup C \setminus A \cup C = B \setminus A$, so we have $A \setminus B \not\cong B \setminus A$. But then, by AP, $A \not\cong B$, a contradiction.

□ Subclaim 7.1

Subclaim 7.2. *Suppose $A \cup C \cong B \cup C$ and $C \cap A = C \cap B = \emptyset$. Then $A \cong B$.*

Proof. By AP, $A \cup C \setminus B \cup C \cong B \cup C \setminus A \cup C$. But since $C \cap A = \emptyset$, $A \cup C \setminus B \cup C = A \setminus B$ and, since $C \cap B = \emptyset$, $B \cup C \setminus A \cup C = B \setminus A$, so $A \setminus B \cong B \setminus A$. So, by AP, $A \cong B$.

□ Subclaim 7.2

We now prove Claim 5.10. The idea is to look at all the non-empty intersections among A, B, A', B' and express everything in terms of pairwise disjoint subsets. This leads to the following definitions:

$$A_0 = A \setminus (A' \cup B')$$
$$A'_0 = A' \setminus (A \cup B)$$
$$D = A \cap A'$$
$$A_1 = A \setminus (A_0 \cup D)$$
$$A'_1 = A' \setminus (A'_0 \cup D)$$
$$B_0 = B \setminus (A \cup A' \cup B')$$
$$B'_0 = B' \setminus (A' \cup A \cup B)$$
$$E = (B \cap B') \setminus (A \cup A').$$

Below is a diagram for the case when $A \cap B = A' \cap B' = \emptyset$.

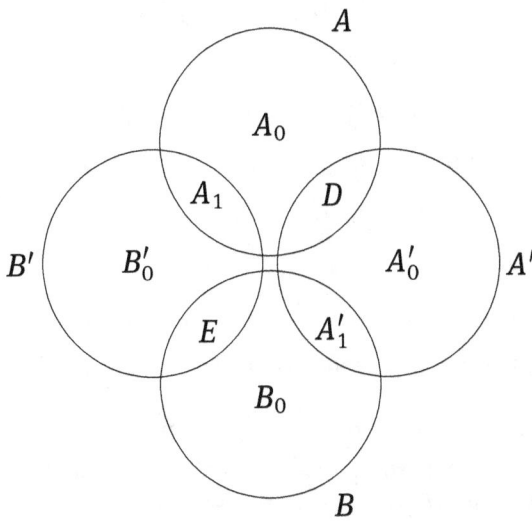

We can describe the relevant sets in terms of these definitions.

$$A = A_0 \cup A_1 \cup D$$
$$A' = A'_0 \cup A'_1 \cup D$$
$$(A \cup B) \setminus (A' \cup B') = A_0 \cup B_0$$
$$(A' \cup B') \setminus (A \cup B) = A'_0 \cup B'_0$$
$$B \setminus A = B_0 \cup E \cup A'_1$$
$$B' \setminus A' = B'_0 \cup E \cup A_1$$

We first show that AP implies SP. Let A, A', B, B' be such that $A \cap B = A' \cap B' = \emptyset$, so the above diagram applies. Assume $A \cong A'$ and $B \cong B'$. We want to show $A \cup B \cong A' \cup B'$. By AP this is equivalent to $A \cup B \setminus A' \cup B' \cong A' \cup B' \setminus A \cup B$. But since $A \cup B \setminus A' \cup B' = A_0 \cup B_0$ and $A' \cup B' \setminus A \cup B = A'_0 \cup B'_0$, it will be sufficient to prove

$$A_0 \cup B_0 \cong A'_0 \cup B_0 \qquad (1)$$

We know $A \cong A'$ and $B \cong B'$. By AP we also know $A \setminus A' \cong A' \setminus A$ and $B \setminus B' \cong B' \setminus B$. But $A \setminus A' = A_0 \cup A_1$ and $A' \setminus A = A'_0 \cup A'_1$, so we have

$$A_0 \cup A_1 \cong A'_0 \cup A'_1 \qquad (2)$$

and $B \setminus B' = B_0 \cup A'_1$ and $B' \setminus B = B'_0 \cup A_1$, so

$$B_0 \cup A'_1 \cong B'_0 \cup A_1. \qquad (3)$$

By Subclaim 7.1, we can add B_0 to both sides of 2 to get

$$B_0 \cup A_0 \cup A_1 \cong B_0 \cup A_0' \cup A_1', \tag{4}$$

and add A_0' to both sides of 3 to get

$$B_0 \cup A_1' \cup A_0' \cong B_0' \cup A_1 \cup A_0'. \tag{5}$$

From 4, 5, since $\cong$ is transitive, we get

$$B_0 \cup A_0 \cup A_1 \cong B_0' \cup A_1 \cup A_0'.$$

By AP we obtain

$$B_0 \cup A_0 \cup A_1 \setminus B_0' \cup A_1 \cup A_0' \cong B_0' \cup A_1 \cup A_0' \setminus B_0 \cup A_0 \cup A_1.$$

But the left hand side is $A_0 \cup B_0$ and the right hand side is $A_0' \cup B_0'$, so we've shown 1, as desired.

Now we show AP implies DP. Assume A, A', B, B' are such that $A \cong A'$ and $A \cup B \cong A' \cup B'$. We want to show $B \setminus A \cong B' \setminus A'$. The left hand side is $B_0 \cup E \cup A_1'$ and the right hand side is $B_0' \cup E \cup A_1$, so we want to show $B_0 \cup E \cup A_1' \cong B_0' \cup E \cup A_1$. By AP, this is equivalent to

$$B_0 \cup E \cup A_1' \setminus B_0' \cup E \cup A_1 \cong B_0' \cup E \cup A_1 \setminus B_0 \cup E \cup A_1'.$$

So, we need to show

$$B_0 \cup A_1' \cong B_0' \cup A_1. \tag{6}$$

We know $A \cong A'$ and $A \cup B \cong A' \cup B'$. We have already seen in proving SP that from $A \cong A'$ we get

$$A_0 \cup A_1 \cong A_0' \cup A_1' \tag{7}$$

By AP, $A \cup B \cong A' \cup B'$ is equivalent to $A \cup B \setminus A' \cup B' \cong A' \cup B' \setminus A \cup B$. Since $A \cup B \setminus A' \cup B' = A_0 \cup B_0$ and $A' \cup B' \setminus A \cup B = A_0' \cup B_0'$, we have

$$A_0 \cup B_0 \cong A_0' \cup B_0' \tag{8}$$

By Subclaim 7.1, we can add A_1' to both sides of 8 to get

$$A_0 \cup B_0 \cup A_1' \cong A_0' \cup B_0' \cup A_1' \tag{9}$$

and add B_0' to both sides of 7, to get

$$B_0' \cup A_0 \cup A_1 \cong B_0' \cup A_0' \cup A_1' \tag{10}$$

We see that the right hand sides of 9 and 10 are the same, so that we get $A_0 \cup B_0 \cup A_1' \cong B_0' \cup A_0 \cup A_1$. So, by AP, $A_0 \cup B_0 \cup A_1' \setminus B_0' \cup A_0 \cup A_1 \cong B_0' \cup A_0 \cup A_1 \setminus A_0 \cup B_0 \cup A_1'$. The left hand side is just $B_0 \cup A_1'$ and the right hand side is $B_0' \cup A_1$, so we have 6.

□ Claim 5.10

A.2 Proof of Lemma 5.18

Lemma 7.3. Π_1^1-CA $\vdash$ CWO

Recall $R \cong S$ and $R < S$ are the Σ_1^1 formulas expressing that R and S are isomorphic (as binary relations) and that R is isomorphic to a proper initial segment of S, respectively. We'll need the following additional notation. We let $R \leq S =_{def} R < S \vee R \cong S$. We'll also need the formulas $f : R < S$ and $f : R \cong S$ which express "f is an order embedding from R onto a proper initial segment of S" and "f is an order isomorphism between R and S", which both have no second-order quantifiers. We define $f : R \leq S =_{def} (f : R < S \vee f : R \cong S)$.

Proof. For $x \in \text{Field}(S)$ for a binary predicate S, recall that S_x denotes the initial segment of S below x, i.e. $S_x(y, z) \Leftrightarrow (S(y, z) \wedge S(y, x) \wedge S(z, x))$ (which exists by Π_1^1-CA). We'll need the following fact.

Claim 7.4. *For R, S well-orders, if $R \leq S$, then there is a unique f such that $f : R \leq S$.*

Proof of Claim Let $f, g : R \leq S$. Suppose $f \neq g$. Let $X(x) \Leftrightarrow f(x) \neq g(x)$ (this predicate exists by Π_1^1-CA). Since $f \neq g$, X is nonempty, so let x be the R-least element of X. Then $f(y) = g(y)$ for all y such that $y \neq x$ and $R(y, x)$. But then $S_{f(x)} = S_{g(x)}$, since $f \upharpoonright R_x = g \upharpoonright R_x$ and these functions are onto $S_{f(x)}$ and $S_{g(x)}$, respectively. So $f(x)$ and $g(x)$ are both the S-least upper bound of Field($S_{f(x)}$), so $f(x) = g(x)$, a contradiction. □ Claim 7.4

Towards proving the Lemma, let R, S be well-orders. Claim 7.4 immediately implies that if $R < S$, then $R \not\cong S$ (if $f : R < S$ and $g : R \cong S$, then both $f, g : R \leq S$, so $f = g$ by Claim 7.4, which is a contradiction since f is not onto Field(S)). Suppose $R \not\leq S$. Note that R must be non-trivial since the trivial well-order, i.e. the unique binary predicate with empty field, is an initial segment of every well-order. If S is trivial, then $S < R$, as desired. So suppose S is non-trivial. Define the predicate F as follows.

$$F(x, y) \Leftrightarrow \exists U, g\, (x \in \text{Field}(R) \wedge y \in \text{Field}(S) \wedge g : U < S \wedge y \notin \text{Range}(g) \wedge$$
$$\forall z ((z \neq y \wedge S(z, y)) \to z \in \text{Range}(g)) \wedge$$
$$\forall u, v(U(u, v) \leftrightarrow [R(u, v) \wedge R(v, x) \wedge v \neq x])).$$

Notice that $F(x, y) \Leftrightarrow R_x \cong S_y$. The above formula just witness that this is Σ_1^1, so F exists by Π_1^1-CA. We check that F is an order isomorphism from an initial segment of R onto an initial segment of S. First we show F is the graph of a function. Suppose $F(x, y)$, so that $R_x \cong S_y$. Then, by Claim 7.4, the unique map $f : R_x \leq S$ is an isomorphism $f : R_x \cong S_y$. So, for any $z \neq y$, there is no map $g : R_x \cong S_z$ (since g would be a map $R_x \leq S$, so $g = f$, contradiction).

Also, F is an order embedding, since if $R(x, y)$ and $g : R_y \cong S_z$, then $g \upharpoonright R_x : R_x \cong S_z$. So, $R_x \not\cong S_z$, hence $\neg F(x, z)$. Next we check that F is onto an initial segment of S. Suppose $F(x, y)$ and $S(z, y)$ with $z \neq y$ (so $z \in \text{Field}(S_y)$). Let $f : R_x \cong S_y$ and let u such that $R(u, x)$ and $f(u) = z$. Then $f \upharpoonright R_u \cong S_z$, so $F(u, z)$, i.e. $z \in \text{Range}(F)$. Similarly, F has domain the field of an initial segment of R. So F is an isomorphism from an initial segment of R onto an initial segment of S.

If F is onto S, then $F^{-1} : S \leq R$ (and this exists by Π_1^1-CA), so we're done. So suppose F is *not* onto S. Note, then, that F has domain a proper initial segment of R, since $R \triangleleft S$ by hypothesis. So we may let u be the R-least element not in the domain of F, and v the S-least element not in the range of F. Let $F'(x, y) \Leftrightarrow F(x, y) \vee (x = u \vee y = v)$ (this exists by Π_1^1-CA). It's easy to see that F' is still an isomorphism from an initial segment of R onto an initial segment of S. Again, if F' is *not* onto, then F' has domain a proper initial segment of R (since $R \triangleleft S$). But then $F' : R_u \cong S_v$, so $F(u, v)$, contradicting choice of u, v. So F' is onto S and so $F'^{-1} : S \leq R$, as desired.

□ Lemma 5.18

Bibliography

Benci, Vieri and Di Nasso, Mauro (2003): "Numerosities of Labeled Sets: A New Way of Counting." In: *Advances in Mathematics*. Vol. 173, 50 – 67.

Benci, Vieri, Di Nasso, Mauro and Forti, Marco (2006): "An Aristotelean Notion of Size." In: *Annals of Pure and Applied Logic*. Vol. 143, 43 – 53.

Benci, Vieri, Di Nasso, Mauro and Forti, Marco (2007): "An Euclidean Measure of Size for Mathematical Universes." In: *Logique et Analyse*. Vol. 50, 43 – 62.

Blass, Andreas, Benci, Vieri, Di Nasso, Mauro and Forti, Marco (2012): "Quasi-Selective Ultrafilters and Asymptotic Numerosities." In: *Advances in Mathematics*. Vol. 231, No. 3–4, 1462 – 1486.

Boolos, George (1997): "Constructing Cantorian Counterexamples." In: *The Journal of Philosophical Logic*. Vol. 26, 237 – 239. Reprinted in Boolos 1998, 339 – 341.

Boolos, George (1998): *Logic, Logic, and Logic*. Cambridge: Harvard University Press.

Burgess, John (2005): *Fixing Frege*. Princeton: Princeton University Press.

Cook, Roy (2012): "Conservativeness, Stability, and Abstraction." In: *British Journal for the Philosophy of Science*. Vol. 63, 673 – 696.

Ebels-Duggan, Sean (2018): "Deductive Cardinality Results and Nuisance Principles." https://www.researchgate.net/publication/323509434_Deductive_Cardinality_Results_and_Nuisance_Principles, visited on October 22, 2018.

Ebert, Phillip and Rossberg, Marcus (2016): *Abstractionism. Essays in Philosophy of Mathematics*. Oxford: Oxford University Press.

Frege, Gottlob (1884): *Die Grundlagen der Arithmetik*. Breslau: Koebner. Austin, J. (trans.). In: *The Foundations of Arithmetic*. 2nd edition, New York: Harper, 1960.
Frege, Gottlob (1893): *Grundgesetze der Arithmetik I*. Jena: H. Pohle. Reprint in: *Basic Laws of Arithmetic*. Ebert, P. A., Rossberg, M. and Wright, C. (eds. and trans.). Oxford: Oxford University Press, 2013.
Frege, Gottlob (1903): *Grundgesetze der Arithmetik II*. Jena: H. Pohle Reprintin: *Basic Laws of Arithmetic*. Ebert, P. A., Rossberg, M. and Wright, C. (eds. and trans.), Oxford: Oxford University Press, 2013.
Hale, Robert and Wright, Crispin (2001): *The Reason's Proper Study: Essays Towards a Neo-Fregean Philosophy of Mathematics*. Oxford: Clarendon Press.
Hazen, Allen (1985): "Review of C. Wright, 'Frege's Conception of Numbers as Objects'." In: *Australasian Journal of Philosophy*. Vol. 63, 251 – 254.
van Heijenoort, Jean (ed.) (1967): *From Frege to Gödel. A Source Book in Mathematical Logic, 1897–1931*. Cambridge: Harvard University Press.
Hodes, Harold (1984): "Logicism and the Ontological Commitments of Arithmetic." In: *The Journal of Philosophy*. Vol. 81, No. 3, 123 – 149.
Kanamori, Akihiro (1997): "The Mathematical Import of Zermelo's Well-Ordering Theorem." In: *The Bulletin of Symbolic Logic*. Vol. 3, No. 3, 281 – 311.
Kanamori, Akihiro (2004): "Zermelo and Set Theory." In: *The Bulletin of Symbolic Logic*. Vol. 10, No. 4, 487 – 553.
Mancosu, Paolo (2009): "Measuring the Size of Infinite Collections of Natural Numbers: Was Cantor's Theory of Infinite Number Inevitable?" In: *The Review of Symbolic Logic*. Vol. 2, 612 – 646.
Mancosu, Paolo (2015): "In Good Company? On Hume's Principle and the Assignment of Numbers to Infinite Concepts." In: *The Review of Symbolic Logic*. Vol. 8, No. 2, 370 – 410.
Mancosu, Paolo (2016): *Abstraction and Infinity*. Oxford: Oxford University Press.
Schweber, Noah (2015): "Transfinite Recursion in Higher Reverse Mathematics." In: *The Journal of Symbolic Logic*. Vol. 80, No. 3, 940 – 969.
Paseau, Alexander (2015): "Did Frege Commit a Cardinal Sin?" In: *Analysis*. Vol. 75, No. 3, 379 – 386.
Shapiro, Stewart (1991): *Foundations without Foundationalism: A Case for Second-Order Logic*. Oxford: Oxford University Press.
Walsh, Sean (2016): "Fragments of Frege's *Grundgesetze* and Gödel's Constructible Universe." In: *The Journal of Symbolic Logic*. Vol. 81, No. 2, 605 – 628.
Wright, Crispin (1983): *Frege's Conception of Numbers as Objects*. Aberdeen: Aberdeen University Press.
Zermelo, Ernst (1904): "Beweis, dass jede Menge wohlgeordnet werden kann." In: *Mathematische Annalen*. Vol. 59, 514 – 516. English translation in: van Heijenoort 1967, 139 – 141.

William Tait
What Hilbert and Bernays Meant by "Finitism"

Abstract: "Finitism" (Tait 1981) presents an argument that finitist number theory is primitive recursive arithmetic (PRA). The argument is based on taking seriously the "finite" in "finitism". But the question remained: what did Hilbert (and Bernays) mean in the early 1920's through the early 1930's by "finitism" and in particular, did they restrict finitist number theory to PRA. In his dissertation (Zach 2003), Richard Zach pointed out that Hilbert endorsed results as finitist that require more than PRA for their proofs. Tait 2002 and tait2005 argue that it is not clear that Hilbert was aware that these results go beyond PRA. But that view is challenged in more recent times in Sieg/Ravaglia 2005 and by the editors of (the invaluable!) *David Hilbert's Lectures on the Foundations of Arithmetic and Logic 1917–1933* (Hilbert 2013). I will survey the old ground and then discuss the new challenge, which claims that, from the early 1920's on, Hilbert accepted as finitist an enumeration function of the primitive recursive functions (which of course is not primitive recursive). The grounds for this are a reading of a passage in §7 of *Grundlagen der Mathematik I* and an argument for the consistency of PRA which goes back to 1922–1923 and is elaborated again in §7 of *Grundlagen der Mathematik I*. I will argue that their reading of the passage in question is a misreading and that the argument for the consistency of PRA uses, not an enumeration function for the primitive recursive functions, but rather mathematical induction on a Π_2^0 predicate (i.e. of the form $\forall x \exists y \phi(x, y)$), which was explicitly rejected by Hilbert as finitist – e.g. notably in Hilbert 1926.

The aim of my talk will be primarily historical: I'll discuss what was meant by the term "finitism" (the *finit, finiter Standpunkt*) in the Hilbert school in the 1920's and early 1930's. I discussed this briefly and tangentially in my paper "Finitism" (Tait 1981: 524 – 556) and in further detail in "Remarks on finitism" (Tait 2002: 410 – 419), and *The Provenance of Pure Reason*, Appendix to Chapters 1 and 2 (Tait 2005: 54 – 60).

But there has been more recent literature on the subject that, I believe, needs further comment. Moreover, further thought about *Grundlagen der Mathematik I* (1934) leads me to strengthen the somewhat agnostic position that I took in those earlier writings: I will now argue that, up to the writing of that work, at least, Hilbert

---

William Tait, University of Chicago, USA, williamtait@mac.com

and Bernays regarded primitive recursive arithmetic (PRA) as a formalization of finitist arithmetic.

A challenge to that position arises from an alleged finitist proof of consistency of PRA. The claim is that the acceptance of that proof implied the assumption that there is an evaluation function for the primitive recursive functions. But, independently of what stance one takes on the main issue of whether Hilbert's finitism extended beyond PRA, we will see that the poof, though classically correct, uses complete induction on a Π_2^0 property, which is explicitly rejected by Hilbert. Moreover, the proof makes no use of an evaluation function for the primitive recursive functions.

I begin with some background of the whys and wherefores of finitist proof theory – – why proof theory and wherefore finitary.

The story, vividly recounted in detail, for example, in Wilfried Sieg's 1999 paper "Hilbert's programs" (Sieg 1999), starts around the beginning of the 20th century, when Hilbert first proposed the problem of proving the consistency of arithmetic. But there were two initial problems with the proposal:

The first was the problem of precisely defining the notion of "consistency" of a theory. A theory T is consistent if there is no proof of some $A \wedge \neg A$. But what is a proof?

Other mathematicians of the nineteenth century, for example Dedekind and Cantor, had focused on consistency as the criterion for admitting ideal structures into mathematics; but the notion of the consistency of the theory T of the structure remained imprecise. There were consistency proofs, but they were proofs only of emphrelative consistency, by interpreting T in the theory of some given structure. But relative consistency proofs must come to an end. And, in the case of a theory of an infinite set, say the theory of a Dedekind infinite set, we are at the end. A syntactical proof was needed, requiring a purely syntactical notion of proof in a theory.

The second problem, posed by Poincaré around 1905, was that a proof of consistency for number theory, however defined, would inevitably involve complete induction in some form or other and so would be circular.

With the help of earlier work by Frege and by Russell and Whitehead, Hilbert solved the first problem in his 1917–1918 lectures on logic, which established the frameworks of first and higher order predicate logic and so gave a precise content to the notion of consistency. Incidentally, these lectures, published in 2013 by William Ewald and Wilfried Sieg in the 1062 page *David Hilbert: Lectures on the Foundations of Arithmetic and Logic 1917–1933* (Hilbert 2013), contain, sometimes verbatim, a substantial part of the content of Hilbert and Ackermann's later *Foundations of Theoretical Logic*. A lot of the history of logic in the 1920's needs rewriting.

As for Poincaré's challenge, the answer was finally attempted by Hilbert and Bernays in a series of papers starting in 1922 and culminating in *Grundlagen der Mathematik I* in 1934 – by which time it was known that their answer to the challenge would at the least need revision.

Their answer was, of course, *finitism*. No precise definition of "finitism" was offered, but judging by the discussion of what counted as finitist and what did not, it was Kronecker's conception of mathematics; but more deeply it was a turning away from the developing conception of mathematics as the study of ideal structures to an older, 18th century conception of mathematics, according to which it is concerned with constructing and computing. It was the conception embodied in Kant's *Critique of Pure Reason*, a fact to which Hilbert and Bernays many times alluded.

But of course Hilbert's proposal wasn't Kronecker's – namely to restrict all mathematics to finitary reasoning. Rather it was to prove the consistency of infinitary mathematics by finitary means in order to free us to use infinitary methods. Mathematical induction would indeed be involved in the consistency proof, but only in the context of finitary reasoning.

It should be mentioned, however, that Hilbert's progress in foundations of mathematics, from the turn of the century up to the finitism of 1922, was not entirely so linear as I have just portrayed it – and this is the point of the plural in Sieg's title "Hilbert's program*s*." There was some flirting with the logicism of Russell and Whitehead, an entirely different approach to foundations, and then a retreat to a very strict form of constructivism, which appears not to admit any general propositions, such as statements of consistency, at all.

The subject matter of Hilbert's finitism included the natural numbers, but it wasn't simply or even primarily about that, of course: its primary aim was a proof of the consistency of formal systems, and so its subject matter included the syntax of such systems. But *we* know that syntax can be coded in the arithmetic of the natural numbers, and so we can restrict ourselves to that. So when I speak of finitism now, I will mean finitist number theory. On the other hand, essentially following Hilbert, we could identify the natural numbers with the *numerals*

$$0, S0, SS0, \ldots$$

making number theory a part of syntax. (Hilbert liked *I, II, III*, ….)

The question, though, is what exactly constitutes finitist reasoning.

In the paper "Finitism", taking the "finite" in "finitism" seriously, I raised the problem of how one can understand a general proposition

$$\forall x\, \phi(x)$$

where x ranges over all the (natural) numbers, such as the consistency statement for a formal system, without the presupposition of an infinite totality. I proposed as an answer that we can understand the notion of an *arbitrary* or *generic number* X and prove

$$\phi(X)$$

from which $\phi(n)$ will follow for each number n.

Incidentally, this was not a new idea, although I didn't recognize that at the time: Kant's idea in the *Critique of Pure Reason* of demonstration of geometric propositions

$$\forall x\, \phi(x)$$

where ϕ is quantifier-free or at most purely existential. Say it is about all triangles. The demonstration begins with 'constructing' the concept ⟨*triangle*⟩ in pure intuition, which meant to construct in imagination a generic triangle X, and then to carry out constructions on X – the *synthetic* part of the demonstration – from which $\phi(X)$ follows analytically. From this it would then follow that $\phi(t)$ would hold for all triangles t.

I will not discuss the details of Kant's theory or its relation to finitism now. A rather unsatisfactory account of it is given in "Kant and Finitism" (Tait 2016). But I aim to do better.

But it is worth mentioning Kant, though, given Hilbert's repeated appeal to his authority. On the other hand, what Hilbert wrote about Kant does not reveal, at least to me, an understanding of the connection I just described. He speaks, repeatedly, of the concrete objects that are given in intuition, but not of objects of *pure* intuition, which was Kant's entryway to synthetic *a priori* geometric truth. Bernays comes closer when he writes of "thought experiments". There are extensive discussions of the relation between Kant's critical philosophy and its descendants and Hilbert's finitism in Volker Peckhaus' *Hilbertprogramm und Kritische Philosophie* (Peckhaus 1990) and in Paolo Mancosu's *Adventure of Reason* (Mancosu 2010).

Returning to finitism, I argued that it is what is formalized in *Primitive Recursive Arithmetic*, PRA, a system first singled out by Skolem in 1923 (Skolem 1923) – although, interestingly, *GdM* I makes no reference to him in this respect.

Here is a brief description of PRA for those not familiar with it:

- Terms: $x, 0, St, f(s, \ldots, t)$
- Formulas: $s = t$ $s \neq t$ where s and t are terms.

- Complement: $\overline{s = t} := s \neq t$ $\overline{s \neq t} := s = t$.
- Definitions: For each function constant f there is a unique definition: an Explicit Definition
$$f(x, \ldots, y) = t$$
where $x, \ldots, y$ are all the distinct variables in t or a Primitive Recursive Definition
$$\begin{cases} f(x, \ldots, y, 0) = g(x, \ldots, y) \\ f(x, \ldots, y, Sz) = h[x, \ldots, y, z, f(x, \ldots, y, z)] \end{cases}$$
The function constants can be ordered $f_0, f_1, \ldots$ so that f_i is defined in terms of f_j only if $j < i$.
- Deductions are of finite sets $\Gamma, \Delta, \ldots$ of formulas understood disjunctively.
- $\Gamma, A := \Gamma \cup \{A\}$.
- Axioms:
$$\Gamma, t = t, \qquad \Gamma, 0 \neq St, \qquad \Gamma, f(s, \ldots, t) = r$$
where the latter is a substitution instance of a defining axiom for f.

- Rules of Inference
  - Successor
  $$\frac{\Gamma, Ss = St}{\Gamma, s = t}$$
  - Substitution
  $$\frac{\Gamma, s = t \qquad \Gamma, A(s)}{\Gamma, A(t)}$$
  - Cut
  $$\frac{\Gamma, s = t \qquad \Gamma, s \neq t}{\Gamma}$$
  - Mathematical Induction
  $$\frac{\Gamma, A(0) \qquad \Gamma, \overline{A(x)}, A(Sx)}{A(t)}$$
  (x does not occur in $\Gamma, A(0)$.)

This is a stripped down formulation of PRA. For example, it doesn't explicitly contain propositional logic. But formulas
$$\phi(x, \ldots, y)$$
of propositional logic whose atoms are equations in PRA can be coded by equations
$$f(x, \ldots, y) = 0$$
of PRA so that the laws of propositional logic are derivable.

Notice that

- If $\Gamma(x)$ is derivable, so is $\Gamma(t)$ for any term t. (Can assume there are no critical variables in the deduction that occur in t.)
- if Γ is derivable, then so is Γ, A for any formula A. (Can assume there are no critical variables in the deduction that occur in A.) Simply add A to each step in the deduction.
- When t is a numeral, then the application of mathematical induction

$$\frac{\Gamma, A(0) \qquad \Gamma, \overline{A(x)}, A(Sx)}{A(t)}$$

can be eliminated in favor of a sequence of t cuts: If t is 0, we already have the deduction. If t is Sn and we assume that we have a deduction of $\Gamma, A(n)$, substitute n for x in the deduction of the premise $\Gamma, \overline{A(x)}, A(Sx)$ and apply the "cut"

$$\frac{\Gamma, A(n) \qquad \Gamma, \overline{A(n)}, A(Sn)}{\Gamma, A(Sn)}$$

The core of the argument that finitism contains PRA is that the generic number X represents in fact a generic finite iteration

$$0, S0, SS0, \ldots, X$$

and, having constructed a $f(Y)$ from the generic number Y, one can transfer the iteration X, obtaining

$$Y, f(Y), f(f(Y)), \ldots, f^X(Y)$$

It is this that enables the derivation of definition of functions by primitive recursion and of proofs by mathematical induction of formulas of PRA.

I don't intend to defend or further discuss this approach. I will note that in the *Tractatus*, 6.02, Wittgenstein explicitly identifies the numbers as iterators. So does Alonzo Church in *The Calculi of Lambda-Conversion* (Church 1941).

The argument that finitism is *exactly* PRA assumes that, for a finitist, introducing a new function f must involve introducing a system e of equations and proving that there is a computation y of a value of f for each x, i.e. in Kleene's notation, proving $\exists y T(e, x, y)$. But one thing that we can all agree on is that finitism does not admit the unbounded existential quantifier – that is stated over and over again by Hilbert and by Bernays. Thus, one needs to put a prior bound on y, so that what has to be proved is the primitive recursive formula

$$\exists y < g(x) T(e, x, y)$$

for some given g. But f is primitive recursive if (and only if) there is a system e which defines it and a primitive recursive such bound g. So this construction will not lead out of the class of primitive recursive functions.

This argument (perhaps) establishes that the constructions $f(X, \ldots, Y)$ of number that are admissible on finitist grounds correspond exactly to the primitive recursive functions f.

But the argument that finitist *reasoning about* these constructions is limited to what is formalizable in PRA is a bit more complicated and I won't discuss it now.

Incidentally, as Michael Friedman has pointed out (Friedman 1992), there is textual evidence from Kant's discussion of the notion of the schema of a concept that for him the construction of the concept ⟨whole number⟩ would also be an arbitrary finite iteration. So, if he had chosen to speak about the epistemology of (whole) number theory, it might have looked very like this proposal for understanding finitism. The difference would be that his notion of iteration is not our abstract notion, but rather it is iteration of constructions in time. Just as the synthetic element of reasoning in geometry was for him construction in space, the synthetic element of reasoning in number theory would be construction in time.

But, anyway – another time!

Kreisel (Kreisel 1960) argued that finitism *should* extend beyond PRA on the ground that, when we reflect on PRA, we are led to new truths, e.g. its consistency. The argument is worth mentioning because it shares a fallacy with other applications of reflection principles, such as those involved in the Turing or Feferman hierarchies of arithmetics based on reflection: to hold for each theorem of a theory T that it is true is not to hold that T is valid. The finitist will accept each theorem of PRA, but (s)he is not in position to say that PRA is valid.

There have also been arguments published that finitism is weaker than PRA; but the more serious ones simply make the point that there is a very natural proper subclass of the primitive recursive functions, namely the Kalmár elementary functions. Indeed, it is a very natural class of functions. But my argument is that, while still taking the "finite" in finitism seriously, we can go further and accept all primitive recursive functions.

I don't know whether the elementary functions satisfy the closure condition mentioned above for the primitive recursive functions, namely whether when there is an elementary bound g on the length of the computations:

$$\exists y < g(x) T(e, x, y).$$

then $\{e\}$ is in fact an elementary function.

But in any case, Hilbert and Bernays explicitly accepted PRA as finitist. The question – I finally come to it – is, whether they were committed to more than that.

The question however is ambiguous: Hilbert did indeed accept as finitist results that cannot be proved in PRA. In his dissertation Richard Zach (Zach 2003) pointed out such a case, namely Hilbert's endorsement of Ackermann's dissertation (Ackermann 1924) as finitist even though it invokes quantifier-free induction up to ω^{ω^ω}. However, as I pointed out in Tait 2005: 56, Ackermann's paper actually *derives* induction only up to ω^2 which, like induction up to ω^n for any $n < \omega$, is derivable in PRA. It is plausible that Hilbert assumed that the stronger principle would also be derivable using arguments that are formalizable in PRA.

The post-Zach question is whether the Hilbert school accepted any principles as finitist that they knew not to be derivable in PRA. I believe the answer is NO.

Here is what is written in *GdM* I on the relation between finitism and PRA:

> Der Unterschied der recursiven Zahlentheorie gegenüber der anschaulichen Zahlentheorie besteht in ihrer formalen Gebundenheit; sie hat als einzige Methode der Begriffsbildung, ausser der expliziten Definitionen, das Rekursionschema zur Verfügung und auch die Methoden der Ableitung sind fest umgrenzt. (*GdM* I: 334/330)

In English:

> The distinction of recursive number theory from intuitive number theory consists in its formal constraints; its only method of concept formation, aside from explicit definition, is the schema of recursion, and also the methods of deduction are strictly circumscribed. (334/330, my translation)

"Intuitive number theory" here means finitist number theory and "recursive number theory" means PRA. So I take this passage to imply that PRA is simply a formalization of finitist number theory.

Bernays continues:

> To be sure we can admit certain *extensions of the schema of recursion* as well as of the induction schema without taking away what is characteristic of the method of recursive number theory. (*GdM* I: 334/330)

He then works through the reductions of multiple recursion, simultaneous recursion, course-of-value recursion, etc., to primitive recursion and then writes

> The question arises whether it might be that all such recursions, in which a procedure of stepwise computation of one or more functions is formalized and which can be presented without the addition of ja new sort of variable can be reduced to primitive recursions. (*GdM* I: 334/330)

By "a new sort of variable" he is referring to variables for functions of higher type. Non-primitive recursive functions of type 1 (i.e. $N^n \longrightarrow N$), can be defined by means of (impredicative) primitive recursive definition of higher type functions.

E.g. in "On the Infinite", Hilbert defined the Ackermann function in this way. Gödel, in the *Dialectica* paper, showed that every function definable by recursion on an ordinal $< \epsilon_0$ can be defined by allowing functions of arbitrary finite type over $\mathbf{N}$ to be defined by primitive recursion. In any case, admitting variables for functions of higher type allows the definition of recursive functions that are not primitive recursive. Bernays is excluding this case, as we would expect of someone who respects the "finite" in "finitism".[1]

Bernays' answer to his question above is of course *no* and he goes on to give examples of non-primitive recursions, such as the definition of the Ackermann function. Thus, "what is characteristic of the method of recursive number theory" can lead out of the class of primitive recursive functions. The question is: what is this characteristic? One proposed answer is that what is characteristic is being finitist. Of course, if that is what Bernays meant, the discussion is over.

But, besides the fact that that answer contradicts the above quote (at least as I read it), to the effect that PRA is a formalization of finitism, the context suggests a different and quite natural answer: what is characteristic is that the functions in question are all defined by – and I again quote – "recursions, in which a procedure of stepwise computation of one or more functions is formalized and which can be presented without the addition of a new sort of variable". In other words, what they have in common is that they are *general recursive* (now called "computable") *functions*.

Finally, I want to discuss another argument to the effect that Hilbert and Bernays intended something stronger than PRA by "finitism". I quote Wilfried Sieg in "Hilbert's Proof Theory":

> From a contemporary perspective the arguments reveal something very important: as soon as a formal theory contains a class of finitist functions it is necessary to appeal to a wider class of functions in this kind of consistency proof. An *evaluation function* is needed to determine uniformly the numerical value of [closed] terms, and such a function is no longer in the given class. As the formal system considered in the above consistency proof includes [PRA], the consistency proof goes beyond the means available in [PRA]. Finitist mathematics is consequentially stronger than [PRA] at this early stage of proof theory. (Sieg 2009: 163)

This is repeated in Sieg and Ravaglia in "David Hilbert and Paul Bernays, *Grundlagen der Mathematik* I and II":

> From a contemporary perspective the arguments reveal something very important: as soon as a formal theory contains a class of finitist functions it is necessary to appeal to a wider class

---

[1] The finitist conception of a first-order function f is of the construction $f(X, \ldots, Y)$ from the generic numbers $X, \ldots, Y$. But how would a finitist understand a second-order function?

of functions in this kind of consistency proof. An *evaluation function* is needed to determine uniformly the numerical value of [closed] terms, and such a function is no longer in the given class. As the formal system considered in the above consistency proof includes [PRA], the consistency proof goes beyond the means available in [PRA]. Finitist mathematics is consequentially stronger than [PRA] at this early stage of proof theory. (Sieg/Ravaglia 2005)

With a slight change of wording, it is repeated again in the editors' introduction in the Ewald-Sieg edition of Hilbert's lectures on logic and arithmetic.

What is supposed to reveal this is an argument partially sketched in Hellmuth Kneser's *Mitschrift* on lectures Hilbert delivered in 1922–1923 for the consistency of PRA, i.e. that there is no deduction of $0 = 1$.

The crux of the argument is showing that, for every closed term t of PRA, there is a deduction, using just the axioms of definition and the rule of substitution, of $t = \bar{t}$, where $\bar{t}$ is a numeral. Call such a deduction a *computation* of t. Given the assumption that there is always such a computation, it is easy to eliminate mathematical induction from any deduction: suppose

$$\frac{\Gamma, A(0) \qquad \Gamma, \overline{A(x)}, A(x')}{A(t)}$$

is an instance of induction with no induction below it. Substituting 0 for every free variable in t if necessary, we can assume that t is closed. Replace the induction by

$$\frac{\Gamma, A(0) \qquad \Gamma, \overline{A(x)}, A(x')}{A(\bar{t})}$$

from which, together with a computation of of t, $t = \bar{t}$, we obtain a deduction of $\Gamma, A(t)$ by a cut. But we have noted that this instance of induction can be eliminated, since $\bar{t}$ is a numeral.

Having eliminates all inductions from the proof, we can eliminate all variables. Apply $\bar{\;}$ to each formula in the deduction, so that defining equations $s = t$ become equations $\bar{s} = \bar{s}$, etc. It is easy to see that every step Γ in the deduction contains a true equation or inequation.

So what is needed for the consistency proof is a construction, for each closed term t, of a computation of t. Notice that the Sieg/Ravaglia condition, that there is a valuation function $t \mapsto \bar{t}$ is insufficient: it is the *computation* that is needed. And that is not being assumed: it is what the Kneser notes aim to construct. Here is what they say:

– Recall that $f_0, f_1, \ldots$ is a fixed enumeration of all the function constants (other than the $'$), ordered so that, if f_i occurs in the definition of f_j, then $i \leq j$.

- If t is a closed term other than a numeral, it contains a subterm $f_i(k_1, \cdots, k_{n_i})$ where the k_j are all numerals. Call such a subterm *critical*. Let t^* result from replacing a critical subterm $f_i(k_1, \cdots, k_{n_i})$ in t by s, where s is is obtained by successively applying he definitional axioms to eliminate f_i.
- E.g. if f_i is defined by

$$f_i(x, 0) = f_j(x) \qquad f_i(x, y') = f_h[x, y, f_i(x, y)]$$

then t^* is obtained by replacing $f_i(k, m)$ in t by

$$f_h(k, m, f_h(k, m-1, \cdots f_h(k, 1, f_j(k)) \cdots)).$$

- *und so weiter* !!

But *und so weiter* what? We have a deduction of $t = t^*$ but we need a deduction of $t = \bar{t}$ for some numeral $\bar{t}$.

There is no attempt to fill in the *und so weiter* that I know of prior to GdM I (2nd ed., 290f.) and there what we have is the following:

i) Prove by induction on $m < \omega$ that f_m is numeral-wise computable, i.e. if k is an argument for f_m consisting of numerals, then $f_m(k)$ is computable. Thus, in the example above

$$f_i(k, m) = f_h(k, m, f_h(k, m-1, \cdots f_h(k, 1, f_j(k)))$$

the induction hypothesis can be applied successively to obtain $\bar{s}_l$ for

$$s_0 = f_j(k), s_1 = f_h(k, 1, \bar{s}_0), \ldots, s_m = f_h(k, \bar{s}_{m-1}).$$

So we have $f_m(k, m) = \bar{s}_m$: Every critical term is computable.

ii) Now, substituting $\bar{s}$ for a critical term s in the closed term t, we obtain a closed term with fewer occurrences of function constants than t. So, by induction on the number of occurrences of function constants, we conclude that every closed term is computable.

It is indeed, a fine argument; but it involves the proof by mathematical induction of numeral-wise computability. That is a Π_2^0 property and can hardly be called finitist.

Going back to the closed term t with the critical subterm $f_i(k)$ and the term t^* obtained by replacing $f_i(k)$ by $f_h(k, m, f_h(k, m-1, \cdots f_h(k, 1, f_j(k))$ where $j, h, i < i$, there is this:

- Let $j_1 > \cdots > j_m$ be all the $j < \omega$ such that f_j occurs in t and let n_h be the number of occurrences of f_{j_h} in t. Set

$$\|t\| = \omega^{j_1} n_1 + \cdots + \omega^{j_m} n_m.$$

Then k

$$\|t^*\| < \|t\| < \omega^\omega.$$

- So, $t \mapsto \bar{t}$ can be defined by by transfinite recursion on $\|t\| < \omega^\omega$, which is equivalent to a double nested recursion of the sort discussed in *GdM* I, §7.

So, for one who believes that Hilbert and Bernays accepted such forms of recursion as finitist, Knesser's argument can be completed. But again, there is no call for an evaluation function for the primitive recursive terms in this argument. (The evaluation function itself can be defined using double nested recursion.)

But, to repeat, I think there are good grounds for thinking that is a misreading of the text and that Hilbert and Bernays did not intend to include those examples of non-primitive recursive functions as finitist.u

I don't know what to make of Bernays' argument in *GdM* I. The attempt at a consistency proof for PRA in the lectures of 1922–1923 is understandable. Gödel had yet to prove the second incompleteness theorem and so it would have been possible to think that consistency of PRA could be proved in PRA. Besides, it was early days in proof theory and the literature through the 1920's and early 1930's was littered with false arguments in proof theory – Hilbert's argument for *CH* in "On the Infinite", Ackermann's and von Neumann's consistency proofs, Herbrand's Lemma.

But the puzzle about the argument in *GdM* I does not concern the question of what Hilbert and Bernays thought finitism was; they certainly excluded from it mathematical induction on Π_n^0 formulas for $n > 0$. I think we are just looking at what is, from a finitist point of view, an invalid argument.

Bibliography

Ackermann, Wilhelm (1924): "Begründung des „tertium non datur" mittels der Hilbertschen Theorie der Widerspruchsfreiheit." In: *Mathematische Annalen*, Vol. 93, 1 – 36.
Church, Alonso (1941): *The Calculi of Lambda-Conversion*. Princeton: Princeton University Press.
Friedman, Michael (1995): *Kant and the Exact Sciences*. Cambridge (MA): Harvard University Press.
Gödel, Kurt (1958): "Über eine bisher noch nicht benutzte Erweiterung des finiten Standpunktes." In *Dialectica*, Vol. 12, 280 – 287. Reprinted with an English translation in Gödel 1990: 240–252. Gödel 1972 is a revised version.

Gödel, Kurt (1972): "On an Extension of Finitary Mathematics Which Has Not Yet Been Used." In: *Collected Works*, Vol. II (Gödel 1990), 271 – 280. Revised version of Gödel 1958.

Gödel, Kurt (1990): *Collected Works*, Vol. II, Oxford: Oxford University Press.

Hilbert, David (1926): "Über das Unendliche." In: *Mathematische Analen*, Vol. 95, 161 – 190.

Hilbert, David; Bernays, Paul (*GdM* I): *Grundlagen der Mathematik; Vol. 1*. Berlin: Springer, 1968² (1934).

Hilbert, David (2013): *Lectures on the Foundations of Arithmetic and Logic 1894 – 1917*. Hallett, Michael et al. (eds.). Dordrecht: Springer.

Kreisel, Georg (1960): "Ordinal Logics and the Characterization of Informal Notions of Proof." In: Todd, J. A. *Proceedings of the International Congress of Mathematicians, Edinburgh*, Cambridge: Cambridge University Press, 289 – 299.

Mancosu, Paolo (2010): *The Adventure of Reason: Interplay Between Philosophy of Mathematics and Mathematical Logic, 1900 – 1940*. Oxford: Oxford University Press.

Peckhaus, Volker (1990): *Hilbertprogramm und Kritische Philosophie*. Göttingen: Vandenhoek und Ruprecht.

Sieg, Wilfried (1999): "Hilbert's Programs: 1917 – 1922." In *Bulletin of Symbolic Logic*, Vol. 5, 1 – 44.

Sieg, Wilfried (2009): "Hilbert's Proof Theory." In: Gabbay, D.; Woods, J. (eds.) *Handbook of the History of Logic 5: Logic from Russell to Church*. Amsterdam: Elsevier, 321– 384.

Sieg, Wilfried; Ravaglia, Mark (2005): "David Hilbert and Paul Bernays, *Grundlagen der Mathematik*." In: Grattan-Guinness, Ivor (ed.). *Landmark Writings in Western Mathematics: Case Studies, 1640 – 1940*. Amsterdam: Elsevier, 981–999.

Skolem, A. T. (1923): *Begründung der elementaren Arithmetik durch die rekurrierende Denkweise ohne Anwendung scheinbarer Veränderlichen mit unendlichem Ausdehnungsbereich*. Videnskapsselskapets. Skrifter. 1. Vol. 6. Reprinted in Skolem 1970, 153 – 188. Translated in van Heijenoort, Jean (ed.) *From Frege to Gödel: A Source Book in Mathematical Logic, 1879 – 1931*. Cambridge (MA): Harvard University Press, 1967, 302 – 333.

Skolem, A. T. (1970). In: Fenstad, J. E. (ed.). *Selected Works in Logic*. Oslo: Universitetsforlaget.

Tait, William (1981): "Finitism." In: *Journal of Philosophy*, Vol. 78, 1981, 524 – 556.

Tait, William (2002): "Remarks on Finitism." In: Sieg, Wilhelm et. al. (eds.). *Reflections on the Foundations of Mathematics: Essays in Honour of Solomon Feferman*. Natick (MA): A. K. Peters/CRC Press, 410 – 419.

Tait, William (2005): *The Provenance of Pure Reason: Essays in the Philosophy of Mathematics and Its History*. Oxford: Oxford University Press.

Tait, William (2016): "Kant and Finitism." In: *The Journal of Philosophy*, Vol. CXIII, No. 5/6, 261 – 273.

Zach, Richard (2003): "The Practice of Finitism: Epsilon Calculus and Consistency Proofs in Hilbert's Program." In *Synthese*, Vol. 137, 211 – 259.

Juliet Floyd
Wittgenstein and Turing

Abstract: A Just-So story, intended as plausible philosophical reconstruction, of the mutual impact of Wittgenstein and Turing upon one another. Recognizably Wittgensteinian features of Turing's diagonal argumentation and machine-model of human computation in "On Computable Numbers, with an Application to the Entscheidungsproblem" (OCN) and his argumentation in "Computing Machinery and Intelligence" (Turing 1950) are drawn out, emphasizing the anti-psychologistic, ordinary language and social aspects of Turing's conception. These were indebted, according to this story, to exposure to Wittgenstein's lectures and dictations. Next Wittgenstein's manuscripts on the foundations of mathematics 1934–1942 are interpreted in light of the impact of Turing's analysis of logic upon them. Themes will include the emergence of rule-following issues, the notion of *Lebensform*, a suggestion about a strand in the private language remarks, and anti-psychologism. The payoff is a novel and more adequate characterization, both of Turing's philosophy of logic and of Wittgenstein's.

1 Introduction

Three assumptions about Wittgenstein and Turing should be surrendered, and it is the argument of this essay that they should be rejected as a whole. First, it is usually assumed that Wittgenstein and Turing were mutually "alien" to one another, standing on opposite sides of a dichotomy between methods of ordinary language and methods of formal logic.[1] Second, it is assumed that in his later philosophy Wittgenstein was concerned to reject Turing's machine model as an analysis of logic: witness the criticisms of talk of processes, states, and experiences in Wittgenstein's famed discussion in *Philosophical Investigations* of "the machine symbolizing its own modes of operation".[2] Third, it is assumed that Turing himself was a computational reductionist, that is, a mechanistic functionalist about the mind. Although Kripke 1982 does not argue for the last two points explicitly – in fact in a long footnote Kripke says he would like to return to this point (Kripke 1982: 35 – 37, n. 24) – his arguments assume that the "dispositionalist" model of the mind

---

[1] Monk 1990.
[2] *Philosophical Investigations*, PI §§193ff.

Juliet Floyd, Boston University, USA, jfloyd@bu.edu

is under attack by Wittgenstein in the famed remarks (PI §§193 – 194), a view promulgated, but then later rejected, by Putnam.[3]

My main claims are these:

- Wittgenstein and Turing shared a matrix of foundational ideas about the nature of logic.
- They also discussed the nature, limits, and foundations of logic over several years.
- They drew from one another, as they both recognized, developing a confluence of ideas forged over many years, not a conflict.

Given current scholarly understandings, I have to make the case in two directions, Wittgenstein → Turing, and Turing → Wittgenstein. The latter is more difficult, and I will merely aim to briefly sketch my story here, relying on previously published papers for details of the arguments.[4]

My story will be justified by appealing to background features of the Cambridge context of, and argumentation in, Turing's great paper "On Computable Numbers, with an Application to the "*Entscheidungsproblem*" (OCN) and Turing's subsequent writings 1937–1954, as well as considerations based on Wittgenstein's construction of the rule-following passages and the emergence of his later style of writing.

The latter came into view beginning in the fall of 1936, with Wittgenstein's failed revision of *The Brown Book* (EPB). Wittgenstein would have learned of Turing's result before leaving Cambridge for Norway in summer 1936. The impact of Turing reached through Wittgenstein's subsequent development, culminating in an explicit remark from 1947, as we shall see in Section 4.2 below.

In the spring of 1937 there was, I shall argue, an especially important series of reactions Wittgenstein had to Turing's paper, as indicated by the fact that the themes of *Regelmässigkeit*, rule-following, technique (*Technik*), and especially forms of life (*Lebensformen*) appear for the first time at this point. They are embedded in Wittgenstein's signature interlocutory style, emerging also at just this time.

In turn, as we shall see, Turing's paper was indebted to the Cambridge milieu in which Wittgenstein's *The Blue and Brown Books* (BB) were handed around and discussed among the mathematics students.

---

[3] Putnam's early functionalist theories (Putnam 1960, Putnam 1967) rejected logical behaviorism and endorsed computationalism, but his subsequent doubts were voiced in Putnam 1988b and Putnam 2009. See Floyd 2017a: 108 and Shagrir 2005 for discussions of the evolution of Putnam's own views.

[4] Floyd 2012b, Floyd 2013, Floyd 2016, Floyd 2017c, Floyd 2018b.

As a package, these issues show us much about Wittgenstein's later conception of philosophy, and the stimulus we may see him having received from reacting to Turing's work. "Forms of life" emerge as fundamental and ubiquitous, but only *after* Wittgenstein read Turing's "On Computable Numbers" (OCN) in the spring of 1937. This chronology mirrors a kind of conceptual regression to what is most fundamental, what is "given" in logic (and philosophy).

The story I shall tell is forwarded as a plausible analytical and philosophical account. It requires us to regard Wittgenstein differently, but also Turing. Encapsulated, the proposed Wittgensteinian re-reading of Turing is this.

1. Turing's philosophical attitude has been distorted by controversies in recent philosophy of mind (Putnam): computationalist and behaviorist reductionisms, functionalism and the idea of an era in which machines will inevitably become the primary drivers of cultural change and creativity. (Of course this is not to deny that Turing pioneered philosophical discussion of computational explanation and modeling in such far-flung fields as cognitive science, artificial intelligence, neurological connectionism.)
2. Turing was neither a behaviorist nor a reductive mental mechanist. Philosophy of logic, not philosophy of mind, was central for his work on foundations. A Cartesian/behavioristic reading of the "Turing Test" (1950) for over 50 years focused on the individual mind at the expense of the social, despite the fact that for Turing it was the delicate, meaning-saturated human-to-human relations in the presence of machines that was fundamental to the test, not human-machine interface per se. Turing himself regarded intelligence as an "emotional" concept, one that is irreducible, response- and context-dependent, socially embedded and driven by human communicative evolution on a global scale.[5]
3. Turing learned from Wittgenstein that the evolution of our symbolic powers, individual and collective, lies within the forms of life and contingencies of contexts in which words are repeatedly embedded in life, types and categories evolving under the pressure of speech and action. To this end, in all his work Turing focussed on taking what we *say* and *do* with words seriously, and on the *limits* of formal methods, not only their power.
4. Everyday language, including our "typings" of objects as they occur naturally in science and everyday life, are an evolving framework or technology. Influenced in part by Wittgenstein, Turing stressed human conversation,

---

[5] See Turing 1969, Proudfoot 2017, Floyd 2017c.

"phraseology", and "common sense", as foundational. In this sense he was a Cambridge philosopher of his time, as well as a pragmatist (Misak 2016).

The structure of the paper that follows is this.

First, in Section 2 we reconstruct the evolution of Wittgenstein's thought, focussing on the key transitions that were made in 1937–1939, as part of his response to Turing's OCN. We use the notion of *simplicity* as a thread through this story.

Next, in Section 3 we explain the importance of Wittgensteinian aspects of Turing's analysis that have been widely appreciated. We draw out first the history of Turing's engagement with Wittgenstein (Section 3.1) and the distinctive nature of Turing's analysis of what a formal system (in the relevant Hilbertian sense) is, emphasizing its philosophical aspects (Section 3.2).

Finally, in Section 4 we consider a distinctive form of diagonal argumentation that both Turing, and then later Wittgenstein – responding to Turing – emphasize. We treat first Turing's own version (Section 4.1) and then Wittgenstein's rendition of the proof (Section 4.2). The latter draws Turing's argument into the orbit of Wittgenstein's mature philosophy quite explicitly.

2 Wittgenstein

2.1 Wittgenstein on Simplicity

To achieve a synoptic overview let us first consider Wittgenstein's development to have taken place in four stages, driven forward by a signal concept for him (and for Turing): the notion of *simplicity*. This notion took on a variety of forms in Viennese philosophy and philosophy of science in the wake of Mach's emphasis on the importance of "economy" in mathematics and logic.[6] Simplicity is not a simple notion.[7] However, roughly but not too controversially, we may regard Wittgenstein's thinking about the role of simplicity in logic as having unfolded in four roughly distinct phases:

 – Simplicity as an absolute ideal (1914-1921)
 – Simplicity as relative to *Satzsystem* (1929-1932)
 – Simplicity given in language-games (1933-1936)

---

[6] Stadler 2018 gives a nice overview of this principle's influence on much subsequent philosophy of science.
[7] Floyd 2017b.

– Simplicity as fluid and ubiquitous (1937-1951)

What I shall argue is that the final step, earmarked by what we may think of as Wittgenstein's *mature* conception of simplicity, was secured by his reading of Turing's OCN.[8]

In the **first** stage (1914–1927), contained in the *Tractatus* (TLP), simplicity in logical analysis is an *absolute* ideal. All propositions are truth-functions of elementary propositions. Objects are simple and indefinable. They show forth in our picturing of possible situations. There is a "calculus of indefinables".[9] The totality of what can be said may be presented (schematically) *via* a "form series" variable, expressing the form of a well-founded ordering of propositions according to a rule, collected by a form-series (step-by-step symbolically specified) rule utilizing truth functions:

$$[\bar{p}, \bar{\xi}, N(\bar{\xi})]^{10}$$

In the **second** stage (1929–1933), the "Middle Wittgenstein" reacts against this absolutist ideal, surrendering the general form of proposition and becoming a relativist about analysis. On his new view, a kind of compromise between the *Tractatus* and what would come later on, simplicity is *relative* to a grammatical "*Satzsystem*". Thus it is no longer essentially truth-functional. For there are many different *Satzsysteme*, or "calculi", each with their own simples (indefinables). These are relative to our forms of representation. The perspective remains a hybrid with the earlier *Tractatus* view, however, for within each *Satzsystem* simplicity is still absolute.

The idea of "aspects" of grammar enters as a newly-centered focus in this relativized conception of simplicity: logical "features" are not merely *Züge* in the sense of formal truth-functional operations on propositions, as in the earlier view, but grammatical features of uses of language. They are drawn out in "perspicuous representations" of grammar.

Influenced by Ramsey and pragmatism about logic, Wittgenstein construes beliefs as purpose-relative hypotheses, open generalizations, tools for organizing expectations. Grappling with Hilbert, Brouwer, Weyl, and Waismann, Wittgenstein develops the idea that generality in mathematics uses templates, schemata, step-by-step "logic-free" definitions. Proofs offer decision procedures, determining the "meaning" of mathematical propositions in particular "spaces" of grammar.

---

[8] For further detail on this framework of analysis, see Floyd 2016, Floyd 2018b.
[9] Wittgenstein MS 111: 31; cf. Engelmann 2013: 128.
[10] TLP 5.2522, TLP 6. For detailed reconstructions see Leblanc 1972, Ricketts 2014, Weiss 2017.

In the **third** stage (1933–1936), Wittgenstein reaches the view expressed in *The Blue and Brown Books* (BB). In these texts it is language-games that are the seat of simplicity: stepwise-embedded, linearly ordered, and anthropologically cast. To imagine a language is to imagine a "culture" (*Kultur*). Simplicity in analysis is comparative, analogical, and evolutionary. Rules are given by tables followed step-by-step by humans. Humans may amalgamate, share, and hand off procedures. There are no longer any "indefinables".

A Spenglerian flavor haunts this stage of Wittgenstein's thought: an additive, linear structure is used to present differing language-games in a quasi-evolutionary way. There is no sharp or general distinction between "automatic" and "non-automatic" behavior: all is cast anthropologically. And Wittgenstein's remarks about the question, "Can a machine think?" – likely read by the undergraduate Turing[11] – treat it as a grammatical or conceptually analogical issue. There remains a contextually important emphasis on the distinction between a "calculation" and an "experiment": the contrast between necessary, internal relations and those that are empirical.

Most importantly, in §41 of *The Brown Book* Wittgenstein broaches the idea of what he calls "general training": the idea that we could teach a person to follow *any* rule couched in terms of symbols and stepwise directional movements. The idea of a rule as a table; the problem of how determinate this "general training" might be, what the scope of this image of logic is – all these things are very close to what Turing would clarify in OCN, as we shall canvas it below.

But there are clear problems, both with the vagueness in Wittgenstein's remarks here, and in his way of presenting language-games. In his mature period (1936–1946) the text of the *Philosophical Investigations* (PI) emerges, beginning in the autumn of 1936.

What are the hallmarks of this mature, **fourth**-stage view?

– Wittgenstein's ideal of simplicity is "domesticated" and the notions of "culture" [*Kultur*] and "common sense" are eliminated in favor of rule-following and simplicity as embedded in environments and "forms of life" [*Lebensformen*]. The term "*Kultur*" is deleted from the manuscript of PI in the fall of 1936, and never returns to any further version of the manuscript.
– Simplicity is now fluid and ubiquitous, achieved, then contested, still comparative, but dynamic and complex. There are analyses, but they are conducted in "investigations", partial searches testing "harmonies" among us. These are then embedded in further searches, moved, separated, amalgamated, etc.

---

11 See Floyd 2017c for the arguments.

– Wittgenstein now conceives applications of the notion of "simplicity" to be "home-spun" [*hausgebachen*] (MS 152: 96): they are woven out of the embeddings of words in forms of life, revealed in what is taken to be meaningful in everyday or ordinary life.

This gives us a way to think about the Turing → Wittgenstein direction of influence, bearing in mind Turing's philosophical achievement in his OCN. For in Wittgenstein's mature philosophy there remains a unity and robustness in the logical, responding to the generality and mathematical robustness of Turing's analysis of what it is to take a "step" in a formal system of logic (see section 3 below). For this is conceived by Wittgenstein in terms of step-by-step, partially-defined, rule-governed, symbolically articulated procedures and their backdrop in interlocutory exchanges and forms of life. This recovered, *realistic* unity, a kind of norm of elucidation for philosophy – the embedding of language-games in forms of life – is what prevents Wittgenstein's mature idea of logic from hardening into a dogmatically asserted totality of propositions, a static, divided archipelago of conventional schemes, or an artificially ordered series of games.

Wittgenstein's conception of the logical after 1937 exhibits certain particularly striking features. We can explain how he got to his mature philosophy by noticing several things connected, I believe, with his response to Turing's OCN.

1. It was first in the spring of 1937 that Wittgenstein revisited themes of the *Tractatus* and of philosophical method.[12]
2. At this time, for the first time, he turned concertedly toward a detailed investigation of the idea of rule-following and *Regelmässigkeit*.
3. Wittgenstein's famed remarks about the machine that "symbolizes its own modes of operation" (PI §§193ff) are first written down in the fall of 1937.
4. For the first time Wittgenstein investigated the shading off of "calculation" and "experiment" in everyday life.
5. Wittgenstein drew in, for the first time, the notion of a form of life (*Lebensform*).[13]
6. Perhaps surprisingly, the term "*Technik*" first occurs in Wittgenstein's writings only in 1937.[14] It is explored thereafter in his writings as a notion and as an object of reflection.

---

[12] Cf. Engelmann 2013 for a discussion.
[13] See Engelmann 2013, Floyd 2016, and Floyd/Mühlhölzer forth.
[14] See MS 118: 874.

When it first enters into his writing, the notion of "technique" is marked by a reference to "Watson".[15] This is an allusion to Wittgenstein's summer 1937 discussions with Alister Watson and Turing, an important fact that connects the notion to his deepening reflections on the idea of *following a rule* and the notion of *Regelmässigkeit*, and regularity.[16] (We shall discuss this below in Section 2.) After this point this term fills the pages of his writings and lectures, becoming a signature notion of his mature philosophy (it occurs in his *Cambridge Lectures on the Foundations of Mathematics* (LFM) 117 times).

As I see it, Turing's analysis of a logical "step" in OCN got Wittgenstein to see a "dynamic" perspective as a way to conceive the nature and limits of the logical, and the notion of a "technique", devised to mark the moment in which a routine is embedded in ordinary life, reflects this.

This chronology is made sense of by the analysis I shall give, and the chronology makes sense of how my analysis works. Let us review some of the key moments in this unfolding of thought.

2.2 The *Urfassung* of PI: 1936–1937

At the end of summer 1936, his Cambridge fellowship over, Wittgenstein went to Norway and attempted to turn the dictated *Brown Book* into a book manuscript (MS 115, EPB). The first appearance of "*Lebensform*" in Wittgenstein's corpus occurs in the fall of 1936 (EPB: 108). It occurs in a discussion of a language or "culture" where there is an environment, and words for color, that are very different from our own. He struggles a bit with the idea of what it is to "think of a use of language or a language" fixing "gaps" in grammar, and after several variants ("life form"/"form of life") he settles on his mature language, the language that remains in PI: To imagine a language is to imagine, not a *Kultur* (as it was in BB), but rather a *Lebensform* (cf. PI §19).

By p. 118 of EPB, Wittgenstein drew a line through the page, writing "This whole 'attempted revision' is *worthless*". After some difficult days, he began a new manuscript (MS 142, see BEE). This would become the so-called "Original Version" of *Philosophical Investigations*, the so-called *Urfassung* (UF in KgE). Seventy-six pages of the *Urfassung* were done by Christmas 1936. Several features are especially important:

---

[15] See the so-called "early version" of PI, the so-called "*Frühfassung*" FF §322 (KgE: 396 = RFM I §133).
[16] On these discussions, see Floyd 2001, Floyd 2017c and Section 2 below.

- Wittgenstein's remarks on Plato's *Theaetetus* and simples are added in more or less their final position (cf. PI §§46ff).
- "Forms of life" enter the manuscript concertedly (PI §§19, 23 – 25), though this key term, being primordial and normative, only occurs five times altogether in PI.[17]
- "Culture" [*Kultur*] and "common sense" are dropped from the manuscript, never entering again.
- Wittgenstein's remarks about Ramsey, logic as a calculus, and logic as a "normative science" (cf. PI §§81ff) are written down.
- The rule-following remarks are broached, but the notion of "technique" is altogether absent.
- The manuscript stops with the question, which remains as yet unanswered, as if a task Wittgenstein leaves himself for the spring: "In what sense is logic something sublime?" (UF §86 = KgE: 130)

Turing sent Wittgenstein an offprint of "On Computable Numbers, with an Application to the *Entscheidungsproblem*" before 11 February 1937. Throughout the spring, in his notebooks Wittgenstein struggles with the notion of simplicity, which he says must be domesticated: "The simple as a sublime term and the simple as an important *form of representation* [*Form der Darstellung*] but with homespun [*hausbackener*] application" (MS 152: 96). Our argument is that it is Turing who showed that analysis in the sense of formal logic, the very idea of "simplicity" of formal steps, their transparency and gap-free character, *must* have a "homespun" use. The terms "simple" and "simplest", explicitly thematized and relied upon, occur 10 times in Turing's OCN.

2.3 From the *Urfassung* of PI to the *Frühfassung*: 1937–38

Wittgenstein completed the *Urfassung* before leaving Norway on May 1st, 1937. During the spring there is substantial development of his mature philosophy of logic: the ideal of the "sublimity" of logic is reworked. Now its "sublimity" lies precisely in our everyday applications of it, what at first seem like "rags and dust" but which allow logic the friction and sensitivities of use it requires (cf. PI §§52, 107). The themes of rule-following and *Regelmässigkeit* are worked through and developed for the first time. And Wittgenstein begins to reconsider the very idea of a "foundation" of mathematics.

---

[17] For discussion see Floyd 2016, Floyd forth.

Completed in spring 1937, the *Urfassung* is the manuscript source of PI §§1–189. It is surely significant that the manuscript ends with what Wittgenstein will set himself to clarify over his summer break: "But are the steps then *not* determined by the algebraical formula?" — The question contains a mistake." (UF §189 = KgE: 204)

And indeed this question *does* contain a mistake, if we think of Turing's way of analyzing the idea of "determining the steps" in something other than a miraculous or "purely formal" way. This we shall discuss below in Section 3: the *Entscheidungsproblem* shows that the demand for a free-standing answer, Yes or No, cannot be made unequivocally.

Back in Cambridge in the summer of 1937, Wittgenstein had a typescript made of the *Urfassung* (TS 220). He showed it to Moore, who noticed the introduction of the new remarks about simples alluding to Plato. According to Rhees, Wittgenstein told Moore that in *The Brown Book* he had used a "false method" (*falsche Methode*), but that now he had found the "right" or "correct" method (*die richtige Methode*). Moore told Rhees that he did not understand what this meant.[18] But I think we can, with the power of hindsight.[19]

It was the *Urfassung's* closing question about steps being or not being determined by an algebraical formula that may have inclined Wittgenstein to join Alister Watson and Turing in a summer discussion group at Cambridge that was devoted to discussing the philosophical significance for foundations of mathematics of the recent undecidability results of the 1930s, including Turing's OCN.[20] Wittgenstein had known both of these Kingsmen since their undergraduate days: so it was not the first time they had talked. But the context was new, and they were each thinking about how to characterize it. After all, these undecidability results show that a naïve conception of "determining the steps" algorithmically has its provable limitations.

Alister Watson's *Mind* paper (Watson 1938) was one result of these discussions. Watson explicitly thanked Turing and Wittgenstein, particularly for discussions of how best to represent the philosophical significance of Gödel's incompleteness theorems.[21] He closed with the thought that we are not much further along in the foundations of mathematics from the ancient Greeks, with their puzzles about the continuum.

Another immediate result of the 1937 discussions was Wittgenstein's turn toward writing numerous remarks on the foundations of mathematics. This,

---

18 See Rhees's introduction to EPB: 12 – 13; the editors of PI disagree with Rhees's claim that Wittgenstein brought both TS 220 and TS 221 with him to Cambridge in the summer of 1937.
19 For more detailed discussion of my claims and the dates see Floyd 2016: 21, Floyd 2018b: 72ff.
20 See Floyd 2001, Floyd 2017c.
21 Watson 1938: 445.

long planned as part of his envisioned book, began in earnest in the autumn of 1937 with remarks echoing those in Watson's "Mathematics and Its Foundations" (Watson 1938). Wittgenstein discussed, not only rule-following and Gödel, but the whole idea of a machine that "symbolizes its own modes of operation". Wittgenstein's focus on the foundations of mathematics lasted through 1944. Floyd/Mühlhölzer forth. discusses the non-extensionalist conception of the real numbers that Wittgenstein developed, focussing on Wittgenstein's responses to Hardy's *A Course of Pure Mathematics*. It is significant that already in the spring of 1937, in light of issues about the unique representability of real numbers, Wittgenstein was turning toward ideas about the differing ways we have of thinking about irrationality, infinity and the continuum.[22] These are themes with which Turing is struggling in OCN.

In the autumn of 1937–1938, right after the discussion with Watson and Turing, Wittgenstein's *Urfassung* of PI was immediately extended to become the so-called "Early Version", the *Frühfassung* (FF) of PI.[23] Here the mature perspective developed in the *Urfassung* is applied to logic the foundations of mathematics. This extension is the basis for what was later excised from PI, and published as RFM I.

In this manuscript we see the first occurrences of Wittgenstein's remarks about our conception of "the machine as symbolizing its own ways of operating" (PI §§193ff). It is we humans who are living creatures who self-conceive *as* machines: we know what it is to "reckon without thinking" according to a rule. The significance of this will become clearer below in Section 3, when we discuss Turing's analysis of a "step" in a formal system.

Drawing out the importance of contrasting varieties of "technique", is what, on Wittgenstein's mature view, *allows* a variety of aspects of numbers to be seen. Aspects are discovered. Techniques are by contrast invented.[24] This is a form of realism, understood in the sense of Diamond's realistic spirit: the *fitting* of concepts

---

[22] As I discuss in Floyd 2016: 21ff., Wittgenstein's Notebook 152, written in the spring of 1937, not only concerns the themes of simplicity and sublimity, but also begins with warmup exercises in the theory of continued fractions, in which the real numbers receive unique decimal representations (unlike our decimal sequence representations). We know Turing was concerned about the implications of this for his analysis of "computable" real numbers; on this see Floyd 2017c: 125, n. 64.

[23] Published in KgE: 205 – 446.

[24] In Floyd/Mühlhölzer forth., chapter 8, I explain that "techniques" are invented, whereas "aspects" are, for Wittgenstein, discovered. Textual evidence may be found at BT §134; RFM II §38, RFM III §§46ff; MS 122: 15, 88, 90; PI §§119, 124 – 129, 133, 222, 262, 387, and 536; xi: 196; PPF xi, §130. Floyd 2018a analyzes this distinction, while Kanamori 2018 applies it to the real numbers.

to reality in forms of life.[25] The idea of a "technique" is designed to register the activity of our designing the "fitting" that goes on.

Wittgenstein lectured at Cambridge in early 1938 on Gödel in an exploratory vein, focussing on the role of negation and the concept of "provability" in Gödel's proof.[26] Oddly it seems he was anticipating questions about the range of proofs about provability only later rigorized.[27] His investigations focussed on the borders of incompleteness, looking at what would be required to establish that they must exist.

Finally, Wittgenstein submitted the *Frühversion* of PI to the Cambridge Press in September 1938[28] with a Preface emphasizing that the method is *not* "gap free" [*lückenlose*], it doesn't run along one "track" (cf. PI, Preface). This apt metaphor, explored in his manuscripts in the period 1937–1939, squares with Turing's analysis of logic as well as Wittgenstein's mature view of formal logic. For Turing shows that it is the partially, and not the totally defined function that must be taken as the basic notion in analyzing the idea of a logical "step". Given this, the embedding of routines in *Lebensformen* – where there may be drift, misunderstanding, and contingencies of application – is inevitable. We shall clarify this point in what follows.

3 Turing

3.1 Turing and Wittgenstein

As is well known, Turing attended Wittgenstein's 1939 lectures at Cambridge on the foundations of mathematics (LFM). Their discussions of contradictions are often regarded as expressing fundamental philosophical or ideological disagreements.[29]

---

[25] See Diamond 1991.
[26] See WCL: 50 – 57.
[27] Henkin 1952: 160 asked a question not too far from some of the questions Wittgenstein raised:

$$\text{If } \phi \text{ is } Bew(\ulcorner\phi\urcorner), \text{ Does } \Sigma \vdash \phi?$$

Löb 1955 then showed:

$$\text{If } \Sigma \vdash Bew(\ulcorner\phi\urcorner) \to \phi, \text{ then } \Sigma \vdash \phi.$$

[28] See Monk 1990: 413.
[29] See Monk 1990.

But what is less emphasized is that Turing's attendance, taking place during just the time he was beginning at Bletchley, was a continuation of earlier discussions. They reflect Turing's even earlier engagement with Wittgenstein as an undergraduate, engagement I shall argue left its imprint, not only on Turing's general philosophical views about logic, but on the precise argumentation he gives in his OCN.[30] Turing's implementation of diagonal argumentation, later revisited by Wittgenstein, will be interpreted below in Section 4.1. It has a Wittgensteinian flavor, one related importantly to the later 1939 discussions between Wittgenstein and Turing.

For now the important point is to note that there was a general Cambridge context, associated Wittgenstein, Whitehead, Russell, Ramsey, Nicod and others, in which foundational issues about logic in general, and types and recursion in particular, were avidly discussed.[31] Turing was an undergraduate 1931–34, and a King's Postgraduate Fellow 1934–36; Alister Watson was "Kingsman" as well, an undergraduate 1926–1933 and then a Postgraduate Fellow 1933–1939.

In the spring of 1932 in his Cambridge course of lectures "Philosophy" Wittgenstein came up with an original analysis of equational recursive specifications in which the need for a uniqueness rule was made explicit.[32] In the autumn of 1932 he began teaching a second, separate course called "Philosophy for Mathematicians" to hash the ideas out further. He argued there that

> What counts in mathematics is what is written down: if a mathematician exhibits a piece of reasoning one does not inquire about a psychological process.[33]

In the autumn of 1933 this course was taught again, and over forty students showed up to the first few lectures.[34] Seeking dialogue and discussion, Wittgenstein dismissed the class, stating that instead of offering lectures he would dictate ideas and distribute the transcriptions to the class. This was the context in which *The*

---

30 See Floyd 2017c for details. Hodges 1983 reports Turing engaged with Alister Watson in discussion of methods of diagonal argumentation in 1935, perhaps before Turing's idea of a "machine" had occurred to him. As I explain in Floyd 2018b: 73, n. 19, the presence of a 0-1 array to present Cantor's method of argument in a recursive, constructive vein was already present in Wittgenstein's MS 157a, written by hand in either 1934 or 1937, and possibly in 1935. This is a precursor to Wittgenstein's presentation of the diagonal argument in RPP 1 §§106ff (MS 135: 118, TS 229, §1764), discussed below in section 4.2.
31 See Floyd 2017c for a detailed argument.
32 See Marion/Okada 2018 for details.
33 AWL: 225.
34 Notes of these lectures have been published in AWL. Recently other transcriptions of these and related discussions taken down by Francis Skinner have been found, including an alternative, longer version of *The Brown Book* and lectures on the nature of logic; these will be edited and published: see Gibson 2010.

Blue Book (1933–1934) and *The Brown Book* (1934–1935) were dictated: mathematics students were the desired audience.

There is good reason to find it plausible that Turing was exposed to these dictations, either by attending the 1933 autumn lectures or reading the dictated notes of them. It is also possible that he attended Wittgenstein's 1932–33 version of the course. For by March of 1933 we know that Turing had avidly read Russell's *Introduction to Mathematical Philosophy* (Russell 1920), in which Wittgenstein's view that logic is tautologous was discussed. And in December 1933 Turing gave a talk to the Moral Sciences Club, arguing that

> ... the purely logistic view of mathematics is inadequate; mathematics has a variety of interpretations, not just one ...[35]

We have here a view orbiting in the circle of Wittgenstein's ideas, quite different from the conception of logic being promulgated at that time by Carnap, in his logical syntax phase. Turing regards this conception as "inadequate".[36]

Whatever the case before 1939, in 1939 Wittgenstein and Turing were continuing conversations in the classroom during that spring in a cooperative, rather than an antagonistic vein. Wittgenstein knew about Turing's famous paper, and they were continuing to discuss the implications of Wittgenstein's new-found focus on rule-following. Each learns from the other, as is evident from the very first lecture where Wittgenstein makes an inside joke with Turing about the distinction between signs and symbols.

It is also clear that Turing continued working on philosophical aspects of logic afterwards, while at Bletchley. He explicitly states that his unpublished paper "The Reform of Mathematical Notation and Phraseology" (Turing 2001b, 1942–44) was influenced by Wittgenstein's lectures, in particular (as he says) the idea of handling types with ordinary ways of speaking.[37] He argues here that what needs to be taken seriously is the end-user, the ordinary "phraseology" of mathematics, rather than the "anti-democratic" ideal of a single, overarching formalism, which would serve as a kind of "straightjacket" to thought.[38]

---

[35] See Hodges 1999: 6, discussed in Floyd 2017c: 126.
[36] See Floyd 2012a and Floyd 2017c for arguments to this effect.
[37] see Floyd 2012b.
[38] I discuss Turing 2001b in my Floyd 2013.

Moreover, in notebooks from the early 1940s Turing continued taking what he called "Notes on Notations". He made analyses and investigations of the specific symbolic devices worked with by Leibniz, Boole, Peano, and others.[39]

These facts serve to correct the portrait of Wittgenstein and Turing as "alien" to one another, or engaged in ideological discussion for and against the use of mathematical logic in philosophy. Instead, they are thinking through foundational issues about logic *with* one another.

But what about the well-known dispute between Turing and Wittgenstein over contradictions in Wittgenstein's 1939 *Cambridge Lectures on the Foundations of Mathematics*?

As is well-known, Wittgenstein insists in LFM on a non-extensional view of contradiction in conversation with Turing. The presence of a formal contradiction allows, by the rules of classical logic, the problem of cascading or explosion: anything becomes derivable in the system. A non-extensional view allows that when formal contradictions are found, one can put them to the side and move elsewhere in the system, giving one or another practical, purposeful reason for so doing.

Wittgenstein is concerned to emphasize with Turing that it is the *uses* of the system that matter to foundations, not only and primarily the ultimate classical logical properties of the sentences of the language with their formal deductive consequences treated ideally, apart from this. This is the idea of the "homespun" character of formal logic discussed in section 2 above. It is instanced today in our hand calculators: punching in a large enough number will cause the addition program to fail. But we still regard the calculator as "adding".

Although we should see Wittgenstein working up a philosophical view that is largely congenial to Turing's OCN, Turing of course pushes back in LFM. Classical logic has its uses, especially in complex empirical situations: there may be situations where these dropping-to-the-side of formalisms would be dangerous, if we are embedding software in powerful and complicated technological projects (such as building bombs, bridges or airplanes). But what will guide us, in addition to issues of consistency and explosion, are approximations, decisions as to scope and probabilities of failure, values about what matters for the purpose at hand.

Wittgenstein's response, then, is that formal issues of consistency in the sense of classical deductive logic are not necessarily the primary, sole foundation of what matters to the objectivity of applications of arithmetic in everyday life. This tells us something important about foundations. The "homespun" idea is that indeed,

---

39 These notebooks, from the estate of Robin Gandy, were sold at Bonham's in 2016 in New York into private hands; see Hodges/Hatton 2015 and Floyd 2017c: 140 n. 100.

for certain purposes and in certain situations, a contradiction is something we may wish to eliminate. But not because it violates an eternal law of logic that is irreversible or somehow set in abstract stone; instead as a matter of technique, a matter of adapting our formalism to actual cases and situations. – The point, actually fully consistent with Turing's argumentation in his OCN, is to reconstrue what debates over the "reality" of the law of excluded middle come to.

Wittgenstein's notebooks from 1939 include much exploration of the method of diagonalization as a technique that reveals new aspects of concepts. He is interested in exploring the differing guises under which we represent, both diagonalization itself as a method and the real numbers. As we shall see in Section 4.1, this reflects an engagement with Turing over the method of diagonal argumentation that Turing himself used in his OCN. Wittgenstein explicitly takes his own philosophical perspective to be reflected in this (see Section 4.2 below). Indeed, Turing's proof has a distinctly Wittgensteinian flavor, as we shall now argue. In particular, Turing sidesteps debates over the general applicability of the law of excluded middle when he frames his argument resolving the *Entescheidungsproblem*.

3.2 Articulations of the *Entscheidungsproblem*

Wittgenstein was perhaps the earliest person to frame the general decision problem for logic.[40] For he wrote to Russell in 1913:

> The big question now is, How must a system of signs be constituted in order to make every tautology recognizable as such *IN ONE AND THE SAME WAY*? This is the fundamental problem of logic! [*Grundproblem der Logik*]![41]

In terms of an overarching conception of logic, Wittgenstein had already begun to forward the following ideas, characteristic of his philosophy throughout his life:

– The propositions of logic are tautologies (or contradictions), "senseless" (*sinnlos*) but not "nonsense" (*unsinnig*), evincing the limits of true-false talk, i.e., sentences with sense (*Sinn*).
– There are no fundamental axioms ("laws") of logic in the sense that axiomatization does not in and of itself reveal to us what is fundamental to logic itself.

---

40 See Dreben/Floyd 1991 for a discussion.
41 Wittgenstein to Russell November or December 1913, see letter 30 in WC: 56ff.

- Logic is to be understood symbolically, in terms of step-by-step procedures that can be written down and recognized by us.
- Philosophy, a part of logic, reflects on the character and limits of this perspective.

The *Entscheidungsproblem* asks whether there exists a definite method that can determine, for every statement of mathematics expressed formally in an axiomatic system (using first-order logic), whether or not that statement can be deduced from the axioms. Hilbert believed in 1930 that the answer would be positive, that there would be no such thing as an "unsolvable" problem.

In 1935 Turing took Newman's course covering the open problems of metamathematics, including the *Entscheidungsproblem*.[42] We know that he was reported discussing diagonal arguments with Alister Watson and Braithwaite at this time. By May 1936 he had resolved the *Entscheidungsproblem* in the negative. It has been an outstanding question how it was that Turing so quickly resolved the question analyzing the notion of a formal system in terms of his "machines". Emphasizing the backdrop to his work in the Cambridge philosophical tradition of discussing the *nature* of logic helps us make clearer sense of this.

The heart of the *Entscheidungsproblem* involved answering the question, What *is* a "definite method"? To satisfactorily resolve it in the negative, one would ultimately have to analyze what is meant *in general* by a "formal system" and a "step" in a formal system in the relevant Hilbertian sense. (Had the problem been answered positively, one would simply have exhibited a Decision Procedure for first-order logical validity). It is crucial that the required general analysis could not be accomplished by simply writing down just another formal system. Nor could it be done by setting out in the metalanguage various kinds of different axiomatic systems. This is why the ("logic-free" versions of) λ-definability and the Herbrand-Gödel-Kleene equational systems were used in the earliest work attempting to clarify what is meant by an "effective" calculation. It is also why Turing devised his machines with command-tables, in a "logic-free", i.e., non-formalized-system-of-logic way. His point was to avoid entanglement with the vagaries of this or that formalization of logic, in order to get to the essence of what a "step" in a formal system *is*.

As is well-known, in 1935 Church, Kleene and Rosser showed that the class of functions calculable in the Herbrand-Gödel-Kleene equational calculus is co-extensive with the class of λ-definable functions.[43] In his "Note on the Entscheidungsproblem" (Church 1936) Church, building on Gödel (Gödel 1931), demon-

---

[42] Hodges 1983.
[43] See Kleene 1981a, Gandy 1988, Sieg 2009.

strated that there is no "effectively calculable" function which decides whether two λ-definable expressions are equivalent. This resolved the *Entscheidungsproblem* in the negative. Next, Turing showed, independently of Church, that no "machine" of the type set out in his OCN can "compute" the desired general procedure as an "application" of his wholly novel analysis, also resolving the *Entscheidungsproblem* in the negative. Faced with having been scooped by Church on the result, Turing nevertheless was able to publish his paper because of its conceptual novelty. (In an Appendix he showed that the functions his "machines" can "compute" are just those that are λ-definable.)

It was the clarification of what a formal system or an algorithm or computation *is* that was new in what Turing achieved. As Kleene later put it,

> Turing's computability is intrinsically persuasive, but λ-definability is not intrinsically persuasive and general recursiveness scarcely so either (its author Gödel being [in 1934] not at all persuaded [that it analyzed the idea of "effective calculability" or "calculation in a logic"]).[44]

As Turing's student Gandy wrote of Turing's way of thinking,

> The approach is novel, the style refreshing in its directness and simplicity. The bare-hands, do-it-yourself approach does lead to clumsiness and error. But the way in which he uses concrete objects such as exercise books and printer's ink to illustrate and control the argument is typical of his insight and originality. Let us praise the uncluttered mind ...
>
> What Turing did, by his analysis of the processes and limitations of calculations of human beings, was to clear away, with a single stroke of his broom, this dependence on contemporary experience, and produce a characterization which – within clearly perceived limits that will stand for all time.[45]

The point is that Turing's particular way of resolving the *Entscheidungsproblem* was *not* the application of a preexisting blueprint of ideas and methods in the metamathematics literature. When he first handed it to Newman, Newman thought it too elementary and nearly discarded it.[46] Instead, Turing offered – in contrast to Gödel, Kleene and Rosser – a philosophically informed, analytic exercise. What he achieved was an intuitively satisfying simplification of ... *simplicity*! (Here of course we mean "simplicity" in the logician's sense of a transparent, unproblematic simplest *step* in a formal system.) He did so by picturesquely drawing in the idea of a human being operating with a table of rules according to a certain routine.

---

44 Kleene 1981b: 49; compare the discussion in Kennedy 2017.
45 Gandy 1988: 78, 93.
46 See Hodges 1983: 112.

This last point has been widely acknowledged.⁴⁷ So is the fact that E.L. Post's analysis of logic in terms of "workers" (drawing in the human element as well) is more or less equivalent, an independent achievement.⁴⁸ What I am arguing is that Turing's deployment of his central argument also bears the stamp of Wittgenstein's way of thinking about logic "anthropologically", rather than "metamathematically": the idea of simplicity as something "homespun", rather than sublime.

Turing analyzed what a step in a formal system *is* by thinking through what it is *for*, i.e., what is *done* with it. The comprehensiveness of his treatment – its lack of "morals" – lies here. Turing made the very idea of a formal system *plain*, unvarnishing it. It is this, I believe, that Wittgenstein responded to beginning in the spring of 1937. Turing took up a "form of life" or "language-game" stance, not an ideological or metaphysical perspective: he *de*-psychologized the notion of "logic". Unlike Post 1936 and Gödel 1972, Turing did not take his analysis to rest on or even necessarily apply to limits of the human mind *per se*. This was part of his Wittgensteinian inheritance.

Differently put, Turing made the notion of a formal system (or definite method) *surveyable* (*übersichtlich, überschaubar*), "open to view". This in turn makes the very idea of surveyability ... surveyable! And this would explain as well why it is that the very notion of "surveyability" becomes such a focus in Wittgenstein's manuscripts in 1939.⁴⁹ Wittgenstein is exploring, in the wake of his discussions with Turing, what it means to say that a proof is "surveyable", "reproducible", "communicable" and so on.

In the end, to clarify the foundations of logic one must draw in the notion of *a human calculator*. This requires, not a psychological account, but a *logical* one: the idea of a *shareable* human calculating procedure that may be offloaded to a machine or another human prover or calculator.

As Sieg puts it,

> Most importantly in the given intellectual context [the move from arithmetically motivated calculations to general symbolic processes that underlie them] has to be carried out programmatically by human beings: the *Entscheidungsproblem* had to be solved by *us* in a mechanical way; it was the normative demand of radical intersubjectivity between humans that motivated the step from axiomatic to formal systems . . . .
>
> It is for this very reason that Turing most appropriately brings in human computers in a crucial way and exploits the limitations of their processing capacities, when proceeding mechanically.⁵⁰

---

47 For a discussion see Kennedy 2017.
48 See Post 1936 and Sieg/Mundici 2017.
49 For a detailed commentary and explication of RFM III, from 1939, see Mühlhölzer 2010.
50 Sieg 2006: 200, my emphasis.

Turing's comparison, in analyzing the idea of a "simplest step" in a formalism, is that:

- (OCN) §9 I: A human computor works locally, step-by-step, and can only take in a certain number of symbols at a glance.
- (OCN) §9 I: The computor takes in "simple operations ... so elementary that it is not easy to imagine them further divided".
- (OCN) §9 III: As Turing himself puts it, we "avoid introducing the notion of a 'state of mind' by considering a more physical and definite counterpart: it is always possible for the computor to break off from his work, to go away and forget all about it, and later to come back and go on with it. If he does this he must leave a note of instructions (written in standard form) explaining how the work is to be continued. This note is the counterpart of the 'state of mind'.

This last point makes very clear that Turing is *not* relying on any theory of mentality, but only presupposing the human *communicability* of a "step" in calculation. The notion of a shareable routine of reckoning-according-to-a-rule is taken as basic in his model.

4 The Diagonal Argument

Why, on our story, would Wittgenstein have been so struck by Turing's 1936 paper?

It is important here to understand certain philosophical aspects of Turing's method of proof in "On Computable Numbers" (OCN). As we have just argued, what Turing offered was a remarkable analysis of our very idea of a "step" in a formal system. And he did this by embedding the idea of "calculation-in-a-logic" in a shared human world: an analogical simplification.

His analysis does *not* turn on a theory of mental states, mathematics, or logic, but instead on the idea that logic is *written down*, just as Wittgenstein had argued it should be in his Cambridge lectures 1932–1935.[51] Turing takes the everyday human ideas of a "command" and a "calculation" as basic elements of logic and works out a (mathematically robust) "comparison" between the activities of a human and that of a machine. In other words, like Wittgenstein Turing takes the *human*

---

[51] In addition to BB there is the so-called "Yellow Book" and transcriptions of the 1932–33 "Philosophy for Mathematicians" (cf. AWL: 43 – 73, 205 – 225).

notion of calculation as basic or simple, and builds his analogy with machines from there.[52]

In effect, Turing used the method of what Wittgenstein called *Vergleichsobjekte* (cf. PI §130), objects of comparison.[53] He states explicitly that we may *compare* the activities of a human computor[54] and a machine (OCN, §1). This was a distinctive move, one that probably would not have been made by a mathematician such as Gödel, Church, Rosser or Kleene: it is remarkably simple, down-to-earth, everyday.

This is why, revisiting Turing's paper in a remark written in 1947, subsequently published in *Remarks on the Philosophy of Psychology*, Vol. I (RPP 1 §1096), Wittgenstein says: "These machines [Turing's 'Machines'] are *humans* who calculate."

4.1 Turing's Diagonal Argument

Let us next turn to the actual diagonal proof used in Turing's OCN to resolve the *Entscheidungsproblem* in the negative. I have made a careful reconstruction of the proof elsewhere (Floyd 2012b) and will simply give an overview of the salient philosophical points here.

It is philosophically crucial that OCN does *not* rely fundamentally on the now readily applied "Halting Argument" in order to show that there is no decision procedure for pure logic. Instead, Turing constructs an idiosyncratic machine, utilizing a kind of *positive* argument that does not turn on the production of a contradiction, or the construction of a machine capable of negating the behavior of another machine, as the Halting Argument does.[55]

Instead, Turing's argument turns on the fact that his machine turns up something analogous to the following command, as I have argued elsewhere:[56]

Do What You Do

This expresses a rule that *cannot* be followed. This makes its point deeply philosophical, not only logico-mathematical. For the fact that we can see that this command is, without further supplementation, unuseable demonstrates that the

---

[52] Floyd 2012b.
[53] This reading is laid out in Floyd 2012b, Floyd 2017c, Floyd 2018b.
[54] Until the late 1940s "computer" referred to a person, often a woman, who carried out calculations and computations in the setting of an office or research facility. Nowadays "computor" is used to make the human user explicit.
[55] Floyd 2012b reconstructs the argument carefully; cf. Floyd 2016, Floyd 2018b.
[56] Floyd 2012b, Floyd 2017c.

human interface, the human context of a shareable command, is fundamental to the nature of logic.

For "Do What You Do" tells you nothing without a specific context of application. It is like a pair of fingers pointing straight at one another. Of course, in an ongoing stream of life, embedded in a conversation or activity with a purpose (e.g., I am showing you how to ride a bike or type the return key on a keyboard repeatedly) "Do What You Do" makes perfect sense, indicating perhaps that you should continue on, doing the same as what you are doing now. Without being embedded in a form of life, however, "Do What You Do" does not issue a command that can be followed (imagine drawing a card in a game with this printed on it). This is what Turing's proof ultimately reveals. The machine he constructs is not contradictory, and does not generate an infinite regress. Rather, we must *see* that such a machine, imagined put into service of a Decision Method for determining first-order logical validity, must stop in the face of its own tautology-like self-inscription. This shows the fundamental need for a context, that is, a form of life in which words and symbols are being embedded.[57]

Right at the beginning of OCN, §9, anticipating his application of the diagonal "process" (as he calls it), Turing notes that he *could* have run his argument differently, by way of contradiction in the manner of the Halting Problem:

> The simplest and most direct proof ... is by showing that, if this general process [of determining whether a machine is "circle free"] exists, then there is another ["contradictory"] machine β. This proof, although perfectly sound, has the disadvantage that it may leave the reader with a feeling that "there must be something wrong".

What might be "wrong" is a concern that Turing has assumed, against the intuitionist, that the law of excluded middle applies univocally to all specifications of all Turing Machines. So Turing says,

> The proof which I shall give has not this disadvantage, and gives a certain insight into the significance of the idea "circle-free". It depends not on constructing β [the "Contrary" machine familiar from the Halting Argument, in which machines that halt are changed to those that do, and vice versa, along the diagonal], but on constructing β', whose nth figure is $\phi_n(n)$.

Turing's β' machine is constructed so as to follow its own commands perfectly, without any difficulty, through a series of stages. The difficulty comes when it reaches the particular stage that embodies the machine that it itself is. At this

---

[57] See Floyd 2012b, Floyd 2016, Floyd 2018b for further discussion of the "Do What You Do" argument of Turing.

point, it comes to the command to do what it itself does: and then it cannot do anything.[58]

An analogy would be with the "positive" Russell Paradox, that is, the issue of the set of all sets that *are* members of themselves. This is the exact complement, so to speak, of the usual Russell set of all sets that are *not* members of themselves. Think of it as the *positive* Russell set. In a certain sense, S "comes before" Russell's set, is more primordial, for there is no use of negation within its definition. And it is not contradictory.

Define

$$S = \{x \mid x \in x\}.$$

Now ask

$$\text{Is } S \in S?$$

And the answer is:

$$\text{If Yes, then } S \in S.$$
$$\text{If No, then } S \notin S.$$

So we have that:

$$S \in S \iff S \in S.$$

There is no inconsistency or paradox here. But there is a problem. For all that we can deduce here is that:

$$S \in S \iff S \in S, \text{ and also } S \notin S \iff S \notin S.$$

We are caught in a kind of circular thought of the form, "it is whatever it is". This is surely not incoherent or inconsistent. The trouble is deeper: the thought cannot be *implemented* or applied.

We have here what might be regarded, following Turing and Wittgenstein, as a kind of performative or empty rule. You are told to do something depending upon what the rule tells you to do, but you cannot do anything, because you get into a loop or tautological circle. This set membership question cannot be a question

---

58 I explain the argument in detail in Floyd 2012b.

that can be applied, because one cannot apply the set's defining condition at every point.

An analogous line of reasoning may be applied to, e.g., "autological" in the Grelling paradox if we ask, "Is autological autological?". Without using negation, one does not get a contradiction. But one may generate a question with the concept that may be sensibly answered with either Yes or No. And in this sense it is an unanswerable question. The trouble is, one cannot get to a decision point here. One cannot *play* the game of Yes and No. "Falls under the concept" and "$\in$" cannot be *used* if they are directly equated.

In the above argument an apparently unproblematic way of thinking is applied, but two different ways of thinking about S are involved. For there is the thinking of S as an object or element that is a member of other sets, and the thinking of S in terms of a concept, or defining condition. Similarly, in Turing's OCN proof, there is the unproblematic characterization of a particular machine, and then there is the difficulty that it must, at one precise point or another, get stuck in a loop, confronted with the command to do what it does.

What is important here is that Turing crafts his argument in OCN carefully, in several respects:

– Even an intuitionistic logician who rejects the law of the excluded middle in infinite contexts can accept his analysis of the idea of a "step" in a formal system: "Do What You Do" is not a contradiction so that the proof is not an indirect one.
– Turing does not build into his notion of a "machine" that it must utilize negation, or change halting to non-halting behavior, in its specification.
– Turing's proof demonstrates clearly that is not part of our notion of "following a rule step-by-step" that we do or do not obey the law of excluded middle.
– More generally, Turing's analysis of a "step" in a formal system is altogether independent of *which* formal system we are speaking of, or which particular "states of mind" are actually used, so that the particular choice of formalism or formalized language is not at issue.
– The internal consistency or precise strength of a command structure is not at issue, nor is the internal coherence or strength of a metastance at issue.

In general, Turing is exploiting the fact that formalization alone doesn't settle the analysis. He refuses to ascend to a "metalevel" in a general way, and instead takes on the needed analogy with human activity, working it out mathematically.

Gödel also resisted the idea that the undecidability results tell us anything general about "human reason", holding instead that they reveal something about "the potentialities of pure formalism in mathematics".[59] What we learn is something about what formal systems *cannot* do. But the idea of a human being and what he or she can take in as "simple", "gap-free" or "transparent" is at the heart of our very idea of a formal system, and it is this that Turing, and not Gödel, was able to draw out.

To be clear, Gödel was unstinting in his praise of Turing's analysis of the general notion of "formal system". He argued that the precise scope of his own 1931 incompleteness result was only determined by Turing's work, writing:

> The precise and unquestionably adequate definition of the general concept of formal system [made possible by Turing's work allows the incompleteness theorems to be] proved rigorously for *every* consistent formal system containing a certain amount of finitary number theory.[60]

The point here was that a kind of potential "gap" remained in our understanding of the scope of applicability of Gödel's 1931 paper until Turing clarified what we mean *in general* by a "formal system of the relevant kind".[61]

Moreover, Gödel argued, the universality of Turing's analysis made it special, freeing it of entanglement with this or that particular formalism:

> With Turing's analysis of computability one has for the first time succeeded in giving an absolute definition of an interesting epistemological notion, i.e., one not depending on the formalism chosen . . . . In all other cases treated previously, such as demonstrability or definability, one has been able only to define them relative to a given language, and for each individual language it is clear that the one thus obtained is not the one looked for. For the concept of computability, however, although it is merely a special kind of demonstrability or definability, the situation is different. By a kind of miracle it is not necessary to distinguish orders, and the diagonal procedure does not lead outside the defined notion.[62]

I would argue that it is hardly a "miracle" that Turing's analysis dodges the issue of relativity-to-language in the way Gödel suggests. Rather, it is a by-product of his starting point. As to what Gödel means by calling Turing's analysis "absolute": unsurprisingly this remark has been much discussed, since this notion is notoriously difficult to make sense of.[63] However, if we focus on the details of Turing's OCN diagonal argument with Wittgenstein mind, I think what it comes to in this context becomes clearer.

---

59 Gödel 1964: 370, discussed in Webb 1990: 292ff.
60 Gödel 1964: 369.
61 Compare Kennedy 2017 for a discussion.
62 Gödel 1946: 1.
63 See however Kennedy 2017 for a recent discussion of "formalism freeness" as a wide-ranging logical phenomenon.

First of all, note that Turing demonstrates that the partially defined, and not the totally defined function, is the basic and more general notion. He does this by framing his Universal Computing Machine U, arguing that one machine can do the work of all, suitably alphabetized in a series of finite coded sequences of particular Turing Machines (see OCN, §6). Given U, we see that if we suppose we have a total listing of all the machines that compute real decimal expansions, given those machines that are undefined on certain inputs, we cannot diagonalize out à la Cantor. In Table 4.1, the downward arrows act like holes in Swiss cheese: they prevent the diagonal method from being applied in such a way that the enumeration may be said to fail:

Table 1: Turing's Partial Functions Prevent Diagonalization à la Cantor

↓	1	1	0	↓	...
1	0	0	0	1	...
0	1	↓	0	0	...
1	1	0	↓	0	...
1	1	1	1	1	...

Turing shows that an analysis of formal logic cannot be "gap free".

4.2 Wittgenstein's Diagonal Argument

In 1947 Wittgenstein wrote down the following remark, subsequently published in RPP 1 §1096ff:

> Turing's "Machines". These machines are *humans* who calculate. And one might express what he says also in the form of games. And the interesting games would be such as brought one *via* certain rules to nonsensical instructions [*unsinnigen Anweisungen*]. I am thinking of games like the "racing game". One has received the order "Go on in the same way" when this makes no sense, say because one has got into a circle. For that order makes sense only in certain positions. (Watson.)[64]

Wittgenstein is remembering or alluding to his 1937 discussions with Watson and Turing here. And his remark makes it clear that he is fully aware of the distinctive argument that lies at the heart of Turing's negative resolution of the *Entscheidungsproblem* in §9 of his OCN.

---

64 RPP I §1098 (MS 135: 117, 1947).

This is clear, because what Wittgenstein does next is to write down an "everyday", "language-game", "forms of life"-embedded version of Turing's proof in OCN. This reformulation casts Turing's argument and its result in a more general manner, one suited to Wittgenstein's mature conception of rule-following and simplicity. On Wittgenstein's view of Turing's argument the idea of a shareable command is shown to be fundamental, and with it the need for techniques and the embedding of words in forms of life. The idea of a rule that is partial, i.e., not everywhere defined, is the basic notion, and not the idea of a rule everywhere defined.

Wittgenstein considers first a list or series of rules – or, as he also say, "laws" – for the expansion of forms of decimal representations of "computable" real numbers

$$\ldots . a_{k1} a_{k2} a_{k3} \ldots$$

He calls this list $\phi(k, \ldots)$. According to his notation, $\phi(k, n)$ is the nth decimal place determined by the kth rule in the list.

He then argues as follows:

A variant of [C]antor's diagonal proof:
Let $v = \phi(k, n)$ be the form of the laws for the expansion of decimal fractions. $\underline{v}$ is the nth decimal place of the $\underline{k}$th expansion. The law of the diagonal then is:

$$v = \phi(n, n) =^{def.} \phi'(n).$$

It is to be proven that $\phi'(n)$ cannot be one of the rules $\phi(k, n)$. Assume it is the 100th. Then we have the formation rule

of $\phi'(1)$: $\phi(1, 1)$
of $\phi'(2)$: $\phi(2, 2)$
etc.,

But the rule for the formation of the 100th place of $\phi'(n)$ becomes $\phi(100, 100)$, that is, it tells us only that the 100th place is supposed to be equal to itself, and so for $n = 100$ is *not* a rule.

[I have always had the feeling that the Cantor proof did two things, while appearing to do only one].

The rule of the game runs "Do the same as..." – and in the special case it becomes "Do the same as you are doing".[65]

---

65 MS 135: 118; the square brackets indicate a passage later deleted when the remark made its way into TS 229/§1764, published as RPP I §1097. As I explain in my Floyd 2012b, in *Zettel* §694 only

In order to understand this proof, we need to read the law $\phi'(n)$ as an instruction or command, in the way that Turing reads his quintuples specifying his "machines" in his 'On Computable Numbers'. For $n = 1$ it says: calculate the first decimal place provided by the law $\phi(1, ...)$; for $n = 2$: calculate the second decimal place provided by the law $\phi(2, ...)$; ....

There will be no trouble at all until we try to say *which* rule on our list, in particular, this instruction is. Suppose (without loss of generality) that it is the 100th. Then at $n = 100$ we have the following command: calculate the 100th decimal place provided by the law $\phi(100, ...)$. But we just presupposed that the law $\phi(100, ...)$ is the *same* as $\phi'(n)$! Therefore, this instruction, namely "Calculate $\phi'(100)$ by calculating $\phi(100, 100)$", is identical with the instruction: "Calculate $\phi(100, 100)$ by calculating $\phi(100, 100)$", which is empty. It is not a rule that we can follow as we can the others on the list, and in that sense it is "*not* a rule", as Wittgenstein says.

This is what I called in the last section the "Do What You Do" argument. It is evidently drawn from Turing's argument in OCN, §9. It is free of any tie to a particular formalism or picture or diagramming method or way of representing decimal expansions or rules. And, since it doesn't use negation to formulate the appeal to the diagonal method, it depends upon no restrictions or extensions of the application of any particular logical law.

What Wittgenstein's version of Turing's diagonal argument proves is that there is a new rule (or command) that is not like the other rules on the list, in that it cannot be followed, because it is quasi-tautologous. In this sense his old view of logic holds up: as shown by Turing, the "limits of logic" lie in rules or instructions that *cannot* be applied. Differently put, the idea of a routine everywhere defined from all perspectives is in a sense incomplete.[66]

The mechanism of the argument clearly depends upon our ability to *see* that a rule cannot be followed, rather than our getting one another to agree or disagree about the status or scope of the law of the excluded middle, or a general point of view on negation or contradictions. In this sense Wittgenstein's diagonal argument draws out something fundamental also to Turing's diagonal argument: that it is fundamental to our very idea of logic – more fundamental, in fact than the idea of any particular logical law holding or not holding – that we have a hold on

---

this second remark concerning the proof is published, thereby separating it from the mention of Turing and Watson – one reason that the close connection with Turing's (OCN) was not noticed by scholars before me.

66 Kreisel later reported (Kreisel 1950: 281 n.) that Wittgenstein's remark about Turing offers a "neat" way of looking at incompleteness, the limitative result being reachable by a command of the form "write what you write".

everyday ways of applying rules, rule-following, and shareable commands. Logic does not need to depend upon community-wide agreement on philosophical theses or conventions about what is to count as a correct logical "law". It is not a question of consensus, but of forms of life.

For this reason Wittgenstein's argument does not work if one considers the decimal expansions *extensionally*, that is, if one severs the results of the expansion rules from the rules themselves. Then all the expansions are pictured as simply spread out before us, and nothing seems to prevent the unaltered diagonal $\phi'(n)$, $n = 1, 2, \ldots$, of the given series from occurring somewhere in the series itself. Yet as soon as one thinks of the rules as genuine commands, i.e., instructions or procedures given that are to be followed in everyday life, the situation changes radically, as Wittgenstein's argument shows. And this draws out in a beautiful way the richness of Wittgenstein's remarks about rules and rule-following.

It is clear that Wittgenstein was not in any way aiming to *refute* the extensional, completed infinite here. There is nothing wrong with it, intrinsically. But it is not adequate on its own to reveal the foundations of logic. And we get into conceptual trouble when we try to think that it is. Instead, Wittgenstein is emphasizing that there are two different points of view that may be taken up on Cantor's diagonal argument. From the extensional point of view, Cantor is showing us something about the limited nature of a list of sequences to catch (and number) the real numbers. From the non-extensional point of view Cantor has given us a "positive recipe" for constructing more and more sequences. Both points of view are valid in their way. But the nature of the limits of each differ.

This may be seen if we imagine a first-person version of the argument. Consider

I Do What I Do

Bernhard Ritter has suggested a remarkable connection between the Do What I Do argument and the private language argument in Wittgenstein at PI §258. Ritter points to Wittgenstein's "Motor Roller",[67] a story of a steamroller Wittgenstein's father once conceived without seeing at first that turned out to be unable to work. The inner and outer sides of the roller of this "machine" have no friction, the machine, as Wittgenstein says, "admits everything" or is "always right". This is an analogy for the idea that however one behaves, what is going on "privately" "inside" one is somehow metaphysically independent of this.

Ritter's suggestion is that as in the case of the diagonal arguments we have considered, Wittgenstein's point is not to emphasize the need for stage-setting

---

[67] Cf. Ritter forth., ch. 18; MS 131: 219 – 222 from 8–9 September 1946.

and context in the use of language (as he does in other remarks on "privacy"), but rather to argue that the "private" diarist cannot use his sensation *itself* to say or explain *which* sensation in particular he is having. As in the positive diagonal argument we have considered, the conclusion must be seen directly, not indirectly, in the very attempt to apply itself to itself.

That a connection is to be drawn with the "vanishing" of the "I" is clear from MS 157a: 17r. This diagonal argument, written in Wittgenstein's hand (in 1934–1937), embeds the usual form of Cantor's diagonal argument, where the numbers along the diagonal are altered, directly in considerations about the vanishing of the "I". "I do", Wittgenstein remarks, has "no volume of experience" but rather "seems like a pointless point, the tip of a needle", something "detached" from phenomena of agency when regarded arbitrarily.

In the context of Turing's OCN, we have seen that there is no diagonalizing out of the class of computable numbers. In this sense the class is robust: Turing's parameter of taking a "step" in a calculation impervious to the vagaries of any particular system of representing them, just as Gödel noted. And yet this "absoluteness" is relative to something else, on the view Wittgenstein thinks Turing's analysis is driven to, in the end: our ability to take in, follow, and recognize one another *as* taking steps in calculation. It is not part of our concept of what it is to follow a rule that we do or do not always follow the law of excluded middle. It *is* part of our concept of following a rule that we can communicate and reach consensus on what *in particular* to *do* with it in a given situation.[68]

Bibliography

Church, Alonzo (1936): "A Note on the Entscheidungsproblem." In: *Journal of Symbolic Logic*, Vol. 1, No. 1, 40 – 41.
Diamond, Cora (1991): *The Realistic Spirit: Wittgenstein, Philosophy, and the Mind*. Cambridge (MA): MIT Press.
Dreben, Burton; Floyd, Juliet (1991): "Tautology: How Not to Use A Word." In: *Synthèse*, Vol. 87, No. 1, 23 – 50.
Engelmann, Mauro Luiz (2013): *Wittgenstein's Philosophical Development: Phenomenology, Grammar, Method, and the Anthropological View*. Basingstoke (UK): Palgrave Macmillan.
Floyd, Juliet (2001): "Prose versus Proof: Wittgenstein on Gödel, Tarski and Truth." In: *Philosophia Mathematica*, Vol. 3, No. 9, 280 – 307.

---

[68] I am tremendously grateful to the organizers of the 41st ALWS meeting in Kirchberg am Wechsel for the opportunity to present and discuss this work. Special thanks are due to Gabriele Mras and Bernhard Ritter for their patience and ideas, as well as Volker Munz, Mathieu Marion and all the students who attended the ALWS Summer School Marion and I taught before the conference.

Floyd, Juliet (2012a): "Wittgenstein, Carnap, and Turing: Contrasting Notions of Analysis." In: Wagner, Pierre (ed.): *Carnap's Ideal of Explication and Naturalism*. Basingstoke (UK): Palgrave Macmillan, 34 – 46.

Floyd, Juliet (2012b): "Wittgenstein's Diagonal Argument: A Variation on Cantor and Turing." In: Dybjer, P. et al. (eds.): *Epistemology versus Ontology, Logic, Epistemology: Essays in Honor of Per Martin-Löf*. Dordrecht: Springer, 25 – 44.

Floyd, Juliet (2013): "Turing, Wittgenstein and Types: Philosophical Aspects of Turing's 'The Reform of Mathematical Notation' (1944 – 5)." In: Turing 2013, 250 – 253.

Floyd, Juliet (2016): "Chains of Life: Turing, *Lebensform*, and the Emergence of Wittgenstein's Later Style." In *Nordic Wittgenstein Review*, Vol. 5, No. 2, 7 – 89.

Floyd, Juliet (2017a): "Positive Pragmatic Pluralism." In: *Harvard Review of Philosophy*, Vol. 24, 107 – 115.

Floyd, Juliet (2017b): "The Fluidity of Simplicity: Philosophy, Mathematics, Art." In: Kossak, Roman; Ordling, Philip (eds.): *Simplicity: Ideals of Practice in Mathematics and the Arts*. New York: Springer, 155 – 178.

Floyd, Juliet (2017c): "Turing on 'Common Sense' Cambridge Resonances." In: Floyd/Bokulich 2017: 103 – 152.

Floyd, Juliet; Bokulich, Alisa (eds.): *Philosophical Explorations of the Legacy of Alan Turing – Turing 100*. New York: Springer.

Floyd, Juliet (2018a): "Aspects of Aspects." In: Sluga, Hans; Stern, David (eds.): *The Cambridge Companion to Wittgenstein*. New York: Cambridge University Press, 2018^2 (2009), 361 – 388.

Floyd, Juliet (2018b): "*Lebensformen*: Living Logic." In: Martin, Christian (ed.): *Language, Form(s) of Life, and Logic: Investigations after Wittgenstein*. Berlin: De Gruyter, 59 – 92.

Floyd, Juliet (forthcoming, expected 2019). "Wittgenstein on Ethics: Working through *Lebensformen*." In: *Philosophy and Social Criticism*.

Floyd, Juliet; Mühlhölzer, Felix (forthcoming, expected 2019): "Wittgenstein on the Real Numbers: Annotations to Hardy's A Course of Pure Mathematics." In *Nordic Wittgenstein Studies*.

Gandy, R. O. (1988): "The Confluence of Ideas in 1936." In: Herken, Rolf (ed.): *The Universal Turing Machine: A Half-Century Survey*. New York: Oxford University Press, 55 – 112.

Gibson, Arthur (2010): "Francis Skinner's Original Wittgenstein *Brown Book* Manuscript." In: Munz, Volker; Puhl, Klaus; Wang, J. *Language and World*, Part One: *Essays on the Philosophy of Wittgenstein*. Vienna: Ontos Verlag, 351 – 366.

Gödel, Kurt (1931): "Über formal unentscheidbare Sätze der Principia Mathematica und verwandter Systeme I." In *Monatshefte für Mathematik und Physik*, Vol. 38, 173 – 198. Reprinted and translated in Gödel 1986a: 144 – 195.

Gödel, Kurt (1946): "Remarks Before the Princeton Bicentennial Conference on Problems in Mathematics." In: Gödel 1986b: 150 – 153.

Gödel, Kurt (1964) "Postscriptum to the 1934 Princeton Lectures." In: Gödel 1986a: 369f.

Gödel, Kurt (1972): "Some Remarks on the Undecidability Results." In: Gödel 1986b: 305f.

Gödel, Kurt (1986a): *Collected Works*, Vol. I: *Publications 1929–1936*. Feferman, Solomon et al. (eds.). Oxford: Clarendon Press.

Gödel, Kurt (1986b): *Collected Works*, Vol. II: *Publications 1938–1974*. Feferman, Solomon et al. (eds.). Oxford: Clarendon Press.

Henkin, Leon (1952): "A Problem Concerning Provability." In: *The Journal of Symbolic Logic*, Vol. 17, 160.

Hodges, Andrew (1983): *Alan Turing: The Enigma of Intelligence*. New York: Touchstone.

Hodges, Andrew (1999): *Turing. The Great Philosophers*. New York: Routledge.
Hodges, Andrew; Hatton, Cassandra (2015): "Turing Point." In *Bonham's Magazine*. Vo. 42, Spring 2015, 18 – 21.
Kanamori, Akihiro (2018): "Aspect-Perception and the History of Mathematics." In: Beaney, Michael; Harrington, Brendan; Shaw, Dominic (eds.): *Aspect Perception After Wittgenstein: Seeing-As and Novelty*. New York: Routledge, 109 – 132.
Kennedy, Juliette (2017): "Turing, Gödel and the 'Bright Abyss'." In: Floyd/Bokulich 2017: 63 – 92.
Kleene, Stephen (1981a): "Origins of Recursive Function Theory." In: *Annals of the History of Computing*, Vol. 3, No. 1, 52 – 67.
Kleene, Stephen (1981b): "The Theory of Recursive Functions, Approaching Its Centennial." In: *American Mathematical Society Bulletin; New Series*, Vol. 5, No. 1, 43 – 61.
Kreisel, Georg (1950): "Note on Arithmetic Models for Consistent Formulae of the Predicate Calculus." In: *Fundamenta Mathematicae*, Vol. 37.
Kripke, Saul (1982): *Wittgenstein on Rules and Private Language: An Elementary Exposition*. Cambridge (MA): Harvard University Press.
Leblanc, Hugues (1972): "Wittgenstein and the Truth-Functionality Thesis." In: *American Philosophical Quarterly*, Vol. 9, 271 – 274.
Löb, M. H. (1955): "Solution of a Problem of Leon Henkin." In: *Journal of Symbolic Logic*, Vol. 20, No. 2, 115 – 118.
Marion, Mathieu; Okada, Mitsuhiro (2018): "Wittgenstein, Goodstein and the Origin of the Uniqueness Rule." In: Stern, David G. (ed.):*Wittgenstein in the 1930s: Between the* Tractatus *and the* Investigations. Cambridge (UK): Cambridge University Press, 253 – 271.
Misak, Cheryl (2016): *Cambridge Pragmatism: From Peirce and James to Ramsey and Wittgenstein*. New York: Oxford University Press.
Monk, Ray (1990): *Ludwig Wittgenstein: The Duty of Genius*. New York/London: Free Press/Jonathan Cape.
Mundici, Daniel; Sieg, Wilfried (2017): "Turing, the Mathematician." In: Floyd/Bokulich 2017: 39 – 62.
Mühlhölzer, Felix (2010): *Braucht die Mathematik eine Grundlegung? Ein Kommentar des Teils III von Wittgensteins BEMERKUNGEN ÜBER DIE GRUNDLAGEN DER MATHEMATIK*. Frankfurt a. M.: Vittorio Klostermann.
Post, Emil L. "Finite Combinatory Processes, Formulation 1." In: *Journal of Symbolic Logic*, Vol. 1, 103 – 105.
Proudfoot, Diane (2017): "Turing and Free Will: A New Take on an Old Debate." In: Floyd/Bokulich 2017: 305 – 322.
Putnam, Hilary (1960): "Minds and Machines." In: Hook, Sidney (ed.): *Dimensions of Mind*. New York: New York University Press, 138 – 164. Reprinted in Putnam 1975: 206 – 214.
Putnam, Hilary (1967): "The Nature of Mental States." In: *Philosophical Papers*; Vol. 1: *Mind, Language and Reality*. Cambridge (UK): Cambridge University Press, 429 – 440. Originally titled "Psychological Predicates." In: Capitan, William H.; Merrill, Daniel D. (eds.): *Art, Mind, and Religion*. Pittsburgh: University of Pittsburgh Press, 1967, 37 – 48. Reprinted in: Putnam 1988a: 150 – 161
Putnam, Hilary (1975): *Philosophical Papers*; Vol. 2: *Mind, Matter and Method*. Cambridge (UK): Cambridge University Press.
Putnam, Hilary (1988a): *The Many Faces of Realism*. Cambridge (MA): MIT Press.
Putnam, Hilary (1988b): *Representation and Reality*. Cambridge (MA): MIT Press.

Putnam, Hilary (2009): "On Computational Psychology." In *Journal of Philosophy*, Vol. 4, 55.
Ricketts, Thomas (2014): "Analysis, Independence, Simplicity and the General Sentence-Form." In: *Philosophical Topics*, Vol. 42, No. 2, 263 – 288.
Ritter, Bernhard (forthcoming). *Kant and Post-Tractarian Wittgenstein: Transcendentalism, Idealism, Illusion*. Palgrave Macmillan, London.
Russell, Bertrand (1920): *Introduction to Mathematical Philosophy*. London/New York: Allen and Unwin/Macmillan, 1920^2 (1919).
Shagrir, Oran (2005): "The Rise and Fall of Computational Functionalism." In: Ben-Menahem, Yemima (ed.): *Hilary Putnam*. Cambridge (UK): Cambridge University Press, 220 – 250.
Sieg, Wilfried (2006). "Gödel on Computability." In: *Philosophia Mathematica*, Vol. 14, No. 2, 189 – 207.
Sieg, Wilfried (2009): "On Computability." In: Irvine, Andrew (ed.): *Handbook of the Philosophy of Science: Philosophy of Mathematics*. Amsterdam: Elsevier BV, 535 – 630.
Stadler, Friedrich (2018): "The Principle of Economy, Simplicity and Parsimony From Mach to the Vienna Circle and Beyond." Paper read at the *First International Workshop INTEREPISTEME*, 13 December 2018.
Turing, Alan (1936): "On Computable Numbers, with an Application to the Entscheidungsproblem." *Proceedings of the London Mathematical Society*, Vol. 2, No. 42, 230 – 265. Corrections: "On Computable Numbers, with an Application to the Decision Problem: a Correction." In: *Proceedings of the London Mathematical Society*, Vol. 2, No. 43, 1937, 544 – 546.
Turing, Alan (1969 [1948]). "Intelligent Machinery." In: Meltzer, B.; Miche, D. (eds.): *Collected Works of A. M. Turing: Mechanical Intelligence*, 106 – 113. Edinburgh: Edinburgh University Press. Originally an unpublished report for the National Physical Laboratory; reprinted in Turing 2013: 501 – 516, and in Turing 2004: 410 – 423.
Turing, Alan (1950): "Computing Machinery and Intelligence." In: *Mind*, Vol. 59, 433 – 460.
Turing, Alan (2001a): *Collected Works of A. M. Turing: Mathematical Logic*. Gandy, R. O.; Yates, C. E. M. (eds.). Amsterdam, etc.: Elsevier.
Turing, Alan (2001b): "The Reform of Mathematical Notation and Phraseology [1944–45]." In: Turing 2001a: 211 – 222; reprinted with commentary in Turing 2013: 245 – 249; original online at http://www.turingarchive.org/search/, King's College Online Archive.
Turing, Alan (2013): *Alan Turing – His Work and Impact*. Amsterdam/Burlington (MA): Elsevier. Cooper, Barry S.; van Leeuwen, Jan (eds.); with commentary on Turing's papers by many experts.
Turing, Alan; Copeland, B. Jack (eds.) (2004): *The Essential Turing: Seminal Writings in Computing, Logic, Philosophy, Artificial Intelligence, and Artificial Life*. Oxford: Oxford Clarendon Press.
Watson, A. G. D. (1938): "Mathematics and Its Foundations." In: *Mind*, Vol. 47, No. 188, 440 – 451.
Webb, Judson C. (1990): "Remark 3; Introductory Note to *Gödel 1972a* [= "Some Remarks on the Undecidability Results"]. In Gödel 1986b: 292 – 304.
Weiss, Max (2017). "Logic in the *Tractatus*." In: *Review of Symbolic Logic*, Vol. 10, No. 1, 1 – 50.
Wittgenstein, Ludwig (TLP): *Tractatus Logico-Philosophicus*. London/New York: Routledge and Kegan Paul. Ogden, C.K. (transl.). First German edition in *Annalen der Naturphilosophie*, Ostwald, Wilhelm (ed.), Vol.14, Leipzig, 1921, 12 – 262. Reprint, second impression with a few corrections, 1931. Republished 1981. Several online free versions of the text are available with hypertext and translation comparison capability. See Kevin Klement's "Side-by-Side" edition, available at http://people.umass.edu/klement/tlp/, the University of Iowa

Tractatus Map http://tractatus.lib.uiowa.edu/, and Luciano Bazzocchi's multiple language site http://www.bazzocchi.com/wittgenstein/.

Wittgenstein, Ludwig (PI): *Philosophische Untersuchungen / Philosophical Investigations*. Anscombe, G. E. M.; Hacker, P. M. S.; Schulte, J. (eds. and transl.); Wiley-Blackwell, Chichester (UK), 2009^4 (1953).

Wittgenstein, Ludwig (KgE): *Philosophische Untersuchungen: Kritisch-genetische Edition*. Suhrkamp Verlag, Frankfurt am Main. Schulte, J.; Nyman, H.; von Savigny, E.; von Wright, G. H. (eds.).

Wittgenstein, Ludwig (BB): *Preliminary Studies for the 'Philosophical Investigations': Generally Known as The Blue and Brown Books*. New York: Harper and Row.

Wittgenstein, Ludwig (RFM): *Remarks on the Foundations of Mathematics*. Anscombe, G. E. M. (ed. and transl.); von Wright, Georg H.; Rhees, Rush (eds.); Blackwell, Oxford, 1978^2 (1956).

Wittgenstein, Ludwig (RPP 1): *Remarks on the Philosophy of Psychology*; Vol 1. Oxford: Blackwell. von Wright, G. H.; Nyman, Heikki (eds.); Anscombe, G.E.M. (transl.).

Wittgenstein, Ludwig (Z): *Zettel*. Oxford: Blackwell, 1981.

Wittgenstein, Ludwig (EPB): *Eine Philosophische Betrachtung (Das Braune Buch)*. *Werkausgabe*; Vol. 5, Rhees, Rush (ed.), Frankfurt a. M.: Suhrkamp, 1984, 117 – 282. Wittgenstein's enlargement of *The Brown Book* (MS 115, TS 310). German translation by von Petra von Morstein. EPB is based on Wittgenstein's manuscript revision of TS 310 contained in MS 115. The revision is not complete and the published EPB concludes with a German translation of the remainder of TS 310.

Wittgenstein, Ludwig (MS / TS): *Wittgenstein's Nachlass: The Bergen Electronic Edition*. Charlottesville (VA)/Oxford: Intelex Corporation/Oxford University Press, 2000. Edited by the Wittgenstein Archive, University of Bergen, under the direction of Claus Huitfeldt. Incorporates Wittgenstein-Source Bergen Facsimile Edition, ed. Alois Pichler, with the assistance of H. W. Kruger, D. C. P. South, T. M. Bruvik, A. Lindebjerg, V. Olstad, published online 2009 (www.wittgensteinsource.org) in a web edition of about 5000 pages of facsimile, diplomatic and normalized transcriptions. CD-ROMs were originally distributed by Intelex Corporation in their Past Masters Series.

Wittgenstein, Ludwig (AWL). *Wittgenstein's Lectures, Cambridge, 1932 – 1935: From the Notes of Alice Ambrose and Margaret Macdonald*. Chicago: University of Chicago Press, 1979.

Wittgenstein, Ludwig (LFM): *Wittgenstein's Lectures on the Foundations of Mathematics: Cambridge, 1939, from the Notes of R.G. Bosanquet, Norman Malcolm, Rush Rhees and Yorick Smythies*. Hassocks, Sussex/Chicago: The Harvester Press/University of Chicago Press, 1976.

Wittgenstein, Ludwig (WC): *Wittgenstein in Cambridge: Letters and Documents, 1911–1951*. McGuinness, Brian (ed.). Malden (MA), Oxford: Blackwell, 2008^4 (1995).

Wittgenstein, Ludwig (MWL): *Wittgenstein: Lectures, Cambridge 1930–1933, From the Notes of G. E. Moore*. Stern, David G.; Rogers, Brian; Citron, Gabriel (eds.) Cambridge (UK): Cambridge University Press, 2016. Originals of the notes are located at the University Library, Cambridge University. Moore published a three part excerpted series from these notes, "Wittgenstein's Lectures in 1930–33" in *Mind*, New Series, Vol. 63, 1954, Part I: 1 – 15; Part II: 289 – 316; Vol. 64, 1955, Part III: 1 – 27, with two corrections on page 254; these were reprinted in his *Philosophical Papers*, New York, Humanities Press, 1959, 252 – 324.

Wittgenstein, Ludwig (WCL): *Wittgenstein's Whewell's Court Lectures, Cambridge, 1938–1941: From the Notes by Yorick Smythies*. Munz, Volker A.; Ritter, Bernhard (eds.); Wiley-Blackwell, 2017.

Charles Parsons
Remarks on Two Papers of Paul Bernays

Abstract: The paper comments on two papers in French that Paul Bernays derived from lectures at a conference on mathematical logic held in Geneva in June 1934. The first, the well-known "On Platonism in mathematics," sets forth a methodological version of Platonism and observes that it can be implemented in some branches of mathematics and not others. He notes that by his definition Brouwer's intuitionism rejects all Platonism, while Hermann Weyl's reconstruction of analysis retains it for generalizations about natural numbers but not for generalizations about real numbers or higher-type objects. He rejects the then widespread idea of a crisis of foundations and argues that the questions raised are philosophical. Bernays' second paper, "Some observations on metamathematics," is a technical sequel to "On Platonism." It describes some basic points in the Hilbert school's proof theory and sketches some results, such as a simple application of Herbrand's theorem, Gödel's proof that if intuitionistic first-order arithmetic is consistent, then so is classical, and Gödel's second incompleteness theorem. A full proof of the latter and a correct proof of Herbrand's theorem only appeared in 1939, in volume II of Hilbert and Bernays, *Grundlagen der Mathematik*.

In this essay I will single out for comment a pair of papers by Paul Bernays. As the reader is likely to know, Bernays was engaged by David Hilbert in 1917 to serve as his assistant (better, collaborator) in the proof-theoretic program that was Hilbert's main professional activity in the later years of his career. A reason why Hilbert chose Bernays was that he had some philosophical training, chiefly through his discipleship in his student years to Leonard Nelson (1882–1927), a broadly neo-Kantian philosopher, although not a member of either of the leading neo-Kantian schools. Bernays worked and taught in Göttingen from late 1917 until he was dismissed in 1933 because he was Jewish. Hilbert was by then retired, and

Charles Parsons, Harvard University, Cambridge (MA), parsons2@fas.harvard.edu

the result was that the Hilbert school effectively dissolved.¹ In 1934 Bernays moved to Zurich and lived there for the rest of his life.²

Nearly all of the writing of the two-volume treatise *Grundlagen der Mathematik* (Hilbert/Bernays 1934/39³) was done by Bernays. Volume 1 was completed (even up to proofreading) before Bernays left Germany. After he moved to Zurich, Bernays had no scientific contact with Hilbert, who seems anyway to have become inactive scientifically after about 1933. However, during the work on volume 2, Bernays did have scientific correspondence with Ackermann and Gentzen.⁴

Early philosophical papers of Bernays were written while he was a student and reflect the views of Nelson. Those of the 1920s articulate the position of the Hilbert school. He probably played a central role in formulating the school's philosophy. The long paper Bernays 1930⁵ may be the most sophisticated articulation of that philosophy. Bernays wrote later that in it he repudiated the formalistic interpretation of the Hilbert school's position, which had been attributed to it for many years.⁶

2. "Sur le Platonisme dans les mathématiques." This paper is one of Bernays's best known philosophical papers.⁷ It resulted from a lecture given at the University of Geneva on 18 June 1934, probably only a couple of months after Bernays's move to Switzerland. The lecture was his first public performance in French. The opening

---

1 Hilbert retired in 1930 and was succeeded by Hermann Weyl. The two other full professors in mathematics, Edmund Landau and Richard Courant, were Jewish and were dismissed by the Nazis in 1933. The attack on Göttingen mathematics was swift even by Nazi standards. Weyl stayed on as director of the institute until the fall of 1933, but then he accepted an appointment at the Institute for Advanced Study. In addition to the problems of functioning under the Nazi regime, Weyl may have doubted how secure his own position was, since he had a Jewish wife.
2 See his "Kurze Biographie," in Müller 1976: xiv – xvi. This volume contains an English version that is somewhat incomplete. A fuller translation, of which I can supply a copy on request, is to appear in the collection of Bernays's philosophical essays edited by Wilfried Sieg and others. However, at present this collection does not have a publisher.
3 These volumes will be cited as HB I and II.
4 I have seen it claimed that during the Nazi period Germans were forbidden to correspond with Jewish emigrés. This example conflicts with that claim. However, it is likely to be true for the period after the war started. I recall seeing in the Bernays papers a letter of Ackermann to Bernays, written shortly after the war, in which he said that he had been unable to write during the war.
5 I do not venture to comment on this paper because of the excellent and thorough (though still unpublished) introduction to it by Wilfried Sieg and W. W. Tait, prepared for the collection cited in note 2 above.
6 Letter to Kurt Gödel, 7 September 1942, in Gödel 2003: 139. On Bernays' philosophical writing after he left Göttingen, see Parsons 2008.
7 An English translation appeared in Benacerraf/Putnam 1964 and Benacerraf/Putnam 1983. A German translation (not by Bernays) appeared in Stegmüller 1978 and also in Bernays 1976.

footnote states that it was in a series *Conférences internationales des sciences mathématiques,* in a subseries on mathematical logic. The general series was a two-year affair with lectures on a number of subjects in mathematics and physics between the fall of 1933 and the fall of 1935. The lectures on mathematical logic were in effect a six-day conference, 18–23 June 1934.[8] Bernays gave in addition to "Sur le Platonisme" four mathematical lectures in this subseries, which were distilled for publication into the paper "Quelques points essentiels de la métamathématique" (Bernays 1935a), to be discussed in §3 below. "On Platonism" is probably Bernays' best known philosophical essay, and for its length it is quite wide-ranging and rich in ideas. The theme of "Platonism" structures most of it but not all.

Near the end of his life Bernays made two statements about what led to changes in his philosophical views:

> I had come close to the views of [Ferdinand] Gonseth on the basis of the engagement of my thinking (*meinen gendanklichen Auseinandersetzungen*) with the philosophy of Kant, Fries, and Nelson, and so I attached myself to his philosophical school (Müller 1976: xvi, my translation).

> ...during the period in which these articles were published, my views on the relevant questions have changed almost exclusively in response to new insights gained from research in the foundations of mathematics. (Bernays 1976: vii)

These statements evidently point in different directions. But the second of the two kinds of influence mentioned here is much more in evidence in the present paper.[9] Gödel's incompleteness theorems are the development in foundational research in the immediately preceding period that had the greatest impact on Bernays. Another significant result that he mentions is Gödel's consistency proof of classical first-order arithmetic relative to intuitionistic. But it is only near the end of the paper, in a discussion of Hilbert's program, that these results are mentioned. In view of the role played in the paper by differences of strength of "Platonistic" conceptions, it is likely that Gödel's theorems played a more fundamental role beneath the surface of Bernays's reflections.[10]

Although the paper is structured by a philosophical conception of "Platonism", Bernays stays close throughout either to questions of mathematical method or to

---

8 This information and information about lectures by others comes from *L'enseignement mathématique* 34 (1935), 116 – 120.

9 On these two different pressures, see Parsons 2008, in particular 131 – 132. Much more could be said on the matter.

10 Cf. Gödel 1932, which observes that the ascent to a stronger system allows the proof of previous undecidable sentences, but the stronger system will have its own undecidable sentences.

mathematical logic. At the outset he describes the paper as concerned with "the present situation in research in the foundations of mathematics" (Bernays 1935: 52). He immediately deflates the then common talk of a crisis of foundations and says that "the mathematical sciences are growing in complete security and harmony" and that "it is only from the philosophical point of view that objections have been raised."

At this point we need to inquire what Bernays means by "Platonism." The term is introduced after he mentions two tendencies of modern mathematics: First, viewing the objects of a theory so that one can apply the law of excluded middle to quantified statements about them:

> For each property expressible using the notions of a theory, it is an objectively determinate fact whether there is or there is not an element of the totality which possesses this property. (Bernays 1935: 52 – 53).

Second, viewing the objects as "existing from the outset," as in Hilbert's geometry, rather than to be constructed, as in Euclid. [11]

Then he describes the tendency as "viewing the objects as cut off from all links with the reflecting subject" (Bernays 1935: 53). He connects this tendency with the philosophy of Plato and so calls it Platonism. The connection with Plato may be tenuous; it seems to be more a modern opposition of realism to idealism that underlies his classification. He might have done better to use the term "realism."[12]

Two important points become clear as the paper goes on: First, Platonistic assumptions can be made in specific branches of mathematics and can differ in strength. Second, as the Euclid example already indicates, Platonism is primarily contrasted with constructivism.

The second point has to be emphasized in view of the post-war usage inaugurated by Goodman/Quine 1947, who characterize Platonism in such a way that it contrasts primarily with nominalism. In the well known Goodman/Quine 1947, the authors begin by saying, "We do not believe in abstract entities," and later they write that they use "'Platonistic' as the antithesis of 'nominalistic'," so that any theory that admits abstract entities is Platonistic (Goodman/Quine 1947: 110 n.). That has been the dominant use of the terms "Platonistic" and "Platonist" in post-war American philosophy of mathematics.

---

[11] These points echo the description of the "existential form," characteristic of the axiomatic method as Hilbert understood it, in HB I 1–2. Elsewhere Bernays uses the probably more memorable phrase "existential axiomatics," e.g. Bernays 1976: 326.

[12] In his post-war writings Bernays seems largely to avoid the term "Platonism." The emphasis on the law of excluded middle as a criterion of Platonism suggests the later development by Michael Dummett of the applicability of bivalent logic as a criterion of realism.

One might see the origin of this difference in the fact that Quine, Goodman, and others who have followed them take classical logic, at least first-order logic, for granted. From the beginning Bernays regards applying classical quantificational logic in mathematics as an assumption that can be contested. Although this view is associated with Brouwer, it is also to be found in Hilbert's papers of the 1920s.[13] But Bernays also does not mention nominalism in this paper (or, so far as I know, elsewhere).

Another relevant point is that in spite of the philosophical gloss in terms of realism, Platonism for Bernays is primarily a methodological stance, which can be taken in some contexts and not in others. To explain this more fully, we need to look at some of the cases Bernays discusses.

Bernays notes that the weakest Platonistic assumption is that of the totality of integers,[14] in particular admitting classical logic when quantifying over numbers. It also implies some instances of excluded middle in the arithmetic of real numbers. But he finds that analysis "is not content with this modest variety of Platonism" (Bernays 1935: 54). At this point Bernays introduces his well known "quasi-combinatorial" picture of arbitrary functions on integers and sets of integers. By analogy of the infinite to the finite,

> ...we imagine functions engendered by an infinity of independent determinations which assign to each integer an integer, and we reason about the totality of these functions (Bernays 1935: 54).

Although this is a picture rather than a worked-out conception, it has the virtue of posing an alternative to the conception of sets of integers as extensions of predicates true or false of integers, about which it is difficult to avoid the conclusion that impredicative reasoning involves a vicious circle, as Weyl famously argued.

Bernays says that the conception is iterated in set theory. Since he regards the quasi-combinatorial view of sets as motivating impredicative reasoning and its iteration as giving rise to full set theory, the question of paradoxes naturally arises.

Bernays calls what he has set forth "only a restricted Platonism" and no more than "an ideal projection of a domain of thought" (Bernays 1935: 56). Whatever he means by the latter characterization, he goes on to describe briefly "absolute Platonism," which he claims is what is threatened by the paradoxes:

---

[13] For further discussion see Parsons 2015.
[14] One might argue that the law of excluded middle is applicable in arithmetic but that this does not commit us to the existence of a *totality* of integers. Bernays does not make this distinction, and I will not make use of it here.

> Several mathematicians and philosophers interpret the methods of Platonism in the sense of conceptual realism, postulating the existence of a world of ideal objects containing all the objects of mathematics. It is this absolute Platonism that has been shown untenable by the antinomies (Bernays 1935: 56).

This formulation might exclude the later views of Gödel, but a sharper formulation is given two paragraphs later, when he says that the antinomies show

> ...the impossibility of combining the following two things: the idea of the totality of all mathematical objects and the general concepts of set and function; for the totality itself would form a domain of elements for sets, and arguments and values for functions (Bernays 1935: 56).

The next part of the essay is devoted to possibilities of eliminating Platonist methods or restricting them more strongly than he has indicated so far. He describes two steps toward such elimination: (1) giving up the (quasi-combinatorial) concepts of set and function and replacing them with "constructive concepts," for example viewing an infinite sequence as given by an arithmetical law; (2) "renouncing the idea of the totality of integers" (57).

Weyl in *Das Kontinuum* (1918) takes the first step and not the second. About this Bernays makes the interesting remark that it is "adapted to the tendency toward a complete arithmetization of analysis," which he says "is not carried through to the end by the usual method" (Bernays 1935: 57). He also remarks elsewhere that the usual set-theoretic construction of analysis is not a complete arithmetization, which he believes Weyl and also Brouwer aspired to (Bernays 1976: viii).

Bernays attributes the second step to Kronecker and Brouwer; he evidently thinks they take both. He could well have regarded Brouwer's theory of choice sequences as a substitute for the quasi-combinatorial conception, but he does not venture to explain it. He is led into a discussion of Kronecker's and Brouwer's views. I will concentrate on what he says about intuitionism. Although he is very clear that intuitionist methods are stronger than finitist, some of his explanations assimilate them. He gives a basic (constructive) explanation of quantification over numbers that echoes Hilbert's explanation in the finitist context.[15] He says that "the negation of a general or existential proposition about integers does not have a precise sense" (Bernays 1935: 58). However, he then describes intuitonist negation as a "strengthened negation" and says that for that negation "the law of the excluded middle is no longer applicable" (Bernays 1935: 58). This is the source of the "characteristic complications" of Brouwer's intuitionistic methods, but they can be avoided in the theory of integers and algebraic numbers.

---

[15] Hilbert 1926: 171 – 173. Hilbert may have been influenced by Weyl 1921, as suggested by van Dalen 1995: 138.

Bernays claims that "roughly" intuitionism is adapted to number theory, the "semiPlatonistic" method admitting only the totality of integers to the arithmetic theory of functions, and the "usual Platonism" to the geometric theory of the continuum (Bernays 1935: 59). He finds "nothing astonishing" in this, since it is usual to restrict one's assumptions to those essential in the relevant domain of science.

At this point Bernays turns to Brouwer's critical case against classical methods. He represents Brouwer as appealing to evidence and as claiming that "the basic ideas of intuitionism are given to us in an evident manner by pure intuition" (Bernays 1935: 60). However, he does not offer much explanation of what he means by "intuition" or what Brouwer meant. He agrees with Brouwer that the concept of number is of intuitive origin and that "one ought not to make arithmetic and geometry correspond in the manner that Kant did," although as elsewhere he finds it "a bit hasty to deny completely the existence of a geometrical intuition" (Bernays 1935: 60). But this is not the point that he presses. Rather, he asks first whether the boundary of intuitive evidence in arithmetic coincides with that of intuitionistic arithmetic, and whether evidence itself has an exact boundary.

On the first point he makes the frequently cited observation that "for very large numbers, the operations required by the recursive method of constructing numbers can cease to have a concrete meaning" (Bernays 1935: 61), pointing out that exponentiation gives rise to numbers far larger than any occurring in experience. He doubts that the evidence for the existence of an Arabic numeral for the number $67^{257^{729}}$ is really intuitive; he asks whether it is not "rather an application of the general method of analogy, consisting in extending to inaccessible numbers the relations which we can completely verify for accessible numbers" (Bernays 1935: 61 – 62).[16] He then suggests, without endorsing it, that one might restrict recursive definitions to those that are "practicable" (*effectuable*; we might now say feasible). This is an early suggestion of a strict finitist view.[17] But the point of making it is to

---

[16] Bernays does not explain the notion of accessibility or say whether it would play a role in the more restrictive view he sketches very briefly in the next paragraph (of the French).

[17] Sieg 1999: 24, uses the term "strict finitist number theory" to describe a point of view expressed in part III of Hilbert's lectures *Logik-Kalkül* of 1920. What Hilbert presents, and Sieg describes, is in one respect less strict and in another respect stricter than what usually goes by that name or what Bernays seems to have in mind here. It is less strict in that Hilbert neither suggests any restriction on introducing what would naïvely be called terms for primitive recursive functions nor mention the idea of feasible or practicable computability. It is stricter in that generalizations are introduced formally; they are not to be taken literally as statements about all numbers but to be made only if a formal proof of a definite kind has been produced. To reinforce this, he does not allow the inference from $A(a)$ to $A(n)$ for particular n, although a proof of $A(a)$ does offer a method of constructing a proof of $A(n)$.

argue that intuitionism "takes as its basis propositions that one can doubt and in principle do without" (Bernays 1935: 62). Although Bernays does not say so, this remark also applies to Hilbert's and his finitary method, as he seems to admit in the postscript in Bernays 1976 to Bernays 1930, where he writes:

> ... the sharp distinction between the intuitive and the non-intuitive, which was employed in the treatment of the problem of the infinite, apparently cannot be drawn so strictly (Bernays 1976: 61).

Bernays goes on to remark that, because quantifications can occur as antecedents of conditionals, and such constructions can be iterated, intuitionist reasoning depends on "abstract reflections" (Bernays 1935: 62 – 63). Thus it is by systematic application of "abstract forms of reasoning" that one can have intuitionistic logic.[18] He concludes that what is characteristic of intuitionism is not that it is "founded on pure intuition" but rather that it is "founded on the relation of the reflecting and acting subject to the whole development of science" (Bernays 1976: 63). He considers that an extreme position, and contrary to the customary procedure of mathematics, which is in his terms Platonist. But with characteristic moderation he writes, "Keeping both possibilities in mind, we shall rather aim to bring about in each branch of science, an adaptation of method to the character of the object investigated" (Bernays 1935: 63 – 64).

Against logicism, he maintains that the intuitive method is most suitable for the theory of numbers. His most distinctive direct objection to logicism, that mathematical abstraction "does not have a lesser degree than logical abstraction, but rather a different direction" (Bernays 1935: 65) alludes rather obscurely to a discussion in Bernays 1930 I, §2.

Bernays goes on to defend the classical set-theoretic treatment of the continuum against intuitionism and, more briefly, the treatment in Weyl 1918. He clearly has also in mind Weyl's enthusiasm in Weyl 1921 for Brouwer's treatment of the continuum. The issue for both is the relation between the singling out of individual points on the continuum and the general conception, say of the real line. Bernays is satisfied with the classical approach:

> The fact is that for the usual method there is a completely satisfying analogy between the manner in which a particular point stands out from the continuum and the manner in which a real number defined by an arithmetical law stands out from the set of all real numbers, whose elements are in general only implicitly involved, by virtue of the quasi-combinatorial conception of sequence. (Bernays 1935: 65)

---

18 Cf. the citation of these remarks in Gödel 1958: 280.

He relates the issue of intuitionism and Platonism to the traditional duality of arithmetic and geometry:

> The concept of number appears in arithmetic. It is of intuitive origin, but then the idea of the totality of numbers is superimposed. On the other hand, in geometry the Platonist idea of space is primordial, and it is against this background that the intuitionist procedures of constructing figures take place. (Bernays 1935: 66).

He concludes that both tendencies are necessary, but the idea of number is "more immediate to the mind than that of space" and "the assumptions of Platonism have a transcendent character which is not found in intuitionism" (Bernays 1935: 66).[19]

The latter consideration leads into his discussion of Hilbert's program. He stresses Hilbert's intention to restrict himself to "intuitive and combinatorial considerations" in consistency proofs and the limited results achieved, falling short of the "axiomatic theory of numbers," that is, classical first-order arithmetic PA.[20] In explaining this he mentions Gödel's theorem, from which he infers:

> It is impossible to prove by elementary combinatorial methods the consistency of a formalized theory which can express every elementary combinatorial proof of an arithmetical proposition (Bernays 1935: 68).

He goes on to say that attempts so far made have not offered any example of an elementary combinatorial proof that cannot be expressed in first-order arithmetic. With some hesitation he concludes that in order to prove the consistency of the latter, a more powerful method will be needed. From a footnote, it appears that one difficulty is the problem of "delimiting precisely the domain of elementary combinatorial methods."

These remarks shed little light on the question debated in recent years of what was the precise extent of finitary arithmetic as conceived by the Hilbert school.[21] If anything, Bernays' footnote suggests that he himself was not sure. However, more that is relevant can be found in the companion paper Bernays 1935a, which will be discussed in §3 below.

Returning to the present paper, Bernays finds the more powerful method he is seeking suggested by the proof of the consistency of PA relative to that of intuitonistic first-order arithmetic HA, which had been obtained by Gödel and Gentzen

---

19 In Bernays 1922: 10, Bernays already described what he later called existential axiomatics (see note 4 above) as "as it were transcendent for mathematics."
20 In what follows I use the contemporary designation PA for this theory, which Bernays here consistently calls the axiomatic theory of numbers. In Bernays 1935a he uses the symbol N, in Hilbert/Bernays 1934/39 of course Z.
21 For a view on this point, see Tait 2019.

(see Gödel 1933). He concludes that "intuitionism, by its abstract arguments, goes essentially beyond elementary combinatorial methods" (Bernays 1935: 69). He raises the question whether a strengthening of the method along these lines would make possible a proof of the consistency of analysis, by which he means full second-order arithmetic. In spite of the considerable achievements of post-war proof theory, that question has not been answered in the affirmative, and some have argued that the limits encountered in proof-theoretic work imply a negative answer.[22]

3. "Quelques points essentiels de la métamathématique." This paper, which followed "Sur le Platonisme" in the issue of *L'enseignement mathématique* containing papers from the conference on mathematical logic in Geneva referred to above, is almost certainly a distillation of the mathematical lectures that Bernays gave there. It may seem to be a technical appendix to "Sur le Platonisme." The early part of the paper is a condensed exposition of concepts and results from the Hilbert school's work in the 1920s. But the paper is more than that. Bernays goes on to sketch some developments that only appear in volume 2 of *Grundlagen der Mathematik* in 1939. The paper is divided into four sections, which pretty clearly correspond to the four lectures from which it is derived. I will assume this and divide my comments into Bernays's sections, headed by Roman numerals.

I. It is shown how in certain cases one can prove the consistency of first-order theories by means of Herbrand's theorem. His example is a toy one, where the axioms simply describe a given predicate as irreflexive and transitive (so that they have only infinite models). Thus they might be

$\forall x \neg Fxx$
$\forall x \forall y \forall z (Fxy \land Fyz \rightarrow Fxz)$

If this theory is inconsistent, then the negated conjunction of these axioms is derivable in pure predicate logic. Then by Herbrand's theorem, its Herbrand expansion is a tautology. Bernays 1935a: 73 – 74 argues that it is not.

It was not suspected at the time that Herbrand's proof of his theorem had serious flaws. Hilbert and Bernays give a correct proof, but it only appears in §3.3 of volume II of HB (see below). In the 1940s Gödel discovered the error in Herbrand's proof and worked out a correction, but he did not publish on the subject. The error

---

[22] See Martin-Löf 2008. Rathjen 2005 also proposes a bound on the constructive reasoning available for proof theory, higher than Martin-Löf's but still within second-order arithmetic. (I am indebted here to Peter Koellner.)

was rediscovered in 1962 by Peter Andrews. In Dreben et. al. 1963 it is proved that some lemmas in Herbrand's proof are false. A corrected proof on Herbrand's lines occurs in Dreben/Denton 1966. The history is narrated in Andrews 2003.[23]

II. The second lecture/section deals with the ε-symbol. Its introduction is preceded by discussion of definite descriptions. Although he says that definite descriptions can be eliminated by a procedure essentially Russell's, more emphasis is placed on a rule by which the description $\iota\, xA(x)$ can be introduced if it has been proved that there is a unique x satisfying $A(x)$. This treatment of descriptions was carried out at length in §8 of HB I. Bernays describes the proof as "un peu pénible."

$\varepsilon xA(x)$ is intended to denote some object satisfying $A(x)$, if there is one, and to *denote some unspecified "throwaway" object otherwise. It proved useful* for metamathematical purposes. In particular, first-order quantificational logic can be derived from the single simple axiom $A(t) \to A[\varepsilon xA(x)]$, using the definitions

$\exists xA(x) \leftrightarrow A[\varepsilon xA(x)]$
$\forall xA(x) \leftrightarrow A[\varepsilon x \neg A(x)]$

Bernays introduces the ε-symbol by way of an intermediate symbol η, which is like the description symbol except that the uniqueness condition is dropped. Then using the somewhat counterintuitive validity

$\exists x[\exists yA(y) \to A(x)]$,

$\varepsilon xA(x)$ can be defined as $\eta x[\exists yA(y) \to A(x)]$. It is then straightforward to prove the rules for quantifiers by the above definitions, although for the generalization rule substitution of terms for free variables is used (Bernays 1935a: 79).

In the number-theoretic context, induction is replaced by the axiom

$\varepsilon xA(x) = Sb \to \neg A(b)$.

Bernays remarks that the formulation of first-order logic by the ε-symbol simplifies metamathematical considerations but is less suitable for actual deductions (Bernays 1935a: 80)

The resulting formalism for number theory was used for the consistency proofs of Ackermann 1924 and von Neumann 1927. However, Bernays concedes that these

---

23 Andrews 2003: 179 asserts that the correction worked out by Gödel is essentially the same as that of Dreben and Denton.

proofs succeed only for the theory with restricted induction. However, he says that a proof of Herbrand's theorem can be obtained by these methods.[24]

III. The following section turns to Gödel's incompleteness theorems. Referring to the limitations of the consistency proofs cited in the previous lecture, he remarks that Gödel's theorem shows that there is an essential obstacle. He says that Gödel's reasoning "is inspired by the idea that leads to Richard's paradox" (Bernays 1935a: 81), which, however, he explains only very briefly (ibid.).[25] He states that Russell's paradox refutes "absolute Platonism" (see Bernays 1935: 56), while the semantical paradoxes rule out the Leibnizian idea of a language that is simultaneously exact and universal (Bernays 1935a: 81).

The treatment of Gödel's first theorem is largely conventional and somewhat sketchy. It assumes that primitive recursive functions are numeralwise expressible (Kleene's term) in the relevant formalism F, from which, together with Gödel's assignment of numbers to expressions, it follows that a formula that says of itself that it is unprovable is, if F is consistent, not provable in F. In contrast, in Gödel 1931 it is *proved* that the numeralwise expressibility of primitive recursive functions holds for his system P (arithmetic with the simple theory of types as underlying logic), and he goes on to show that this holds also for PA.

For the second theorem, Bernays observes that mathematical induction must be able to be carried out in F. But in the sketch of the proof, he hardly gives more detail than was given in Gödel 1931 §4. It was not until HB II that a full proof was published, in particular with a statement of assumptions about derivability that are sufficient for the proof.

Bernays ends the section with the remark:

> To all appearances, the framework in which Hilbert confined the methods inspired by the "finitary point of view" is not wide enough for a theory of proof. The question is then to find out whether this framework can be enlarged without abandoning the goal pursued by metamathematics. We shall see that that is indeed the case. (Bernays 1935a: 88)

IV. The title of the last section is "The relation of the axiomatic theory of numbers and intuitionistic arithmetic." You will recall that by "the axiomatic theory of numbers" Bernays means PA. He begins by citing Gödel's second theorem, which implies that if PA is consistent, the formula stating that it is consistent is not provable in PA. He says that we are led "by various tests" to believe that any proof of a theorem of arithmetic satisfying the exigences of the finitary point of view

---

[24] Apparently he means von Neumann's. As noted above, the proof appears in §3.3 of HB II.
[25] In Bernays' motivation in HB for the argument for Gödel's first incompleteness theorem, it is the liar paradox that plays a larger role (HB II, §5.1.a).

can be formalized in PA. Thus there is not a finitary proof of the consistency of PA. This section is the first place where Bernays explores the idea that PA and possibly stronger systems can be proved consistent by methods that go beyond finitism but are still constructive.

Bernays makes some brief remarks about the finitary method and why it would not be strong enough to prove the consistency of PA. He notes that some mathematicians have thought of it as not different from the intuitionistic method of Brouwer.[26] In this connection he remarks:

> What is in favor of that interpretation is that the restrictions made by the intuitionistic method were just those necessary for metamathematics, because that method is *fully characterized* by the exigency to avoid suppositions resting on the analogy of the infinite to the finite, in particular that of the totality of whole numbers. (Bernays 1935a: 88 – 90, emphasis mine)

This remark makes one wonder how much Bernays had studied Brouwer's writings., in particular those that set forth the foundation of intuitionistic analysis.[27] But he goes on to say that in metamathematical proofs one has always held to a narrower framework, reasonings that can be formalized without using bound variables. He gives an example of a "non-elementary" recursion that is not captured by PA.[28] A recursion not capturable within PA is needed to prove transfinite induction on an ordering of type ε_0. But he notes that such an induction is provable in intuitionistic mathematics. It could be that one would find a proof of the consistency of PA that would be intuitionistic and in which such an induction would be the only assumption going beyond PA. (Bernays 1935a: 91) However, he says that "for the moment that is only a possibility" (ibid.), although it was realized not long after in the proof of Gentzen 1936.

In Gödel 1933 it is proved that PA can be interpreted in intuitionistic arithmetic HA, so that if the latter is consistent, so is the former.[29] That seems to have convinced everyone that intuitionistic methods went beyond the finitary method as conceived by Hilbert. Bernays sketches briefly a finitistic proof, based on Gödel's result, of the consistency of PA relative to that of HA. To prove the consistency of

---

[26] He mentions von Neumann, Kalmár, and Herbrand. He does not say whether that was his own view before learning of Gödel's theorems. I will not address the question what he or other members of the Hilbert school thought the limits of the finitary method were before the publication of HB I.
[27] I am thinking particularly of Brouwer 1925, Brouwer 1925a, Brouwer 1926 and Brouwer 1927.
[28] He says nothing about the formula B that figures in the recursion on p. 90, but it seems clear that if it is primitive recursive or Σ_1, the recursion will not go beyond PA. On the other hand, Π_1 is sufficient.
[29] Gerhard Gentzen proved the same result but withdrew his paper when he learned that he had been anticipated by Gödel. See Gödel 1986: 284.

PA by this means, one needs to prove the consistency of HA. Bernays does not address this issue, but it is likely that he assumes that the consistency of HA is established intuitionistically by a truth-definition.

In his final remark, Bernays concedes that this approach to the consistency of arithmetic does not extend to analysis.

In his informal "lecture at Zilsel's" in Vienna early in 1938, Gödel is very critical of using intuitionistic methods, taken at face value, for consistency proofs.[30] He looks for a framework that is closer to finitism but still strong enough for interesting consistency proofs.

Bibliography

Ackermann, W. (1924): "Begründung des *tertium non datur* mittels der Hilbertschen Theorie der Widerspruchsfreiheit." In: *Mathematische Annalen* 93, 1 – 36.

Andrews, Peter (2003): "Herbrand Award Acceptance Speech." In: *Journal of Automated Reasoning*, Vol. 31, 169 – 187.

Benacerraf, Paul; Putnam, Hilary (1964): *Philosophy of Mathematics: Selected Readings*. Englewood Cliffs, NJ: Prentice-Hall.

Benacerraf, Paul; Putnam, Hilary (1983): Second edition of *Benacerraf and Putnam 1964*, with a somewhat revised selection. Cambridge University Press.

Bernays, Paul (1922): "Über Hilberts Gedanken zur Grundlegung der Arithmetik." In: *Jahresbericht der Deutschen Mathematiker-Vereinigung* 31, 1, 10 – 19

Bernays, Paul (1930): "Die Philosophie der Mathematik und die Hilbertsche Beweistheorie." In: *Blätter für deutsche Philosophie* 4, 326 – 367.

Bernays, Paul (1935): "Sur le Platonisme dans les mathématiques." In: *L'enseignement mathématique*, Vol. 34 (1935), 52 – 69.

Bernays, Paul (1935a): "Quelques points essentielles de la métamathématique." In: *L'enseignement mathématique*, Vol. 34, 70 – 95.

Bernays, Paul (1976): *Abhandlungen zur Philosophie der Mathematik*. Darmstadt: Wissenschaftliche Buchgesellschaft.

Brouwer, L. E. J. (1925): "Zur Begründung der intuitionistischen Mathematik I." In: *Mathematische Annalen*, Vol. 93, 244 – 257.

Brouwer, L. E. J. (1925a): "Zur Begründung der intuitionistischen Mathematik II." In: *Mathematische Annalen*, Vol. 95, 453 – 472.

Brouwer, L. E. J. (1926): "Zur Begründung der intuitionistischen Mathematik III." In: *Mathematische Annalen*, Vol. 96, 451 – 488.

Brouwer, L. E. J. (1927): "Über Definitionsbereiche von Funktionen." In: *Mathematische Annalen*, Vol. 97, 446 – 463.

Dreben, Burton; Andrews, Peter; Anderaa, Stal (1963): "False Lemmas in Herbrand." In: *Bulletin of the American Mathematical Society* 69, 699 – 706.

---

30 Gödel, "Vortrag bei Zilsel," transcribed and translated in Gödel 1995.

Dreben, Burton; Denton, John (1966): "A Supplement to Herbrand." In: *The Journal of Symbolic Logic* 31, 393 – 398.
Gentzen, Gerhard (1936): Die Widerspruchsfreiheit der reinen Zahlentheorie. *Mathematische Annalen*, Vol. 112 (1936), 493 – 565.
Gödel, Kurt (1931): "Über formal unentscheidbare Sätze der Principia Mathematica und verwandter Systeme." In: *Monatshefte für Mathematik und Physik*, Vol. 38, 173 – 198. Reprinted with translation in Gödel 1986.
Gödel, Kurt (1932b): "Über Vollständigkeit und Widerspruchsfreiheit." In: *Ergebnisse eines mathematischen Kolloquiums*, Vol. 3, 12 – 13. Reprinted with translation in Gödel 1986.
Gödel, Kurt (1933): "Zur intuitionistischen Artithmetik und Zahlentheorie." In: *Ergebnisse eines mathematischen Kolloquiums*, Vol. 4, 34 – 38. Reprinted with translation in Gödel 1986.
Gödel, Kurt (1958):" Über eine bisher noch nicht benützte Erweiterung des finiten Standpunktes." In: *Dialectica*, Vol. 12, 280 – 287. Reprinted with translation in Gödel 1990.
Gödel, Kurt (1986): *Collected Works; Vol. I: Publications 1929–1936*. Solomon Feferman et al., eds. New York and Oxford: Oxford University Press.
Gödel, Kurt (1990): *Collected Works; Vol. II: Publications 1938–1974*. Solomon Feferman et al., eds. New York and Oxford: Oxford University Press.
Gödel, Kurt (1995): *Collected Works; Vol. III: Unpublished Essays and Lectures*. Solomon Feferman et al., eds. New York and Oxford: Oxford University Press.
Gödel, Kurt (2003): *Collected Works; Vol. IV: Correspondence A–G*. Solomon Feferman et al., eds. Oxford: Clarendon Press.
Goodman, Nelson; Quine, W. V. (1947): "Steps Toward a Constructive Nominalism." In: *The Journal of Symbolic Logic* 12, 105 – 122.
Hilbert, David (1926): "Über das Unendliche." In: *Mathematische Annalen*, Vol. 95, 161 – 190.
Hilbert, David; Bernays, Paul (1934/39): *Grundlagen der Mathematik*. 2 vols. Berlin: Springer. 2nd ed. Berlin: Springer, 1968, 1970. Cited as HB, according to 2nd ed. unless otherwise indicated.
Martin-Löf, Per (2008). "The Hilbert-Brouwer Controversy Resolved?" In: van Atten, Mark; Boldini, Pascal; Bourdeau, Michel; Heinzmann, Gerhard (eds.), *One Hundred Years of Intuitionism (1907–2007): The Cerisy Conference*, 243 – 256. Basel, Boston, Berlin: Birkhäuser.
Müller, Gert H. (ed.) (1976): *Sets and Classes: On the Work by Paul Bernays*. Amsterdam: North-Holland.
Parsons, Charles (2008): "Paul Bernays' Later Philosophy of Mathematics." In: Costa Dimitracopoulos, Ludomir Newelski, Dag Normann, and John R. Steel (eds.), *Logic Colloquium* 2005, 129 – 150. Lecture Notes in Logic 28. Urbana, Association for Symbolic Logic, and Cambridge University Press. Reprinted in Parsons 2014.
Parsons, Charles (2014): *Philosophy of Mathematics in the Twentieth Century. Selected Essays.* Cambridge (MA): Harvard University Press.
Parsons, Charles (2015): "Infinity and a Critical View of Logic." In: *Inquiry*, Vol. 58, 1 – 19.
Rathjen, Michael (2005): "The Constructive Hilbert Program and the Limits of Martin-Löf Type Theory." In: *Synthese*, Vol. 147, No. 1, 81 – 120.
Sieg, Wilfried (1999): "Hilbert's Programs, 1917–1922." In: *The Bulletin of Symbolic Logic*, Vol. 5, 1 – 44.
Stegmüller, Wolfgang (ed.) (1978): *Das Universalienproblem*. Darmstadt: Wissenschaftliche Buchgesellschaft.
Tait, W. W. (2019): "What Did Hilbert and Bernays Mean by Finitism?" In this volume.

van Dalen, D. (1995): "Hermann Weyl's Intuitionistic Mathematics." *The Bulletin of Symbolic Logic*, Vol. 1, 145 – 169.
von Neumann, J. (1927): "Zur Hilbertschen Beweistheorie." *Mathematische Zeitschrift*, Vol. 26, 1 – 46.
Weyl, Hermann (1918): *Das Kontinuum. Kritische Untersuchungen über die Grundlagen der Analysis*. Leipzig: Veit.
Weyl, Hermann (1921): "Über die neue Grundlagenkrise der Mathematik." In: *Mathematische Zeitschrift*, Vol. 10, 39 – 79.

Richard Zach
The Significance of the Curry-Howard Isomorphism

Abstract: The Curry-Howard isomorphism is a proof-theoretic result that establishes a connection between derivations in natural deduction and terms in typed lambda calculus. It is an important proof-theoretic result, but also underlies the development of type systems for programming languages. This fact suggests a potential importance of the result for a philosophy of code.

1 Introduction

Many results of mathematical logic are thought to be of philosophical significance. The most prominent examples are, perhaps, Gödel's completeness and incompleteness theorems, and the Löwenheim-Skolem theorem. The completeness theorem is thought to be significant because it establishes, in a mathematically rigorous way, the equivalence between syntactic and semantic definitions of logical consequence. The incompleteness theorem, on the other hand, is significant because it shows the *in*equivalence of syntactic and semantic definitions of mathematical truth. The Löwenheim-Skolem theorem is considered philosophically significant especially because of the use Putnam put it to in his model-theoretic argument (Putnam 1980). I will argue below that the Curry-Howard isomorphism is a logical result that promises to be philosophically significant, even though (among philosophical logicians, at least) it is little known and if it is, is often considered a mere curiosity. That result, in brief, is this: to every derivation in natural deduction there corresponds a term in a lambda calculus, such that transformation of the derivation into a normal form corresponds to evaluating the corresponding lambda term.

Of course, results of mathematical logic, just like mathematical results more generally, are often thought to be significant not just for philosophy, but more generally. The question of what makes a mathematical result significant is a fruitful topic for philosophical investigation, and I will not be able to do it justice here. But a few things can be said. Mathematical results can be significant for a number of different reasons. Significance can arise from the result's theoretical importance. It may, for instance, elucidate a concept by relating it to others. A prime example of this is again the completeness theorem: it elucidates semantic and proof-theoretic

Richard Zach, University of Calgary, Canada, rzach@ucalgary.ca
DOI 10.1515/9783110657883-18

definitions of consequence by showing them to be equivalent (in the case of first-order logic). A result can also be considered significant because it is fruitful in proving other results; it provides methods used in further proofs. The compactness theorem is significant for this reason: it allows us to show that sets of sentences are consistent (and hence satisfiable) by showing that every finite subset is consistent (or satisfiable).

Another dimension along which the significance of results could be measured is the breadth of fields for which it is significant, either by elucidating notions, or by providing proof methods, or by paving the way for "practical" applications. By a "practical" application, I mean a method for finding answers to specific questions or solving particular problems. The line between what counts as a theoretical application and what as a practical application is of course not a clear one. But a paradigm case, I take it, of a practical application is, for instance, determining if a specific inference is valid or finding an interpretation for a specific set of sentences. The specific inferences or sets of sentences here often and importantly may not be problems in mathematical logic itself, but lie in other areas of mathematics or even outside mathematics. Inferences from databases, say, or finite models for formal specifications of circuits can be seen as specific questions that can be solved with methods – proof and model-building methods – from mathematical logic. That such proof methods exist and that they are sound and complete, again, is one reason they are significant. Here, of course, the proof methods are ones that are amenable to implementation in software, such as resolution, and not those commonly used or studied in mathematical logic itself.

Soundness, completeness, and compactness results for various logical systems are easily seen to be significant for these reasons and along these dimensions. Proof-theoretic results such as cut-elimination and consistency proofs more generally are harder to certify as significant. Of course, consistency proofs are one of the more prototypical philosophically significant results of mathematical logic, since they arose directly out of philosophical concerns, viz., Hilbert's program. Normalization of natural deduction plays an important role in the formulation of proof-theoretic semantics.

Consistency proofs are also often practically significant. In the absence of a semantics for a logical formalism, for instance, a consistency proof (e.g., a cut-elimination result) establishes a kind of safety of the formalism. Historically, such results established the safety of various non-classical logics. They also paved the way for the development of formalisms that were amenable to software implementation, i.e., automated theorem proving. Proof search is only feasible if the search space can be sufficiently restricted, and cut-elimination guarantees this. If the cut rule were not eliminable, proof search using analytic calculi such as the sequent calculus would not be feasible – and cut elimination proofs establish this even in the absence of a proof of cut-free completeness.

Strengthenings of consistency proofs also have theoretical significance in mathematical logic. Here I have in mind the kinds of results arising out of the work of consistency proofs which show that there are procedures that transform proofs in one system into proofs in another, weaker system (at least for certain classes of theorems). These results provide proof theoretic reductions of one system to another (a theoretically significant result inside mathematical logic), provide the basis for what Feferman 1992 has called "foundational reductions" (a philosophical payoff), but also sometimes provides methods for extracting bounds from proofs of existential theorems in various systems of arithmetic or analysis (a mathematical payoff). They also measure (and hence elucidate) the strength of axiom and proof systems in more fine-grained ways than simple consistency strength, as when the proof-theoretic reduction shows a speed-up of one system over another.

The significance of the Curry-Howard isomorphism has been hard to assess. It started off as a mere curiosity, when Curry 1934 observed a perhaps surprising but possibly merely coincidental similarity between some axioms of intuitionistic and combinatory logic. Following Howard 1980 (originally circulated in 1969) and Reynolds 1974, it became clear that the isomorphism applied in a wider variety of cases, and it took on both theoretical and philosophical significance. Its theoretical significance lies in the fact that it can be used to prove strong normalization of natural deduction, i.e., the result that normal derivations do not just exist but that they are unique. Its philosophical significance arose originally out of the claim that the lambda term assigned to a derivation can be taken to be the "computational content" of the derivation. But in what sense this does provide a "content" in any kind of clear and robust sense was never really elucidated. More work has to be done to justify calling the lambda term assigned to a derivation its "content," and to explain in what sense that content is "computational."

In the past two decades or so, however, it has become clear that the Curry-Howard isomorphism is of very clear practical significance: it now forms the basis of type systems of programming languages and the natural deduction systems on the proof side of the isomorphisms are the basis of automatable and indeed automated type checkers and type inference systems for real-life programming languages. I will argue that this practical significance allows for an overdue philosophical study of programming languages.

2 Natural Deduction

The Curry-Howard isomorphism is a correspondence between proofs in natural deduction systems and terms in lambda calculi. The correspondence is explained much more easily if natural deduction is formulated not as it was originally by Gentzen 1934 and Prawitz 1965 as inferring formulas from formulas, but as proceeding from sequents to sequents. In classical natural deduction, initial formulas in a proof are assumptions, some of which may be discharged by certain inference rules. A proof is thought of as establishing that the end-formula follows from the set of assumptions that remain undischarged or "open." In "sequent style" natural deduction, each sequent $\Gamma \vdash A$ in the proof records the set of assumptions that a formula follows from at every step in the proof. So an assumption by itself is recorded as $A \vdash A$, and generally a proof with end-formula A from open assumptions Γ would be represented, in sequent-style natural deduction, as a tree of sequents with end-sequent $\Gamma \vdash A$. In this setup, the inference rules of natural deduction become:

$$\frac{A, \Gamma \vdash B}{\Gamma \vdash A \to B} {\to}\mathrm{I} \qquad \frac{\Gamma \vdash A \to B \quad \Gamma \vdash A}{\Gamma \vdash B} {\to}\mathrm{E}$$

$$\frac{\Gamma \vdash A \quad \Gamma \vdash B}{\Gamma \vdash A \wedge B} \wedge\mathrm{I} \qquad \frac{\Gamma \vdash A \wedge B}{\Gamma \vdash A} \wedge\mathrm{E}$$

$$\frac{\Gamma \vdash A \wedge B}{\Gamma \vdash B} \wedge\mathrm{E}$$

Of course, this is just the fragment involving the conditional and conjunction, for simplicity. We're also ignoring, for the time being, the subtleties of keeping track of which assumptions are discharged where, but this will be fixed later on. Here is an example derivation of the theorem $(A \wedge B) \to (B \wedge A)$:

$$\frac{\dfrac{\dfrac{A \wedge B \vdash A \wedge B}{A \wedge B \vdash B} \wedge\mathrm{E} \quad \dfrac{A \wedge B \vdash A \wedge B}{A \wedge B \vdash A} \wedge\mathrm{E}}{A \wedge B \vdash B \wedge A} \wedge\mathrm{I}}{\vdash (A \wedge B) \to (B \wedge A)} {\to}\mathrm{I}$$

It corresponds to the following derivation in the original formulation of natural deduction:

$$\frac{\dfrac{\dfrac{[A \wedge B]^x}{B} \wedge\mathrm{E} \quad \dfrac{[A \wedge B]^x}{A} \wedge\mathrm{E}}{B \wedge A} \wedge\mathrm{I}}{(A \wedge B) \to (B \wedge A)} {\to}\mathrm{I}\ x$$

The x labelling the $\to$I inference indicates that the assumption labelled x is discharged at that inference. In the sequent-style derivation, all sequents above the

→I inference depend on the assumption $A \wedge B$ and so $A \wedge B$ appears on the left of the turnstile; in the final sequent, that assumption has been discharged and so $A \wedge B$ no longer appears on the left of the turnstile.

Derivations in natural deduction *normalize*, i.e., there is always a sequence of reduction steps which transforms a derivation into a derivation in normal form. A derivation is in normal form if no introduction rule for a connective is immediately followed by an elimination rule for the same connective. This result plays a similar role for natural deduction as the cut-elimination theorem plays for the sequent calculus. In particular, it establishes the subformula property for natural deduction derivations. As noted above, the subformula property is essential for the practical implementation of proof search algorithms, since it limits the search space and so eliminates one source of infeasibility in actually carrying out the search for a derivation.

Normalization is carried out by removing "detours" – applications of introduction rules immediately followed by elimination rules – one by one. In the sequent-style natural deduction formalism sketched above, such a replacement would be, for instance, the following:

$$\cfrac{\cfrac{\overset{\pi}{A, \Gamma \vdash B}}{\Gamma \vdash A \to B} \to I \quad \overset{\pi'}{\Gamma \vdash A}}{\Gamma \vdash B} \to E \quad \mapsto \quad \cfrac{\pi[\pi'/A]}{\Gamma \vdash B}$$

Here, $\pi[\pi'/A]$ is the result of replacing, in the subderivation π, every initial sequent of the form $A \vdash A$ by the subderivation π' ending in $\Gamma \vdash A$. Note that no natural deduction rules apply to the left side of sequents (the so-called context). Hence, any occurrence of A in the contexts of sequents in π is replaced by Γ, and the A's in the contexts in π disappear. Thus, if the derivation π has end-sequent $A, \Gamma \vdash B$, then $\pi[\pi'/A]$ has end-sequent $\Gamma \vdash B$.

In the case of a ∧E rule following a ∧I rule, the reduction step is even simpler. Here, nothing has to be done to the original (sub-)derivations.

$$\cfrac{\cfrac{\overset{\pi}{\Gamma \vdash A} \quad \overset{\pi'}{\Gamma \vdash B}}{\Gamma \vdash A \wedge B} \wedge I}{\Gamma \vdash A} \wedge E \quad \mapsto \quad \overset{\pi}{\Gamma \vdash A}$$

Of course, if the conclusion of ∧E had been $\Gamma \vdash B$, the derivation would instead be reduced to just the derivation π' of the sequent $\Gamma \vdash B$.

3 The Typed Lambda Calculus

The Curry-Howard isomorphism is a correspondence between derivations in natural deduction and terms in a typed lambda calculus. But let's begin with the untyped lambda calculus. It is a term calculus; terms are built up from variables. If t is a term, so is $\lambda x.t$, called a lambda abstract. Intuitively, it represents a function (a program) that takes x as argument and returns a value specified by the term t (in which x occurs free). In $\lambda x.t$, x is bound. So terms can represent functions; applying a function to an argument is simply represented by a term (ts), where s is another term representing the argument to the function represented by t. Our toy calculus also contains operations working on pairs: If t and s are terms, then $\langle t, s \rangle$ is the pair consisting of t and s. If t is a term representing a pair, then $\pi_1 t$ represents its first component and $\pi_2 t$ its second.

The lambda calculus is a very simple programming language, in which terms are programs. *Execution* of a program for an input is the conversion of terms to one that cannot be further evaluated. Such terms are said to be in normal form; they represent outcomes of computations. The conversion of terms proceeds according to the following *reduction rules*:

$$(\lambda x.t)s \mapsto t[s/x]$$
$$\pi_1 \langle t, s \rangle \mapsto t$$
$$\pi_2 \langle t, s \rangle \mapsto s$$

The notation $t[s/x]$ means: replace every free occurrence of x in t by s. Terms of the form given on the left are called *redexes*; they are the kinds of (sub)terms to which reduction can be applied.

The reduction rules apply not just to entire terms, but also to subterms. For instance, here is a very simple program that inverts the order of the elements of a pair:

$$\lambda x.\langle \pi_2 x, \pi_1 x \rangle$$

If we apply this term to a pair $\langle u, v \rangle$, conversion will produce $\langle v, u \rangle$:

$$(\lambda x.\langle \pi_2 x, \pi_1 x \rangle)\langle u, v \rangle \mapsto \langle \pi_2 \langle u, v \rangle, \pi_1 \langle u, v \rangle \rangle$$
$$\mapsto \langle v, \pi_1 \langle u, v \rangle \rangle$$
$$\mapsto \langle v, u \rangle$$

In the untyped lambda calculus, it is allowed to apply terms to terms to which they intuitively shouldn't be applied. In the simple example above, the program on the left expects the argument x to be a pair, but the formation rules don't prohibit applying it, e.g., to another function $\lambda y.t'$. In the *typed* lambda calculus, the syntax prohibits this. Now, every variable comes with a type, denoted by an uppercase letter: x^A means that the variable x only takes values of type A. One can think of types as objects of a certain sort (e.g., truth values or natural numbers) but there are also function types (e.g., functions from numbers to truth values). The type $A \to B$ are the functions with arguments of type A and values of type B. There may be other types as well. In our example language, for instance, we have product types: if A and B are types, then $A \wedge B$ is the type consisting of pairs where the first element is of type A and the second of type B. So our example program now becomes $\lambda x^{A \wedge B}.\langle \pi_2 x, \pi_1 x \rangle$: the variable x is restricted to objects of type $A \wedge B$. It is then easy to see that the term represents a function from pairs to pairs, but while arguments are of type $A \wedge B$, values are of type $B \wedge A$. So our term is of type $(A \wedge B) \to (B \wedge A)$. This is represented in a *type judgment*

$$\lambda x^{A \wedge B}.\langle \pi_2 x, \pi_1 x \rangle : (A \wedge B) \to (B \wedge A)$$

Now the point of the typed lambda calculus is to disallow terms in which the types don't match up. For instance, we should not allow the formation of the term

$$(\lambda x^{A \wedge B}.\langle \pi_2 x, \pi_1 x \rangle)(\lambda y^A.y)$$

because the term on the left is a function of type $(A \wedge B) \to (B \wedge A)$ and the term $\lambda y^A.y$ on the right is a term of type $A \to A$, i.e., certainly not a pair. To do this, we give formation rules for terms that take the types of variables and subterms into account. They operate on sequents $\Gamma \vdash t : A$ where Γ is now a set of type judgments for the free variables in the term t. For instance,

$$x : A, y : B \vdash \langle x, y \rangle : A \wedge B$$

means that if x is of type A and y of type B, then the term $\langle x, y \rangle$ is of type $A \wedge B$. So the rules for forming terms now become:

$$\frac{x : A, \Gamma \vdash t : B}{\Gamma \vdash \lambda x^A.t : A \to B} \to\!I \qquad \frac{\Gamma \vdash t : A \to B \quad \Gamma' \vdash s : A}{\Gamma, \Gamma' \vdash (ts) : B} \to\!E$$

$$\frac{\Gamma \vdash t : A \quad \Gamma \vdash s : B}{\Gamma \vdash \langle t, s \rangle : A \wedge B} \wedge\!I \qquad \frac{\Gamma \vdash t : A \wedge B}{\Gamma \vdash \pi_1 t : A} \wedge\!E$$

$$\frac{\Gamma \vdash t : A \wedge B}{\Gamma \vdash \pi_2 t : B} \wedge\!E$$

Of course, $x : A \vdash x : A$ is always true, and so sequents of this form serve as axioms.

4 The Curry-Howard Isomorphism

The first part of the Curry-Howard isomorphism consists in the observation that the right-hand sides of the type judgments in the rules of term formation for the typed lambda calculus given in the previous section are exactly the introduction and elimination rules of natural deduction. When considered as term formation and type inference rules, the →E rule, for instance, tells us that if s is a term of type $A \to B$ and t is a term of type A, then (st) is a term of type B. However, these same rules also allow us to systematically assign terms to sequents in a natural deduction derivation (once variables are assigned to the assumptions). If we focus not on the term side of the type judgments but on the type (formula) side, we can read the inference rules another way. For instance, in the case of →E, they say that if s has been assigned to the premise $A \to B$, and t to the premise A, then we should assign (st) to the conclusion B. In this way, we get that for every derivation in natural deduction we can assign terms from the typed lambda calculus to each sequent (or rather, the formula on the right of the sequent). For instance, here is a derivation of $(A \land B) \to (B \land A)$ with the corresponding terms assigned to each formula on the right; the assignment is determined once we pick variables to assign to the assumptions (in this case, we assign x to the assumption $A \land B$):

$$\cfrac{\cfrac{\cfrac{x : A \land B \vdash x : A \land B}{x : A \land B \vdash \pi_2 x : B} \land E \quad \cfrac{x : A \land B \vdash x : A \land B}{x : A \land B \vdash \pi_1 x : A} \land E}{\cfrac{x : A \land B \vdash \langle \pi_2 x, \pi_1 x \rangle : B \land A}{\vdash \lambda x^{A \land B}.\langle \pi_2 x, \pi_1 x \rangle : (A \land B) \to (B \land A)} \to I} \land I$$

The second part of the Curry-Howard isomorphism extends this observation. Applying the normalization procedure to derivations in natural deduction corresponds to evaluation (reduction) of the corresponding terms. In the case where we remove a →I/→E detour, the term assigned to the conclusion is of the form $(\lambda x.t)s$ and the term assigned to the conclusion of the derivation after applying the normalization step is $t[s/x]$, i.e., the result of a lambda calculus reduction step:

$$\cfrac{\cfrac{\begin{array}{c}\pi\\ x : A, \Gamma \vdash t : B\end{array}}{\Gamma \vdash \lambda x^A.t : A \to B} \to I \quad \begin{array}{c}\pi'\\ \Gamma \vdash s : A\end{array}}{\Gamma \vdash (\lambda x^A.t)s : B} \to E \quad \mapsto \quad \begin{array}{c}\pi[\pi'/A]\\ \Gamma \vdash t[s/x] : B\end{array}$$

The same happens when we remove a ∧I/∧E detour:

$$\cfrac{\cfrac{\begin{array}{c}\pi\\ \Gamma \vdash t : A\end{array} \quad \begin{array}{c}\pi'\\ \Gamma \vdash s : B\end{array}}{\Gamma \vdash \langle t, s \rangle : A \land B} \land I}{\Gamma \vdash \pi_1 \langle t, s \rangle : A} \land E \quad \mapsto \quad \begin{array}{c}\pi\\ \Gamma \vdash t : A\end{array}$$

Note in particular that a derivation in natural deduction is normal (contains no detours) if and only if the term assigned to it is in normal form: any $\to$I/$\to$E detour would correspond to a redex of the form $(\lambda x^A.t)s$, and any $\wedge$I/$\wedge$E detour would correspond to a redex of the form $\pi_i\langle s, t\rangle$. (A redex, again, is a subterm to which a reduction can be applied; a term is normal if it contains no redexes.)

More detailed expositions of the Curry-Howard isomorphism for more comprehensive systems can be found in Girard et al. 1989 and Sørensen/Urzyczyn 2006.

5 The Significance of the Curry-Howard Isomorphism

In its logical form, the Curry-Howard isomorphism consists of the following facts:

1. Natural deduction proofs have associated proof terms.
2. Normalization corresponds to reduction of the corresponding proof terms.

We have only seen it at work in the toy case of the $\to$/$\wedge$-fragment of minimal logic, but similar proof term assignments can and have been given for a wide variety of natural deduction systems and logics. In this version, the Curry-Howard isomorphism is theoretically important for one main reason: it allows us to prove what's called strong normalization. It is relatively easy to prove that there always is a sequence of normalization steps that results in a derivation in normal form. Strong normalization is the claim that *any* sequence of reduction steps (a) terminates and (b) results in the same normal form. Via the Curry-Howard correspondence, this statement about derivations and normalization is a consequence of a corresponding result about lambda calculus and evaluation of typed lambda terms: any sequence of evaluation steps (a) terminates in a term in normal form (a value) and (b) all possible sequences of evaluation steps result in the same value.

The Curry-Howard isomorphism, in its computational version, is the basis for a very important body of work developed over the last 30 years, in the area of theorem proving and programming language theory. The computational version results from considering the term (program) side of the isomorphism as primary:

1. Well-formed lambda terms have associated type derivations.
2. Evaluation of terms corresponds to normalization of the corresponding type derivations.

Its importance for the theory of programming languages lies in the fact that it provides a model for performing type checking for programs in various programming languages. In our toy example, the typed lambda calculus plays the role of the programming language. A term of the form $\lambda x^A.t$ is a program which takes inputs of type A. Its outputs are of type B just in case the term itself is of type $A \to B$. To type check the program means to ensure that the term has this type. It obviously corresponds exactly to the proof-theoretic question of whether there is a derivation of $\vdash \lambda x^A.t : A \to B$, i.e., whether there is a derivation of $A \to B$ which has the type $\lambda x^A.t$ assigned to it as a proof term. The Curry-Howard isomorphism establishes that well-formed terms can always be type checked by a derivation in normal form. Since derivations in normal form have the subformula property, searching for a derivation that type checks a given program can (often) be done effectively.

A programming language is called type safe if it prevents programmers from writing programs that result in type errors. A type error occurs if a program of type $A \to B$, say, is applied to an argument which is not of type A, or when its evaluation for an argument of type A results in something that is not of type B. The Curry-Howard isomorphism, and properties like it for actual programming languages, guarantees type safety. This has tremendous theoretical and practical importance: programs in type-safe languages cannot hang: there is always a way to continue the execution until a value is arrived at. The Curry-Howard isomorphism has this consequence because it implies two things: If a term is not already a value (i.e., it is not in normal form), then evaluation can continue because it contains a redex that can be reduced. This property is called "progress." But more importantly – and here the Curry-Howard isomorphism comes in – when a term is reduced, its type stays the same (and there is a corresponding natural deduction derivation which verifies this). This property is called "type preservation." Together, these two properties establish that if t is a term (program) of type $A \to B$, and s is a term (argument) of type A, (ts) will always evaluate to a normal term (value) of type B. The strong normalization property for the typed lambda calculus establishes that evaluation always terminates and that any order of evaluation steps results in the same value. The Curry-Howard isomorphism, via its consequence of type preservation, establishes that each intermediate step and the final result, will be a term of type B.

The ability to type check programs, and the property of type safety, are kinds of completeness and soundness properties for programs. Like consistency in the foundations of mathematics, type safety provides a minimal safety guarantee for programs in type-safe languages. Type safety does not, of course, guarantee that any program will output the specific desired result – it is still possible that there are programming errors – but it does guarantee that the program will not produce a result of the wrong type. It won't hang, and it won't produce, say, a number when the program is of type, say, natural numbers to Boolean values.[1]

6 Toward a Philosophy of Code

For a long time, debates in the philosophy of mathematics were dominated by questions of little or no significance or even relation to mathematical practice, e.g., the realism/anti-realism debate. Concurrently, philosophy of science, and philosophies of the special sciences, concentrated on questions that were of quite central and immediate relevance to their respective areas. At the same time, computer science developed as an independent discipline. Philosophers of mathematics, however, have focused on just a very narrow set of questions related to computability – mainly those related to the limits of computability, the status of the Church-Turing thesis, philosophical analyses of the notion of mechanical computation, and the relation between formal and physical models of computability. Philosophy of programs and higher level programming languages is just in its infancy.

By analogy with the philosophy of mathematics, logical results like the Curry-Howard isomorphism promise to play a similar role in such a nascent philosophy of code as earlier results like Gödel's incompleteness theorems, the Löwenheim-Skolem theorem, and consistency proofs and proof-theoretic reductions play in the philosophy of mathematics. It can help frame and solve philosophical questions that naturally arise about programming languages. One such question, for instance, is this: Although several foundational frameworks are available for mathematics, one of them, Zermelo-Fraenkel set theory, has dominated both in mathematics and in the philosophy of mathematics. By contrast, many more frameworks for specifying algorithms have been and are being proposed by computer scientists. What explains this? What are the fundamental differences between programming languages and according to which criteria should we compare them? There are obvious candidates, such as efficiency or suitability for specific applications. But to differentiate programming languages and programming paradigms it will be important to consider the conceptual differences between languages, such as the presence or absence, and the power and structure of their type systems.

Recent work in the philosophy of mathematics has made headway on some aspects of mathematics that were bracketed by traditional mid-century philosophical engagement with mathematics, and which are more closely related to the

---

[1] For more detail on the importance of the Curry-Howard isomorphism and type systems for programming languages, see Cardelli 2004, Pierce 2002, Wadler 2015.

practice of mathematics and to methodological considerations made by mathematicians themselves. For instance, philosophers have considered the questions of what makes a proof explanatory (Mancosu 2018), the notions of elegance and beauty as employed by mathematicians (Montano 2014), differences in mathematical style (Mancosu 2017), and the question of when proof methods count as "pure" (e.g. Detlefsen/Arana 2011). By analogy, methodological considerations made by computer scientists in proposing, designing, revising, criticizing, and choosing between programming languages provide an area suitable for philosophical analysis and reflection – which may well be of interest to computer scientists the way, say, some work in the philosophy of biology has been of interest in biology itself. Computer scientists also talk about programs and programming languages in aesthetic (elegance), cognitive (readability), or practical terms (efficiency and speed, ease of maintenance, security) which play a role in the above-mentioned comparison between and design of programming languages. Type-safe languages, for instance, have been claimed to be superior to non-type-safe and untyped languages along these lines. They have been claimed to be easier to read, easier to maintain, easier to debug, and they are, demonstrably, more secure. It is also claimed that typed languages are more abstract, i.e., that they enable programming at a higher level of abstraction. This notion of abstraction itself could be subjected to philosophical analysis. In all of this, a logical result – the Curry-Howard isomorphism – plays a fundamental role, and it would be important to understand this role better, and to make use of it for philosophical discussions of code and programming languages.

Bibliography

Cardelli, Luca (2004): "Type Systems." In: *Computer Science and Engineering Handbook*. Tucker, Allen B. (ed.), 2nd edition, chapter 97. Boca Raton, FL: CRC Press
Curry, Haskell B. (1934): "Functionality in Combinatory Logic." In: *Proceedings of the National Academy of Sciences of the United States of America*. Vol. 20, No. 11, 584 – 590.
Detlefsen, Michael and Arana, Andrew (2011): "Purity of Methods." In: *Philosopher's Imprint*. Vol. 11, No. 2, 20.
Feferman, Solomon (1992): "What Rests on What? The Proof-Theoretic Analysis of Mathematics." In: *Akten des 15. Internationalen Wittgenstein-Symposiums: 16. Bis 23. August 1992, Kirchberg Am Wechsel*. Vol. 1, 147 – 171. Vienna: Hölder-Pichler-Tempsky. Reprinted in: Feferman 1998, chapter 10, 187 – 208.
Feferman, Solomon (1998): *In the Light of Logic*. New York and Oxford: Oxford University Press.
Gentzen, Gerhard (1934): "Untersuchungen über das logische Schließen I–II." In: *Mathematische Zeitschrift*. Vol. 39, 176 – 210, 405 – 431. English translation in Gentzen 1969, 68 – 131.

Gentzen, Gerhard (1969): *The Collected Papers of Gerhard Gentzen*. Szabo, Manfred E. (ed.). Amsterdam: North-Holland.

Girard, Jean-Yves, Taylor, Paul, and Lafont, Yves (1989): *Proofs and Types*. New York: Cambridge University Press.

Howard, William A. (1980): "The Formulae-As-Types Notion of Construction." In: *To H. B. Curry: Essays on Combinatory Logic, Lambda Calculus and Formalism*. Seldin, Jonathan P.; Hindley, J.R. (eds.). London and New York: Academic Press, 480 – 490.

Mancosu, Paolo (2017): "Mathematical Style." In: *The Stanford Encyclopedia of Philosophy*. Zalta, Edward N. (ed.). Metaphysics Research Lab, Stanford University, fall 2017 ed. https://plato.stanford.edu/archives/fall2017/entries/mathematical-style/.

Mancosu, Paolo (2018): "Explanation in Mathematics." In: *The Stanford Encyclopedia of Philosoph*. Zalta, Edward N. (ed.). Metaphysics Research Lab, Stanford University, summer 2018 ed. https://plato.stanford.edu/archives/sum2018/entries/mathematics-explanation/.

Montano, Ulianov (2014): *Explaining Beauty in Mathematics: An Aesthetic Theory of Mathematics*. Berlin: Springer.

Pierce, Benjamin C. (2002): *Types and Programming Languages*. Cambridge (MA): MIT Press.

Prawitz, Dag (1965): *Natural Deduction*. Stockholm Studies in Philosophy 3. Stockholm: Almqvist & Wiksell.

Putnam, Hilary (1980): "Models and Reality." In: *Journal of Symbolic Logic*. Vol. 45, No. 3, 464 – 482.

Reynolds, John C. (1974): "Towards a Theory of Type Structure." In: *Programming Symposium*. Robinet, B. (ed.), Lecture Notes in Computer Science 19. Berlin and Heidelberg: Springer, 408 – 425.

Sørensen, Morten Heine and Urzyczyn, Pawel (2006): *Lectures on the Curry-Howard Isomorphism*. In: *Studies in Logic and the Foundations of Mathematics*. Vol. 149. New York: Elsevier.

Wadler, Philip (2015): "Propositions as Types." In: *Communications of the ACM*. Vol. 58, No. 12, 75 – 84.

Felix Mühlhölzer
Reductions of Mathematics: Foundation or Horizon?

Abstract: The usual reductions of large parts of mathematics to much more restricted parts, with the reduction to set theory as a sort of paradigm, are virtually uncontroversial from a purely mathematical point of view. But what is their point? According to the standard answer, they are important because they provide *foundations* for mathematics. What that precisely means, however, can be explained and also be criticised in quite different ways. There is a Wittgensteinian way of criticism that proves to be particularly instructive and that is summed up in the following beautiful passage in (RFM VII, §16): "The *mathematical* problems of the so-called foundations are no more at the basis of mathematics for us than the painted rock is the support of a painted castle." There is another answer given by Bourbaki: such a reduction provides an *horizon* for mathematics. This is a totally different idea from the idea of a foundation. The horizon of mathematics is understood as a perfect formalization that lies *in front* of us and that guides us, but it is not *beneath* us like a foundation, i.e. a sort of rock that supports the edifice of mathematics. However, Claude Chevalley, the Bourbakist who had actually developed this idea, later discarded it, and his criticism is in line with a Wittgensteinian perspective. So the question remains: what is the real point of the reductions?

Mathematics can be reduced to set theory. By "mathematics" I mean what some people call "classical mathematics"[1], which is mathematics as taught in the usual university courses and which comprises the overwhelming part of existing mathematics. By set theory I mean "axiomatic set theory" as presented in one of its familiar forms. The set theorist Kenneth Kunen describes the reduction of mathematics to set theory as follows:

> All mathematical concepts are defined in terms of the primitive notions of set and membership. In axiomatic set theory we formulate [...] axioms about these primitive notions [...]. From such axioms, all known mathematics may be derived. (Kunen 1980: xi)

This reduction is a clear and indisputable mathematical fact. Kunen, however, is not content with simply stating and describing this fact but preludes his de-

---

[1] For example Maddy in Maddy 2017.

Felix Mühlhölzer, Georg August University Göttingen, fmuehlh@gwdg.de

scription with the claim: "Set theory is the foundation of mathematics". What *this* means, however, and in what sense it may be true, is anything but clear. Kunen certainly wants to say that the foundational role of set theory is important, and many people would agree, but again the precise nature of this importance needs to be clarified. In what follows, I will present some thoughts concerning the role and importance of such reductions of mathematics, thoughts that are oriented towards Wittgensteinian considerations. It is true that Wittgenstein himself did not think about the reduction of mathematics to set theory. He was concerned with the reduction to logic as carried out by Frege or Russell, but his critical remarks about the alleged foundational role of logic can in most cases be applied to set theoretic reductions as well.

There are different ways in which mathematics can be reduced to set theory, but this variety is of mathematical and hardly of philosophical interest. Irrespective of this mathematical variety, however, it is the concept *foundation* that can be understood in very different ways, and it is these ways in which we should be interested in philosophy. It would go beyond the scope of this paper to list, let alone to discuss, all the possible meanings of the word "foundation" with respect to mathematics, as it is attempted in a recent paper by Penelope Maddy (Maddy 2017). Here I want to concentrate on merely two of these meanings. They are very prominent and are explicitly inspected by Wittgenstein. Both come along with characteristic pictures which make them particularly suitable for a Wittgensteinian treatment.

Perhaps what is most in keeping with the word "foundation" is the idea of a *solid basis* on which our mathematical edifice *rests*. Bourbaki, at the very beginning of his book on set theory, explicitly says that the main purpose of his many volumes about different mathematical theories is "to provide a solid foundation for the whole body of modern mathematics" (Bourbaki 1968: v), and the prevalent picture is the picture of "a rock on which to build the edifice" (Guedj 1985: 18). This solidity may prevent us from running into contradictions (although this is actually not the main purpose of the Bourbakian endeavour itself), it should lead to the mathematicians' consent, guarantee agreement among them, and similar things.

Wittgenstein never mentions Bourbaki, but what he says about 'foundations of mathematics' in §16 of Part VII of his *Remarks on the Foundations of Mathematics* can be immediately applied to Bourbaki's and other people's idea of a solid rock. This is the relevant Wittgensteinian passage:

> The *mathematical* problems of what is called foundations are no more at the basis of mathematics for us than the painted rock is the support of a painted castle.

We must understand what precisely Wittgenstein has in mind here.

According to the idea of a foundation just described, set theory is seen as being 'under' mathematics in the sense of being *at the bottom* of it. According to the *second* idea that I want to discuss, set theory is 'under' mathematics in the sense of being under the usual mathematical *clothes*. These clothes disguise the really important forms that are underneath, which are the set-theoretic forms wherein the reduction of mathematics ends. An idea like this was quite common in logicism when people described the logicist reductions as showing that mathematics (or at least arithmetic) is *logic in disguise,* and it is then simply transmitted to set-theoretic reductions when mathematics is now seen as *set-theory in disguise.*

This is an idea, and a picture, rather different from the first one (the idea of a rock at the bottom), and also this second idea and picture is explicitly picked up by Wittgenstein. Wittgenstein actually talks about logicism in the form presented by Russell, but what he says can be applied to set-theoretic reductionism as well. The relevant passage is in RFM III, §25:

> The Russellian signs veil the important forms of proof as it were to the point of unrecognizability, as when a human form is wrapped up in a lot of cloth.

Again, we must understand what precisely Wittgenstein has in mind here.

In RFM, and more so in his manuscripts from 1929 to 1944, Wittgenstein scrutinizes quite different traits of mathematical reductionism. One is the following: Consider, for a start, elementary arithmetic dealing with the natural numbers only, that is, the numbers 1, 2, 3, 4, ... (From now on I will call the natural numbers simply "numbers"; and for the sake of simplicity I ignore the number 0.) We can easily reduce the familiar numerals as just used to numerals using only the symbols "1" and "+", producing the following sequence:

(H) $1, 1+1, 1+1+1, 1+1+1+1, \ldots$

(Again for the sake of simplicity, I leave out brackets.) A real *set*-theoretic reduction, of course, would go much further, until only set-theoretic symbols are used – which is in fact very much further.[2] But already at this intermediate stage, the stage (H), we should pause.

---

[2] To be precise, a reduction to ZFC, say, would not lead to expressions replacing our familiar *numerals* but to expressions replacing our *sentences* about numbers because the language of ZFC does not allow for singular terms. In this case it is the set theoretic wrapping up of our sentences we should be concerned with. (I'm grateful to Volker Halbach for calling my attention to this point.) In the Bourbakian version of set theory, however, there *are* expressions replacing our numerals and they are of incredible complexity; see Mathias 2002.

Prima facie it may sound quite plausible to say at this stage that the new symbols are now '*nearer* to the numbers themselves', that they are, so to speak, more '*tailored* to the numbers themselves' in comparison to our familiar symbols, which are based on something mathematically arbitrary like the decimal system. And aren't these new symbols actually *underlying* the old ones? Aren't *they* the really important ones we should have in mind when trying to *understand* arithmetic? And don't the old, familiar symbols actually *disguise* this underlying structure? Aren't they – the old, familiar ones – actually wrapping up the really important ones in a lot of cloth? Hilbert, for example, when presenting his so-called 'finitary' arithmetic, can be interpreted as claiming something like that, either with respect to the symbols in (H) or, equivalently, with respect to strings of strokes like |, ||, |||, ... Hilbert even treated these symbols as numbers themselves – as 'anschauliche Zahlen', one might say – and not merely as numerals.

However, one can see the situation also in reverse, and this is what Wittgenstein did already in 1929. The first, rather straightforward problem with notations like $1 + 1 + 1 + \ldots + 1$ or strings of strokes is that for long expressions of this sort we simply can no more recognize *what number* is meant. And this unrecognizability becomes much worse, of course, when our reduction approaches the ground level formulated with the primitive notions of pure set theory. A problem of this sort – the unanswerability of the question "What number is this?" – is hinted at in RFM III, §§7 & 8 and it is thoroughly treated by Kripke in several unpublished lectures.[3] It allows a good interpretation of the end of RFM III, §25, which I already quoted and in which Wittgenstein says: "The Russellian signs [or the signs of set theory (as we should add)] veil the important forms of proof as it were to the point of unrecognizability, as when a human form is wrapped up in a lot of cloth." According to this point of view, the unrecognizability now pertains to signs like "$1 + 1 + 1 + \ldots + 1$", and when there are, say, 31.746 "1"s in such a sign, the 'human form' – namely: the numeral "31.746" – really appears to be wrapped up in the accumulation of "1"s. Furthermore, an unrecognizability of this sort becomes worse and worse when the ground level of the reduction is approached.

One might say that this problem is rather unimportant with regard to the actual aim of the reduction at hand and that we shouldn't make much fuss about it. After all, for these reductions individual numbers are not of interest (with the exception perhaps of Zero and One). What one is aiming at are more general insights. But there is another problem, a more serious one which not only concerns the question of what number is *meant* by a sign like "$1+1+\ldots+1$", but which targets the *identity* of

---

[3] See Kripke (Kripke 1992, Kripke 1997 and Kripke 2011).

the sign itself. Wittgenstein was posing this sort of problem since 1929[4], and I think that it belongs to the various impulses initiating his post-*Tractatus* philosophy. It is simply the problem that the identity of symbols like "1 + 1 + ... + 1" or of strings of strokes is blurred when these symbols are too long. In 1929 Wittgenstein says this:

> A fundamental question: How can I know that "|||||||" and "|||||||" are the *same sign*? After all, it is not enough that they look similar?
> For it is not the rough sameness of the gestalt that should constitute the identity of the sign, but rather the sameness of numbers. (MS 106: 22f.)[5]

Of course, this not only concerns strings of strokes but also signs like "1 + 1 + 1 + 1 + 1 + 1 + 1" and then the bulk of *all* the expressions that are near to the basis of the reduction, including the *proofs* formulated at this basis. Wittgenstein calls this the *problem of the unsurveyability* of many mathematical expressions. It is thoroughly investigated in RFM III.[6]

This problem is not taken seriously by the typical mathematician, or perhaps I should not say that it is 'not taken seriously' but that it is actually ignored by most mathematicians. Otherwise the widespread opinion that our usual mathematical expressions are merely *abbreviations* of the set-theoretic expressions at the ground-level of the reduction couldn't be explained. This opinion obviously takes for granted that the identity of the expressions at the ground level is settled. The list of people who talk that way, emphasizing the word "abbreviation", is long, and I could mention and quote many.[7] But of course there cannot be a question about *really* devising abbreviations of the unsurveyable expressions, which typically are

---

4 See, e.g., MS 106: 22f. (1929); MS 111: 156f. (1931); MS 112: 15 (1931).
5 Two years later, in 1931, he explicitly emphasizes the importance of this problem: "The problem of the distinction between 1 + 1 + 1 + 1 + 1 + 1 + 1 and 1 + 1 + 1 + 1 + 1 + 1 + 1 is much more important / fundamental than appears at first sight" (MS 112: 15).
6 I deal with this in Mühlhölzer 2006 and Mühlhölzer 2010.
7 Let me give one example that is particularly funny. It occurs in Alonzo Church's classical book *Introduction to Mathematical Logic* where we read the following about the definitions that typically occur in reductions of mathematics and that are also used in his own book: "[They] are concessions in practice to the shortness of human life and patience, such as in theory we disdain to make. The reader is asked, whenever we write an abbreviation of a [well-formed formula], to pretend that the [well-formed formula] has been written in full and to understand us accordingly. [...] Indeed we must actually write [well-formed formulas] in full whenever ambiguity or unclearness might result from abbreviating. And if any one finds it a defect that devices of abbreviation [...] are resorted to at all, he is invited to rewrite this entire book without use of abbreviations, a lengthy but purely mechanical task." (Church 1956: 75f.)

hugely unsurveyable. To talk about 'abbreviations' is, literally understood, utter nonsense.

So, we are in a strange situation now: On the one hand, the unsurveyability phenomenon emphasized by Wittgenstein is not taken seriously by mathematicians, on the other hand their own talking about 'abbreviations' cannot be taken seriously as well. Obviously, their attitude towards unsurveyability and abbreviation cannot refer to our actual practice of using mathematical symbols. It must be of a more *theoretical* nature. The question then is what this theoretical nature and what the relevant theory might be.

I see two possibilities here. The first is to say that our practical limitations are of no importance for mathematics as such. I think that at the basis of such a view must be a certain picture, or a certain theory, of an *ideal subject*, an *ideal mathematician*, who is not limited by human restrictions. Even Philip Kitcher, when holding an empiricist, partly Millian view of mathematics – which he did for a long time – found himself forced to advocate a specific theory of an ideal mathematician in order to accommodate the indefinite iterations of mathematical operations.[8] However, in all views of this sort it remains unclear what precisely *are* the super-human powers that should be attributed to such a fictional ideal subject, and all these views are uncoupled from our actual practice of referring to mathematical objects, not only with respect to unsurveyable notations but quite generally. We should not take them seriously, and Philip Kitcher, for example, has in fact retracted his own former theory of an ideal mathematician.[9] To my mind, the essential mistake of such a theory is to *first* talk about the ideal subject and *then* to say with its help how to deal with the unsurveyable expressions. I think it should be the other way round. There is a beautiful, pithy statement by Hilary Putnam concerning the idea that the notion of an *ideal machine* might help us to understand what mathematics is about. This is what Putnam says: "Talk of what an ideal machine could do is talk *within* mathematics; it cannot fix the interpretation *of* mathematics" (Putnam 1979: 119). Analogously I would say: Talk of what an ideal mathematician could do is to a significant extent talk *within* mathematics; it cannot fix the interpretation *of* mathematics, and in particular, it cannot say how to deal with unsurveyable mathematical expressions. I think this is the right diagnosis, and I will not go deeper into this issue here.

It immediately leads, however, to the second possibility to understand the theoretical nature of the mathematicians' attitude towards unsurveyability and

---

[8] See Kitcher 1984: 107 – 148, and Parsons 1986: 133f.
[9] Kitcher presents his reasons for his retraction in Kitcher 2012: 168 and 185; see also the reasons presented in Emödy 1994, inspired by Kripke 1982.

abbreviation, namely: not to downgrade our practical limitations but to consider the unsurveyable expressions as *mathematical* entities; as mathematical entities referred to in *metamathematics*. In fact, in metamathematics the so-called "signs" are considered as on a par with numbers. Gödel is rather clear about that when in the Introduction to his incompleteness paper he writes:

> Of course, for metamathematical considerations it does not matter what objects are chosen as primitive signs, and we shall assign natural numbers to this use. Consequently, a formula will be a finite sequence of natural numbers, and a proof array a finite sequence of finite sequences of natural numbers. (Gödel 1931: 147)

What Gödel here envisages is a syntactical structure considered as a genuine *mathematical* structure. When the situation is seen in that way, the problem concerning the identity of the unsurveyable expressions is circumvented in the same way it is circumvented with respect to numbers; or at least we can say that this identity-problem is not a special one but the familiar identity-problem we know from mathematical entities in general. According to my view, the objects of metamathematics are genuine mathematical objects, like numbers. Their identity can be settled purely axiomatically, and they are *not used*.

Is this really true? Don't we make use of the numbers *themselves* when counting, computing, measuring, when *applying* mathematics in everyday life and in science? – Yes, we speak that way, but this manner of speaking blurs an important difference with regard to the criteria of identity that are assumed. If we compute with numbers, say, we use specific representations of them, and it is the identity of these representations that is relevant. The respective criteria of identity then in fact are dependent on the use that is made. But in the case of the numbers themselves the criteria of identity are totally different, and the same is true in the case of *syntactical objects*, considered as *mathematical* objects.[10] When syntactical objects are used they appear in the form of specific notations, and if these notations become too complicated they cannot be used anymore (even in machines, of course). Whereas the genuine syntactical objects, considered as mathematical objects, are independent of specific notations. It is irrelevant, say, whether we mark the successor function in Peano arithmetic with the letter "S" or with a little punctuation mark ("'"). And the properties of these syntactical objects can also be described by referring only to certain relations between them, quite independent of the specific notational representatives, in complete analogy to our dealing with numbers *via* axiom systems. Of course when actually writing down these axioms systems, we

---

[10] In what follows I draw upon the nice paper "Structuralism and Meta-Mathematics" by Simon Friederich (Friederich 2010).

use specific notations. But we are totally aware of the irrelevance with respect to the *mathematical* situation. See what Gödel said about metamathematics.

The idea to regard syntactical expressions as genuine mathematical entities is particularly evident when the allegedly 'abbreviating' definitions are considered not as producing *enlargements* of the basic, primitive language[11] – enlargements in order to make the language 'less cumbersome', as people often say – but as *referring* to the expressions of the primitive language. What is less cumbersome, then, is a *new* language devised in order to represent the primitive language. This is the view of definitions favored by Bourbaki, for example. Bourbaki calls the expressions of the primitive language "assemblies" and he says this:

> To simplify the exposition [of a mathematical theory] it is convenient to denote [its] assemblies by less cumbersome symbols. We shall use, especially, [...] bold-face italic letters [...]. We shall often say that [these] symbols *are* assemblies, rather than that they *denote* assemblies: expressions such as "the assembly A" or "the letter x" [...] should therefore be replaced by "the assembly denoted by A" or "the letter denoted by x". (Bourbaki 1968: 16f.)

This view has the following consequence: Take the sign "2", that is, the numeral normally used to refer to the number 2. In the Bourbakian setting it is *defined* as a really cumbersome assembly formulated in the formal language of pure set theory, and furthermore, according to Bourbaki, it *denotes* this assembly. That is, it doesn't denote a number but an expression in a formal language. But doesn't this expression for its part refer to the number 2? No, because it is not a *used* expression, and "reference" is a word that makes sense only with respect to used expressions. So, according to the Bourbakian view, the sign "2" – which normally *is* a used sign, of course – doesn't refer to a number but is defined to be a certain set-theoretic expression. This is Bourbaki's brand of formalism.

To consider reference as essentially tied to use is, of course, a Wittgensteinian thought, but it seems to be rather common. In the literature one comes across the so-called *use thesis* concerning reference.[12] It says that in the determination of the reference of our terms the use we make of the terms must play an essential role, and this thesis has been widely adopted. If I see it correctly, the term *use thesis* was originally introduced by Stewart Shapiro, not, however, as a thesis about reference but about *understanding* a language.[13] Nevertheless, Shapiro seems to

---

[11] This is the view adopted, for example, in Hilbert et al. 1968: 292.
[12] I discuss this thesis in detail in Section 3 of Mühlhölzer 2014b.
[13] He explains it thus: "The claim is that understanding should not be ineffable. One understands the concepts embodied in a language to the extent that she knows how to use the language correctly. Call this the use thesis." (Shapiro 1990: 252; similarly in Shapiro 1991: 211 – 214, and Shapiro 1997: 204 – 206)

connect this thesis with "reference", at least implicitly. It is Hilary Putnam who in a more explicit way connects use with reference. This is what he says: "On any view, the understanding of the language must determine the reference of the terms, or, rather, must determine the reference given the context of use." (Putnam 1980: 24)

So, in order to be precise we should say that the so-called *signs* at the fundamental level are actually no 'signs' at all because they are not used, and for the same reason this so-called primitive *language* is, strictly speaking, not a genuine language. It consists of mathematical entities similar to numbers – in fact, these entities *can* be numbers – which we may *denote* in our usual way of speaking, but they themselves do not denote anything. This, of course, is the prevalent stance in metamathematics.

We are now in the following situation: There is our familiar practice of using mathematical expressions. Via appropriate definitions these expressions can be reduced to more complex ones, these to still more complex ones, and so on. In reality this step by step process of reduction can be performed only for the first few steps, and at later steps the so-called expressions are no more used ones but theoretically postulated mathematical entities. They are entities like numbers, and they in fact can simply be considered as numbers or as sequences of numbers. In this transition from used to not-used entities there is a certain grey area, but it is relatively small and the large area beyond it consists of purely mathematical expressions, with the primitive expressions of axiomatic set theory at the very end.

How does this situation go with the idea of a *foundation*? – I mentioned two characteristic shapes this idea may take. The second one makes use of the picture of the clothing: our real mathematics consists of clothes that disguise the proper mathematical forms, which are the set-theoretic forms wherein the reductions of mathematics end. But should we really say that our familiar mathematical practice with its used signs *jackets* the abstract mathematical structure to which it is reduced? I do not think that this was the intended picture. It does not go with the ubiquitous saying that our familiar signs are *abbreviations* of set-theoretic expressions – used signs do not 'abbreviate' non-used abstract entities –, and it does not go with the quite common idea (as adopted by Bourbaki, for example) that our used signs *refer* to such entities. Clothes do not 'refer' to the bodies they cover. The picture of the clothing does not seem adequate.

The other idea of a 'foundation' makes use of the picture of the 'solid basis', of the 'rock' on which mathematics is built. But should we really say that our mathematical *practice*, that our familiar *use* of mathematical symbols, is *based* on the abstract, non-used expressions of axiomatic set theory? Is our use of the numeral "3", say, 'based' on the set-theoretic substitute of this number? Is it based on the corresponding Bourbakian 'assembly'? All these things are too far away from our actual practice and in fact categorically different from it, and what is meant by saying that our practice is *based* on them remains utterly unclear.

One may ask whether the idea of a foundation really aims at founding our actual mathematical *practice*? Doesn't it aim at giving an *abstract model* of this practice? – But what people say doesn't sound like that. For there remains the ubiquitous saying that our familiar signs are *abbreviations* of set-theoretic expressions, and many even claim – like Alonzo Church, for example – that we should "actually write [the formulas] in full whenever ambiguity or unclearness might result from abbreviating" (Church 1956: 76). It doesn't make sense, however, to say things like that of abstract models of our practice. So, the picture of the solid rock is beset with confusions.

In the light of what I've said now, I want to come back to the Wittgensteinian passage I cited at the beginning:

> The *mathematical* problems of what is called foundations are no more at the basis of mathematics for us than the painted rock is the support of a painted castle. (RFM VII, §16)

According to what I've said so far, the following interpretation of this passage would make sense: The original idea of the 'rock' aims at giving our mathematical practice sufficient solidity. But what we then do is devising a reduction to a purely mathematical structure that we now regard as our basis. It is a structure far away from our practice, in particular from the use of symbols which is essential to this practice. Wittgenstein's talking about the *mathematical* problems that are involved may lead us, then, to think of abstracting from this use and of transforming our familiar practice into a purely mathematical structure as well. But with this move we have abandoned our idea of a solid rock. We then have reached the painted castle (our practice modelled as a mathematical structure) in its relation to painted rock (which is set theory). This relation no more involves a foundation as originally conceived because it is no more a genuine relation of *support*.

Is this the thought Wittgenstein had in mind? Perhaps not. The quoted passage is about paintings and paintings are not non-used things. Nevertheless, the crucial thought is the same: As in the case of paintings, so also according to my interpretation just given there cannot be any question of support. Furthermore, as I said before, also when *not* transforming our practice into something purely mathematical, it remains unclear what it might mean to say that the basic mathematical structure 'supports' this practice. The mathematical structure itself does not influence us and our practice. So it cannot support us and our practice. This supposed support rather looks like a fake support. The painted castle is not supported by the painted rock, and the relation between the abstract basis of set theory and our actual mathematical practice cannot be a relation of support either. I read

Wittgenstein's quoted remark simply as a warning: What may look like a support might be no support at all; it may be a fake support.

The question then is: what *is* the relation between the signs used in our mathematical practice and the purely mathematical, syntactic structure at the end of a reduction? Perhaps it is wrong or confused to call it "foundational" – but how should it be called instead? One may be tempted to call the transformation that leads from our used signs to these non-used signs *idealization*, but this would be totally inappropriate. I would propose to call it *petrification*. Our normal processes of idealization, as common in empirical science, consist in disregarding certain untidy aspects of real situations that we deem inessential in the context at hand, as when we leave aside small but inevitable occurrences of friction in physics. But when petrifying signs, as it is done in metamathematics and on the fundamental level of a mathematical reduction, we disregard *what is essential* to them *as signs* – what gives *life* to them, as Wittgenstein sometimes says[14] –, namely their being used, and seen in that way, petrification appears to be almost the opposite of idealization.

Let me mention another idea connected with the reduction of mathematics to set theory, an idea expressed by Bourbaki. It is the idea of an *horizon*. Sometimes people, and even Bourbakists themselves, mix it up with the idea of a foundation in the sense of a solid rock, but it is actually quite different. The 'horizon' of mathematics, as understood by Bourbaki, is a perfect formalization of mathematics that lies *in front of us* and that guides us; it is not *beneath* us like a rock that supports the mathematical edifice. An important part of the idea of such a mathematical horizon is that when actual mathematical thoughts and texts create doubts or suggest rectifications, then (as Bourbaki says)

> the process of rectification, sooner or later, invariably consists in the construction of texts which come closer and closer to a formalized text until, in the general opinion of mathematicians, it would be superfluous to go any further in this direction. (Bourbaki 1968: 8)

With "this direction" is meant the direction towards the horizon, and the horizon itself is considered to be the ideal endpoint of this process, when no further rectification is envisaged. This end has the form of axiomatic set theory.

At this point, however, Bourbaki isn't careful enough. The horizon in his sense should be a *text*, as he says, but this so-called text is very different from the texts actually produced by mathematicians. The latter ones are used texts whereas the horizon, as envisaged by Bourbaki, is not suitable for any use. It is, once more,

---

14 See, for example, BB: 3 – 5, and PI §§432 & 454.

a purely *mathematical structure* – a *metamathematical* structure – which only mimics a real text but is actually categorically different from it.

What is the purpose of the horizon? According to Bourbaki, the horizon provides the *standard of perfect rigour* which mathematicians have, or should have, before their eyes.[15] This idea mainly stems from Claude Chevalley, a Bourbakist of the first hour. Later, however, Chevalley retracted it, and it is instructive to quote him in this respect. His retraction can be found in a beautiful interview with Denis Guedj, given in 1985:

> At the level of mathematical logic, there's a point on which I am [now] totally separated from [the Bourbakists]. [...] It's what in Bourbaki one called the horizon; you describe the formal rules, but there's no way you can apply them systematically because it would take up too much space. However, these rules can at least ideally describe "a horizon", a perfect text from the standpoint of rigour. Now, in my opinion, that's not possible. It was in reading Castoriadis that I understood this impossibility. For example, the idea of a symbol which is "the same", although written in different places and at different times, is not at all an idea that stands by itself. But it must stand by itself if one has this conception, even purely theoretically, of mathematics. Not only can this idea not possibly be realized, but its content is absurd. A symbol cannot possibly be "the same" if it does not have an aura of signification. There, there is an appeal to something human that contradicts the idea of a perfect "horizon". (Guedj 1985: 22)

This passage is not perfectly clear, and in texts by Castoriadis I couldn't find satisfying clarifications. Nevertheless, what Chevalley says doesn't seem to me especially difficult. As I read him, he originally wanted this so-called 'horizon' to be a *perfect* text from the standpoint of rigour, and as such it should not be subject to the human aspects of our human practice. In Chevalley's own words: "the idea of a symbol which is 'the same', although written in different places and at different times" – that is, the idea concerning the *identity of the symbol* – should be "an idea that stands by itself" in order to belong to a perfect text. However, he now sees that this doesn't seem to be possible. He observes that "a symbol cannot possibly be 'the same' if it does not have an aura of signification" and that this sort of "appeal to something human contradicts the idea of a perfect 'horizon'". This is Chevalley's own, peculiar way of conveying his thought. It can be expressed, however, also in a more Wittgensteinian manner, for example as follows: Real signs belonging to real texts must be subject to the human aspects of our human practice because their identity depends on the actual use we make of them within this practice. Therefore

---

15 Referring to his whole series of volumes, his so-called *Elements of Mathematics*, Bourbaki expresses it thus: "written in accordance with the axiomatic method and keeping always in view, as it were on the horizon, the possibility of a complete formalization, our series lays claim to perfect rigour" (Bourbaki 1968: 12).

they cannot have the non-human 'perfection' Chevalley originally wanted from them.

What, then, about the idea of the horizon as a mathematical structure, that is, as a syntactical structure in the sense of metamathematics? I cannot imagine that Chevalley did not think of this possibility, but he doesn't mention it. As I understand him, he discards it because it is incapable of providing the desired "perfect text from the standpoint of rigour", as he says. For, what can be meant by that? According to Bourbaki, this text should serve as the *perfect standard for the correctness of what we do* in mathematics, a standard on which mathematicians should agree. But this cannot be accomplished by a mathematical structure as such. The horizon, as a mathematical structure, is 'too far way', so to speak, from our actual practice. To be more precise: it is categorically different from this practice, and we get into many of the difficulties uncovered by Wittgenstein in his rule-following considerations. Presumably, Chevalley himself didn't know these considerations, but Chevalley's *attitude*, as shown in the passage I quoted, can very well be supported by Wittgenstein's reflections. When, in PI §201, Wittgenstein talks about the "way of grasping a rule [...] which, from case to case of application, is exhibited in what we call 'following the rule' and 'going against it'", then this is precisely such an "appeal to something human that contradicts the idea of a perfect 'horizon'", as Chevalley says.

So, the idea of a perfect horizon is totally in order if it is meant in the sense of a syntactical structure belonging to genuine mathematics, to which our actual mathematics can be reduced. But it cannot serve the purpose of providing an ultimate standard for perfect rigour, as imagined by Bourbaki.

In the end, my result is that the idea of the horizon, beautiful as it may appear at first sight, suffers the same fate as the two ideas of a foundation that I discussed before. The envisaged aims of these foundations *and* of Bourbaki's horizon prove to be dubious. This is an end very much in accordance with Wittgenstein's considerations as published in his *Remarks on the Foundations of Mathematics*. I interpret the expression "Foundations of Mathematics" occurring in this title as referring mainly to the mathematicians' attempts at 'founding' mathematics, and what Wittgenstein says is mainly critical of these attempts.

What then *is* the point of the reductions of mathematics to set theory (and of other sorts of reductions that are on the market)? And what point may be seen from a Wittgensteinian perspective? These are questions to be dealt with at other occasions, and I'm not sure that there are Wittgensteinian answers to them.

Bibliography

Bourbaki, Nicolas (1968): *Elements of Mathematics, Theory of Sets*. Paris: Hermann/Reading, Mass.: Addison-Wesley.

Church, Alonzo (1956): *Introduction to Mathematical Logic*. Revised and enlarged edition. Princeton: Princeton University Press.

Emödy, Marianne (1994): "Empirismus und die Bedeutung mathematischer Zeichen." In: *Dialektik*. 1994/3, 43 – 58.

Friederich, Simon (2010): "Structuralism and Meta-Mathematics." In: *Erkenntnis*. Vol. 73, 67 – 81.

Gödel, Kurt (1931): "On formally undecidable propositions of *Principia mathematica* and related systems I.". Reprinted in: *Collected Works, Volume I, Publications 1929–1936*. Feferman, Solomon et al. (eds.). Oxford University Press, 1986, 145 – 195; page numbers according to this edition.

Guedj, Denis (1985): "Nicholas Bourbaki, Collective Mathematician: An Interview with Claude Chevalley." In: *The Mathematical Intelligencer*. Vol. 7, 18 – 22.

Hilbert, David and Bernays, Paul (1968): *Grundlagen der Mathematik I*. Berlin, Heidelberg, New York: Springer.

Kitcher, Philip (1984): *The Nature of Mathematical Knowledge*. Oxford: Oxford University Press.

Kitcher, Philip (2012): "Mathematical *Truth*?" In: *Preludes to Pragmatism*. Oxford: Oxford University Press, 166 – 191.

Kripke, Saul (1982): *Wittgenstein on Rules and Private Language*. Oxford, New York: Basil Blackwell.

Kripke, Saul (1992): "Logicism, Wittgenstein, and *de re* Beliefs about Numbers." Typescript of Kripke's *Alfred North Whitehead Lectures*. Harvard University, May 4–5, 1992.

Kripke, Saul (1997): "Logicism, Wittgenstein, and the Identification of Numbers." Typescript of Kripke's *Rosenkrantz Lectures*. UCLA, November 19 – 20, 1997.

Kripke, Saul (2011): "Wittgenstein, Logicism, and 'Buck-Stopping' Identifications of Numbers." https://www.youtube.com/watch?v=KE9m6BuORGI.

Kunen, Kenneth (1980): *Set Theory: An Introduction to Independence Proofs*. Amsterdam: Elsevier.

Maddy, Penelope (2017): "Set-theoretic Foundations." In: *Foundations of Mathematics*. Caicedo, E.; Cummings, J.; Koellner, P.; Larson, P.B. (eds.). American Mathematical Society, 289 – 315.

Mathias, A.R.D. (2002): "A Term of Length 4 523 659 424 929." In: *Synthese*. Vol. 133, 75 – 86.

Mühlhölzer, Felix (2006): "'A mathematical proof must be surveyable' - What Wittgenstein meant by this and what it implies." In: *Grazer Philosophische Studien*. Vol. 71, 57 – 86.

Mühlhölzer, Felix (2010): *Braucht die Mathematik eine Grundlegung? Ein Kommentar des Teils III von Wittgensteins BEMERKUNGEN ÜBER DIE GRUNDLAGEN DER MATHEMATIK*. Frankfurt am Main: Vittorio Klostermann.

Mühlhölzer, Felix (2012): "Wittgenstein and Metamathematics." In: *Wittgenstein: Zu Philosophie und Wissenschaft*. Stekeler-Weithofer, Pirmin (ed.). Hamburg: Felix Meiner, 103 – 128.

Mühlhölzer, Felix (2014a): "On Live and Dead Signs in Mathematics." In: *Formalism and Beyond. On the Nature of Mathematical Discourse*. Link, Godehard (ed.). Berlin: Walter de Gruyter, 183 – 208.

Mühlhölzer, Felix (2014b): "How Arithmetic Is About Numbers: A Wittgensteinian Perspective." In: *Grazer Philosophische Studien*. Vol. 89, 39 – 59.

Parsons, Charles (1986): "Review of Philip Kitcher's *The Nature of Mathematical Knowledge*." In: *The Philosophical Review*. Vol. 95, 129 – 137.

Putnam, Hilary (1979): "Analyticity and apriority: beyond Wittgenstein and Quine." Reprinted in: Putnam 1983, 115 – 138.

Putnam, Hilary (1980): "Models and reality." Reprinted in: Putnam 1983, 1 – 22.

Putnam, Hilary (1983): *Realism and Reason, Philosophical Papers, Vol. 3*. Cambridge: Cambridge University Press.

Shapiro, Stewart (1990): "Second-Order Logic, Foundations, and Rules." In: *The Journal of Philosophy*. Vol. 87, 234 – 261.

Shapiro, Stewart (1991): *Foundations without Foundationalism: A Case for Second-Order Logic*. Oxford: Oxford University Press.

Shapiro, Stewart (1997): *Philosophy of Mathematics: Structure and Ontology*. Oxford: Oxford University Press.

Wittgenstein, Ludwig (1969): *The Blue and Brown Books*. Oxford: Blackwell.

Wittgenstein, Ludwig (1978): *Remarks on the Foundations of Mathematics*. von Wright, G. H.; Rhees, Rush; Anscombe, G. E. M. (eds.), Anscombe, G. E. M. (trans.), revised edition. Oxford: Blackwell.

Wittgenstein, Ludwig (2000): *Manuscripts from Wittgenstein's Nachlass, published in the Bergen Electronic Edition*. Oxford: Oxford University Press. Cited with manuscript and page number according to this edition; translations are my own.

Wittgenstein, Ludwig (2009): *Philosophical Investigations*. (the German text, with an English translation by Anscombe, G. E. M., Hacker, P. M. S. and Schulte, Joachim), rev. 4th edition by P. M. S. Hacker and Joachim Schulte. Chichester: Wiley-Blackwell.

Jan von Plato
What Are the Axioms for Numbers and Who Invented Them?

Abstract: The Peano axioms of arithmetic were first published in 1889, in a rare Latin tract *Arithmetices Principia*. This original source became widely known by its inclusion fifty years ago in the collection *From Frege to Gödel*, edited by Van Heijenoort (cf. Heijenoort 1967). Ever since, the belief has been held by which "Peano (1891b) acknowledges that his axioms come from Dedekind (1888)," as the editor put his words. The ordering problems of who found the axioms are here resolved in Peano's favour.

1 Introduction

It is now fifty years since the publication of Jean van Heijenoort's *From Frege to Gödel: A Source Book in Mathematical Logic, 1879–1931*. That collection of original works with editorial comments was the result of a project that has shaped the understanding of the development of logic and foundations of mathematics for several generations of logicians and others. The only comparable contribution is the collection *The Undecidable: Basic Papers on Undecidable Propositions, Unsolvable Problems and Computable Functions*, prepared by Martin Davis and published two years earlier, but with a very specific scope as is seen already from the subtitle (cf. Davis 1965).

Van Heijenoort's editorial introductions of the work of Giuseppe Peano (1858 – 1932) are written under convictions we formulate succinctly as follows:

1. Dedekind invented the Peano axioms. Moreover, he invented the recursive definitions of the basic arithmetic operations.
2. Peano had no rules of inference. Moreover, the recursive definitions of the arithmetic operations in Peano's work are circular.

It is easy to find any amount of repercussions of these Van Heijenoortian theses in the literature. We shall give just one example, on the provenance of the Peano axioms.

Jan von Plato, University of Helsinki, jan.vonplato@helsinki.fi

DOI 10.1515/9783110657883-20

2 Peano's Deductive Machinery in the *Arithmetices Principia*

Peano published in 1889 a separate little treatise, the 36-page *Arithmetices Principia, Nova Methodo Exposita*, or "The principles of arithmetic, presented by a new method," (Arith. Princ.) in what follows. It was written in Latin and the earlier parts got an English translation in the Van Heijenoort collection in 1967. The original is readily available online and one sees that this booklet consists of a 16-page preface and presentation, and a 20-page systematic development that begins with § 1: On numbers and on addition. Peano writes in the introductory part (Van Heijenoort's translation, Heijenoort 1967: 85):

> I have denoted by signs all ideas that occur in the principles of arithmetic, so that every proposition is stated only by means of these signs. ...
> With these notations, every proposition assumes the form and the precision that equations have in algebra; from the propositions thus written other propositions are deduced, and in fact by procedures that are similar to those used in solving equations.

Peano's signs are, first of all, *dots* that are used in place of parentheses, and then: P for *proposition*, $a \cap b$, even abbreviated to ab, for *the simultaneous affirmation of the propositions a and b*, $-a$ for *negation*, $a \cup b$ for *or*, V for *truth*, and the same inverted for *falsity* Λ. The letter C inverted has a double use, one for *containment* between two classes, the other for consequence, as in today's stylized implication sign $a \supset b$. Even if it is read "deducitur" (one deduces) when used for consequence, it is clearly a connective, because it is found iterated. For example, Peano's second propositional axiom is:

$$a \supset b . b \supset c : \supset . a \supset c$$

There is also the connective of *equivalence*, $a = b$, definable through implication and conjunction as $a \supset b . \cap . b \supset a$.

Peano writes in the preface that he has followed in logic amongst others Boole, and for proofs in arithmetic Herman Grassmann's 1861 book *Lehrbuch der Arithmetik*, "in arithmeticae demonstrationibus usum sum libro: H. Grassmann." Grassmann's book presents an abstract approach to natural numbers, by a "basic sequence" that is generated from a ground element through the addition of a unity. The arithmetic operations are defined through recursion and their properties proved by induction, such as the associativity and commutativity of addition. Grassmann's approach was endorsed in textbooks, the best-known being Ernst Schröder's *Lehrbuch der Arithmetik und Algebra* of 1873. The "stroke symbol" for successor, as in the recursive definition $a + b' = (a + b)'$, was taken into use by Schröder.

There is no evidence of the influence of Frege, even if some of the initial statements about the ambiguity of language and the necessity to write propositions only in signs are very close to those in Frege's *Begriffsschrift*.

Peano says of definitions (Heijenoort 1967: 93):

> A *definition*, or *Def.* for short, is a proposition of the form $x = a$ or $\alpha \supset . x = a$, where a is an aggregate of signs having a known sense, x is a sign or aggregate of signs, hitherto without sense, and α is the condition under which the definition is given.

Pure logic is followed by a chapter on *classes*, or sets as one could say. The notation is $a \varepsilon b$ for *a is a b*, and $a \varepsilon K$ for *a is a class*.

When Peano proceeds to arithmetic, he first adds to the language the symbols *N* (*number*), 1 (*unity*), $a + 1$ (*a plus 1*), and $=$ (*is equal to*). The reader is warned that the same symbol is used also for logic. Next he gives the famous Peano axioms for the class *N* of natural numbers, nine in the original formulation. Four of the axioms are general principles about equality, namely axioms 2 to 6, the other five the essentially arithmetic axioms:

Table 1. Peano's Axioms for Natural Numbers
1. $1 \varepsilon N$
2. $a \varepsilon N . \supset . a = a$
3. $a, b \varepsilon N . \supset : a = b . = . b = a$
4. $a, b, c, \varepsilon N . \supset \therefore a = b . b = c : \supset . a = c$.
5. $a = b . b \varepsilon N : \supset . a \varepsilon N$.
6. $a \varepsilon N . \supset . a + 1 \varepsilon N$.
7. $a, b \varepsilon N . \supset : a = b . = . a + 1 = b + 1$.
8. $a \varepsilon N . \supset . a + 1 - = 1$.
9. $k \varepsilon K \therefore 1 \varepsilon K \therefore x \varepsilon N . x \varepsilon k : \supset_x . x + 1 \varepsilon k :: \supset . N \supset k$.

These axioms use the same sign for equality of numbers and equivalence of propositions, with axiom 7 standing out as one that has both in one formula. They also use the same symbol for implication and class containment, the latter as in the last inverted *C* of axiom 9.

Peano's basic notions give the natural numbers in a unary system that contains just the symbols 1 and +. The list of axioms is followed by a definition:

10. $2 = 1 + 1; 3 = 2 + 1, 4 = 3 + 1$; and so forth.

Peano's suggested definition contains the same defect as that of Grassmann, revealed by the elliptic "etc" or similar: namely, Peano is very clear about his basic concepts of unit 1 and successor +1 that produce expressions of the form $1 + \cdots + 1$, yet no way is given for inductively producing arbitrary decimal expressions from such expressions in the unary successor form, what would today be called a conversion algorithm from base one to base ten.

Now follows a list of theorems, the first one with a detailed proof:

11. $2 \, \varepsilon \, N$.

Proof.

P 1 . ⊃ :	$1 \, \varepsilon \, N$	(1)
1 [a] (P 6) . ⊃ :	$1 \, \varepsilon \, N . \supset . 1 + 1 \, \varepsilon \, N$	(2)
(1) (2) . ⊃ :	$1 + 1 \, \varepsilon \, N$	(3)
P 10 . ⊃ :	$2 = 1 + 1$	(4)
(4).(3).(2, 1 + 1) [a, b] (P 5): ⊃ :	$2 \, \varepsilon \, N$	(Theorem).

The justifications for each step are written at the head of each line so that they together imply the conclusion of the line. The derivation begins with P 1 in the antecedent, justification part of an implication, and $1 \, \varepsilon \, N$ in the consequent as the conclusion. The meaning is that from axiom P 1 follows $1 \, \varepsilon \, N$. The second line has similarly that from axiom P 6 with 1 substituted for a follows $1 \, \varepsilon \, N . \supset . 1 + 1 \, \varepsilon \, N$. The next line tells that from the previous lines (1) and (2) follows $1 + 1 \, \varepsilon \, N$. The following line tells that definition 10 gives 2=1+1. The last line tells that lines (4) and (3) give, by the substitution of 2 for a and 1+1 for b in axiom P 5, the conclusion $2 \, \varepsilon \, N$. The order in which (4) and (3) are listed is 2=1+1 and $1 + 1 \, \varepsilon \, N$. The instance of axiom P 5 is $2 = 1 + 1 . 1 + 1 \, \varepsilon \, N : \supset . 2 \, \varepsilon \, N$. Thus, we have quite formally in the justification part the expression:

$$(2 = 1 + 1) . (1 + 1 \, \varepsilon \, N) . (2 = 1 + 1 . 1 + 1 \, \varepsilon \, N : \supset . 2 \, \varepsilon \, N).$$

Line (3) is similar: It has two successive conditions in the justification part:

$$(1 \, \varepsilon \, N) . (1 \, \varepsilon \, N . \supset . 1 + 1 \, \varepsilon \, N)$$

There are altogether two instances of logical inference, both written so that the antecedent of an implication as well as the implication itself is in the justification part, and the consequent of the implication as the conclusion of the line. Each line of inference in Peano therefore has one of the two forms, with b a substitution instance of axiom a in the first:

$a \supset b.$
$a . a \supset b :\supset b.$

After the above theorem, there follow other very simple consequences about the equality relation, numbered 12–17. Then follows a section with theorems proved by induction, clearly ones suggested by those in Grassmann. Number 19 shows that natural numbers are closed with respect to addition, number 22 is a principle of replacement of equals in a sum, $a = b \supset a + c = b + c$, and 23 is the associative law. To arrive at commutativity, Peano proves first as 24 the lemma $1 + a = a + 1$, then finishes with 25 that is commutativity of addition to 28 that is replacement at both arguments in sums, $a = b \,.\, c = d \supset a + c = b + d$. The part on natural numbers is finished by sections on the recursive definition and basic properties of subtraction, multiplication, exponentiation, and division, all of it following Grassmann's order of things in definitions and theorems to be proved (§§ 2 – 6).

3 Dedekind and the Peano Axioms

Richard Dedekind (1831–1916) is known for two foundational contributions, the 1872 booklet *Stetigkeit und irrational Zahlen* (Continuity and Irrational Numbers), and another of 1888, the *Was sind und sollen die Zahlen?* (What Are the Numbers and What Are They For?). The latter is written in set-theoretic terms that have become standard in mathematics, even if Dedekind himself was thoroughly idealistic in his mathematical philosophy: he writes that sets are "collections of things" and the latter in turn "completely determined by everything that can be stated or thought about them" (Dedekind 1888: 1). Of the natural numbers, he writes in the preface as an answer posed by his title that "the natural numbers are creations of the human spirit" and that they "serve as means to conceive the distinctness of things [Verschiedenheit der Dinge] more easily and sharply."

The preface of Dedekind's second booklet is somewhat apologetic: he had worked on the foundations of arithmetic in the 1870s but had other duties. In the meanwhile, works by others on the natural numbers appeared of which he mentions Schröder's book of 1873, Kronecker's works, and von Helmholtz' essay of 1887 on counting and measuring, with the unreserved addition that his own approach had been "formed since many years and without any influence from whatever side." Four topics are listed and claimed as his proper main contributions (Dedekind 1888: viii, numbering added):

1. The sharp distinction into the finite and infinite.
2. The concept of the number of things [Anzahl].
3. That complete induction... really proves things.
4. That the definition by induction (or recursion) is determinate and consistent.

Dedekind's book introduces an abstract set-theoretic mode of thinking, with the basic notions of objects, sets, and mappings, conceived independently of Georg Cantor's set theory as he suggests.

As concerns the natural numbers, Dedekind's abstract way of conceiving them gives the following: Any set S with a mapping φ into S and a "ground element," designated by 1, generates a sequence of natural numbers by iteration, $1, \varphi(1), \varphi(\varphi(1)), \ldots$. Natural numbers are the "chain" or closure of the ground element relative to the mapping, designated by N. The properties Dedekind lists are four, here in a modern terminology (Dedekind 1888: 16):

Table 2. Dedekind's Four Postulates of 1888
α $\varphi(N)$ is a subset of N.
β   The closure of the iteration gives N.
γ   The ground element 1 is not in $\varphi(N)$.
δ   The mapping φ is injective.

The set N is infinite by the two criteria γ and δ Dedekind had laid down, what came to be known as "Dedekind's infinity axioms." These properties for Peano's primitive, the successor mapping, are covered by axioms 7 and 8 of Peano's list: One direction of axiom 7 can be seen as a principle of replacement of equals $a = b$ in the successor function $a + 1$. The remaining direction expresses the injectivity of the successor function. Peano's axiom 8 states that $a + 1 \neq 1$ by which the image of N under the successor function mapping is a proper subset of N.

In the preface part of Arith. Princ., Peano mentions Dedekind's *Was sind und was sollen die Zahlen?* of 1888 as a "recent script in which questions that pertain to the foundation of numbers are acutely examined."

In Van Heijenoort's words (Heijenoort 1967: 83):

> Peano (1891b) acknowledges that his axioms come from Dedekind (1888), §71, definition of a simply infinite system.

What are we to make out of this suggested acknowledgement in Peano's 1891 paper "Sul concetto di numero"? Here are some facts and observations:

Dedekind's 1888 booklet has a preface dated in October 1887. One can presume that it appeared some time in the earlier parts of 1888, and that it took some time for Peano to have the tract available. Peano's first writings on logical matters were:

1. The logical preliminaries to his book on the *Calcolo Geometrico* of 1888. The book itself was a contribution to Hermann Grassmann's vectorial way of looking at geometry. The logical preliminary consists of some twenty pages of equivalences in classical propositional logic, say, De Morgan's laws, and the predicament that logic consists in such equivalences and their continuous use in rational reasoning.

2. The next year, 1889, Peano published two booklets, the first one the *Arithmetices Principia*, published by the Bocca brothers of Turin (Fratelli Bocca Editrice di Torino). The second booklet is the Italian *I Principii di Geometria, Logicamente Esposta*, or "The Principles of Geometry, Logically Exposed." This latter booklet

has a preface dated June 1889 and the admission that Arith. Princ. was a "preceding opuscule." A barest look at Arith. Princ. shows that its main text of 20 pages after the introduction contains hardly a single word; it's all a strictly formal symbolic development of elementary arithmetic. This formal development is, compared to the verbose logical prose in the *Calcolo Geometrico* of 1888, a very strict endeavour, and one that must have taken all of Peano's time.

3. In 1891, Peano started the journal *Rivista di Matematica*, meant at least initially mainly for school teachers. The first issue begins with his article *Principii di logica matematica* that explains the symbolic notation, propositional logic, and the idea of expressing mathematics in the symbolism. There follows another article with the title *Formole di logica matematica* that contains an improved axiomatization of propositional logic. A third article is the "Sul concetto di numero" to which Van Heijenoort refers.

Dedekind, in contrast to Peano, was an idealist to whom the objects of mathematics existed in the mathematician's mind. His scheme of things had little place for a language, not to speak of a strict formal syntax as in Peano. The places are numerous in which Peano expresses his faith in the use of symbols: A mathematical notation wrought purely in symbols is the only way of divesting mathematics from the equivocality of ordinary language. Peano, much more than Frege or Hilbert, endorsed a formalistic approach in which there is a language of symbols, operations, and relations, with a formal development of the combinatorial structure that results from the syntactic stipulations. Nothing of the kind is found in Dedekind.

Here is a concrete expression of the difference between an idealist with a direct access to the realm of the objects of mathematics, and the formalist who has to take recourse to mere expressions in an artificially constructed language: We consider the seventh Peano axiom. In Peano, it is an equivalence by which $a + 1 = b + 1$ if and only if $a = b$. The two parts are that $a + 1 = b + 1$ implies $a = b$ and that $a = b$ implies $a + 1 = b + 1$. We don't find the latter in Dedekind, and why is that? It is natural for one concerned with a language to see to it that two different expressions for the same object can be replaced everywhere. It is equally natural for a pure extensionalist like Dedekind to think that no such explicit principle is needed.

Now we come to Peano's article "Sul concetto di numero" and Peano's reference to Dedekind. In the article, the Peano axioms are given as (Peano 1891c: 90):

Table 3. Peano's 1891 Axioms for Natural Numbers
1. $1 \varepsilon N$
2. $+ \varepsilon N \backslash N$
3. $a, b \varepsilon N . a+ = b+ : \supset . a = b$
4. $1 - \varepsilon N+$
5. $s \varepsilon K . 1 \varepsilon s . s+ \supset s : \supset . N \supset s.$

The infix backslash notation in axiom 2 is suggested, as Peano explains, by Dedekind's general theory of operations: + belongs to the class of operations from N to N. With operations, "some properties of numbers can be made dependent on more general ones and treated in a more concise form" (Peano 1891c: 87). The fourth axiom states that 1 is not a member in the class $N+$. The fifth shows how Peano uses the same symbol for implication and class containment, with the class $s+$ contained in s, and in the consequent, N contained in s, to be read "every N is an s." The notation also shows the novelty mentioned above, the result of the successor operation written as $a+$, with the equation $a+ = a + 1$ coming out as a theorem from a proper recursive definition of sum.

The passage to which Van Heijenoort refers is Heijenoort 1967: 93:

> The preceding primitive propositions [axioms of table 3] are due to Dedekind, op. cit. n. 71; there is, however, a mild difference in the enunciation of our proposition 5 (which is Dedekind's β) on which we don't stop here. They are identical in substance to those I had posed in the *Arith. Princ.*, save that the introduction of the sign \ allows to simplify their form.

Initially, the timing of things, and the labour of working out the formal details of *Arith. Princ.*, left me very dubious about any direct use of Dedekind's axioms in Peano. In the passage of 1891, Peano is very generous toward Dedekind, as a comparison of Tables 2 and 3 at once shows. Peano reformulated in 1891 his axioms in terms of Dedekind's theory of operations and as he writes, this formulation led him to prove the independence of the axioms. The above passage continues, (Heijenoort 1967: 93):

> These propositions express the necessary and sufficient conditions for the objects of a system to correspond univocally to the series of the N; and they can be enunciated also as follows:
> 1. The name 1 is given to a particular object of the system.
> 2. Let an operation be defined for which there corresponds to every object a of the system another, $a+$, even that in the system.
> 3. And that two objects, the correspondents of which are equal, be equal.
> 4. The object called 1 shall not be the correspondent of any.
> 5. And finally that it be the class common to all classes s that contain the individual 1 and that, when they contain an individual, they also contain its correspondent.
> It is easy to see that these conditions are independent.

Browsing further in the *Rivista*, one finds in volume VI of 1899 Peano's notes on the *Formulario* project with the following passage (Peano 1899: 85):

> The composition of my work of the year 1889 was still independent of the mentioned script of Dedekind; I had, before the printing, the moral proof of the independence of the primitive propositions from which I began, those with the substantial coincidence with the definitions of Dedekind. Later I succeeded in proving the independence.

In the 1889 booklet on the foundations of geometry, we find a similar admission that the independence of the geometric axioms is a "moral certainty."

Peano's last exposition of his arithmetic was in the fifth edition of the *Formulario Mathematico* of 1908, written in his own invented language "Latino sine Flexione." On page 15, he explains:

> We prove that a system of primitive propositions is mutually independent, in an absolute way, if we adduce, for each proposition, an interpretation of the system of primitive ideas that satisfies each primitive proposition except the one considered.

Such proofs for the Peano axioms are given on page 27 of the *Formulario*.

Dedekind's contribution to Peano's developments was to make him see how the axioms he had invented in 1889 can be formulated abstractly in a general framework of operations, with the possibility to interpret the formalism in whatever way, and the consequent possibility of interpretations that validate all of the axioms except a chosen one. The mentioning of Dedekind's 1888 script in the *Arith. Princ.* was just a late addition to its preface without effect on its content.

To end this discussion of the provenance of Peano's axioms, I shall now give the promised example of the effect of Van Heijenoort's reading of Peano:

> We know so well the natural number system from Peano's five axioms, published by him in [1889] and [1891]. In fact, as Peano acknowledged in [1891, 93], these axioms come from the definition of a simply infinite system in Dedekind, "Was sind und was sollen die Zahlen?" [1888].

This passage is found among the opening sentences of Kleene's 1981 essay *The theory of recursive functions approaching its centennial*. There are reasons to think it very unlikely that Kleene had read Peano's article, among them that it was published in late 19th century in an obscure journal in Italian. It would have been useful if Kleene had stated where his assessment came from, but there is no mentioning of Van Heijenoort.

Being influenced by someone's reading is one thing. Kleene's total ignorance of the development of arithmetic on the basis of recursive definitions prior to Peano and Dedekind, instead, is hard to fathom (Kleene 1981: 44):

> Under Dedekind and Peano's treatment, the natural numbers constitute the system of objects obtained by starting with an object 0 ("zero"), and repeatedly generating a next object by an operation ' ("successor" or "+").

Dedekind refers on his first page to Schröder and von Helmholtz; what they contain we shall soon see. In some fifty years of work with recursiveness, Kleene does not seem to have looked at these sources, nor at Grassmann whom Peano identifies as the one who originated recursive arithmetic. Had Kleene read the passage on Grassmann in Peano (Peano 1819b: 96), three pages ahead from the page he cites, or pretends to cite, he would have found out the true prehistory of computability. Now instead he presented a presumed historical fact out of context, as if he had discovered it by himself; an attempt by a logician to contribute to the history of the field without proper training in the methodology of the history of exact sciences.

4 Recursive Definitions in Dedekind

It is common to read that Dedekind invented the recursive definition of the arithmetic operations in his 1888 booklet. Next to Kleene, one example is Van Heijenoort's 1985 essay on Herbrand's logical work, by which "at the beginning of the 1920s primitive recursive functions, introduced by Dedekind as early as 1888, had become the very paradigm of a computable function" (Heijenoort 1985: 113). One arrives at such a view, apart from nationalistic tendencies that hardly apply in this case, from Dedekind's claim that he has "not been influenced by anyone in any way," in combination with the presentation without any reference to anyone of the recursive definition of addition, multiplication, and exponentiation towards the end of the booklet.

The list of Dedekind's essential novelties of 1888, given above, does not contain the foundation of arithmetic on recursive definitions. Indeed, Dedekind refers on the first page of the preface for his book to two sources that contain it: Hermann von Helmholtz' article *Zählen und Messen* of 1887, and Ernst Schröder's *Lehrbuch der Arithmetik und Algebra* of 1873 mentioned already above. Dedekind's stroke notation for the successor comes from Schröder. Behind these presentations there is Hermann Grassmann who in his textbook of 1861 developed arithmetic extensively on the basis of recursive definitions. All of this is detailed out in my "In search of the roots of formal computation" (von Plato 2016, also von Plato 2017).

The first steps in Grassmann consisted of inductive proofs of the associativity and commutativity of addition. These properties had been postulated as axioms in the earlier literature. Discussing this matter, Peano writes in his 1891 essay on the concept of number (Peano 1891c: 96):

> The rigorous proofs of these properties that we have reported are due to Grassmann (1861). They were then repeated by Hankel 1867, Peirce, Dedekind, etc, the last one of which enunciated even the principle of mathematical induction, of which the others made use of "as a known way of inference," without explicit enunciation.

Hankel was the first one to explain Grassmann's method, followed by Schröer 1873 and Helmholtz 1887, but neither Peirce 1881 nor Dedekind 1888 mention Grassmann. In the former, recursion works on the first term of a sum, something Dedekind proves as a lemma, but Peano states very clearly that these two authors were just repeating Grassmann. Peano is still giving generous credit to Dedekind, here for his abstract formulation of the induction principle freed of the specifics of the sequence of natural numbers.

The recursive definition of functions gets in Dedekind the following general formulation, with Z_n the set of the first n natural numbers, as in the above generation process (§9): Given any set Ω with an element ω and a mapping θ to Ω, there is for every n a mapping ψ_n from Z_n to Ω such that:

  I $\psi_n(Z_n)$ is a subset of Ω,
  II $\psi_n(1) = \omega$,
  III $\psi_n(t') = \theta\psi_n(t)$ for $t < n$.

This scheme for ψ_n is formulated as a theorem with an inductive proof by n. Next a more general formulation is given, the "theorem of definition by induction" numbered as the paragraph 126 that for any mapping of a set Ω to itself and distinguished element states the existence of a unique function from a chain N to Ω with the above three properties.

Dedekind had written a first version of his 1888 tract in the 1870s and distributed it to some extent. One such early version has been reproduced in Dugac 1976. Large parts of the published form of *Was sind und was sollen die Zahlen?* are found included verbatim there, but there is no trace of the recursive definition of the arithmetic operations.

5 Rules of Inference and Recursive Definitions in Peano

From the derivations in Peano's treatise, the following structure emerges:

Peano's formal derivations consist of a succession of formulas that are:

(i) *Implications in which an axiom implies its instance.*
(ii) *Implications in which previously derived formulas a and a ⊃ b imply b.*

Russell took over *verbatim* this structure of formal derivations in his 1906 article 'The Theory of Implication'.

Peano likened his propositions to the equations of algebra and his deductions to the solving of the equations. Rather startlingly, Van Heijenoort, instead of figuring out what Peano's notation for derivations means, claims in his introduction (Heijenoort 1967: 84) that there is "a grave defect. The formulas are simply listed, not derived; and they could not be derived, because no rules of inference are given... he does not have any rule that would play the role of the rule of detachment." Had he not seen the forms $a \supset b$ and $a \mathbin{.} a \supset b \mathbin{:}\supset b$ in Peano's derivations, the typographical display of steps of axiom instances and implication eliminations with the conclusion b standing out at right, and the rigorous rule of combining the antecedent of each two-premiss derivation step from previously concluded formulas?

Van Heijenoort's unfortunate assessment, and it becomes much worse if one reads further, has undermined the view of Peano's contribution for a long time, when instead Peano's derivations are constructed purely formally, with a notation as explicit as one can desire, by the application of axiom instances and implication eliminations. These rules hit the eye of anyone who reads Peano with any attention. One witness is Kurt Gödel, a meticulous reader of Peano. His notes on the *Formulaire* are found in one of his "Excerptenhefte" where he lists Peano's propositional axioms, then writes: "Rules: implication and substitution of equals (not formulated but used)."[1]

Van Heijenoort writes about Peano's definition of addition and multiplication (Heijenoort 1967: 83):

> Peano ... puts them under the heading "Definitions" although they do not satisfy his own statement on that score, namely, that the right side of a definitional equation is "an aggregate of signs having a known meaning".

When introducing his primitive signs for arithmetic, Peano enlisted *unity*, notation 1, and *a plus* 1, notation $a + 1$. Thus, the sum of two numbers was not a basic notion, but just the successor, and Peano's definition 18 lays down what the addition of a successor $b + 1$ to another number means, in terms of his primitive notions:
    18. $a, b \, \varepsilon N. \supset . a + (b + 1) = (a + b) + 1$.

Peano notes (Heijenoort 1967: 95):

> *Note.* This definition has to be read as follows: if a and b are numbers, and if $(a + b) + 1$ has a sense (that is, if $a + b$ is a number) but $a + (b + 1)$ has not yet been defined, then $a + (b + 1)$ signifies the number that follows $a + b$.

---

[1] Found on frame 515 left part, reel 20 of the Gödel microfilm collection, written in a mixture of German and Gabelsberger shorthand. Transcription in German is: Regeln: Implikation und Einsetzung für gleiche (nicht formuliert aber angewendet).

If, as Peano assumes, $a + b$ is a number, i.e., if $a + b \, \varepsilon \, N$, then even $(a + b) + 1 \, \varepsilon \, N$, so the definiens "has a meaning" as Peano writes, and one really wonders what Van Heijenoort may have been thinking here. Peano's definition is followed by his theorem 19 by which the class of natural numbers is closed with respect to sum: $a, b \, \varepsilon N. \supset . a + b \, \varepsilon N$. A detailed proof by induction on the second summand is given. Peano's misfortune was perhaps to use the same notation for the operation of a successor and for an arbitrary sum. This feature was soon corrected, when from 1891 on Peano wrote the successor as $a+$.

Bibliography

Davis, M. (1965): *The Undecidable. Basic Papers on Undecidable Propositions, Unsolvable Problems and Computable Functions.* Hewlett (NY): Raven Press.
Dedekind, R. (1872): *Stetigkeit und Irrationale Zahlen.* Vieweg Reprint, 1969.
Dedekind, R. (1888): *Was sind und was sollen die Zahlen?* Vieweg Reprint, 1969.
Dugac, R. (1976): *Richard Dedekind et les fondements des mathématiques.* Paris: Vrin.
Frege, G. (1879): *Begriffsschrift, eine nach der arithmetischen nachgebildete Formelsprache des reinen Denkens.* Halle: Nebert.
Grassmann, H. (1861): *Lehrbuch der Arithmetik für höhere Lehranstalten.* Berlin: Enslin.
Hankel, H. (1867): *Vorlesungen über die complexen Zahlen und ihre Functionen I.* Leipzig: Leopold Voss.
van Heijenoort, J. (ed.) (1967): *From Frege to Gödel, A Source Book in Mathematical Logic, 1879–1931.* Cambridge (MA): Harvard University Press.
van Heijenoort, J. (1985): "Jacques Herbrand's work in logic and its historical context." In: Van Heijenoort's *Selected Essays.* Naples: Bibliopolis, 99 – 121.
von Helmholtz, H. (1887): "Über Zählen und Messen, erkenntnistheoretisch betrachtet." In: *Philosophische Aufsätze, Eduard Zeller zu seinem fünfzigjährigen Doctorjubiläum gewidmet.* Leipzig: Fues' Verlag, 17 – 52.
Kleene, S. (1981): "The Theory of Recursive Functions Approaching Its Centennial." In: *Bulletin of the American Mathematical Society.* Vol. 5, 43 – 61.
Peano, G. (1888): *Calcolo Geometrico Secondo l'Ausdehnungslehre di H. Grassmann, Preceduto dalle Operazioni della Logica Deduttiva.* Turin: Fratelli Bocca Editori.
Peano, G. (1889): *Arithmetices Principia, Nova Methodo Exposita.* Fratelli Bocca Editrice, Turin. Partial English translation in Van Heijenoort (1967).
Peano, G. (1889): *I Principii di Geometria, Logicamente Esposta.* Turin: Fratelli Bocca Editrice.
Peano, G. (1891): "Principii di logica matematica." *Rivista di Matematica.* Vol. I, 1 – 10.
Peano, G. (1891): "Formole di logica matematica." *Rivista di Matematica.* Vol. I, 24 – 31.
Peano, G. (1891): "Sul concetto di numero." In: *Rivista di Matematica.* Vol. I, 87 – 102.
Peano, G. (1899): "Sul §2 del Formulario. Vol II: *Aritmetica.*" In: *Rivista di Matematica.* Vol. 6, 75 – 89.
Peano, G. (1901): *Formulaire de Mathématiques.* Paris: Carré et Naud.
Peano, G. (1908): *Formulario de Mathematico.* Turin: Fratelli Bocca Editori.
Peirce, C (1881): "On the Logic of Number." In: *American Journal of Mathematics.* Vol. 4, 85 – 95.

von Plato, J. (2016): "In Search of the Roots of Formal Computation." In: *History and Philosophy of Computing*. Gadducci, F.; Tavosanis, M. (eds.), Cham (Switzerland): Springer, 300 – 320.

von Plato, J. (2017): *The Great Machinery Works: Theories of Deduction and Computation at the Origins of the Digital Age*. Princeton, London: Princeton University Press.

Russell, B. (1906): "The Theory of Implication." In: *American Journal of Mathematics*, Vol. 28, 159 – 202.

Schröder, E. (1873) *Lehrbuch der Arithmetik und Algebra für Lehrer und Studirende. Erster Band. Die Sieben algebraischen Operationen*. Leipzig: Teubner.

Part III: **Wittgenstein**

Mathieu Marion and Mitsuhiro Okada
Following a Rule: Waismann's Variation

Abstract: We reconstruct a variation by Friedrich Waismann of Wittgenstein's rule-following argument, based on what we call the 'guessing game' (BB: 112 and PI§§151 and 179), and contrast it with Kripke's case of a deviant pupil (PI §§143 and 185). Our reconstruction follows Waismann's reliance on the cause-reason distinction, and it is completed by an explanation of what it means for the 'chain of reasons' to have an end, beyond which one can only appeal to causes. To conclude, we identify the contemporary debate on 'blind reasoning' as an area where Waismann's variation could play a role.

In this paper, we wish to present a reconstruction of what we call a 'variation' by Friedrich Waismann of Wittgenstein's arguments on following a rule, based on chapter VI of *The Principles of Linguistic Philosophy* (PLP) and §§10 – 11 of his lectures on causality (C). Although we quote Wittgenstein extensively for obvious reasons, especially *The Blue and Brown Books* (BB), on which Waismann relies heavily, our aim is not to set up this variation in order to provide a reading of *Philosophical Investigations* (PI §§143 – 242). We will at any rate say a few words about this in sections 3 and 4. We therefore avoid discussing the secondary literature, using only on occasions Saul Kripke's *Wittgenstein on Rules and Private Language* (Kripke 1982) as a foil. Our goal is, minimally, to provide the *prima facie* case for extracting a variation from PLP and C.

We motivate our approach in section 1 and then present the variation as such in section 2, as based on the distinction between causes and reasons, and centred on what we call the 'guessing game'. In section 3 we complete our presentation, with an argument based on the idea that 'the chain of reasons has an end'. We conclude with brief remarks in section 4 intended to identify an area, within contemporary debates on the epistemology of logic, where Waismann's variation could potentially play a role.

Mathieu Marion, Département de philosophie, Université du Québec à Montréal, marion.mathieu@uqam.ca
Mitsuhiro Okada, Department of philosophy, Keio University, mitsu@abelard.flet.keio.ac.jp

DOI 10.1515/9783110657883-21

1 Wittgenstein and Waismann: Co-operation and Variation

The story of Waismann's co-operation with Wittgenstein is sufficiently well known, we need only recall a few points.[1] As early as 1929, possibly earlier, Waismann was asked to write an account of Wittgenstein's philosophy entitled *Logik, Sprache, Philosophie*, which was later on intended to appear as the first volume of *Schriften zur wissenschaftlichen Weltauffassung*. Wittgenstein was for a while meant to co-author the book, and gave Waismann access to his manuscripts and typescripts, including BB – an important fact given that our case largely relies on parallels between BB and C. After Schlick's death in 1936, that project was abandoned and Waismann decided to finish the book on his own, with Margaret Paul translating it into English prior to the war. Alas both originals (German and English translation) were lost, but an edition based on the galley proofs of the English translation appeared six years after his death in 1965 (PLP), while a reconstruction of the German original appeared in 1976 (Waismann 1976).

Co-operation with Wittgenstein generated a lot of material in Waismann's *Nachlass*,[2] from the shorthand notes of conversations with Schlick and Wittgenstein (1929–1932), published in 1967 as *Ludwig Wittgenstein und der Wiener Kreis*[3], and a short account of Wittgenstein's philosophy, *Thesen*, published as Appendix B to that book, to the voluminous dictations now published as *The Voices of Wittgenstein* (Wittgenstein et al. 2003). Waismann also acted as Wittgenstein's spokesman at the well-known Königsberg meeting (Waismann 1982), and he published in the 1930s papers translated and collected in *Philosophical Papers* (Waismann 1977), as well as *Einführung in das mathematische Denken* (1936). Although this book and the papers on probability (1930) and on identity (1936) show Wittgenstein's influence, Waismann nevertheless established his credentials as a philosopher on his own right. This independence became much more pronounced in his post-war writings and lectures at Oxford. The lecture notes on causality discussed here are undated, but internal evidence show that they are from the late 1940s (Marion 2011: 32). Not only PLP cannot be fully identified with a mere presentation of Wittgenstein, but our case relies, as we pointed out, on drawing parallels between BB and C, thus with a text from this later period, where Waismann is very much his own man.

---

[1] For further details, see Hampshire 1960, Quinton 1977, Baker et al. 1976, McGuinness 1979, Baker 1979, Baker 1997, Baker 2003 and McGuinness 2011b.
[2] See Schulte 1979.
[3] Translated in English as Wittgenstein 1979.

The main target of C §§10 – 11 is Wolfgang Köhler's claim in Chapter X of *Gestalt Psychology* – whose title happens to be "Insight" – that one can have an immediate awareness or 'insight' into the connexion between cause and effect, so that one does not need to observe regularities to establish a causal link (Köhler 1930: 270f.).[4] For example, Köhler claims that when listening to an alto at the concert-hall, he was directly aware of – or had an 'insight' into – the cause of his admiration, namely the alto's singing. Against this, Waismann argues in his lectures that Köhler had confused in his description the 'object' of his admiration with the 'cause' of his admiration, and he points out that there is no 'logical connection' between the singing and his admiration, since it may have happened that he took a dose of mescal before going to the concert, and that we might then very well consider this to be the cause of his admiration (C: 169). This is not, as we shall see, entirely unrelated to Wittgenstein's concerns in BB, but the context – a discussion of determinism and causality – is very much Waismann's.

Attitudes towards the material that resulted from co-operation with Wittgenstein range from seeing Waismann as providing an authoritative exposition[5] of Wittgenstein (at least for the period 1928–1936) to scepticism about this, given that a close examination reveals many divergences.[6] We need not take a stance on this issue. On May 19, 1936, Wittgenstein wrote a letter to Waismann remonstrating him for insufficient acknowledgement in his paper on identity, given that its essential idea was Wittgenstein's. The tone of Wittgenstein's letter is nevertheless one of benevolence, as he was anxious also to provide Waismann with room to maneuver:

> When a composer A has written variations on a theme by composer B, he does not write: "for this piece of music I have received valuable suggestions from B", but he writes: "Theme by B". [...] in saying that the theme is by B, nothing is said about the *value* of the variations; it can be as valuable, indeed even more valuable than the theme itself.[7]

We think that this use of the musical concept of 'variation' as a simile also gives us room to think of Waismann as providing what is neither an authoritative interpre-

---

4 Waismann gives numerous examples (C: 164 – 165).
5 See Baker 1997: xix and Baker 2003: xvii, xxxiii & xli.
6 See Schulte's searching analyses in Schulte 2011.
7 The full passage, reproduced from the electronic edition Wittgenstein 1998, reads: "Zuerst ein Gleichnis: Wenn der Komponist A Variationen über ein Thema des Komponisten B geschrieben hat, so heißt es nicht /schreibt er nicht/ : 'zu diesem Musikstück habe ich wertvolle Anregungen von B erhalten', sondern er schreibt: 'Thema von B'. Obwohl man ja das Thema als eine Anregung zu den Variationen bezeichnen kann. – Ferner mit der Feststellung, das Thema sei von B, ist nichts über den Wert der Variationen gesagt; dieser kann ebenso groß, ja größer sein, als der des Themas".

tation, nor something else of little or no value. Saul Kripke's notorious reading of PI §§143 – 242 could also be seen as a variation, with its own intrinsic value, but it is perhaps too far from Wittgenstein's 'theme' to count as a 'variation', while we would claim that Waismann's remains close enough to it.

2 The Cause/Reason Distinction and the Guessing Game

There are strong parallels between a passage at BB: 12 – 15 and both PLP, chapter VI, §5, and C §§10 – 11, that tell us how Waismann understood that passage, as he focussed on the distinction between 'cause' and 'reason' – he also calls the latter 'ground'. After all, that passage is often seen as the *locus classicus* for that distinction within the analytic tradition.[8] In PLP, chapter VI, §5, Waismann uses that distinction to argue against the 'causal interpretation of language' (ostensibly the sort of view argued for by Russell in *Analysis of Mind* or Ogden & Richards in *The Meaning of Meaning*). For example, he argues that giving the reason for an action justifies it, while giving the cause does not (PLP: 121 – 122). But Waismann's argument is part of his defense of the conception of 'language as calculus' – which he introduces immediately afterwards in §6 – and it is trivial to point out that the later Wittgenstein moved away from it. Furthermore, Waismann seems to overstate his case, linking it to what is now known as the 'logical connection argument':[9]

> The causal interpretation of language is due to a confusion of *logical* and *causal* consequences of a command, the expression of a wish, the statement of a fact, etc. Suppose an engine-driver is asked 'Why do you stop here?' and answers 'Because the signal was at "Stop!"'. This answer is mistakenly regarded as stating a cause, when in fact it states a reason. (PLP: 121)

For these reasons, although we will note parallels, it is better to steer clear and rely mainly on the lecture notes on causality.[10] In these lectures, the distinction is illustrated thus:

> Imagine someone writing down various figures while he does a sum. When asked why he wrote just these particular figures, he may reply in two different ways. He may say "You see,

---

8 Davidson is often read as having undermined this distinction in Davidson 1963, but it has recently become the topic of more scrutiny, including from Wittgenstein's standpoint. See Schroeder 2012, Stoutland 2010, Glock 2013, and Sandis 2015.
9 On the 'logical connection argument', see Stoutland 1970.
10 As we just saw, he still remains close to the 'logical connection argument' in his lectures.

I was adding these numbers and, in doing this, I followed such and such a rule". He then states the *reason* for his behaviour. Or he might have said "In my brain processes of such and such a kind were going on which innervated the muscles of my fingers in such a way that they made movements so as to write down these figures." Then he states the *cause* of his action. (C: 171)

In order to argue for the distinction, Waismann musters four points: (1) a causal explanation will appeal to processes situated in time, while an explanation invoking reasons will refer to timeless entities such as rules (C: 170); (2) contrary to grounds, causes cannot be appealed to in order to justify an action (C: 171); (3) contrary to causes, reasons cannot be discovered by observation (C: 172); and (4) causes and reasons are involved in different ways in the process of learning to follow rules (C: 172). We shall not discuss further the first argument, and concentrate on the other three.[11] We just saw that (2) is already in PLP, so is (3) in PLP: 121. Note that in C §§10 – 11, Waismann also distinguishes between 'reasons' and 'motives' in an interesting way, but we need not delve into this.

We should first note that Waismann remains at all times rather close to Wittgenstein's text. We can see this by comparing his argument for (4) with the *Blue Book*:

> A. The teaching is a drill. [...] The drill of teaching could [...] be said to have built up a psychical mechanism. This, however, would only be a hypothesis or else a metaphor. We could compare teaching with installing an electric connection between a switch and a bulb [...]
> In so far as the teaching bring about the association, feeling of recognition, etc. etc., it is the cause of the phenomena of understanding, obeying, etc.; and it is an hypothesis that the process of teaching should be needed in order to bring about these effects [...]
> B. The teaching may have supplied with a rule which is itself involved in the process of understanding, obeying, etc. (BB: 13)

> So we must distinguish between *ground* and *cause*, for we learn of both in different ways. The *cause* for his writing down certain figures may lie in the fact that he was taught so in school and that this teaching has created a disposition, e.g. left definite traces in his nervous system and his brain; the *ground* for his procedure is the *rule* which he states when asked for the ground. (C: 172)

What is of interest to us is the fact that, in parallel with BB: 13 to be quoted below, Waismann introduces into his discussion a case of rule-following in PLP: 120 – 121 and C: 171 – 173 in order to argue for (3):

---

[11] But one should note that Wittgenstein argues in the first part of *Remarks on the Foundations of Mathematics* that internal relations are 'non-temporal' (Wittgenstein 1978 I, §101 – 105).

> Let us imagine that someone writes on a board the numbers 0, 1, 4, 9, 16 in this order. We, watching him, may suppose that, in doing this, he is following a definite rule, e.g., that he is writing down the squares of the integers in order. Have we found out this rule by observation? Not at all; our supposed rule is merely a hypothesis, which would account for the numbers he has actually written down. But the figures written down are always subsumable under an infinite number of mathematical laws. How are we now to tell which rule he in fact followed? By making him continue the figures? But even if he wrote a thousand figures, he still might have been following any one of an infinite number of rules. [...] It is quite different if he tells which rule he has been following. Suppose he says 'I have been using the formula $y = x^2$, and I have substituted for x the first 5 integers 0, 1, 2, 3, 4.' The expression 'the rule he is following' has now altered its meaning. In this latter sense the rule is determined by what the calculator *says*, not by observation of the figures which he is writing down; though these may help us to *guess* the rule. (C: 172)

In order to explain how a variation is implicated here, we would like to point out that in PI §§143 – 242, Wittgenstein discusses rule-following with help of not one but two cases. Kripke relies on the notoriously case at §§143 & 185 of the pupil, who, having been asked to follow the rule '+2', goes on adding '+4' after 1000:

$$2 \ 4 \ 6 \ \ldots \ 996 \ 998 \ 1000 \ 1004 \ 1008 \ \ldots$$

But one should not overlook that there is a second case at PI §§151 & 179, which is in fact the language-game of §§62 – 64 of the *Brown Book*:

> Let the game be this: A writes down a row of numbers. B watches him and tries to find a system in the sequence of these numbers. When he has done so he says: "Now I can go on". This example is particularly instructive because 'being able to go on' here seems to be something setting in suddenly in the form of a clearly outlined event. – Suppose then that A had written down the row 1, 5, 11, 19, 29. At that point B shouts "Now I can go on". What was it that happened when suddenly he saw how to go on? A great many things might have happened. Let us assume then that in the present case, while A wrote one number after the other, B busied himself with trying out several algebraic formulae to see whether they fitted. When A had written "19" B had been led to try the formula $a_n = n^2 + n - 1$. A's writing 29 confirms his guess. (BB: 112)

Let A be called 'Smith' and B called 'Jones'. We can speak here of a 'guessing game': Smith writes down on the blackboard the initial segment of a series and Jones must try and guess which algebraic formula/rule Smith is following. We could express the essential difference between the case of the deviant pupil and this guessing game by stating that the latter involves the second-person standpoint, while Kripke's focus on the former is, in an 'internalist' manner, in terms of first-person epistemology (one's own inability to justify which rule one follows on the basis of one's past intentions, dispositions, etc.). So, we really have two distinct cases, and Waismann's variation is built only on the guessing game.

Let us suppose the game goes on the following way. Smith writes down:

$$1\ 4\ 9\ 16$$

Having observed Smith, Jones deduces that he must have been working the first values of the function:

$$y = x^2$$

But Waismann suggests that Smith might have replied:

> The rule I was using was different; it was only by chance that these first numbers coincided with the beginning of the series of the squares. For instance my rule was $y = \frac{x}{50} \times (24 + 35x^2 - 10x^3 + x^4)$. (C: 172)

Let us call this function F. As Waismann points out:

> These examples will give us some idea how infinitely many possibilities there are, and how unfounded it would be to suppose that we can discover the reason for a man's action by observation. (C: 172)

This is our point (3) above. As Wittgenstein had pointed out, any guess by Jones is an 'hypothesis':

> The proposition that your action has such and such a cause, is hypothesis. The hypothesis is well-founded if one had a number of experiences which, roughly speaking, agree in showing that your action is the regular sequel of certain conditions which you then call causes of the action. In order to know the reason which you had for making a certain statement, for acting a particular way, etc., no number of agreeing experiences is necessary, and the statement of your reason is not a hypothesis. (BB: 15)

It is indeed an 'hypothesis', since it could be falsified by the next number written down by Smith but never fully confirmed. That is, until Smith tells Jones that he had been following F, in which case it is his assertion that justifies his having written these numbers.

The reason for this hypothetical character is, as Waismann points out, an underlying mathematical fact: it is nearly impossible for Jones to discover which function Smith was computing by mere observation because any such initial segment, whatever its length, is the initial segment of an infinite number of mathematical functions, none of them being more privileged than the other. Wittgenstein himself alludes to this point three times, for example in the *Blue Book*:[12]

---

[12] See also PI §§179 & 213.

> Taken an example. Some one teaches me to square cardinal numbers; he writes down the row
>
> $$1\ 2\ 3\ 4,$$
>
> and asks me to square them. [...] Suppose, underneath the first row of numbers, I then write:
>
> $$1\ 4\ 9\ 16.$$
>
> What I wrote is in accordance with the general rule of squaring; but *it obviously is also in accordance with any number of other rules; and amongst these it is not more in accordance with one than with another*. (BB: 13) (Our italics.)

We have thus far argued that in order to defend for the distinction between cause and reason, Waismann deployed many arguments, including (3) above: that contrary to causes, reasons cannot be discovered by observation, and that he defended this claim with the guessing game. To complete our description of his variation, we need to say something concerning the idea that the 'chain of reasons' has an end.

3 The Chain of Reasons

Again, the idea that the 'chain of reasons' has an end is Wittgenstein's. He writes:

> Now there is the idea that if an order is understood and obeyed there must be a reason for our obeying it as we do, and, in fact, a chain of reasons reaching back to infinity. [...]
>   If on the other hand you realize that the chain of *actual* reasons has a beginning, you will no longer be revolted by the idea of a case in which there is *no* reason for the way you obey the order. (BB: 14f.)

We extrapolate a bit here and recall Turing's analysis of human computation in his renowned 1936 paper – after all his 'machines' were not computers but "*humans* who calculate" (Wittgenstein 1980 §1096). Let us then imagine Jones asking further why questions about Smith's following of F, that lead, from one answer to another, to the breaking down of the computation of values of F into elementary steps, arriving at "'simple operations' which are so elementary that it is not easy to imagine them further divided" (Turing 1936: 250), and applying them would involve no guesswork. When prompted about such basic cases, Smith would have no choice but to claim that 'This is simply what I do'. The chain of reasons would come to an end, because there is no further breaking down of the operations into further simpler steps.[13]

---

[13] Wittgenstein argues for that claim from an analogy with the infinite divisibility of the line (BB: 14).

Now, Jones might continue asking questions beyond that point. Wittgenstein sees this as mistaken, given that Smith's answers would then move into causal territory, and he offers a diagnosis:

> Now there is the idea that if an order is understood and obeyed there must be a reason for our obeying it as we do, and, in fact, a chain of reasons reaching back to infinity. [...] If on the other hand you realize that the chain of *actual* reasons has a beginning, you will no longer be revolted by the idea of a case in which there is *no* reason for the way you obey the order. At this point, however, another confusion sets in, that between reason and cause. One is lead into this confusion by the ambiguous use of the word "why". Thus when the chain of reasons has come to an end and still the question "why?" is asked, one is inclined to give a cause instead of a reason.
>
> The double use of the word 'why', asking for the cause and asking for the motive, together with the idea that we can know, and not only conjecture, our motives, gives rise to the confusion that a motive is a cause of which we are immediately aware, a cause 'seen from the inside', or a cause experienced. (BB: 15)

Waismann remains again close to Wittgenstein, in fact so close that he silently lifts the last sentence of that quotation, which we underline here:

> I sum up then: <u>the ambiguous use of the word 'why', asking for the cause and asking for the motive, together with the idea that we can *know*, and not merely *guess*, our motives, gives rise to the confusion that a motive is a cause of which we are immediately aware, a cause 'seen from inside', or a cause directly experienced.</u>
>
> If you now look back on the examples given by Köhler, it will not need much effort to realise that he is constantly taken in by the ambiguities of speech, which make him confuse, on the one hand, the *object* with the *cause*, and, on the other hand, the *motive* with the *cause*. (C: 174)

At this point, we would like to digress and make a small point of scholarship. As explained above, in Waismann's target in his lectures on causality, §§10 – 11 was Köhler's idea of an 'insight'. Now, the locutions in the sentence Waismann took from the *Blue Book*, 'being immediately aware of a cause' and 'a cause seen from the inside', give away Wittgenstein's target, and Waismann's mention of Köhler immediately afterwards confirms this. In this respect, it is worth noticing that revised German translation of *Gestalt psychology* (Köhler 1930), appeared in 1933 under the title *Psychologische Probleme* (Köhler 1933), and that we find during the same period the very first occurrence of this very point against Köhler (and rule-following) in *Philosophical Grammar*, but, as in BB: 15 without mentioning his name:

> If I write '16' under '4' in accordance with the rule, it might appear that some causality was operating that was not a matter of hypothesis, but of something immediately perceived (experienced). (Confusion between 'reason' and 'cause'.) (Wittgenstein 1974 §61)

One usually locates Wittgenstein's reading of Köhler and his discussion at a later period, 1946–51.[14] If the foregoing is right, then Wittgenstein was already acquainted with Köhler's book as early as 1933–35, and many well-known developments would relate to this.

As we already said, it is not our intention to launch into a new re-reading of PI §§143 – 242 on the basis of Waismann's variation, so we limit ourselves to only one small exegetical remark. In a nutshell, the variation amounts to this: on the basis of the guessing game, it can be argued that, contrary to causes, reasons for an action cannot be discovered by observation. Furthermore, in asking why-questions, one soon reaches the end of the 'chain of reasons', where one can only answer 'This is simply what I do', and beyond which, if further why-questions are asked, one will begin to answer with an illicit appeal to causes as if one were still supplying reasons. We view the matter as follows: on the one hand, Waismann read the relevant passages from BB and construed his own very 'Wittgensteinian' argument about rule-following, using the guessing game and relying heavily on the distinction between causes and reasons, while on the other hand Wittgenstein moved forward to develop an intricate set of remarks on following a rule in PI §§143 – 242. Can we find at least echoes of Waismann's variation, within their fabric?

The idea that the 'chain of reasons' has an end is surely related to the ideas of 'reaching bedrock' (§217) and 'obeying the rule blindly' (§219). Re-reading the relevant sections, we see that they are still couched in the language of 'causes' and 'reasons'; we underline here the key words and add the original words in parenthesis:

> 211. No matter how you instruct him in continuing the ornamental pattern, how can he *know* how he is to continue it by himself – Well how do *I* know? – If that means "Have I <u>reasons</u> (*Gründe*)?" the answer is: my <u>reasons</u> (*Gründe*) will soon give out, And then I shall act without <u>reasons</u> (*Gründe*).
>
> [...]
>
> 217. "How am I able to follow a rule?" – If this is not a question about <u>causes</u> (*Ursachen*), then it is about the <u>justification</u> (*Rechtfertigung*) for my acting in this way in complying with the rule.
> Once I have exhausted the <u>justifications</u> (*Begründungen*), I have reached bedrock, and my spade is turned. Then I am inclined to say: "This is simply what I do."
>
> [...]
>
> 219. When I obey a rule I do not choose. I obey the rule blindly.

This is an idea which is likely to be misunderstood, for example when one describes the bedrock as some sort of 'pre-normative' foundation, or when Kripke

---

14 See, for example, Benjafield 2008: 105.

understands 'obeying the rule blindly' (§219) as involving "an unjustified stab in the dark" (Kripke 1982: 16).

4 Conclusion: Blind Rule-Following

To conclude, we would like to look at the argument of the variation as it stands, on its own, and ask into what service might it be pressed? Leaving aside here issues concerning the validity of the distinction between 'causes' and 'reasons' – where obviously more would need to be said to shore up the variation – we would like merely to identify an area in the epistemology of logic where potential contributions could be made, within the recent debate generated by Paul Boghossian's "Blind Reasoning" (Boghossian 2003).

One frequent way to read PI §§143 – 242 consists in seeing these sections as providing a set of interrelated arguments *against* the idea that in order to follow a rule, one must *grasp* it, having it 'in mind' and, as it were, tracking its requirements.[15] Waismann gets essentially the same point across:

> To think of a rule of arithmetic may, indeed, be the cause of its being followed. Notice, however, that the cause of the fact that a rule is being followed may also lie in something different – for instance, in the habit of doing a sum in this way; this habit, in its turn, may be the result of an antecedent process of training. At any rate, *to say that whenever I do something in accordance with a rule, I must have been aware of the rule, or must have rehearsed it for myself, is unrealistic.* A chess player, when he is not a beginner, makes a move without thinking of the rule; his acting in accordance with the rules is just due to habit; and so in other cases. (C: 173) (Our underlining)

The suggestion here is that the idea that one must 'have in mind' the rule in order to follow it is still to speak in causal terms.

An 'internalist' explanation of rule-following of the type rejected by both Wittgenstein and Waismann would start with the assumption a *rule* of inference such as *Modus Ponens*:

$$A, A \rightarrow B \vdash B, \qquad (MP)$$

is somehow 'normatively inert', so that one would need to 'have in mind' the corresponding *implication* – as a logical truth or principle – in order to act according to it:

---

15 See, for example, Wright 2001: 184.

$$(A \& (A \rightarrow B)) \rightarrow B \qquad (P)$$

In Lewis Carroll's paradox of inference, when faced with an instance of *MP*, the Tortoise refuses to infer *B*, and when Achilles proposes to add *MP* to the premises, she still demurs and a regress follows upon repeating the procedure (Carroll 1895). This is often taken to mean that logic alone does not 'move the mind', so that one always needs something else – a desire, disposition, habit, etc. – that *causes* one to infer. As Simon Blackburn put it:

> There is always something else, something that is not under the control of fact and reason, which has to be given as a brute extra. (Blackburn 1995: 695)

According to the above, this is equivalent to not seeing that the chain of reasons has come to an end, and thinking that something else needs to be supplied. Laurence Bonjour has suggested that this extra might be 'rational insight' (Bonjour 1998: 106 – 107),[16] and Elijah Chudnoff argued that 'intuition' is needed (Chudnoff 2013), but this is precisely the sort of move that would be barred by Waismann's variation.

Having now identified this as an area for further work, we would like to conclude with a further comment about Carroll's paradox. Boghossian pointed out that the 'internalist' explanation of rule-following is circular: in order to recognize that one particular instance of *MP* is valid, I must justifiably infer that it is valid in virtue of *P* – the principle that says that all instances of *MP* are valid. But in order justifiably to infer that it is valid from *P*, I must be able justifiably to infer according to *MP*.[17] Now, basic cases of rule-following such as inferring in accordance with *MP* are precisely located where the 'chain of reasons' ends, so to speak 'at the boundary' of our inferential practice.[18] For this reason basic cases of rule-following must be 'blind', simply because no anterior reason could be provided, on pains of circularity. As Crispin Wright put it:

> With respect to a wide class of concepts, a grasp of them is not anterior to the ability to give them competent linguistic expression but rather *resides in* that very ability. [...] the

---

[16] For an effective criticism of Bonjour, see Boghossian 2003: 230f.
[17] This is adapted from Boghossian 2003: 233, see also Wright's critique of the 'Modus Ponens Model' (Wright 2007: 490f.). The 'adoption problem', set up by Romina Padro on the basis of lecture notes by Kripke, also lurks in the vicinity. By 'adoption' she means whether someone who has no prior notion of *P* could come to adopt *MP* not as an inferential practice but on the basis of acceptance of *P* (Padro 2015: 31). Her claim is that this is not possible, as Carroll's paradox points to a similar sort of circularity.
[18] "A *reason* can only be given *within* a game. The links of the chain of reasons come to an end, at the boundary of the game. (Reason and cause.)" (Wittgenstein 1974 §55).

modus ponens model [MPM] *must* lapse for basic cases. Basic cases – where rule-following is 'blind' – are cases where rule-following is *uninformed by anterior reason-giving judgement*. (Wright 2007: 496)

This may simply be the lesson from Carroll's paradox: while it is true that

$$A, A \rightarrow B \vdash B \text{ if and only if } (A \,\&\, (A \rightarrow B)) \rightarrow B,\text{ }^{19}$$

one should beware of the philosophical confusions that this biconditional entails, one of them being precisely that one supposedly needs to 'have in mind' the implication-as-belief P in order to detach B from $A \rightarrow B$ given A.[20] What Waismann's variation would tell us at this stage is that invoking P in this manner is involving a confusion between causes and reasons.[21]

Bibliography

Baker, Gordon (1979): "Verehrung und Verkehrung: Waismann and Wittgenstein." In: *Wittgenstein: Sources and Perspectives*. Luckhardt, C. G. (ed.). Ithaca N.Y.: Cornell University Press, 243 – 285.
Baker, Gordon (1997): "Preface to the Second Edition." In: PLP, xi – xxiii.
Baker, Gordon (2003): "Preface." In: Wittgenstein et al. 2003, xvi – xlviii.
Baker, Gordon and McGuinness, Brian (1976): "Nachwort". In: Waismann 1976, 647 – 662.
Benjafield, John G. (2008): "Revisiting Wittgenstein on Köhler and Gestalt Psychology." In: *Journal of the History of Behavioral Sciences*. Vol. 44, 99 – 118.
Blackburn, Simon (1995): "Practical Tortoise Raising." In: *Mind*. Vol. 104, 695 – 711.
Boghossian, Paul (2003): "Blind Reasoning." In: *Proceedings of the Aristotelian Society, Supplementary Volumes*. Vol. 77, 225 – 293.

---

19 With the proviso that the right-hand side formula is logically true. This formula is what Stewart Shapiro calls a 'transfer principle' (Shapiro 2000: 337).
20 For a detailed discussion, see Marion 2016.
21 The ideas presented here were first presented by Mathieu Marion at the international colloquium on *Waismann – Causality and Logical Positivism* at the Institut Wiener Kreis, Vienna, in October 2010. Prior to the 41st International Wittgenstein Symposium at Kirchberg (August 2018), Mathieu Marion presented them in Brazil at the University of São Paulo (2013), the colloquium *Perspectives on Wittgenstein* (Universidad Federal do Rio Grande do Sul, Porto Alegre, May 2013) and the colloquium *Middle Wittgenstein IV* (Pirenópolis, May 2014), in Japan at Keio University in February 2011 and January 2014, and at the Annual Meeting of the Japanese Association for the Philosophy of Science at the University of Chiba, in June 2018, while Mitsuhiro Okada presented them at a Wittgenstein Meeting, Keio University, Hiyoshi in December 2016. This work was partly supported by the following research grants: MEXT-JSPS KAKENHI Basic Research (B)17H02265, (B)26284005 and (C)23520036. We would like to thank the above audiences and, especially, Mauro Engelmann, Andrew English and Takashi Iida for their comments.

Bonjour, Laurence (1998): *In Defence of Pure Reason.* Cambridge: Cambridge University Press.
Carroll, Lewis (1895): "What the Tortoise said to Achilles." In: *Mind* (n.s.). Vol. 4, 278 – 280.
Chudnoff, Elijah (2013): *Intuition.* Oxford: Oxford University Press.
Davidson, Donald (1963): "Actions, Reasons and Causes." In: *Journal of Philosophy.* Vol. 60, 685 – 700.
Hampshire, Stuart (1960): "Friedrich Waismann 1896-1959." In: *Proceedings of the British Academy.* Vol. 46, 309 – 317.
Glock, Hans-Johann (2014): "Reasons for Action: Wittgensteinian and Davidsonian Perspectives in Historical and Meta-Philosophical Context". In: *Nordic Wittgenstein Review.* Vol. 3, 7 – 46.
Köhler, Wolfgang (1930): *Gestalt Psychology.* London: Bell.
Köhler, Wolfgang (1933): *Psychologische Probleme.* Berlin: Springer.
Kripke, Saul (1982): *Wittgenstein on Rules and Private Language.* Cambridge Mass.: Harvard University Press.
Marion, Mathieu (2011): "Waismann's Lectures on Causality: An Introduction." In: McGuinness 2011a, 31 – 51.
Marion, Mathieu (2016): "Lessons from Lewis Carroll's Paradox of Inference." In: *The Carrollian.* Vol. 28, 48 – 75.
McGuinness, Brian (1979): "Editor's Preface." In: Wittgenstein 1979, 11 – 31.
McGuinness, Brian (ed.) (2011a): *Friedrich Waismann. Causality and Logical Positivism.* Dordrecht: Springer.
McGuinness, Brian (2011b): "Waismann: The Wandering Scholar." In: McGuinness 2011a, 9 – 16.
Padro, Romina (2015): *What the Tortoise Said to Kripke: The Adoption Problem and the Epistemology of Logic.* New York: City University of New York.
Quinton, Anthony (1977): "Introduction." In: Waismann 1977, 9 – 19.
Sandis, Constantine (2015): "One Fell Swoop. Small Red Books Historicism Before and After Davidson." In: *Journal of the Philosophy of History.* Vol. 9, 372 – 392.
Schroeder, Severin (2001): "Are Reasons Causes? A Wittgensteinian Response to Davidson." In: *Wittgenstein and Contemporary Philosophy of Mind.* London: Palgrave, 150 – 170.
Schulte, Joachim (1979): "Der Waismann-Nachlass. Überblick – Katalog – Bibliographie." In: *Zeitschrift für philosophische Forschung.* Vol. 33, 108 – 140.
Schulte, Joachim (2011): "Waismann as Wittgenstein's Spokesman." In: McGuinness 2011a, 9 – 16.
Shapiro, Stewart (2000): "The Status of Logic." In: *New Essays on the A Priori.* Boghossian, Paul; Peacocke, Christopher (eds.), Oxford: Clarendon Press, 333 – 66.
Stoutland, Frederick (1970): "The Logical Connection Argument." In: *Studies in the Theory of Knowledge. American Philosophical Quarterly.* Malcolm, Norman et al. (eds.), Monograph Series No. 4, 117 – 129.
Stoutland, Frederick (2010): "Reasons and Causes." In: *Wittgenstein. Mind, Meaning and Metaphilosophy.* Frascolla, Pasquale; Marconi, Diego; Voltolini, Alberto (eds.). Basingstokes: Palgrave MacMillan.
Turing, Alan Mathison (1936): "On Computable Numbers with an Application to the Entscheidungsproblem." In: *Proceedings of the London Mathematical Society.* (2nd series), Vol. 42, 230 – 65.
Waismann, Friedrich (1936): *Einführung in das mathematische Denken.* Vienna: Gerold und Co.
Waismann, Friedrich (1976): *Logik, Sprache, Philosophie.* Stuttgart: Reclam D. Reidel.
Waismann, Friedrich (1977): *Philosophical Papers.* Dordrecht. D. Reidel.

Waismann, Friedrich (1982): "Über das Wesen der Mathematik: Der Standpunkt Wittgensteins." In: *Lectures on the Philosophy of Mathematics*. Amsterdam: Rodopi, 157 – 167.

Waismann, Friedrich (1997): *The Principles of Linguistic Philosophy*. 2nd edition. Basingstokes/London: MacMillan.

Waismann, Friedrich (2011): "Causality." In: McGuinness 2011a, 91 – 184.

Wittgenstein, Ludwig (1969): *The Blue and Brown Books*. 2nd edition. Oxford: Blackwell.

Wittgenstein, Ludwig (1974): *Philosophical Grammar*. Oxford: Blackwell.

Wittgenstein, Ludwig (1978): *Remarks on the Foundations of Mathematics*, revised edition. Oxford: Blackwell.

Wittgenstein, Ludwig (1979): *Ludwig Wittgenstein and the Vienna Circle*. Oxford: Blackwell.

Wittgenstein, Ludwig (1980): *Remarks on the Philosophy of Psychology*, Vol. 1. Oxford: Blackwell.

Wittgenstein, Ludwig (1998): *Gesamtbriefwechsel/Complete Correspondence. The Innsbruck Electronic Edition*. McGuinness, Brian; Seekircher, Monika; Unterkircher, Anton (eds.). Charlottesville, Va.: InteLex Corporation.

Wittgenstein, Ludwig and Waismann, Friedrich (2003): *The Voices of Wittgenstein. The Vienna Circle*. London, New York: Routledge.

Wittgenstein, Ludwig (2009): *Philosophical Investigations*. revised 4th ed. Oxford: Wiley-Blackwell.

Wright, Crispin (2001): *Rails to Infinity*. Cambridge Mass.: Harvard University Press.

Wright, Crispin (2007): "Rule-Following without Reasons: Wittgenstein's Quietism and the Constitutive Question." In: *Ratio* (n. s.). Vol. 20, 481 – 502.

Michael Potter
Propositions in Wittgenstein and Ramsey

Abstract: In *Begriffsschrift* Frege proposed to ignore the part of content that is irrelevant to logic; what remains he called "conceptual content". In "On Sense and Reference" he renamed this "sense" but failed to stress that it is a notion belonging to the philosophy of *logic*, not of language. Russell seems to have seen the importance of the notion only briefly. Wittgenstein did not make use of the notion until he was in Norway, and only introduced the terminology of "sign" and "symbol" to mark the distinction while composing the *Tractatus*. Ramsey proposed to treat sign and symbol as merely two different ways of typing token inscriptions, but this unduly brushes over the difficulties the notion of a symbol involves. The most striking feature of Wittgenstein's thinking on this is the way that he generalized Frege's argument for the notion of sense so as to bypass his incorrect particularization to the case of identity.

This essay has the same title as the talk I gave in Kirchberg, but its scope is more limited. I began that talk by remarking that the technical notion of a symbol played in the *Tractatus* an analogous role to the one that conceptual content had played in *Begriffsschrift* and sense in Frege's later semantics, namely that of singling out the part of the content of a sign that is relevant to logic. I intended this remark as little more than throat-clearing – an uncontroversial observation that would help my audience to situate the argument I then went on to lay out – but during the question period it was met with consternation. I therefore think it best to devote my written contribution to explaining what I meant by the remark, since until I do that no one is likely to be convinced by an argument that treats it as an obvious background assumption.

1 Begriffsschrift

In the first few sections of *Begriffsschrift* Frege made several important logical distinctions. Right at the beginning, for example, he distinguished between a judgment and its content. He was not, of course, the first philosopher to make this distinction, but he was, as far as I know, the first to do it for the reason – dubbed the "Frege point" by Geach (Geach 1965) – that it is required to explain how in the

---

Michael Potter, University of Cambridge, Cambridge (UK), mdp10@cam.ac.uk

course of a logical argument a single content may occur unasserted and asserted in different occurrences. Even so, to make such a distinction between judgment and content is not yet to say anything illuminating about the nature of the latter. An important step – arguably the most important in all his philosophy of logic – came when he singled out what he called *conceptual* content – the part of content that is relevant to inference. Two sentences have the same conceptual content just in case the same inferences may be drawn from each. He called the conceptual content of a declarative sentence a "judgable content'. The part of the content that this notion of conceptual content excludes he variously called "colouring" (*Färbung*), "illumination" (*Beleuchtung*), or "scent" (*Duft*); in English Dummett (Dummett 1973) called it "tone". In *Begriffsschrift* Frege instanced "and" and "but" as words differing only in tone, not conceptual content; elsewhere he cited various other examples, such as "horse" and "steed", or "dog" and "cur". The fact that he never settled on a single word for tone is symptomatic of his lack of interest in it: he only ever mentioned it in order to set it aside as irrelevant to logic.

The importance of this distinction is that it enabled Frege to make prominent those features of the structure of sentences that are relevant to inference. It is part of the logician's task, as of the grammarian's, to study how sentences reveal the structure of the contents they express, but their differing interests lead to different conceptions of that structure. The two sentences "The Greeks defeated the Persians at Plataea" and "The Persians were defeated by the Greeks at Plataea" have the same conceptual content, because the same consequences may be derived from each, whereas grammar quite properly distinguishes between the active voice of the verb in the former and the passive voice in the latter. The notion of tone is thus applicable not only to the contents of individual words but to the manner in which they are assembled to form sentences: sentences with the same conceptual content may differ in emphasis, and hence in tone, by having different terms as their grammatical subjects.

As is familiar, Frege later realized that his *Begriffsschrift* notion of conceptual content was inadequate and should be replaced. The important point to note here is that the notion he replaced it with, "sense", was still intended to play the role of constituting the part of content that is relevant to logic. Unfortunately, though, this point is obscured by a significant oddity in the way "On Sense and Reference" is written, namely that he chose hardly to mention formal logic there at all. Why not? The only reason I can think of is that he was continuing, as he had in *Grundlagen* to separate the formal from the philosophical in order to increase the chances of being read. At any rate, he deliberately emphasized the non-mathematical applications of his semantic theory and therefore obscured the point that his notion of sense was primarily a contribution to the philosophy of *logic*, not of language. It is uncontentious that there is *some* difference in content between

"Hesperus" and "Phosphorus": what is at issue in "On Sense and Reference" is whether logic should recognize it.

2 Russell

I have stressed the importance of the notion of conceptual content or sense for Frege's philosophy of logic. To see how important it was, we have only to look to the salutary example of Russell. He learned his quantificational logic from Peano, not Frege, and his first reaction to $\phi x_x \psi x$ (what we would now write as $\forall x(\phi x \rightarrow \psi x)$) was that this was plainly a different proposition from "every ϕ is a ψ. He took it to be obvious that the former says something about everything there is (namely that it is either a ψ or not a ϕ), whereas the latter only says something about ϕs (Russell 1903 §41). This is an instance – the *Principles* has many others scattered through it – of his failure to grasp Frege's insight that logicians should ignore the part of content that is irrelevant to inference. The result was that Russell kept treating as logical what were really only grammatical distinctions, and hence devising theories of hopelessly unnecessary complexity.

Not until the autumn of 1902, with the main text of the *Principles* complete, did Russell settle down to study Frege's published papers and write an appendix to his book that summarized them. Over the next two or three years the influence of his reading of Frege was occasionally visible in his own work. For instance, one feature of his 1905 paper that often goes unnoticed is the extent to which Russell's famous theory of descriptions depends on accepting something like the Fregean conception. It is one of the curiosities of "On Denoting" that it slips in this crucial step in a footnote, where, somewhat oddly, Russell attributes it to Bradley (Russell 1905: 481). The importance of focussing on the part of content that is relevant to inference is not the only thing in "On Denoting" that might have a Fregean source. The other central concern of the paper is the relevance of scope in explaining ambiguities. Thus, for instance, Russell used different possible scopes of the quantifier to explain he difference between *de re* and *de dicto* understandings of "George IV wished to know whether Scott was the author of Waverley". (Frege's influence on Russell evidently waned after 1905. When he re-explained his theory of descriptions later, in the Introduction to *Principia*, he did not use scope distinctions to motivate it.)

After 1911 he came under Wittgenstein's influence and became suspicious of his platonistic conception of logic as having a subject matter consisting of entities such as disjunction or negation. From the autumn of 1913 onwards, therefore, he used "proposition" to mean the sentence, rather than what the sentence expresses.

In his post-war work these linguistic items were the subject matter of logic, with the result that his conception of the subject became somewhat psychologistic.

3 Early Wittgenstein

Whereas Frege proposed a "one-kind, two-step" semantics for singular terms, Russell's was a "two-kind, one-step" theory, i.e. one that distinguished two kinds of singular term, definite descriptions and logically proper names, and proposed that the latter referred directly without any need for a third-realm intermediary. It may well be that Wittgenstein at first shared Russell's one-step conception, and hence rejected the need for Frege's distinction between sense and reference. In support of this one might quote a letter Jourdain wrote to Frege in January 1914. Jourdain at that time lived on the outskirts of Cambridge and had got to know Wittgenstein through Russell. (For a time it was proposed that he and Wittgenstein would collaborate on a translation into English of some parts of Frege's writings, but this did not occur, presumably because of the war: when this translation eventually appeared in 1915–17 Jourdain's collaborator was named as Johann Stachelroth.) Jourdain wrote to ask Frege, *inter alia*, "whether, in view of what seems to be a fact, namely, that Russell has shown that propositions can be analyzed into a form which only assumes that a name has a 'Bedeutung', & and not a 'Sinn', you would hold that 'Sinn' was merely a psychological property of a name" (Frege 1980: 78). Perhaps Wittgenstein was behind Jourdain's question. If so, then at this time (or, at any rate, when he departed for Norway the previous October) Wittgenstein still sided with Russell against Frege's sense/reference distinction.

At that time, though, Wittgenstein still hoped to devise a "logically perfect" notation which would make the logical properties of a sign transparent. By the time of the Moore dictation in April 1914, however, his search for such a notation had stalled. He therefore began instead to stress the importance of seeing past the "particular scratches" to the symbolism's "logical properties" (NB: 112). Nonetheless, with no label for the distinction he was compelled to speak opaquely of "seeing the sign in the sign" (NB, 23 Oct. 1914). Not until he had reached page 54 of the *Prototractatus* volume did he introduce a separate word for what it is that we are supposed to see. Even then, his immediate purpose in doing so was only to make a point about the ambiguity of ordinary language, namely that one sign might represent different symbols in different occurrences. Not until later, it seems, did he realize that even if he had, *per impossibile*, found a logically perfect notation, there would still have been a role for the distinction.

4 Sign and Symbol

In the *Tractatus* the distinction between sign and symbol is introduced via that between propositional sign and proposition, and this in turn emerges as a consequence of features of the picture theory of meaning. A picture, according to that theory, is a fact that shares its form with the part of reality that it attempts to represent. A proposition is a picture with two extra features: first, its form of depiction is logical; second, it is "expressed perceptibly through the senses" (TLP 3.1). These two features then give rise to the need for a distinction between the proposition the propositional sign through which it is expressed, since otherwise the picture theory would hold, absurdly, that we can read off the structure of the world from the structure of the signs we use to represent it.

One might be tempted to think that by "propositional sign" Wittgenstein just meant "sentence", except that this word, like "picture", in ordinary usage means a complex, whereas Wittgenstein was explicit (TLP 3.14) that a propositional sign is a fact, i.e. it is already parsed so that its grammatical structure is revealed. So a propositional sign and a proposition are both facts, but their forms are different: perceptible in the former case; purely logical in the latter. It is this difference in form that explains why we cannot simply read off the structure of the world from that of language. Propositions, according to the *Tractatus*, share their form with the parts of the world they represent; propositional signs do not.

Having thus distinguished between propositional sign and proposition, Wittgenstein distinguished analogously between their meaningful constituents, which he called *signs* and *symbols* (or *expressions*) respectively. (The qualifier "meaningful" is needed here to rule out treating "es is mor", for instance, as a sign because it occurs in "Socrates is mortal".) "What is essential in a symbol," Wittgenstein said in the *Tractatus*, "is what all symbols that can serve the same purpose have in common." (TLP 3.341) In the notes dictated to Moore he was slightly more explicit about what kind of "purpose" he had in mind. "What symbolizes in a symbol," he there wrote, "is that which is common to all the symbols which could in accordance with the rules of logic ... be substituted for it." (NB: 117)

We have now arrived at the sense in which Wittgenstein's notion of a symbol in the *Tractatus* is analogous to that of conceptual content in *Begriffsschrift* and sense in *Grundgesetze*. Wittgenstein, like Frege, wanted to focus attention on the features of a word's content that contributes to its logical role. "Seeing the symbol in the sign" consists in seeing past the contingent properties of a sign to these features.

> It can never indicate the common characteristic of two objects that we symbolize them with the same signs but by different methods of symbolizing. For the sign is arbitrary. We could therefore equally well choose two different signs and where then would be what was common in the symbolization? (TLP 3.322)

To say that Tractarian symbols play broadly the same role as Fregean senses is not to say that they are identical, though. Frege's idea was that conceptual content or sense is inferential power. The difficulty comes in how we cash this out. If sentences have the same conceptual content only when they have the same *immediately derivable* consequences, then the notion becomes too sensitive to which consequences we take to be immediate. You may find obvious an inferential step that for me requires explanation, in which case this notion has no place in an account of logic that aspires, as Frege's did, to be independent of psychology. Once a formal system is in place, there will of course be an objective criterion for the immediacy or otherwise of a deduction, but this will depend on the system chosen and hence be unsuitable to serve, as he intended his notion of conceptual content to serve, as part of the grounding for that very system. Wittgenstein accused Frege of having wrongly imported psychological considerations into logic. "It is remarkable," he complained, "that so exact a thinker as Frege should have appealed to the degree of self-evidence as the criterion of a logical proposition." (TLP 6.1271) This is an instance, I think, of Wittgenstein's repeated tendency to take hold of an idea and apply it more resolutely than its originator. Here he resolutely ignored psychological considerations of obviousness, so that what the individual reasoner is irrelevant to "seeing the symbol in the sign"; all that matters is what is said about the world. In particular, then, logically equivalent propositional signs express the same proposition.

5 Ramsey

Wittgenstein famously had a copy of the typescript in his possession at the armistice, and it was this copy that he sent to Keynes in 1919 for onward delivery to Russell. Russell left this copy with Wrinch when he departed for China, and she eventually managed to get Ostwald to publish it as an issue of his monograph series. An advance proof copy of this version of the book reached Ogden in Cambridge at the beginning of November 1921 (letter to Russell, 5 Nov. 1921). It was presumably this copy that Ogden lent to Ramsey (then a second year undergraduate) shortly thereafter and which, as his father later recalled, "interested him greatly".

These details of just when Ramsey first read the book lend an intriguing aspect to the earliest of his philosophical essays to have survived, namely a talk on the nature of propositions that he delivered to the Cambridge Moral Sciences Club at the end of that November. There he proposed a multiple relation theory of judgment which, although certainly not the same as the one in the *Tractatus*, is at least similar in broad outline. What is clear, at least, is that his interest in the *Tractatus* made him an obvious choice as translator. According to his father's memoir he dictated his draft at Miss Pate's typing office in Trinity Street Cambridge in March 1922. The story of how Ogden sent the resulting typescript to Wittgenstein in Trattenbach to correct is now well known.

The book finally appeared in the dual-language edition in the autumn of 1922. Moore (who had been responsible for suggesting the Latin title by which the English version is known) was by then Editor of *Mind* and commissioned a critical notice of the book from Ramsey. He wrote this in August 1923, just after graduating at Cambridge as what was then called a "B* Wrangler' in the Mathematical Tripos. (This meant that he got a First in Part A and a Distinction in Part B of his final exam.) Ramsey's critical notice was the first serious contribution to the secondary literature on the *Tractatus*, and it remains one of the best. For our purposes here what is relevant is an exegetical suggestion he made there. From what was said above we might well be puzzled about the nature of Tractarian symbols. To help us understand the idea, Ramsey proposed to make use of Peirce's terminology of type and token. This is now standard, of course, but when Peirce proposed it in 1906, few others took much notice. Two who did were Ogden and Richards, who quoted the relevant passage of Peirce in their book, *The Meaning of Meaning* (Ogden/Richard 1923: 433 – 434). Ramsey then adopted this terminology. Sign and symbol, he suggested, are not token and type – a sign is already a type, as Wittgenstein's gnomic observation that " 'A' is the same sign as 'A' " (TLP 3.203) was intended to make clear – but rather two ways of typing the tokens, whether according to syntactic properties (signs) or logical role (symbols).

On the face of it, Ramsey's way of putting the matter discourages us from hypostasizing symbols as occupants of a Fregean "third realm" intermediate between language and reality. However, there are two points that tell against such a deflationary view. First, as Ramsey also later noted (*after* he had discussed the *Tractatus* with its author, it should be said), there may well be some types of which there are no token instances. "It cannot be any concern of ours," he said, "whether anyone has actually symbolized" it (Ramsey 1931: 33). Whether we want to say that a type with no tokens inhabits an abstract "realm" is perhaps a relatively minor matter. Second, the transition from a propositional sign to a proposition is not quite as straightforward as Ramsey's exegesis suggests. It is not merely a matter of grouping different signs together, but of changing the form of the fact. The form of the propositional sign might be spatial (if it is written) or temporal (if it is spoken), whereas the form of the proposition is purely logical. So it cannot be said that Ramsey has altogether demythologized the transition.

6 Identity

Wittgenstein used the sign "=" (much as Frege had used "≡" in *Begriffsschrift*), to stand between signs that express the same symbol. In particular, as he explained in the *Prototractatus* (PT 4.2213), he used it between propositional signs that express the same proposition. Oddly, though, he dropped this explanation from the final version of the book, despite continuing there to use it in this way (e.g. 4.0621, 5.51, 5.52). This similarity between the *Tractatus* and *Begriffsschrift* uses of the equality sign invites an obvious question, though. In "On Sense and Reference" Frege rejected the *Begriffsschrift* account of identity. So why does this objection not also apply to Wittgenstein's account in the *Tractatus*? Frege's objection was in effect that his previous account had made identities metalinguistic. (He did not use that word, of course, which did not become current until the 1930s.) In the *Tractatus* Wittgenstein simply accepted this. On his account "=" is a sign of the metalanguage, not the object language.

What, though, of Frege's objection that identity sentences do sometimes express non-trivial information? Frege's central idea was that we need the notion of sense in order to allow for the fact that the objects we refer to may have other aspects of which we are currently unaware, and yet that does not prevent us from referring to them successfully. In his draft reply to the letter from Jourdain quoted earlier, Frege made this point by using the example of a mountain seen from different directions by two explorers, one of whom names it "Aphla", the other "Ateb".

What is striking here is surely the sense of two philosophers taking past each other. Frege's example of Aphla and Ateb certainly would not have impressed Russell, who would have said that "Aphla" and "Ateb" are not logically proper names but disguised descriptions. The place where this point surfaces in the *Tractatus* is in Wittgenstein's stipulation that in the formal language we have no need of a sign of identity, because we can just agree never to refer to the same object with two different simple signs. Frege's point had been that this stipulation is impossible to implement, because there are cases in which we do not yet know that the two signs do refer to the same thing. Wittgenstein's response was that when it comes to Tractarian objects, we do always know: a Tractarian object, unlike Frege's mountain, is not the kind of thing that can have other, as yet unknown aspects to it.

I suggested earlier that it was a feature of Wittgenstein's way of thinking to press an idea more resolutely than its originator had done. Here we have another instance of this. In 1914 Frege still conceived of his sense/reference distinction, as he had when he introduced it in the 1890s, as motivated narrowly by a point concerning *identity*. In this, though, he was mistaken. To see why, we need only notice the difference in inferential power, on Frege's conception, between "Hesperus is a planet" and "Phosphorus is a planet". This difference suffices to show that "Hesperus" and "Phosphorus" have different senses, but does not mention identity at all. This suffices to show that Frege had mislocated the point of the distinction. Frege thought that the distinction between sense and reference was needed so as to leave room for objects to have unknown aspects. On Wittgenstein's view, the distinction is needed in order to leave room, more generally, for the structure of our signs to differ from that of the world.

Bibliography

Dummett, Michael (1973): *Frege: Philosophy of Language*. London: Duckworth,
Frege, Gottlob (1879): *Begriffsschrift: Eine der arithmetischen nachgebildete Formelsprache des reinen Denkens*. Halle a.S.: Nebert.
Frege, Gottlob (1980): *Philosophical and Mathematical Correspondence*. Oxford: Blackwell.
Geach, P. T. (1965): "Assertion." In: *Philosophical Review*, Vol. 74, 449 – 465.
Ogden, C. K. and Richard, I. A. (1923): *The Meaning of Meaning*. London: Kegan Paul, Trench, Trubner & Co.
Ramsey, F. P. (1931): *Foundations of Mathematics, and Other Essays*. London: Kegan Paul, Trench, Trubner & Co.
Russell, Bertrand (1903): *The Principles of Mathematics*. Cambridge (UK): Cambridge University Press.
Russell, Bertrand (1905): "On Denoting." In: *Mind*, Vol. 14, 479 – 493
Wittgenstein, Ludwig (NB): *Notebooks 1914–1916*. Von Wright, G. H.; Anscombe, G. E. M. (eds.). Oxford: Blackwell, 1961.
Wittgenstein, Ludwig (TLP): *Tractatus Logico-Philosophicus*. London: Kegan Paul, Trench, Trubner & Co, 1922.
Wittgenstein, Ludwig (PT): *Prototractatus*. McGuiness, B. F.; Nyberg, T., von Wright, G. H. (eds.). London: Routledge, 1971.

Jean-Yves Béziau
An Unexpected Feature of Classical Propositional Logic in the *Tractatus*

Abstract: We study the relation between classical propositional logic (CPL) as it is nowadays and how it appears in the *Tractatus* focusing on a specific feature expressed in the paragraph 5.141. In a first part we make some general considerations about CPL, pointing out that CPL is difficult to characterize and define, that there is no definite final version of it presented in one given reference book. In a second part we analyze the network of concepts related to paragraph 5.141 of the *Tractatus* involving notions corresponding to what are nowadays called "semantical consequence", "distribution of truth values", "valuations" and "models". We make the link with Tarski's definition of logical consequence in his famous 1936 paper. This leads us to examine in a third part up to which point CPL is in the *Tractatus* considered as a Boolean algebra.

1 Introduction: The Relation Between Two Icons

On the one hand classical propositional logic (hereafter CPL) is the most famous logical system of modern logic, on the other hand the *Tractatus Logico-Philosophicus* (hereafter *Tractatus*, TLP) is one of the most famous books in the history of modern logic. It seems therefore worth to ask the following questions:

- How is CPL in the *Tractatus*?
- What are the differences and similarities of CPL in the *Tractatus* and other versions?
- What is the contribution of the *Tractatus* to CPL?

These three questions are interrelated. It is not that simple to give answers to them and the aim of the present paper is not to give full and final answers to these questions.

Jean-Yves Béziau, University of Brazil, Rio de Janeiro and Brazilian Research Council, jyb@ufrj.br

We will concentrate on the paragraph 5.141 of the *Tractatus*,[1] which is the following single sentence: If p follows from q and q from p then they are one and the same proposition.

We will of course not comment this proposition in isolation, artificially extracted from the *Tractatus*. We will deal with the related network of concepts presented in the *Tractatus*, and compare this framework to CPL. This involves all aspects of CPL: historical, mathematical and philosophical.

2 The Inherent Ambiguity of Classical Propositional Logic

On the one hand we have a book, the *Tractatus*, on the other hand we have a logical system, CPL. One of the reasons why it is difficult to make a comparison between CPL and CPL as it is in the *Tractatus* is that, contrarily to what one may think, CPL is not something directly clear and obvious, precisely and univocally presented or defined.

Let us emphasize that this is the case of any scientific system or theory: the theory of evolution of course, but also the theory of relativity or to take an even simpler example, more directly related to CPL, lattice theory. It would be too naïve to believe that lattice theory reduces to a group of axioms. There are fairly different axiomatizations of lattice theory, moreover everything is in the axioms only potentially.

It is also important to stress that a scientific system or a theory is not codified in one given book. There is no Bible of lattice theory, although this case is the closest we can imagine because Garret Birkhoff's book is one of the most famous books of mathematics of the 20th century (see Birkhoff 1940 and Bennett 1973). But it is still quite different from the Bible ... or the *Tractatus*. The *Tractatus* has a rigid and precise linguistic structure, allowing, like in the Bible case, never ending discussions and interpretations of each sentence.

Furthermore a scientific theory evolves forever, lattice theory as it is nowadays is not the same at it was at the time of Birkhoff. CPL has evolved quite a lot since the time of the *Tractatus*. The situation of CPL is much more complicated than the situation of lattice theory, because (1) it can be presented in even more different ways (proof theory vs. semantics) (2) it is surrounded by philosophical nebulosity

---

[1] We will call the numbered items of the *Tractatus* "paragraphs", following a certain tradition, although they are not always syntactically speaking such entities.

and (3) there is not one specific reference book by a famous author devoted to it, like in the case of lattice theory with Birkhoff.

Many people have the idea that CPL is trivial and simple, logic for babies. But in fact it is not the case. The situation is similar with the one of natural numbers. It is only apparently simple. Number theory is not the simplest mathematical theory, as shown in quite different ways by Gödel and Bourbaki. The similarity is even stronger if we consider that the structure of the set of propositions in propositional logic is an absolutely free algebra, which, like Peano algebra (an absolutely free algebra with only one generator and one function), is not axiomatizable in first-order logic (for details, see Béziau 1999).

CPL was not born in one day, out of the spirit of one man. Before its definitive version there were many drafts. People like Boole, Peirce, Frege, Russell, Wittgenstein made different contributions to it. And to speak of a definitive version of it is quite misleading. However we can say that what we find in the work of Post in 1921 is something close to it (the case of first-order logic is more complicated).

Funny enough Post's work was published the same year as Wittgenstein's *Tractatus*: 1921. Post's work is a seminal work. After Peirce who proved that all the 16 connectives can be defined by only one (joint work with his student Christine Ladd-Franklin, Ladd-Franklin 1882), it is the first work with important mathematical results: completeness, functional completeness and Post completeness. In mathematics results work together with conceptualization. In Post's paper we find for the first time a clear distinction between proof and truth in CPL, distinction on which basis the completeness theorem which is herein presented makes sense.

There are various philosophical interpretations of CPL and the philosophical view is interacting with the formal aspect of CPL. This is in particular the case in the *Tractatus* with the idea of elementary propositions, on the basis of which was promoted "logical atomism" by Russell, a terminology not used by Wittgenstein himself but already introduced in the preface of the *Tractatus* by Bertrand Russell to describe Wittgenstein's theory.

Gödel showed that it is possible to prove the completeness theorem of CPL without considering that there are atomic formulas. Generally CPL is presented with atomic formulas, but it can also be presented without. This paper was commented by Quine (see Gödel 1932).

The same Quine wrote a famous paper in *Mind* in 1934, which was pivotal for the tendency to speak about "Sentential Logic" rather than "Propositional Logic", arguing that it is better to conceive CPL as dealing with sentences than propositions. At the same time, in Poland people were going in the other direction, in particular considering connectives as functions, so that on the one hand we have an algebra of propositions whose operators are connectives (idea due to Lindenbaum) and on the other hand logical matrices (theory developed by Łukasiewicz and Tarski) where there are some operators defined on truth-values corresponding to the connectives (see Béziau 2002). Using this correspondence Lindenbaum proved a famous theorem according to which any logic can be characterized by a matrix, result published by Jerzy Łoś after the war (see Łos 1949).

In Poland the terminology "zero-order logic" was introduced to talk about CPL, which is quite neutral as the nature of the elements dealing with and establishes a correspondence with first-order logic. Another important innovation in Poland was to consider a consequence operator or consequence relation, not only a set of tautologies. This approach of CPL is now quite standard, but few people know that if there is not restriction of finiteness, if we consider a consequence relation as a relation between on the one hand a set of formulas (a theory) of any cardinality and on the other hand a formula (consequence of the theory), then CPL is not decidable, despite compactness (see Béziau 2001). This proof is presented in the book of Enderton (Enderton 1972).

These remarks show that CPL as it is today is necessary quite different from as presented in the *Tractatus*.

3 The Paragraph 5.141 and Related Concepts

We will now focus on paragraph 5.141 of the *Tractatus*. It is as follows:

> 5.141 If p follows from q and q from p then they are one and the same proposition.

For the sake of precision and exactness, below is the German version as printed in the original publication:

> 5.141 Folgt p aus q und q aus p, so sind sie ein und derselbe Satz.

And here is the position of 5.141 in the *Tractatus*, explicitly presented as a tree:[2]

---

[2] The illustration is used with permission from the University of Iowa Tractatus map (tractatus.lib.uiowa.edu). For more on the Tractatus map, see Stern 2016 and Stern 2019.

We will not in general present the full German original of the sentences and paragraphs that we are commenting, but only the German originals of central notions of interest for us here. In general translations from German to English are quite straightforward for the topic of our paper.

The only, but essential, case which is tricky is the one of "Satz", which is very important for us here. Depending on the situation, it can be translated in English, as: sentence, proposition, statement, principle (cf. *Satz vom Grund*, corresponding to *Principle of Reason*). Let us note that on both English translations of the *Tractatus*,[3] "Satz" has been translated by "proposition" and we will also follow here this translation. It seems reasonable to think that Wittgenstein uses "Satz" as corresponding to what is called a "proposition" in *Principia Mathematica*, and in fact he is using the letters "p" and "q" similarly to what is done in Whitehead and Russell's book (Russell 1910). The sign "p", which is the first letter of the word "proposition", is there used as a variable for propositions, due to the intended range of it, like "n" is used as a variable for numbers, and then "q" and "m" respectively follow.

There are two important notions in 5.141: *follow* and *proposition*. To properly understand 5.141 we need to have a correct understanding of these two notions, the two being interrelated: if we want to understand what a proposition in the *Tractatus* is, we need to understand the meaning of "follow". This is a technical notion

---

[3] We will generally follow Ogden's translation, which is less nice than Pears and McGuinness's one, but closer to the German original.

depending on two other technical notions. The meaning of "follow" ("folgen", in German) is presented in the paragraph 5.11:

FOLLOW: If the truth-grounds which are common to a number of propositions are all also truth-grounds of some one proposition, we say that the truth of this proposition *follows* (*folge*) from the truth of those propositions.

As we can see the meaning of "follow" depends on the notion of truth-ground (*Wahrheitsgrund*), which is defined in paragraph 5.101:

TRUTH-GROUNDS: Those truth-possibilities of its truth-arguments, which verify the proposition, I shall call its *truth-grounds* (*Wahrheitsgründe*).

And as we can see this notion depends on the notion of truth-possibilities. (Wahrheitsmöglichkeiten):

We can represent truth-possibilities by schemata of the following kind ('T' means 'true', 'F' means 'false'; the rows of 'T's' and 'F's' under the row of elementary propositions symbolize their truth-possibilities in a way that can easily be understood):

p	q	r
T	T	T
F	T	T
T	F	T
T	T	F
F	F	T
F	T	F
T	F	F
F	F	F

,

p	q
T	T
F	T
T	F
F	F

,

p
T
F

.

(TLP 4.31)

This corresponds to what are called nowadays "distributions of truth-values". Some books follow Wittgenstein's presentation, using "T" and "F" (in the original text we have the German initials: "W" and "F"), other books instead use "1" and "0". Emil Post was using a notation which is rarely used: "+" and "-". The original terminology of Wittgenstein "truth-possibilities" has also not been followed, at least for CPL, but there is a connection with possible worlds in Kripke semantics for modal logic via Carnap (Carnap 1947).

Let us note that nowadays there is no specific or/and standard word in CPL for what is called in the *Tractatus* a truth-ground ("ground" has recently became famous through Kit Fine but with another meaning, see e.g. Fine 2012). However this notion is perfectly clear. This is what can be called a model, but a model in CPL is not a mathematical structure, like in first-order logic, it is a function from the set of propositions into {0,1}. Generally "model" is not used in CPL; but Chang and Keisler (Chang 1973) used it to make a uniform presentation of CPL and First-Order Logic.

Such a function is generally called a "valuation" by contrast to "distribution of truth-values" which are functions defined only on the set of atomic propositions. Wittgenstein is not making the distinction. The fact that valuations can be generated by distributions of truth-values and that a distribution of truth-value has a unique extension which is a valuation, is directly related to the concept of absolutely free algebra. These technicalities were made precisely clear in the Polish school, in particular by Łoś (see Łos 1958).

Using a bit of symbolism, denoting a Truth-Ground as TG and a valuation as v, we can put the definition of Truth-Ground as follows:

$$TG[p] = \{v; v(p) = 1\}$$

And if we replace the terminology "Truth-Ground" by "Model". We have:

$$mod[p] = \{v; v(p) = 1\}$$

We can therefore reformulate 5.141. as follows:

$$\text{if } mod[p] = mod[q], \text{ then } p = q$$

It is worth noting that Tarski in 1936 used the same terminology, "folgen" (his paper was written in German, Tarski 1936a), and a definition similar to the one of Wittgenstein in the *Tractatus*, but more general in two aspects: it does not reduce to propositional logic, it is a relation between theories and propositions:

"The sentence X follows logically from the sentences of the class K if and only if every model of the class K is also a model of the sentence X." (Tarski 1936d)

And Tarski is the guy who proved that CPL is a Boolean algebra (Tarski 1936e), but not on the basis of this notion, nowadays standardly called "semantical" consequence.

4 Is CPL in the *Tractatus* a Boolean Algebra?

First of all this question has not to be confused with the quite funny question "Is the *Tractatus* a Boolean Algebra?". The answer to the latter is: certainly not! This is what Donald Duck would reply, or any rational animal. The (structure of the)

Tractatus is just a tree. However a nice tree with lots of flowers and fruits... But there is a connection between relations of order and Boolean algebra: as Marshall Stone discovered (Stone 1935), a distributed complemented lattice is a Boolean Ring, the two being two equivalent formulations of a Boolean algebra.

And this gives us a clue to the original question, because to answer it we need to have a clear idea of what is a Boolean Algebra. The simplest Boolean algebra is the Boolean algebra on {0,1}. And the simplest way to consider this algebra is to consider the two operations + and × defined on these numbers by the following tables:

×	0	1		+	0	1
0	0	0		0	0	1
1	0	1		1	1	1

We have then what is called an "Idempotent Ring". Something very simple despite this quite poetic name and that may look complicated or/and incomprehensible for non-mathematicians. Someone may think that CPL in the *Tractatus* is not a Boolean Algebra because we cannot find "Idempotent Ring", "0" and "1" and such tables. But of course we have to go beyond appearances. We can rewrite these two tables as follows:

.	F	T		v	F	T
F	F	F		F	F	T
T	F	T		T	T	T

These are exactly the same tables, we just have changed the signs. All these signs are used in the *Tractatus*, but these tables themselves are not presented. Note however that Russell and Wittgenstein were drawing similar tables in the 1910s (before the *Tractatus*).

If we consider these tables are defining operations on {F,T}, i.e. with domain and co-domain {F,T}, then what we have is what is called the semantics of CPL. And the semantics of CPL is nothing else than the Boolean algebra on {0,1}. One could say that CPL is a Boolean algebra because its semantics is a Boolean algebra, but this would be a bit exaggerated not to say confusing. CPL is a Boolean algebra in another way which is different, in particular because it is a different Boolean algebra than this simplest one. And this is this second way which is connected to the paragraph 5.141.

Boole was considering that $x^2 = x$ (where x is a variable for a proposition, and using a notation mimicking arithmetic) is the fundamental law of thought, from which in particular it is possible to derive the law of contradiction (see Béziau 2018). Wittgenstein was less extravagant but nevertheless would have agreed with Boole that p and $p.p$ are identical.

But in CPL p and $p.p$ are considered as two *different* propositions. We are not writing "p" and "$p.p$", because we are considering the objects they refer to. Note also that Wittgenstein is not using quotation marks in the paragraph 5.141.

The two propositions p and $p.p$ are different but they are considered as *logically equivalent*. What does this mean? According to 5.141, p and $p.p$ are one and the same proposition because one follows from the other one and vice versa, because they have the same truth-grounds according to the definition given in 5.101. In CPL they are not the same, but they are equivalent. But considering that logical equivalence is a congruence relation we can "identify" them and this leads us to a Boolean algebra.

Wittgenstein does not make this detour, he is directly considering the algebra that we can get by factoring CPL with logical equivalence. This is generally called a Lindenbaum-Tarski algebra because this methodology can be applied to logics other than CPL, but in case of CPL the so-called Lindenbaum-Tarski algebra is in fact a Boolean algebra. It is not the Boolean algebra on {0,1}, it has in particular much more than two elements. Leibniz is famous for the following definition: "Two terms are the same (eadem) if one can be substituted for the other without altering the truth of any statement (salva veritate)." (Leibniz 1680, ch. 19, def. 1) Leibniz is talking here about terms and gives the following illustration: "For example, 'triangle' and 'trilateral', in every proposition demonstrated by Euclid concerning 'triangle', 'trilateral' can be substituted without loss of truth (salva veritate)." (Leibniz 1680, ch. 20, def. 1) We can generalize this view applying this definition of identity to any objet, including propositions. Now to claim that the two propositions p and $p.p$ are the same, in this Leibnizian sense, because they can be substituted for the other without altering the truth of any statement, we have to prove the so-called replacement theorem. This is what Tarski did and therefore showed that CPL is a Boolean algebra. Wittgenstein did not prove this theorem, so the sameness he is talking about in 5.141 is ambiguous because there is no guarantee that it can work. And moreover Wittgenstein had no idea that this corresponds to what we now call a Boolean algebra.

What we can say is that the *Tractatus*, through 5.141, is aiming at conceiving CPL has a Boolean algebra.

Bibliography

Bennett, Mary K. (1973): Review of *Lattice theory* by Garrett Birkhoff, 3rd edition and *Lattice theory: first Concepts and distributive lattices* by George Grätzer, Bulletin of the American Mathematical Society, Vol. 79, 1 – 5.

Béziau, Jean-Yves (1999): "The mathematical structure of logical syntax." In: *Advances in contemporary logic and computer science*. 3 – 15.

Béziau, Jean-Yves (2001): "What is classical propositional logic?" In: *Logical Investigations*. Vol. 8, 266 – 277.

Béziau, Jean-Yves (2002): "The philosophical import of Polish logic." In: *Methodology and philosophy of science at Warsaw University*. 109 – 124.

Béziau, Jean-Yves (2007): "Sentence, proposition and identity." In: *Synthese*. Vol. 154, 371 – 382.

Béziau, Jean-Yves (2018): "Is the Principle of Contradiction a Consequence of $x^2 = x$?" In: *Logica Universalis*. Vol. 12, 55 – 81.

Birkhoff, Garrett (1940): *Lattice Theory*. New York: American Mathematical Society.

Boole, George (1854): *An Investigation of the Laws of Thought on Which are Founded the Mathematical Theories of Logic and Probabilities*. London: MacMillan.

Carnap, Rudolf (1947): *Meaning and necessity: a study in semantics and modal logic*. Chicago: University of Chicago Press.

Chang, Chen C. and Keisler, Howard J. (1973): *Model Theory*. Amsterdam: North Holland.

Enderton, Herbert B. (1972): *A mathematical introduction to logic*. New York: Academic Press.

Fine, Kit (2012): "The pure logic of ground." In: *Review of Symbolic Logic*. Vol. 12, 1 – 25.

Gödel, Kurt (1932): "Eine Eigenschaft der Realisierungen des Aussagenkalküls." In: *Ergebnisse eines mathematischen Kolloquiums*. Vol. 2, 27 – 28.

Ladd-Franklin, Christine (1882): *On the Algebra of Logic*. PhD thesis directed by Charles Peirce, Baltimore: The Johns Hopkins University.

Leibniz, Gottfried Wilhelm (1680): "On the Universal Science: Characteristic." In: *Monadology and Other Philosophical Essays*. Schrecker, Paul; Schrecker, Anne (eds.). New York: Bobbs-Merrill, 1965, 11 – 21.

Łoś, Jerzy (1949): "O matrycach logicznych." In: *Prace Wrocławskiego Towarzystwa Naukowego*. Seria B, No. 19, 1 – 41.

Łoś, Jerzy and Suszko, Roman (1958): "Remarks on sentential logic." In: *Indigationes Mathematicae*. 177 – 183.

Quine, William v.O. (1934): "Ontological remarks on the propositional calculus." In: *Mind*. Vol. 43, 472 – 476.

Post, Emil (1921): "Introduction to a general theory of elementary propositions." In: *American Journal of Mathematics*. Vol. 43, 163 – 185.

Stern, David G. (2016): "The University of Iowa Tractatus Map." In: *Nordic Wittgenstein Review*, Vol. 5, No. 2, 203 – 220.

Stern, David G. (2019): "The Structure of *Tractatus* and the *Tractatus* Numbering System." In: Limbeck-Lilienau, Christoph; Stadler, Friedrich (eds.): *The Philosophy of Perception and Observation: Proceedings of the 40th International Ludwig Wittgenstein Symposium*. Boston, Berlin: de Gruyter, forthcoming.

Stone, Marshall (1935): "Subsumption of the theory of Boolean algebras under the theory of rings." In: *Proceedings of National Academy of Sciences*. Vol. 21, 103 – 105.

Tarski, Alfred (1936a): "Über den Begriff der logischen Folgerung." In: *Actes du Congrès International de Philosophie Scientifique, VII Logique*. 1 – 11.
Tarski, Alfred (1936b): "O pojęciu wynikania logicznego." In: *Przeglad Filozoficzny*. Vol. 39, 58 – 68.
Tarski, Alfred (1936c): "On the concept of logical consequence." In: *Logic, semantics, metamathematics*.
Tarski, Alfred (1936d): "On the concept of following logically." In: *History and Philosophy of Logic*. (2003), Stroińska, Magda; Hitchcock, David (trans.). Vol. 23, 155 – 196.
Tarski, Alfred (1936e): "Grundzüge des Systemenkalküls." In: *Fundamenta Mathematicae*. Vol. 25, 503 – 526, Vol. 26, 283 – 301.
Whithead, Alfred N. and Russell, Bertrand (1910): *Principa Mathematica*. Edition 1, Cambridge: Cambridge University Press.
Wittgenstein, Ludwig (TLP): *Tractatus Logico-Philosophicus*. Pears, David F.; McGuinness, Brian (trans.). London: Routledge, 1961.

Janusz Kaczmarek
Ontology in *Tractatus Logico-Philosophicus*: A Topological Approach

Abstract: The paper describes some topological tools: the first part defines Wittgenstein's topology and a lattice of situations (as a theorem). Next, there is considered a non-atomistic lattice of situations, sometimes called a hybrid lattice. It allows us to explore the differences between the atomistic and non-atomistic approaches.

1 Some Theses of Wittgenstein's *Tractatus*

The ontology of logical atomism developed by Wittgenstein and Russell[1] is so complex that it is impossible to provide its overall characteristic in this paper. Instead, let us recall the core tenets of Wittgenstein's philosophy of logical atomism made in *Tractatus Logico-Philosophicus*:

1 The world is everything that is the case.
1.1 The world is the totality of facts, not of things.
1.11 The world is determined by the facts, and by these being all the facts.
1.12 For the totality of facts determines both what is the case, and also all that is not the case.
1.13 The facts in logical space are the world.
1.2 The world divides into facts.
1.21 Any one can either be the case or not be the case, and everything else remain the same.
2 What is the case, the fact, is the existence of atomic facts.
2.01 An atomic fact is a combination of objects (entities, things).
2.011 It is essential to a thing that it can be a constituent part of an atomic fact.[2]

The reason that we quote Wittgenstein here is to differentiate situation and elementary situation from varieties of other concepts of logical atomism like fact, independence of state of affairs, case, compatibility of state of affairs; but not to

---

[1] Cf. TLP and Russell 1985.
[2] Cf. TLP.

Janusz Kaczmarek, University of Łódź, Poland, janusz.kaczmarek@filozof.uni.lodz.pl

propose any clear-cut reading of logical atomism as this seems to be a hopeless task. My aim is modest: to propose a plausible interpretation of Wittgenstein's theses applying Wolniewicz's lattices of situation and other topological methods that would render the hybrid and general modelling of situations possible.

2 Wolniewicz's Lattices

Wolniewicz introduced the notion of a lattice of elementary situations (Wolniewicz 1982, Wolniewicz 1985, Wolniewicz 1999) formally investigating Wittgenstein's ontology of logical atomism. The axioms he presented in Wolniewicz 1999 are as follows:[3]

Axiom 1. *Let CES be a set (empty or non-empty) of the so-called contingent (or proper) elementary situations and ES a set of elementary situations. Define: ES = CES ∪ {o, λ}, where o is called empty elementary situation and λ (≠ o) the impossible one. The contingent situations and empty situation are called possible ones.*

Axiom 2. *A pair (ES, ≤), where ≤ is a partial order on ES is a partially ordered set and for any $x \in ES$: $o \leq x \leq \lambda$.*

The fact: $x \leq y$ is read: *x obtains in y*.

Axiom 3. *For any $A \subseteq ES$ there exists $x \in ES$ such that: $x = \sup A$.*

Remark 1. Wolniewicz wrote:
  Thus, ES is a complete lattice, with the join $x \vee y = \sup\{x, y\}$, and the meet $x \wedge y = \inf\{x, y\}$. Thus, for any $x, y \in ES$, the usual equivalences hold:

$$x \vee y = y \text{ iff } x \leq y, \quad x \wedge y = x \text{ iff } x \leq y.\text{[4]}$$

Axiom 4. *For any $x \in CES$ there exists $y \in CES$ such that $x \vee y = \lambda$*

Axiom 5. *For any $x, y, z \in ES$:*

(5a) if $(x \vee y \neq \lambda$ and $x \vee z \neq \lambda)$, then $(x \vee y) \wedge (x \vee z) \leq x \vee (y \wedge z)$,
(5b) if $y \vee z \neq \lambda$, then $x \wedge (y \vee z) \leq (x \wedge y) \vee (x \wedge z)$[5].

---

[3] Cf. Wolniewicz 1999: 20 – 23.
[4] Cf. Wolniewicz 1999: 21.
[5] Of course, the relation ≤ can be replaced by equality (=), because the converse relations in consequents hold in any lattice.

Axiom 6. *For any $x, y \in ES$: if $x \leq y$, then there is an element $x' \in ES$ such that*

$$x \wedge x' = o \text{ and } x \vee x' = y.$$

It means that for any $y \in ES$ the interval $\langle o, y \rangle = \{x \in ES : o \leq x \leq y\}$ is complemented.

Axiom 7. *Let $LS = \{x \in ES : \lambda \text{ covers } x\}$ be called a logical space of possible worlds. There is a non-empty set LS such that for any contingent or empty situation x exists $w \in LS$ and $x \leq w$.*

Of course, if $CES = \emptyset$, then $LS = \{o\}$.

Remark 2. "λ covers x" means that $x \leq \lambda$ and for any $y \in ES$: if $x \leq y$ and $x \neq y$, then $y = \lambda$.

Axiom 8. *Let AES be a set of atomic elementary situations, i. e., $AES = \{x \in ES : x \text{ covers } o\}$. We assume that there is a non-empty set AES such that for any $x \in ES$ there exists $A \subset AES$ such that: $x = \sup A$.*

Remark 3. *If $x = o$, then $o = \sup \emptyset$.*

Axiom 9. *For any $x, y \in ES$ such that neither $x = o$ and $y = \lambda$, nor conversely: if $x \vee y = \lambda$, then there exist $s, t \in AES$ such that $s \leq x$ and $t \leq y$ and $s \vee t = \lambda$.*

Axiom 10. *For any $x, y, z \in AES$: if $x \vee z = \lambda$ and $y \vee z = \lambda$, then $x = y$ or $x \vee y = \lambda$.*

Wolniewicz explains that Axiom 10 "may look odd as an axiom, but to stay in the spirit of the *Tractatus* we want the following relation to be transitive:

if $x, y \in AES$, then (xdy iff ($x = y$ or $x \vee y = \lambda$)),

and **Axiom 10** is just this"[6].

It is evident that the relation d is an equivalence on AES. Any class of the partition $D = AES/d$ is to be called the *logical dimension* of the space LS (as long as ES is not empty).

Axiom 11. *Let "dim LS" mean the number of logical dimensions (as Wolniewicz assumes):*

$$\dim LS = n,$$

where $n \geq 0$ is a natural number and $\dim LS = 0$, if $CES = \emptyset$; $\dim LS = \operatorname{card} D$, otherwise.

---

6 Cf. Wolniewicz 1999: 23.

Wolniewicz concludes:

> The three axioms, **Axiom 8 – Axiom 10**, against the background of **Axiom 1 – Axiom 7**, embody the philosophy of Logical Atomism. Indeed, the "logical atomism" of Russell, whatever they were, had two basic ontological properties: they were simple, and they were mutually independent. Now **Axiom 8** means simplicity of A-situations[7] with regard to their logical space. And in view of the **Axiom 9** and **Axiom 10**, A-situations belonging to different logical dimensions are independent to each other in a Wittgensteinian sense of the term (…).[8]

We can think of a given lattice of elementary situations as the following figures:

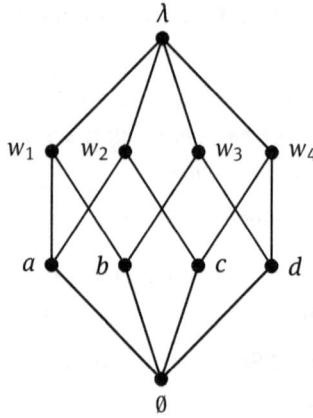

Fig. 1: A lattice with signature (2, 2) refers to dimension $D_1 = \{a, d\}$ with 2 elements and dimension $D_2 = \{b, c\}$ with 2 elements. Source: Wolniewicz 1999: 30.

Example 4. *Now let us make some remarks regarding the lattice. The set $\{a, b, c, d\}$ is a set of atoms, i.e., simple states of affairs (simple and compound states are called elementary situations). If they are, for example, "it's cold", "it's wet", "it's dry" and "it's warm", respectively, then $\{a, d\}$ and $\{b, c\}$ are two logical dimensions of temperature and moisture. Wolniewicz assumes that the number of dimensions is finite but the number of atoms in a given dimension D is arbitrary (finite or infinite). Every lattice of elementary situations SE comprises at least two improper ones: the impossible situation λ and the empty one $\emptyset$ ($\emptyset \neq \lambda$). Wolniewicz assumes, additionally, that the set $ES' = SE - \{\lambda\}$ is a set of possible elementary situations, and CES =*

---

7 The term "A-situation" is used originally by Wolniewicz for "atomic elementary situation".
8 Cf. Wolniewicz 1999: 27.

$ES - \{\emptyset, \lambda\}$ is a set of contingent situations. The set $\{w_1, w_2, w_3, w_4\}$ is a set of possible worlds (a logical space). Obviously, for any situation $x : \emptyset \leq x \leq \lambda$.

Example 5. *A lattice with finite number of dimensions.*

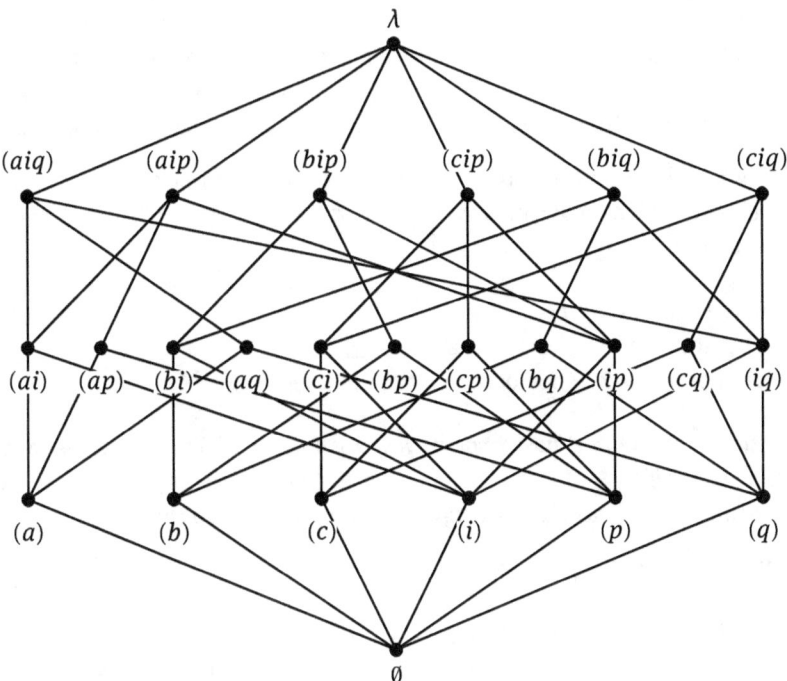

Fig. 2: A lattice with signature (2, 1, 3). Source: The author.

The lattice has 3 dimensions: $D_1 = \{p, q\}$, $D_2 = \{i\}$, and $D_3 = \{a, b, c\}$. Wolniewicz assumes that the number of dimensions is finite but the number of elements in a dimension can be finite or infinite. In the latter case we obtain a lattice of finite height and infinite width. It is easy to prove that **Axioms 1 – 11** are consistent. The lattice presented in **Figure 1** fulfils the axioms.

3 Towards Topological Ontology of Elementary Situations

Now, let us pose the question about topological ontology. Although it belongs to a formal ontology[9], topological concepts, theorems, and structures are normally used and, in result, a topological approach is favoured. I intend to convince the reader about the advantages of modelling ontological concepts and theorems within topological framework. To do it, let me start with some preliminaries.

Definition 1. *Let X be a set (not necessarily nonempty) and τ_X a family of subsets of X. A pair (X, τ_X) is a topology or a topological space on X, if the following conditions are fulfilled:*

  a) $\emptyset \in \tau_X$ and $X \in \tau_X$,
  b) *A union of sets from τ_X is a set of τ_X,*
  c) *A finite intersection of sets from τ_X is a set of τ_X.*

Definition 2. *Let (X, τ_X) be a topological space and $A \subseteq X$. Then (A, τ_X) is called a subspace of X, if $\tau_A = \{A \cap B : B \in \tau_X\}$. τ_A is usually called the subspace topology on A. Sometimes we say that topology τ_A is induced by τ_X on the set A.*

Example 6. *Topologies*

  $\tau 1$. *If $X = \emptyset$, then $(\emptyset, \{\emptyset\})$ is a topological space.*
  $\tau 2$. *If $X = \{1; 2\}$, then $(X, \{\emptyset, \{1\}, X\})$ is a topological space. It is known as the Sierpiński's space.*
  $\tau 3$. *If $X = \mathbb{R}$, $\mathbb{R}$ is the set of real numbers, and any set of $\tau_\mathbb{R}$ is a union of sets in form $(u; v) = x \in \mathbb{R} : u < x < v$, for $u, v \in \mathbb{R}$, then $(\mathbb{R}, \tau_\mathbb{R})$ is a topological space called natural topology on $\mathbb{R}$ (or Euclidean topology).*
  $\tau 4$. *If $X = \mathbb{R}$ and $\emptyset \neq A \subset X$, then $(X, \{\emptyset, A, X - A, X\})$ is a topological space.*
  $\tau 5$. *For any set X the discrete topology on X is the topology τ_d such that $\tau_d = \{U : U \subseteq X\}$, so the collection of sets of τ_d equals the power set of X, i.e. $\tau_d = \mathcal{P}(X)$. Next, the indiscrete topology (or trivial topology) on X is the topology $\tau_{triv} = \{\emptyset, X\}$.*

Definition 3. *Any element of τ_X is called an open set. If A is open set, then $X - A$ is called a closed set of the topological space τ_X. Of course, X and $\emptyset$ are open and closed in each topological space at the same time.*

---

9 Cf. Kaczmarek 2008.

Definition 4. *A collection* $\mathbb{B}$ *of open sets of a topological space* (X, τ_X) *is called a basis of* (X, τ_X), *if each open set in* X *can be represented as a union of elements of* $\mathbb{B}$.

Definition 5. *Let* (X, τ_X) *be a topological space. A collection* $\mathbb{S}$ *of open sets is called a subbasis of* (X, τ_X), *if each open set in a basis of* (X, τ_X) *can be represented as a union of finite intersections of elements of* $\mathbb{S}$.

Definition 6. *For two topologies,* τ *and* τ', *on* X *we say that* τ *is weaker (or coarser) than* τ' *(equivalently: that* τ' *is stronger or finer than* τ) *if* $\tau \subset \tau'$ *(we write:* $(X, \tau) \leq (X, \tau'))$. *It means that each open set of* τ *is also an open set in* τ'. *Of course, for any set* X *and any topology* τ *on* X *we have:*

$$\tau_i \subset \tau \subset \tau_d \text{ (or: } (X, \tau_i) \leq (X, \tau_X) \leq (X, \tau_d)).$$

Remark 4. *For any family* $\tau'_X \subseteq 2^X$, *there exists the least topology* τ_X *having the following properties: 1)* $\mathbb{S} \subseteq \tau_X$, *2) for any* $\tau'_X \subseteq 2^X$, *if* $\mathbb{S} \subseteq \tau'_X$, *then* $\tau_X \subseteq \tau'_X$.

Proof. Really, the least topology τ_X which includes the family $\mathbb{S}$ is a topology defined in 3 steps:

a) subbasis: $\mathbb{S} \cup \{\emptyset, X\}$,
b) basis: $\mathbb{B} = \{A_1 \cap \ldots \cap A_n : A_i \in \mathbb{S}\} \cup \{\emptyset, X\}$, for $i = 1, \ldots, n, n \in \omega$,
c) topology: $\tau_X = \{\bigcup W : W \subseteq \mathbb{B}\}$

Remark 4 is important for understanding Wittgenstein's topology and the ideas standing behind it.

Example 7. *Subbases and bases*

τ6. Let $\mathbb{Q}$, be a set of rational numbers. A family $\mathbb{B}$, = $\{(u; v)$, for $u, v \in \mathbb{B}, u < v\} \cup \{\emptyset\}$ is one of the basis of natural topology.

τ7. A family $\mathbb{S} = \{(-\infty; v), \text{for } v \in \mathbb{B}\} \cup \{(u; \infty)$, for $u \in \mathbb{B}\}$ is one of the subbasis of the natural Euclidean topology.

τ8. Let us consider the subbasis $\mathbb{S} = \{\{1, 2, 3\}, \{2, 3, 4\}, \{3, 4, 5, 6\}\} \cup \{\emptyset\}$. Then a family:

$$\mathbb{B} = \{\{1, 2, 3\}, \{2, 3, 4\}, \{3, 4, 5, 6\}, \{2, 3\}, \{3\}, \{3, 4\}\} \cup \{\emptyset\}$$

is the basis and

$$\tau_{\{1,2,3,4,5,6\}} = \{\emptyset, \{1, 2, 3, 4, 5, 6\}, \{1, 2, 3\}, \{2, 3, 4\}, \{3, 4, 5, 6\}, \{2, 3\},$$
$$\{3\}, \{3, 4\}, \{1, 2, 3, 4\}, \{2, 3, 4, 5, 6\}\}$$

is a topology generated by the given subbasis and basis.

As some connections between topologies and lattices are shown in the sequel, let us remark that the topology $\tau_{\{1,2,3,4,5,6\}}$ can be depicted as the lattice structure:

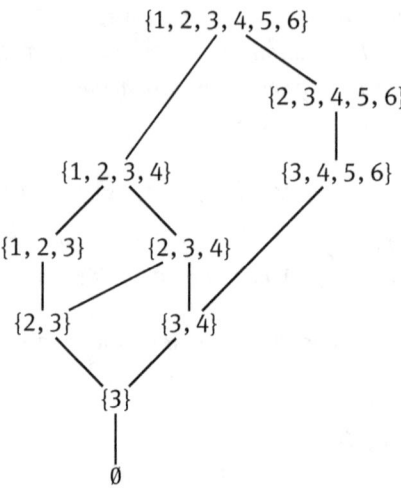

Fig. 3: The topology $\tau_{\{1,2,3,4,5,6\}}$ as a lattice. Source: The author.

4 Discrete Topology of Situations

Now, it turns out that the concept of a topological space can be used to define some classes of lattices proposed by Wolniewicz, and some other classes, too. Let us construct a lattice of signature (2, 1, 3) as in **Figure 2**. Taking into account a basis $B = \{\{a\}, \{i\}, \{q\}\}$, we obtain a topology that is a set of all objects x of **Figure 2** such that $x \in (aiq)$. To be more accurate, we have to change the nodes of the lattice: (a) into $\{a\}$, (aq) into $\{a, q\}$, (aiq) into $\{a, i, q\}$, and so on. The family $\tau_{\{a,i,q\}} = \{\emptyset, \{a\}, \{i\}, \{q\}, \{a, i\}, \{a, q\}, \{i, q\}, \{a, i, q\}\}$ is a topological space on $\{a, i, q\}$. Other topological spaces can be easily found in **Figure 2**. Note that any topological space is discrete and the lattice consists of elements of all topological spaces and of λ.

Now, let us define a lattice of elementary situations in more general terms.

Let $A = \{A_1, A_2, A_3, ...\}$ be any countable family of sets and for any A_i, card$(A_i) \geq 2$ and for any $A_i, A_j, A_i \cap A_j = \emptyset$ (now we are rejecting Wolniewicz's axiom that the number of dimensions is finite (**Axiom 11**). Each A_i is understood as a set of atomic situations and each A_i consists of singletons (i. e. $A_i = \{\{a_{i_1}\}, \{a_{i_2}\}, \{a_{i_3}\}, ...\}$). Put A_i as a dimension of a lattice and a set of incompatible atoms (for any a and b from $A_i\{a\} \cup \{b\} = \lambda$; Cf. Point 5, below). Now let us take into account a function $c : \mathbb{N} \to \bigcup_{k=1}^{\infty} A_k$, $\mathbb{N}$ is a set of natural numbers, such that $c(k) \in A_k$. Then we fix, for the given c:

$$\mathbb{B}_c = \{c(k) : k \in \mathbb{N}\} \cup \{\emptyset\}.$$

Fact 1. *A pair (X, τ_X), where $X = \bigcup_{k=1}^{\infty} c(k)$ and for any $A \in \tau_X$, $A = \bigcup B$ for $B \in \mathbb{B}_c$, is a topological space.*

Proof. $\bigcup B$ means a union of some sets from $\mathbb{B}_c$. Each basis can be treated as a subbasis, thus by **Remark 4**, we obtain the least topological space generated by $\mathbb{B}_c$.

Definition 7. *I propose to call the space fixed by **Fact 1** a Wittgenstein's topology.*

Remark 5. *The name "Wittgenstein's topology" is further extended to refer to the finite cases; thus, $\tau_{\{a,i,q\}}$ defined above is a Wittgenstein's topology.*

So we can obtain:

Fact 2. *A union of all Wittgenstein's topologies is a lattice (with $\emptyset$ as an empty situation and $\bigcup(A_1 \cup A_2 \cup ...)$ as λ); let us call it $\mathcal{LWT}$ ($\mathcal{LWT}$ refers to a lattice of Wittgenstein's topologies), where supremum $\underline{\cup}$ and infimum $\underline{\cap}$ is defined in the following way (S_i, S_j are any elementary situations, i.e., $S_i, (S_j)$ belongs to some topology or is λ):*

$$S_i \underline{\cup} S_j = \begin{cases} S_i \cup S_j, & \text{if } S_i \cup S_j \in \tau_{A_k}, \text{for some } k \in \omega \\ \lambda, & \text{oth.} \end{cases}$$

$$S_i \underline{\cap} S_j = \begin{cases} S_i \cap S_j, & \text{if } S_i \cap S_j \in \tau_{A_k}, \text{for some } k \in \omega \\ \emptyset, & \text{oth.} \end{cases}$$

Proof. Having the definitions of $\underline{\cup}$ and $\underline{\cap}$, it is evident that for any S_i, S_j there exist $S_i \underline{\cup} S_j$ and $S_i \underline{\cap} S_j$. An order $\leq$ in the $\mathcal{LWT}$ is defined in a standard way.

Definition 8. *Elements from $A_1, A_2, A_3, ...$ are called atoms of the lattice (atomic elementary situations). Elements of the form $w = c(\mathbb{N})$, for any functions of type c, are called possible worlds.*

Remark 6. Note that Wittgenstein's topology is maximal in a given lattice $\mathcal{LWT}$. Thus, if (X, τ_X) is a Wittgenstein's topology and $(X', \tau_{X'})$ is a topology such that $X \subset X'$ and $\tau_X \subset \tau_{X'}$, then $(X, \tau_X) = (X', \tau_{X'})$. In a given lattice $\mathcal{LWT}$ we could point to different topologies that are not maximal. Take into account **Figure 2**: the topological space

$$\tau_{\{a,i,q\}} = \{\emptyset, \{a\}, \{i\}, \{q\}, \{a,i\}, \{a,q\}, \{i,q\}, \{a,i,q\}\}$$

is maximal in a given lattice but

$$\tau_{\{a,i\}} = \{\emptyset, \{a\}, \{i\}, \{a,i\}\}$$

is not a maximal topological space.

Fact 3. *All axioms of Wolniewicz's lattice are fulfilled in $\mathcal{LWT}$ (except **Axiom 11** saying that the number of dimensions is finite).*

Proof. Note that each Wittgenstein's topology is discrete and is a Boolean algebra. The relation d can be defined trivially by the condition:

(d) for any $S_i, S_j \in AES : (S_i d S_j$ iff $S_i, S_j \in A_k$ for some $k \in \omega)$.

For example, let us prove **Axioms 8** and **9**.

Let $S \in \mathcal{LWT}$ be any elementary situation. (1) If $S = \emptyset$, then $\emptyset = \sup \emptyset$, (2) if $S = \lambda$, then $\lambda = \sup \{\{a\}, \{b\}\}$, where $\{a\}$ and $\{b\}$ belong to the same dimension, (3) if S is a contingent situation, then S is an element of some Wittgenstein's topology; thus, by **Fact 1**, S is a union of some sets from the basis of this topological space, and hence, there exists a subset B of AES such that $S = \sup B$; thus **Axiom 8** holds.

Next, let S, S' be any elementary situation such that: either ($S \neq \emptyset$ or $S' \neq \lambda$) or ($S \neq \lambda$ or $S' \neq \lambda$). We assume that (*) $S \cup S' = \lambda$, so we should prove three cases. Consider only the case of S and S' being contingent situations. If $S \cup S' = \lambda$, then S and S' do not belong to one Wittgenstein's topology. Assume that $S \in \tau_X$ and $S' \in \tau_{X'}$, and let $\mathbb{B}_c$ and $\mathbb{B}_{c'}$ be two bases for two Wittgenstein's topologies. Then, S and S' are unions of elements from $\mathbb{B}_c$ and $\mathbb{B}_{c'}$, respectively. Suppose, contrarily, that $S \cup S'$ has no elements from one dimension. Thus there exists function $c(k)$ such that $S \cup S' \subset c(\mathbb{N})$, but the family $\{A : A \subset c(\mathbb{N})\}$ is a Wittgenstein's topology and $S \sup S' \in \{A : A \subset c(\mathbb{N})\}$, thus we obtain a contradiction with (*).

Definition 9. *Two given topological spaces, (X_1, τ_{X_1}) and (X_2, τ_{X_2}), are homeomorphic iff there exists a function*

$$f : X_1 \to X_2,$$

such that

(a) f is a "one-to-one" and "on" mapping,
(b) both f and f^{-1} are continuous functions.

Remark 7. *General topology defines a continuous function in the following way. Let (X_1, τ_{X_1}) and (X_2, τ_{X_2}) be two topological spaces. Given a function $f : X_1 \to X_2$ and an element x_0 of the domain X_1, f is said to be continuous at the point x_0 when the following holds: for any neighbourhood C of point $f(x_0)$ from τ_{X_2} there exists some neighbourhood $D \in \tau_{X_1}$ of x_0 such that $f(D) \subset C$. If a given function f is continuous at any point of its domain, then f is said to be continuous.*

Fact 4. *Any two Wittgenstein's topologies in a given $\mathcal{LWT}$ are homeomorphic.*

Proof. A $\mathcal{LWT}$ is defined in **Fact 2**. Take into account any two different Wittgenstein's topological spaces, (X, τ_X), $(X', \tau_{X'})$, where $X = \bigcup_{k=1}^{\infty} c(k)$ and $X' = \bigcup_{k=1}^{\infty} c'(k)$, for some functions c and c' defined above. Thus:
$$\mathbb{B}_c = \{c(k) : k \in \mathbb{N}\} \cup \{\emptyset\} \text{ and } \mathbb{B}_{c'} = \{c'(k) : k \in \mathbb{N}\} \cup \{\emptyset\}$$
are the bases of these topologies, respectively.

Notice that $\mathbb{B}_c$ and $\mathbb{B}_{c'}$ are cardinally equivalent sets and take the function

$$g : \{c(k) : k \in \mathbb{N}\} \cup \{\emptyset\} \to \{c'(k) : k \in \mathbb{N}\} \cup \{\emptyset\}$$

such that $g(c(k)) = c'(k)$ and $g(\emptyset) = \emptyset$. Next, we extend the function g to a function $h : X \to X'$ by the condition: for any $A \in \tau_X h(A) = C$, where $A = \bigcup B$ for some sets $B \in \mathbb{B}_c$ and $C = \bigcup g(B)$. It is easy to show that h is a homeomorphism. It suffices to prove that h is continuous. Let x be any point of X. Consider $h(x)$ and any neighbourhood C of $h(x)$. Of course, $\{x\} \in \mathbb{B}_c$ and $h(\{x\}) \in \mathbb{B}_{c'}$. The set $C \in \tau_{X'}$ and, because $(X', \tau_{X'})$ is a discrete topology and belongs to atomistic lattice (**Axiom 8** is fulfilled), then $C = \bigcup g(B)$, for some sets $B \in \mathbb{B}_c$. So, we can take a neighbourhood $A = \bigcup g^{-1}(g(B))$. It is evident that $h(A) = C$, so $h(A) \subset C$. Thus, h is a continuous function, because we have taken into account any $x \in X$. In the same way we prove that h^{-1} is continuous. Proving the condition (a) of **Definition 9** is trivial. It is enough to see that both topologies are of the same cardinality and that the function g generates one-to-one mapping h.

Remark 8. *To prove that h and h^{-1} is continuous, we can also use some theorems of general topology. Namely:*

Theorem 2. *Let (X_1, τ_{X_1}) and (X_2, τ_{X_2}) be two topological spaces and $f : X_1 \to X_2$. If (X_1, τ_{X_1}) is a discrete topological space, then f is continuous.*

Conclusion. *Any Wittgenstein's topologies in Wolniewicz's lattices are homeomorphic.*

Proof. Take, for example, a lattice described in diagram 2 (**Figure 2**) along with the explanations at the beginning of **Section 4**. Each maximal topology in a Wolniewicz's lattice is finite and function g and homeomorphism h give the expected result.

Fact 5. *There exist many topologies in $\mathcal{LWJ}$ that are homeomorphic with the given Wittgenstein's topology.*

Proof. Take a $\mathcal{LWJ}$ defined in **Fact 2**. Each Wittgenstein's topological space (X, τ_X) in $\mathcal{LWJ}$ has a countable base. Let $\mathbb{B}_c = \{c(k) : k \in \mathbb{N}\} \cup \{\emptyset\}$ be a base of (X, τ_X). Taking into account a base $\mathbb{B}_{2c} = \{c(2k) : k \in \mathbb{N}\} \cup \{\emptyset\}$ we obtain a new topology generated by $\mathbb{B}_{2c}$ that is homeomorphic with (X, τ_X). It is enough to consider the function $g : B_c \to B_{2c}$ defined by $g(c(k)) = c(2k)$ and $g(\emptyset) = \emptyset$.

Remark 9. *Philosophical conclusion.* As we have already mentioned, Wolniewicz assumes a finite number of dimensions but in a given dimension any number of atoms are admitted. So, a width of a lattice can be finite, countable, or uncountable. According to **Conclusion** we obtain the following: we can investigate structural properties of a different maximal topologies in a given Wolniewicz's lattice by investigating one of them only. Next, according to **Fact 5**, it is possible to explore maximal topologies by exploring their subspace (although still of the same cardinality).

5 Dependence and Independence of State of Affairs: Separated Situations

Let us remark that atomic elementary situations belonging to different logical dimensions are independent of each other in the Wittgensteinian sense. Following Wolniewicz, we define[10]

Definition 10. *Let ES be a non-empty set of elementary situations from $\mathcal{LWJ}$. ES is independent iff $\sup ES \neq \lambda$ and for any $A, B \in ES : A = B$ or $A \cap B = \emptyset$*[11].

---

[10] Cf. Wolniewicz 1985, Wolniewicz 1999. Wolniewicz uses a term W-independent (independency in the Wittgensteinian sense). We owe next concepts of compatibility and separation to Wolniewicz, too.

[11] The term *ES* was used earlier (**Point 2**) as a set of all elementary situations in a given Wolniewicz's lattice. Now, in the case of lattices which consist of topological spaces, *ES* is related to any set of situations.

Fact 6. *Any two (different) situations A, B are independent iff $A \cup B \neq \lambda$ and $A \cap B = \emptyset$.*

Proof. The proof is evident by **Definition 10** but notice that situations A and B have to be different. If $A = B$, then $A \cup B \neq \lambda$ and $A \cap B = \emptyset$ only if $A = \emptyset$, so only $\emptyset$ is independent to $\emptyset$. If A or B is equal with λ, then $A \cup B = \lambda$, thus A and B are not independent. An interesting case is when A and B are contingent elementary situations.

Definition 11. *Any two contingent situations, A and B, are called incompatible iff $A \cup B = \lambda$. Of course, two such situations are compatible iff they are not incompatible.*

Conclusion. *A and B are compatible iff there exists a possible world w such that $A \leq w$ and $B \leq w$.*

Fact 7. *Let IN be a set of independent situations and COM a set of compatible ones. Then:*

(i) $IN \cap COM \neq \emptyset$,
(ii) $IN - COM \neq \emptyset$,
(iii) $COM - IN \neq \emptyset$.

Definition 12. *Any two (different) sets A and B of $\mathcal{LWJ}$ are separate iff: if $A \neq B$ then there exists a world w such that $(A \subset w$ and $B \not\subset w)$ or $(A \not\subset w$ and $B \subset w)$.*

Fact 8. *Let D be a dimension of $\mathcal{LWJ}$ and $A, B \in D$. It is evident that if $A \neq B$, then A and B are separate. Moreover, for any $A, B \in \mathcal{LWJ}$, if $A \neq \emptyset \neq B$ and $A \neq B$, then A and B are separate*[12].

Proof. Take any A and B. $A \neq B$, so if $A = \lambda$, then B is a contingent or empty situation and thus, by **Axiom 7**, there exists w such that $B \subset w$ (but, obviously, $A \not\subset w$). If $A \neq \lambda \neq B$, $A \neq B$, then $A - B \neq \emptyset$ or $B - A \neq \emptyset$. Suppose: $A - B \neq \emptyset$. Thus there exists $a \in A$ such that $a \notin B$. We see that $\{a\} \in D$ for some dimension of $\mathcal{LWJ}$. Because $card(D) \geq 2$, then there exists $\{a'\} \in D$ such that $\{a'\} \neq \{a'\}$. Consider a basis $\mathbb{B}$ such that $B \cup \{a'\} \subset \bigcup C$, for $C \in \mathbb{B}$ and a topological space generated by basis $\mathbb{B}$. Of course, $B \subset \bigcup C (= w)$ and $A \not\subset \bigcup C$.

Remark 10. *Note that the lattice given in **Figure 2** is not separate. Consider, for example, the nodes (ap) and (aip).*

---

[12] In Wolniewicz 1985 this fact is an axiom, in Wolniewicz 1999 a theorem.

6 Non-Atomistic Topology and Hybrid Lattices

When Wittgenstein was asked to give an example of a statement that would refer to a simple state of affairs, he was to answer: *I don't know*. This answer suggests that he could have in mind a set of non-atomistic situations (states of affairs)[13]. Let us have a look upon this kind of sets. I propose to define a lattice of Wittgenstein's topological spaces in a more general sense as a lattice which admits topologies with atoms and topologies without atoms.

Definition 13. *Let (X, τ_X) be any topological space, $A_1, A_2, \ldots$, a subset of X such that for any $i, j, i \neq j : A_i \not\subset A_j$ and $A_j \not\subset A_i$, and for any $k \in \omega : (A_k, \tau_{A_k})$ be any topological subspace of (X, τ_X). Then the family $\mathcal{GWL} = \{S : S \in \tau_{A_k}$ for some $k \in \omega\} \cup \{\lambda\}$ with $\underline{\cup}$ as a supremum, $\underline{\cap}$ as a infimum and $\emptyset$ (or zero) and λ (or unit) as the least and greatest element of $\mathcal{GWL}$ is a general Wittgenstein's lattice, where $\underline{\cup}$ and $\underline{\cap}$ are defined by conditions as in **Fact 2**.*

Remark 11. *$\mathcal{GWL}$ can be also called a hybrid lattice because it admits atomic and non-atomic topological spaces. Let us provide some representative examples.*

Example 8. *Let (R, τ_R) be natural topological space on $\mathbb{R}$. Take into account $A_1 = \{1, 2, 3\}$ and $A_2 = (0, 1)$.*

The topological subspace induced by A_1, i.e., (A_1, τ_{A_1}) can be visualized as:

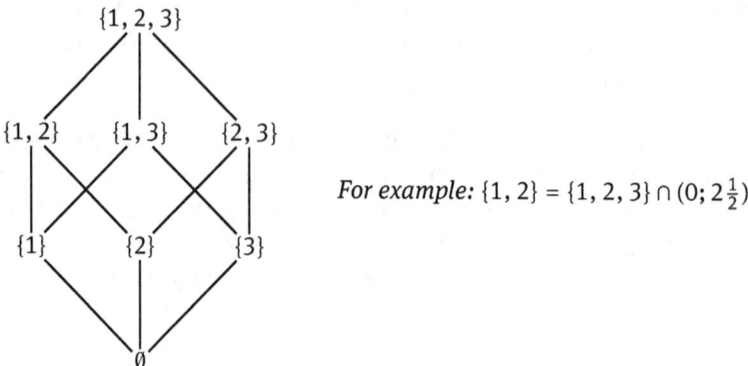

For example: $\{1, 2\} = \{1, 2, 3\} \cap (0; 2\frac{1}{2})$

Fig. 4: The topology $\tau_{\{1,2,3\}}$. Source: The author.

---

[13] Glock emphasizes that Wittgenstein had a problem with an example of a simple sentence or an atomic state of affairs. Cf. the discussion presented in Glock 1996, entries: Fact, Elementary Sentence, Sentence.

and the topological subspace induced by A_2, i.e., (A_2, τ_{A_2}) is more complicated to be visualized:

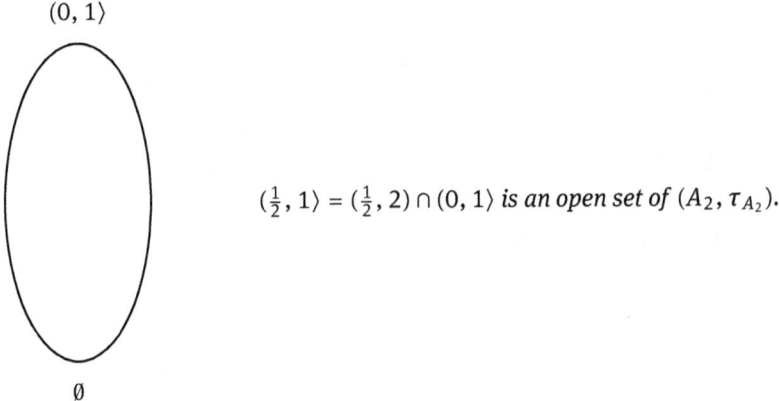

$\langle \frac{1}{2}, 1 \rangle = \langle \frac{1}{2}, 2 \rangle \cap (0, 1)$ is an open set of (A_2, τ_{A_2}).

Fig. 5: The topology $\tau_{(0,1)}$. Source: The author.

Next, consider a set $A_3 = \{\frac{1}{n}; n \in \omega\} \cup \{2\}$, a topological subspace (A_3, τ_{A_3}) induced by natural topology on A_3, and a lattice of (A_k, τ_{A_k}) for $k = 1, 2, 3$. Then we obtain **Figure 6**.

In the lattice below (see **Fig. 6**), there are countable sets of atoms and the topology (A_2, τ_{A_2}) without atoms in the middle of the figure. This means that for any $A \in \tau_{A_2}$ there exists $B \in \tau_{A_2}$ such that $\emptyset \subset B \subset A$ (and, of course, $\emptyset \neq B \neq A$). I hope that the reader is convinced that hybrid lattices can be used to model a (possible) world which is only partially atomistic.

Example 9. Now, take into account $X = (0; 1)$ and any $r \in (0; 1)$. Consider topological subspaces on $X - \{r\}$ induced by natural topology. Let us fix $\lambda = (0; 1)$, the topology on $X - \{r\}$ as τ_r and $\mathcal{GWL}(X - \{r\}) = \bigcup_r \tau_r \cup \{\lambda\}$.

Fact 9. $\mathcal{GWL}(X - \{r\})$ with $\lambda = (0; 1)$ as a unit, $\emptyset$ as zero and any $r \in (0; 1)$ is a $\mathcal{GWL}$.

It is evident that for any r a set $(0; 1) - \{r\}$ plays the role of possible world but in the lattice we have no atoms. Notice also that a set $\{a\} \in (0; 1)$ is not an atom of $\mathcal{GWL}(X - \{r\})$ because for any topological space $\tau_r : \{a\} \notin \tau_r$.

Fact 10. For a given $\mathcal{GWL}(X - \{r\})$, let $A, B \in \mathcal{GWL}(X - \{r\})$. If $A \cup B = (0; 1)$, $r \in A$, where $r \in (0; 1)$ is a certain real number, and $r \notin B$, then A and B are incompatible and separate.

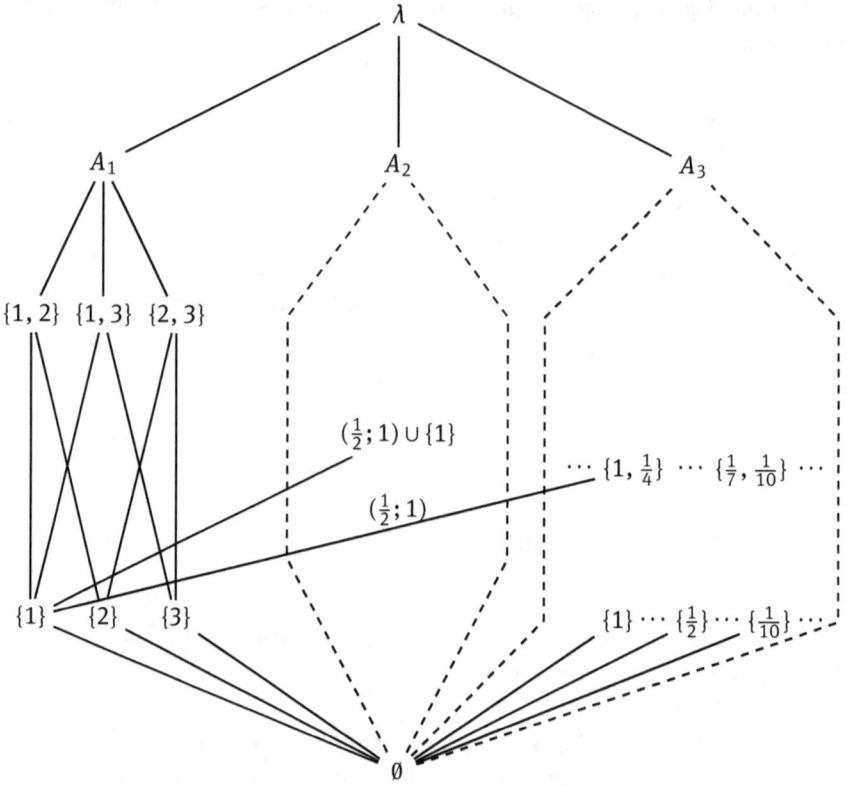

Fig. 6: Hybrid lattice of topologies. Source: The author.

Proof. If $A \cup B = (0; 1)$, then there does not exist r' such that A and $B \in \tau_{r'}$ and thus, the sets are incompatible. If for a certain $r : r \in A$ and $r \notin B$, then $A \notin \tau_r$ and $B \in \tau_r$. Thus, the sets are separate.

7 Negation of an Atomic Elementary Situation

At the 41st International Wittgenstein Symposium, in the discussion, Professor Weingartner raised the question of the nature of negation of atomic elementary situation. I shall try to answer it now. The problem of a negative state of affairs is finely grained so the required topological tools applied to it need to be finely grained, as well. In my opinion, within Wittgenstein's ontology it is impossible to speak about the negation of a given state of affairs since Wittgenstein says:

1.21 Any one (fact, state of affairs, elementary situation) can either be the case or not be the case, and everything else remain the same.

I treat negation as a logical operation characteristic for sentences of logical languages (sentential language, first order logic language, etc.). Therefore, for me, there is no negation present in the Wittgenstein quote. However, we can investigate the so-called negative situations. To this aim, suppose that:

B is a negative situation to an atomic one, A, iff (a) $A \cup B = \lambda$ and (b) $A \cap B = \emptyset$.

Of course, this condition has to be treated as a proposal only, perhaps one of many.

Consider some example given in **Figure 1**. We can notice that $a \vee d = \lambda$, $a \wedge d = o$, or $\{a\} \cup \{d\} = \lambda$, $\{a\} \cap \{d\} = \emptyset$ (in topological terminology). Observe that the same occurs in the case of nodes a and w_3 (w_4). Thus, if our proposal is correct, you can say that the atomic situation like a has two negative situations, d and w_3. The situation d is atomic in the lattice but w_3 is compound. The latter result is parallel with Glock's considerations. He arguments that negation of an atomic fact (or a negative fact) is not atomic but molecular[14].

In the case of non-atomic lattices or hybrid lattices the matter is more complicated but still interesting. For example, taking into account the lattice $\mathcal{GWL}(X-\{r\})$ defined above in **Point 6**, we can observe the following: the set, as for example $(0, \frac{1}{2})$, has no negative situation. To prove this consider topological subspace on the set $(0; 1) - \{\frac{3}{4}\}$ generated (induced) by natural topology. A universe of this topology is: $X = (0; \frac{3}{4}) \cup (\frac{3}{4}; 1)$. In turn $X - (0; \frac{1}{2}) = \langle \frac{1}{2}; \frac{3}{4} \rangle \cup (\frac{3}{4}; 1)$. But this set does not belong to any topology τ_r, for $r \in (0; 1)$. For example, if $r = \frac{3}{4}$, then only the set $A = X - (0; \frac{1}{2})$ fulfils the conditions:

(1) $(0; \frac{1}{2}) \cap A = \emptyset$ and
(2) $(0; \frac{1}{2}) \cup A = \lambda = (0; 1)$,

but A does not belong to any topological space τ_r.

The final emphasis should be put on the fact that topology is seldom applied in the ontological investigations. Nevertheless, I hope that the tools, and some of the results obtained and presented in the paper, open up a new perspective. The majority of axioms proposed by Wolniewicz are fulfilled in $\mathcal{LWT}$ and non-atomistic $\mathcal{GWL}$. I hope that some other topological notions (like separation axioms, dense

---

14 Cf. Glock 1996, entry: Fact.

set, closed set, connected set, derivative set etc.) can be also used to establish novel theorems in terms of which Wittgenstein's *Tractatus* and ontology could be further interpreted.[15]

Bibliography

Glock, Hans-Johann (1996): *A Wittgenstein Dictionary*. Blackwell Publishers, Oxford.

Kaczmarek, Janusz (2008): "What is a Formalized Ontology Today? An Example of IIC." In: *Bulletin of the Section of Logic*. Vol. 37, No. 3 – 4, 233 – 244

Kuratowski, Kazimierz (1977): *Wstęp do teorii mnogości i topologii*, (wraz z dodatkiem R. Engelkinga: *Elementy topologii algebraicznej*), (*Introduction to Set Theory and Topology with a supplement by R. Engelking: Elements of algebraic topology*). PWN Warszawa, 358 (see also: Kuratowski, *Topology*. Vol. I, 1966, Vol. II, 1968).

Russell, Bertrand (1985): *The Philosophy of Logical Atomism*. Open Court, Peru

Wittgenstein, Ludwig (1997), *Tractatus logico-philosophicus*, BKF, PWN, Warszawa,109. B. Wolniewicz (transl. and introd.). First edition. *Logisch-philosophische Abhandlung*. In: *Annalen der Naturphilosophie*. 1921.

Wolniewicz, Bogusław (1982): "A Formal Ontology of Situations." In: *Studia Logica*. Vol. 41, No. 4, 381 – 413

Wolniewicz, Bogusław (1985): *Ontologia sytuacji (Ontology of Situations)*. PWN Warszawa, 134.

Wolniewicz, Bogusław (1999): *Logic and Metaphysics. Studies in Wittgenstein's Ontology of Facts*. Polskie Towarzystwo Semiotyczne (Ed. by Polish Semiotic Association), Warszawa.

---

[15] This paper is supported by the National Science Centre, Poland, No 2017/27/B/HS1/02830.

Franz Berto
Adding 4.0241 to TLP

Abstract: *Tractatus* 4.024 inspired the dominant semantics of our time: truth-conditional semantics. This is focused on possible worlds: the content of p is the set of worlds where p is true. It has become increasingly clear that such an account is, at best, defective: we need an "independent factor in meaning, constrained but not determined by truth-conditions" (Yablo 2014: 2), because sentences can be differently true at the same possible worlds. I suggest a missing comment which, had it been included in the *Tractatus*, would have helped semantics get this right from the start. This is my 4.0241: "Knowing what is the case if a sentence is true is knowing its ways of being true": knowing a sentence's truth possibilities *and* what we now call its topic, or subject matter. I show that the famous "fundamental thought" that "the 'logical constants' do not represent" (4.0312) can be understood in terms of ways-based views of meaning. Such views also help with puzzling claims like 5.122: "If p follows from q, the sense of 'p' is contained in that of 'q'", which are compatible with a conception of entailment combining truth-preservation with the preservation of topicality, or of ways of being true.

> *The way history has unrolled, the old containment idea was beaten out by the notion of truth preservation.*
> – David Kaplan, *The Meaning of "Ouch" and "Oops"*

1 How Did We Get From There to Here?

We've all heard it so many times:

4.024   To understand a proposition means to know what is the case, if it is true.

Franz Berto, University of St Andrews, UK, and University of Amsterdam, NL, fb96@st-andrews.ac.uk

Tractatus Logico-Philosophicus (TLP)'s section[1] 4.024 has been at the core of the dominant semantics of 20th Century: truth-conditional semantics. This has focused, as a matter of historical fact, on truth-at-possible-worlds conditions. The meaning or content of a sentence is given by its intension: a function from possible worlds to truth values or, equivalently, the set of possible worlds where the sentence is true. Call this Standard Possible Worlds Semantics (SPWS).

Possible worlds don't show up explicitly in the TLP. Let's thus start by laying on the table some textbook Tractarian semantics – by which I mean: a minimal account, which doesn't get into big interpretative issues.[2] Say that an atomic sentence is one that doesn't have further sentences as its syntactic constituents. Such a sentence is meaningful by being a picture of a state of affairs (4.032): a possible configuration or combination of objects, perhaps of objects and properties/relations (2.01, 2.031–2.032, 2.202). The state of affairs is the sense of the sentence qua picture (2.221, 3.13), what we know by understanding it (4.021). States of affairs can obtain, or fail to obtain. Obtaining states of affairs are facts (The TLP's terminology is not very uniform here, but let's not quibble over this either). By obtaining, a state of affairs makes true the sentence it is the sense of (4.25). The world is the totality of facts, that is, of obtaining states of affairs (1–1.2, 2.04–2.06, 2.063). Thus, all the true atomic sentences taken together make for a complete description of the world (4.26). Complex sentences are truth-functional compounds of atomic sentences (5): their senses consist of the combinations of the obtaining and non-obtaining of the states of affairs pictured by their atomic constituents, which make them true (4.2, 4.4–4.41, 4.431).

How did we get from (something like) *this* to SPWS? Carnap is usually taken as the main responsible. While introducing his method of intension and extension in *Meaning and Necessity*, he claims that "some ideas of Wittgenstein were the starting-point for the development of this method" (Carnap 1947: 9). Notoriously, he talked of state descriptions, not possible worlds, but a state description is "a class of sentences […] which contains for every atomic sentence either this sentence

---

[1] I will call "sections" the numbered sentences, or groups thereof, composing the TLP (TLP); these are sometimes called "propositions", but I find the terminology a bit confusing. I will use "sentence" as short for "declarative sentence": a linguistic configuration that can be true or false. The TLP uses "Satz" for this, and the Ogden translation has "proposition"; I will only leave the word with this meaning when directly quoting the TLP. I myself will use "proposition" or "(propositional) content" for the meaning or content of a declarative sentence.

[2] Like, What exactly is an elementary or atomic sentence? Does it include predicative terms? What's an atomic fact? How do the objects that constitute an atomic fact hang together? Are properties and relations objects, too? What are Tractarian objects like, by the way? – and so on. There are many excellent guides to the TLP, giving overviews of the debates on these issues. I recommend especially Frascolla 2007 and White 2006.

or its negation, but not both, and no other sentences" (Ibid.). Ignoring various complications, one may take them as close enough to complete descriptions of possible worlds – or, as the worlds themselves, in a linguistic ersatz (Lewis 1986, Ch. 3) account of them (Carnap claims, a bit misleadingly, that "the state-descriptions represent Leibniz' possible worlds or Wittgenstein's possible states of affairs", Ibid). Equivalently, one can take the totality of possible states of affairs, and form maximally consistent combinations of them. These will be, in a combinatorial setting, the possible worlds. Let their collection be W. The actual world is but the totality of states of affairs that obtain. And then we have it: the intension of a sentence p, giving its propositional content (or so-called UCLA proposition), is a function $I_p : W \to \{T, F\}$ mapping each $w \in W$ to its extension: truth, if p is true at w; falsity otherwise.

2 Aboutness, Topics, Subject Matters

The last decades have made increasingly clear that the SPWS account of propositional content is, at best, defective. According to Yablo, we need an "independent factor in meaning, constrained but not determined by truth-conditions" (Yablo 2014: 2). That's because sentences can be differently true – true in different *ways* – at the same possible worlds, and SPWS tells us very little on how and why sentences are true at the relevant worlds. Take for instance:

> Equilateral triangles are equiangular.
> 2 + 2 = 4.

These are true at the same worlds (all of them, if mathematical necessity is unrestricted) but they don't seem to mean or say the same thing: only one is *about* triangles, and made true by how they are.

One may quibble over the use of necessary truths, but take an example made famous by Hempel:

> All ravens are black.
> All non-black things are non-ravens.

These are (classically) logically equivalent, thus true at the same worlds, and arguably contingent, but they cannot quite mean the same if, as it seems intuitive, different pieces of evidence confirm them. Only one is *about* ravens, and made true by how ravens happen to be.

According to Yablo, *aboutness* is the missing ingredient in SPWS. This is "the relation that meaningful items bear to whatever it is that they are *on* or *of* or that they *address* or *concern*" (Yablo 2014: 1): their *subject matter* or, as I shall also call it, their *topic*. I'll say something on what topics or subject matters are, or could be, in a minute. Here's what they are not – or, cannot easily reduce to: the things referred to in a sentence. Topicality also has to do with what is *said about* those things. "Dog bites man" and "Man bites dog" (Yablo 2014: 34) involve the same things: dog, man, perhaps biting – but they don't say the same.

The main alternative to SPWS for an account of propositional content comes from structured propositions. Can't they get it right? In the so-called Russellian view, the content of "John kisses Mary" is a structured object involving John, Mary, and the relation of kissing, in some order. Ordering can tell it apart from the content of "Mary kisses John". There are several problems with structured propositions as an account of content and same-saying at the right level of semantic fine-grainedness. I cannot get into them here but, for an insightful discussion, one can read Ripley 2012. I'll just mention one issue, raised in Yablo 2014: 1, for it will be important in the following: in the Russellian account, "Mary does not kiss John" differs in content from "Mary kisses John", in that the content of the former includes a component, *not*, which the content of the latter lacks. Also, the content of "Mary kisses John and Paul is jealous" includes *and* as a component. This seems wrong for subject matters: there's no sense in which "Mary does not kiss John" talks about negation, or "Mary kisses John and Paul is jealous" talks about conjunction.

Work on subject matter and ways of being true has been flourishing lately. This has revitalized a semantic tradition to some extent alternative to SPWS, and focused on a *mereological* view of meaning, whereby contents can be taken as having parts, as including other contents, as capable of being fused into wholes which inherit the proper features from the parts. Ideas concerning subject matters can be found already in Parry 1933, Ryle 1933, Goodman 1961, Perry 1989. Ways of being true have more recently been understood in terms of partial content (Humberstone 2000), world-partitions (Lewis 1988), world-divisions (Yablo 2014, Osorio-K. 2016), truthmakers (Fine 2014, Fine 2015), and more (Gioulatou 2016, Hawke 2017).

I don't think that the semantic ideas included in the TLP lead *inevitably* to SPWS. I'm not a specialist of Wittgenstein – only an admirer of the TLP since I was an undergrad. Still, it seems to me that there are, in it, traces of an alternative, mereological view of content. Before proposing, in two sections, a "missing gloss" to TLP 4.024 which, had it been there, would have helped 20th Century semantics get things right from the start, in the next section I will dig into a place in the TLP where such traces seem to show up: the Tractarian account of logical consequence.

3 The Uncomfortable 5.122

SPWS can easily make sense of 5.12 and its first subsection:

5.12 In particular the truth of a proposition p follows from that of a proposition q, if all the truth-grounds of the second are truth-grounds of the first.
5.121 The truth-grounds of q are contained in those of p: p follows from q.

The truth-grounds of a sentence are "those truth-possibilities of its truth-arguments, which verify the proposition" (5.101). These can be understood via the notion of Tractarian logical space (1.13, 2.11, 2.202, 3.4, 3.42, 4.463), whose mainstream interpretation is in terms of possible worlds (Stenius 1960) and relies, essentially, on the Carnapian ideas sketched above: each possible world is a combination of the obtaining and non-obtaining of states of affairs depicted by the atomic sentences. Then 5.12-entailment is close enough to truth preservation at all worlds in all interpretations or models, as per SPWS.

The problems come with the second gloss to 5.12:

5.122 If p follows from q, the sense of "p" is contained in that of "q".

Here the direction of containment is the other way around: the conclusion is included in the premise. A related idea pops up in 5.14: "If a proposition [p] follows from another [q], then the latter [q] says more than the former [p], the former [p] less than the latter [q]". The entailing q says at least whatever the entailed p said, thus including it in some sense. As pointed out in the beautiful Negro 2017: 5, it is not straightforward to understand this in terms of SPWS. Negro notices that 5.122-consequence has given headaches to several authoritative interpreters of the *Tractatus*. Here's Ramsey:

> I think this statement is really a definition of containing as regards senses, and an extension of the meaning of "assert" partly in conformity with ordinary usage, which probably agrees as regards $p \wedge q$ and p [...] but not otherwise. (Ramsey 1923: 471, notation adapted for consistency)

Ramsey's way of making sense of sense containment has it that, when by asserting p we happen to assert q, the sense of p has that of q as a part. As an intuitive ("conforming to ordinary usage") example, he points at the sense of "$p \wedge q$"s including that of p. Because $p \wedge q$ entails q according to SPWS, this does not tell sense containment apart from SPWS logical consequence. But there's another intuitive example in the vicinity, which does: p entails $p \vee q$ according to SPWS, but surely "$p \vee q$" can be partly about something p is not about, namely whatever q is about. We'll get back to both entailments later.

Max Black is more outspoken than Ramsey: he calls the idea of sense containment "an impediment to clarity", and glosses on 5.122 by speaking of "a peculiar (and possibly unfortunate) use of this word [scil. 'contained']" (Black 1964: 251 – 252). Fogelin 1976 says that sense containment is no more than a metaphor.

Can we make sense of sense-containment? Negro 2017 and Frascolla 2007 have interesting proposals (differing from each other in ways I won't get into here), which pivot on looking, rather than at truth-grounds, at their complements, namely, falsity-grounds, in a negative or exclusionary view of content (Rumfitt 2008). We can then explain the reversed inclusion direction of sense containment as some kind of inclusion of falsity-grounds.

Now I want to leave Wittgensteinian exegesis behind, though, and broaden the view a bit. There is a less-than-metaphorical tradition in logic and semantics – variously stretching back at least to Kantian ideas on analyticity – of looking for content-preserving entailment relations: relations that hold between p and q only when q introduces no content alien to that of p. Among such relations are those labeled as "tautological entailment" (Van Fraassen 1969), "analytic containment" (Angell 1977), "analytic implication" (Parry 1933, Fine 1986, Ferguson 2014). I will sketch two ways to approach content-containment and ways-based semantics in the coming section, after I've added my proposed gloss to 4.024.

4 4.0241*, and Two Ways to Ways

Here's my main proposed addition to TLP, meant to help semantics move beyond SPWS (let me flag it with an asterisk, to clarify that it's my stuff, not Wittgenstein's):

4.0241* Knowing what is the case if a sentence is true is knowing its ways of being true.

How do we understand such ways of being true? One way to have a ways-based account of propositional content is found in Kit Fine's works, e.g., Fine 2014, Fine 2015. Fine's core idea is very Tractarian in spirit: sentences are made true by states of affairs or situations, rather than by whole possible worlds. His semantics has a set with a partial ordering on it, $\langle S, \leq \rangle$. Each $s \in S$ is a situation, or state of affairs, or a configuration of objects, or of objects and properties. States of affairs are things that can obtain or fail, and are chunks of reality that ground the truth and falsity of sentences, so we're not tremendously far from TLP's own states of affairs.

We are far from possible worlds, however – though one can recover a certain account of worlds by taking them as limit constructions out of states. The ordering $s_1 \leq s_2$ may be read in a metaphysically loaded way: "s_1 is a real part of s_2", thus, states of affairs can be literally included in larger ones (this works well mainly with states of affairs involving concrete entities: the state consisting of St Andrews' being in Scotland is included in the state consisting of St Andrews' being in the UK). I favor a less metaphysically loaded reading of "≤" in terms of information: "s_2 preserves all the information in s_1", or "s_2 supports the truth of anything whose truth is supported by s_1", or so. Then what matters about states of affairs, for semantics, is the information they encode or support.

Either way, worlds would be things which are maximal with respect to ≤: when w_1 and w_2 are worlds, the only way for "$w_1 \leq w_2$" to hold is for w_1 to *be* w_2. Finean states are not like that: like the situations of Barwise/Perry 1983's situation semantics, they are partial and *relevant* for the things whose truth they support. And it's states, not worlds, which are at the core of Kit Fine's semantics – just as they are at the core of the Tractarian theory of representation: it's of facts, i.e., (obtaining) states of affairs, that we make pictures of (2.1). Pictures are facts (2.141): the facts consisting in their elements' being in such-and-such relations with one another (2.14). The elements of the picture stand for (*vertreten*) the objects which are the constituents of the pictured state of affairs (2.131) – and so on, as per the well-known (though variously interpreted) pictorial theory of representation. Thoughts are logical pictures, (3), and sentences are the expression of thoughts, (3.1). The sense of a sentence – its content – is the state of affairs pictured by the sentence. What I understand when I grasp the content of *p*, is the state of affairs *p* represents (4.021).

I will not get into the details of Fine's state-based semantics, however, for I favor a different way of having a way-based semantics. This comes from two-component (2C) accounts of content, which are more friendly to worlds. One account of this kind has been developed in great detail in Yablo 2014, but I'll stick to a more abstract 2C setting I'm working on, together with some friends (Berto 2018a, Berto 2018b, Hawke 2017). The content of a sentence *p* is the *thick proposition* it expresses, [[*p*]]. This has two components: (1) *p*'s intension, |*p*| ⊆ *W*, its truth set as per SPWS (the *thin* UCLA proposition), and (2) *p*'s *topic* or *subject matter*, *t*(*p*). Overall, [[*p*]] = ⟨|*p*|, *t*(*p*)⟩. *p* and *q* say the same, that is, express the same thick proposition, when (1) they are true at the same worlds, and (2) they have the same topic. Given that topics can have parts, we also get a natural view of content inclusion: the content of *p* includes that of *q* when (i) any *p*-world is a *q*-world and (ii) the topic of *q* is part of that of *p*. When both (i) and (ii) obtain, we claim that *p* "says more than (or, at least as much as)' *q*.

Two natural questions now are, Why two components? What are these topics or subject matters? As for the first one, our understanding of propositional contents can be naturally seen as involving two elements. Now I attempt a further envisaged addition to the TLP – a subsection glossing on my proposed 4.0241*, and pushing more in the 2C direction:

4.02411* One knows how a sentence is true by knowing its truth possibilities *and* what it is about.

One understands what *p* means or says, when one understands (1) what the sentence speaks *about*, its topic or subject matter, *and* (2) what it *says* about that topic or subject matter. One understands "La neige est blanche" as soon as one knows that (1) it speaks about *the color of snow, how snow is like, snow's whiteness*, etc. – and, one knows that (2) it says that things are such-and-so with respect to that subject matter, that is, that snow is white.

In a plain state-based account à la Fine, the two components are taken care of by one kind of things: the states. A state of affairs verifying *p* is both what *p* is about, and what truthmakes it. But we may have tentative reasons for splitting the two components. It seems that people can sometimes have a partial understanding of the meaning of a sentence by grasping only its truth conditions, or only its topic. William III could have understood the truth conditions of "Either France will get into a nuclear war with England, or not": he'd know, by only looking at its logical form, that the sentence would be true no matter what. He couldn't have grasped what the sentence is about, for he had no idea of what a nuclear war is. William III could have grasped the topic of "Louis XIV is bald" (say, *Louis' baldness*) without knowing under which conditions the sentence would be true (just in case Louis had *n* hairs – for what *n*? A range between some *n* and some *m*? Which *n* and *m* then? A fuzzy range? etc.).

That we split components doesn't mean that we must treat them as conceptually or metaphysically irreducible to each other. Here our second question comes in: What are topics, or subject matters? One may take them, again, as truthmakers or states of affairs, and reduce worlds to them by taking worlds as maximally informative states, or constructions out of states. But one may also go the other way around. This was the way pursued in Lewis 1988's seminal work on subject matters. Understand subject matters starting from questions: the subject matter of *p* (in context *c*), is given via the question or questions *p* can be taken as answering to (in *c*). This determines a partition of the set of worlds: w_1 and w_2 end up in the same cell when they agree on the answer.[3] "The number of stars is eight" has as its topic *the number of stars*. It can be taken as an answer to the question, How

---

[3] I won't get in detail into the following issue in this paper, *but*: as subject matters are context-dependent, it makes sense to say that the same sentence can get different subject matters in different contexts insofar as it answers to different questions: "Matteo Plebani is a communist' can be paired to, What are *Matteo's political preferences*?, or to Who are the *communists in the*

many stars are there? This splits and groups the worlds depending on how they answer: all the zero-star worlds end up in one cell; all the one-star worlds end up in another; etc.

Yablo 2014 proposes to generalize Lewisian partitions, thus equivalence relations, to divisions: sets of worlds that collectively cover the modal space, but which allow overlap, determined by similarity relations which are reflexive and symmetric, but not transitive. Some questions have more than one correct answer ("Where can I find a B&B in Kirchberg?"), so a world can be in more than one cell with respect to them. In any case, subject matters reduce to ways of splitting and grouping worlds into sets.

Before I give you more 2C details, I need to talk of Wittgenstein's fundamental thought.

5 4.0312, 2C, and Back to 5.122

Whether one takes a Fine-style, a 2C-style, or some other approach, a key principle followed in accounts of subject matter goes hand in hand with what Wittgenstein took as his "fundamental thought" in the TLP: that the logical constants represent nothing, for "the *logic* of the facts cannot be represented" (4.0312).

In a motto: the logical vocabulary contributes nothing to topicality, or subject matter. Let's stick with the Boolean connectives.[4] We saw that there is something wrong in the claim that "Snow is not white" is about negation. "Snow is not white" must be about whatever "Snow is white" is about: that may be *snow's color, the whiteness of snow*, etc. In general, "$\neg p$" must be exactly about what p is about. We can phrase this first guiding principle for a propositional recursion on topics in 2C-terms: $t(\neg p) = t(p)$. Thus, also, $t(\neg\neg p) = t(p)$. Remember Frege on the *Sinn*-preservation of double negation, a view endorsed by Wittgenstein:

> 4.0621 That, however, the signs "p"and "q" *can* say the same thing is important, for it shows that the sign "∼" corresponds to nothing in reality.
> That negation occurs in a proposition, is no characteristic of its sense (∼∼ $p = p$). [...]
> 5.44 [...] And if there was an object called "∼", then "∼∼ p" would have to say something other than "p". For the one proposition would then treat of ∼, the other would not.

Plebani family? – etc. A nice development of subject matter theory in this direction is Plebani/Spolaore 2018.

4 The view could be readily extended to the universal and particular quantifiers, if one took them as generalized conjunctions and disjunctions, respectively; but there are some subtleties involved here I don't want to get into.

As for conjunction and disjunction: "$p \wedge q$" and "$p \vee q$" have the same subject matter: a fusion or merging of the subject matter of p and that of q: "John is tall and handsome" and "John is tall or handsome" are both about the same topic, namely the height and looks of John's. (Again, that doesn't make them express the same thick proposition: their truth sets are distinct.) Where "⊕" stands for topic fusion, the second guiding principle for a propositional recursion on topics is: $t(p \wedge q) = t(p \vee q) = t(p) \oplus t(q)$. Again, the logical vocabulary is topic-transparent.

We can now build a small 2C semantics. Let's have a plain propositional language $\mathcal{L}$ with atoms $\mathcal{L}_{AT}$: p, q, r ($p_1, p_2, \ldots$), negation $\neg$, conjunction $\wedge$, disjunction $\vee$, round parentheses (), as auxiliary symbols. We use A, B, C ($A_1, A_2, \ldots$) as metavariables for formulas. The well-formed formulas are the atoms and, if A and B are formulas: $| \neg A | (A \wedge B) | (A \vee B) |$.

A model for $\mathcal{L}$ is a tuple $\mathfrak{M} = \langle W, T, \oplus, v, t \rangle$, where:

- W is a nonempty set of possible worlds;
- T is a nonempty set of possible topics;
- $\oplus$ is an idempotent, commutative, associative binary operation on T;
- $v: \mathcal{L}_{AT} \to \mathcal{P}(W)$ assigns a truth set $v(p) = |p|$ to each atom p;
- $t: \mathcal{L}_{AT} \to T$ assigns a topic $t(p)$ to each atom p.

Out of fusion we can define what it means that topic x is part of topic y, $x \leq y =_{df} x \oplus y = y$ – making of parthood a partial ordering (with the strict ordering, $<$, defined from the nonstrict $\leq$, the usual way). v is extended to the whole $\mathcal{L}$ via the usual recursive clauses assigning a truth set $|A|$ to each formula A of $\mathcal{L}$:

- $|\neg A| = W - |A|$
- $|A \wedge B| = |A| \cap |B|$
- $|A \vee B| = |A| \cup |B|$

We also extend t to the whole $\mathcal{L}$ following our two guiding principles above:

- $t(\neg A) = t(A)$
- $t(A \wedge B) = t(A \vee B) = t(A) \oplus t(B)$

This double recursion gives us a fully recursive assignment, to each formula A of $\mathcal{L}$, of a thick propositional content $[[A]] = \langle |A|, t(A) \rangle$. A and B say the same, i.e., they express the same content (in model $\mathfrak{M}$), when $[[A]] = [[B]]$, that is, $|A| = |B|$ and $t(A) = t(B)$ (and, one easily defines "saying at least as much as" and "saying strictly more").

This account of propositional content is strictly more fine-grained than what one gets in SPWS, given that any difference in truth set will make for a difference in thick proposition, but there will be additional distinctions warranted by different topic-assignments, e.g., $[[\neg(A \wedge \neg A)]] \neq [[B \vee \neg B]]$ ("It's not the case that the Liar is both true and untrue" doesn't say the same as "Either Goldbach's Conjecture holds, or not"). Still, lots of sentences turn out to express the same thick proposition, e.g., $[[A \wedge B]] = [[B \wedge A]]$ ("It's rainy and cold today" and "It's cold and rainy today" say the same thing); $[[A]] = [[\neg\neg A]]$ ("Mary is on time" says the same as "Mary isn't late"); $[[\neg A \vee \neg B]] = [[\neg(A \wedge B)]]$ ("Either Mary doesn't like John or she doesn't like Paul" says the same as "Mary doesn't like both John and Paul").

Now back to Wittgenstein. We saw that, as pointed out by Negro, one cannot straightforwardly understand sense containment, as per 5.122, merely in terms of SPWS truth sets. But one can understand it by taking propositional contents as more than truth sets. 2C-semantics can get close enough to the spirit of 5.122 via a characterization of logical consequence (say, $A \vDash B$: "B follows from A", or "A entails B"), that embeds topic-containment. We now know what it means, in a 2C-setting, that one thick proposition includes another: $[[A]]$ includes $[[B]]$ (let's write "$[[A]] \trianglerighteq [[B]]$", that is, $\langle |A|, t(A) \rangle \trianglerighteq \langle |B|, t(B) \rangle$) when (1) $|A| \subseteq |B|$, all the A-worlds are B-worlds, and (2) $t(B) \leq t(A)$, that is, the topic or subject matter of B is included in that of A. We can interpret in 2C terms "A proposition affirms every proposition that follows from it" (5.124), and "If one proposition follows from another, then the latter says more than the former, and the former says less than the latter" (5.14): we have precise characterizations of same-saying, saying strictly more, saying strictly less, saying at least as much. In particular, we claim that $A \vDash_{\mathfrak{M}} B$ just in case $[[A]] \trianglerighteq [[B]]$ in $\mathfrak{M}$, that is, A says at least as much as B there: while A's truth set involves no more worlds, its subject matter is at least as big as that of B. And we say that $A \vDash B$ when $A \vDash_{\mathfrak{M}} B$ for all $\mathfrak{M}$.

6 Conclusion: Pulling in Different Directions

I wouldn't want to claim that *all* of the TLP can be made consistent with a topic-sensitive account of propositional content, whether of the 2C kind, or of some other sort – and in particular, that 4.0241*, and especially 4.02411*, don't betray the TLP in any way. On the contrary, the book often pulls in an opposite direction. One of the many things clarified by Negro 2017, is that Wittgenstein himself admitted, some years after the publication of the TLP, that his ideas around sense containment were less than fully clear when he wrote the book. In his *Notebook I*, which dates back to the late Twenties or early Thirties, Wittgenstein says:

> If the proposition q follows from the proposition p, then I thought that $p \wedge \neg q$ has to be a contradiction and this I saw quite rightly. I believed that I had to further infer from this [...] that q in some sense has to be *contained* in p. For if both propositions had nothing to do with each other, how could $p \wedge \neg q$ be a contradiction? In what sense q is supposed to be contained in p does not yet emerge from this [...] and I did not clearly specify it [...]. There is a clear sense in saying: q is contained in the logical product $p \wedge q$. (Baker 2003: 127 and 197, notation adjusted for consistency.)

Perhaps Wittgenstein had clear the obvious case of content containment given by $A \wedge B$'s entailing A, but couldn't readily see how to generalize it to a full-fledged notion of content-preserving entailment: since "he had only intuitively grasped the general case as correct, [this] was thus left completely unexplained in the *Tractatus*" (Negro 2017: 6).

In the TLP, A entails $A \vee B$ in the 5.12 sense of entailment: all the truth-grounds of the former are truth-grounds of the latter or, as we would say today, any possible world in any interpretation or model making A true will also make $A \vee B$ true. But, unlike $A \wedge B \vDash A$, this entailment is not topic-preserving, for B may break the boundaries of A's subject matter: it is not the case that, in general, whatever "$A \vee B$" says was already said by A. These have been taken as paradigmatic cases of content inclusion and noninclusion, and as data a theory of content-preserving entailment must comply with:

> A paradigm of inclusion, I take it, is the relation that simple conjunctions bear to their conjuncts – the relation *Snow is white and expensive* bears, for example, to *Snow is white*. A paradigm of noninclusion is the relation disjuncts bear to disjunctions; *Snow is white* does not have *Snow is white or expensive* as a part. (Yablo 2014: 11)

> A guiding principle behind the understanding of partial content is that the content of A and B should each be part of the content of $A \wedge B$ but that the content of $A \vee B$ should not in general be part of the content of either A or B. (Fine 2015: 1)

Here the hiatus comes to the fore:

> 4.465 The logical product of a tautology and a proposition says the same thing as the proposition. Therefore that product is identical with the proposition.

Thus, in particular, A says the same as $A \wedge (B \vee \neg B)$. If one only looks at truth sets, these coincide. But it needn't be the case that $t(A) = t(A \wedge (B \vee \neg B))$, so the two cannot express the same thick proposition. In fact, I think that the interpretation of TLP sense-containment provided by Negro in terms of inclusion of falsity-grounds (Negro 2017: 17 – 20) does a better job than my little 2C semantics in making sense of the whole stance of the TLP with regards to logical consequence and content inclusion.

However, when it's about characterizing propositional content and same-saying, it's the thick proposition view that gets it right, not the merely truth-conditional view – or so I claim. How can "Jane fell out of bed" say the same as "Jane fell out of bed and either John is 6 feet tall or not"? Only the latter is, partly, about John's heights. Mary may have brought it about that Jane fell out of bed without bringing it about that Jane fell out of bed and either John is 6 feet tall or not (Perry 1989).

It still seems to me reasonable to say, at least, that the direction into which truth-conditional semantics has *de facto* been developed into SPWS and UCLA thin propositions is not straightforwardly mandated by the TLP. One should not be puzzled by the puzzlement of authoritative interpreters like Ramsey or Black: of course they had a hard time coordinating what the TLP says about sense containment in 5.122 with the direction it takes elsewhere. I'm more puzzled by the fact that the authoritative interpreters already saw the merely truth-conditional direction as the default one, and treated the other one, the content-containment direction, as the one which shouldn't have been there, or should have been treated as a mere metaphor.[5]

Bibliography

Angell, R.B. (1977): "Three Systems of First Degree Entailment." In: *Journal of Symbolic Logic*. Vol. 47, 147.
Baker, G. (2003): *The Voices of Wittgenstein: The Vienna Circle – Ludwig Wittgenstein and Friedrich Waismann*. London: Routledge & Kegan Paul.
Barwise, J. and Perry, J. (1983): *Situations and Attitudes*. Cambridge (MA): MIT Press.
Berto, F. (2018): "Aboutness in Imagination." In: *Philosophical Studies*. Vol. 175, 1871 – 1886.
Berto, F. (2018): "Simple Hyperintensional Belief Revision." In: *Erkenntnis*. 1 – 17.
Black, M. (1964): *A Companion to Wittgenstein's Tractatus*. Ithaca (NY): Cornell University Press.
Carnap, R. (1947): *Meaning and Necessity*. Chicago, London: University of Chicago Press.
Ferguson, T.M. (2014): "A Computational Interpretation of Conceptivism." In: *Journal of Applied Non-Classical Logics*. No. 4, Vol. 24, 333 – 367.
Fine, K. (1986): "Analytic Implication." In: *Notre Dame Journal of Formal Logic*. Vol. 27, 169 – 179.
Fine, K. (2014): "Truthmaker Semantics for Intuitionistic Logic." In: *Journal of Philosophical Logic*. Vol. 43, 549 – 577.
Fine, K. (2015): "Angellic Content." In: *Journal of Philosophical Logic*. Vol. 45, 199 – 226.
Fogelin, R.J. (1976): *Wittgenstein*. London: Routledge & Kegan Paul.

---

[5] I am very grateful to Lello Frascolla, Peter Hawke, Levin Hornischer, Diego Marconi, Antonio Negro, Naomi Osorio, Aybüke Ozgün, Matteo Plebani, for lovely conversations and useful comments on the subject matter of this paper.

Frascolla, P. (2007): *Understanding Wittgenstein's Tractatus*. New York: Routledge.
Gioulatou, I. (2016): *Hyperintensionality*. ILLC, Universiteit van Amsterdam.
Goodman, N. (1961): "About." In: *Mind*. Vol. 70, 1 – 24.
Hawke, P. (2017): "Theories of Aboutness." In: *Australasian Journal of Philosophy*. Vol. 96, 697 – 723.
Humberstone, L. (2000): "Parts and Partitions." In: Theoria. Vol. 66, 41 – 82.
Lewis, D. (1986): *On the Plurality of Worlds*. Oxford: Blackwell.
Lewis, D. (1988): "Relevant Implication." In: *Theoria*. Vol. 54, 161 – 174.
Negro, A. (2017): "Making Sense of Sense Containment." In: *History and Philosophy of Logic*. Vol. 38, 1 – 22.
Osorio-Kupferblum, N. (2016): "Aboutness: Critical Notice." In: *Analysis*. Vol. 76, 528 – 546.
Parry, W.T. (1933): "Ein Axiomensystem für eine neue Art von Implikation (Analytische Implikation)." In: *Ergebnisse eines Mathematischen Kolloquiums*. Vol. 4, 5 – 6.
Perry, J. (1989): "Possible Worlds and Subject Matter." In: *Possible Worlds in Humanities, Arts and Sciences*. Sture, Allén (ed.). Berlin: De Gruyter, 173 – 191.
Plebani, M. and Spolaore, G. (2018): "Subject Matter: A Modest Proposal." In: *Unpublished Manuscript*.
Ramsey, F.P. (1923): "Critical Notice." In: *Mind*. Vol. 32, 465 – 478.
Ripley, D. (2012): "Structures and Circumstances: Two Ways to Fine-Grain Propositions." In: *Synthese*. Vol. 189, 97 – 118.
Rumfitt, I. (2008): "Knowledge by Deduction." In: *Grazer Philosophische Studien*. Vol. 77, 61 – 84.
Ryle, G. (1933): "About." In: *Analysis*. Vol. 1, 10 – 12.
Stenius, E. (1960): *Wittgenstein's Tractatus: A Critical Exposition of its Main Lines of Thought*. Oxford: Blackwell.
Van Fraassen, B. (1969): "Facts and Tautological Entailments." In: *Journal of Philosophy*. Vol. 66, 477 – 487.
White, R.M. (2006): *Wittgenstein's Tractatus Logico-Philosophicus: A Reader's Guide*. London: Continuum.
Wittgenstein, L. (1922): *Tractatus logico-philosophicus*. Ogden, C.K. (trans.), London: Routledge & Kegan Paul.
Yablo, S. (2014): *Aboutness*. Princeton, Oxford: Princeton University Press.

Štefan Riegelnik
Understanding Wittgenstein's Wood Sellers

Abstract: In the collection *Remarks on the Foundations of Mathematics* (I, §149) Wittgenstein encourages us to imagine a group of people selling wood at a price relative to the area covered by the pile of wood irrespective of the height of the pile. In "Wittgenstein and Logical Necessity" Barry Stroud argues that Wittgenstein uses this scenario to steer between two untenable positions: (i) Frege's Platonism, according to which the wood sellers must be considered to be insane, and (ii) a version of conventionalism which leaves open the possibility of ways of inferring, counting, and calculating different to ours. At first sight, the behaviour of the wood sellers seems to be comprehensible. But, as Stroud argues, the more we project our grammatical structures and categories into their verbal and non-verbal behaviour, the less intelligible the wood sellers become. In what follows, I discuss Stroud's account of the unintelligibility of the wood sellers and I contrast it with Johan Canfield's critical reading of this verdict.

1 Introduction

If one is asked to add up 5 and 7, the result *must* be 12. A person claiming a different number to be the sum has either misunderstood the question or has made a mistake. In a word, mathematical statements like "5 + 7 = 12" are necessarily true and a denial of them is impossible or unintelligible. One of the main philosophical questions concerning mathematical necessity concerns how we can account for the source of this necessity and in consequence of the unintelligibility of a denial of a true mathematical statements. More specifically, how do we account for the unintelligibility of a statement, or an as-if-statement, which at first appears to be intelligible, but eventually turns out to be not intelligible at all? For obvious reasons, the statement itself cannot be part of an inference. For in order to be a premise or a conclusion, a statement must be intelligible and it would be paradoxical if it turned out that it is not.

In §149 of the *Remarks on the Foundations of Mathematics* (1996)[1] Wittgenstein appeals to the idea of people selling wood relative to the area covered by the pile of wood irrespective of the height of the pile. In an influential article Barry Stroud

---

[1] Henceforth RFM.

Štefan Riegelnik, University of Zurich, Switzerland, stefan.riegelnik@philos.uzh.ch

DOI 10.1515/9783110657883-26

(Stroud 1965) argues that Wittgenstein's appeal to this scenario (and others) is supposed to show that while we might acknowledge the possibility of alternative mathematical systems, but at the same time we do not understand them, i. e. their applications are unintelligible. According to Stroud, this allows Wittgenstein to steer between the Scylla of Platonism and the Charybdis of conventionalism as accounts of the source of mathematical necessity. John Canfield (Canfield 1975) counters this line of reasoning by arguing that we can complete the scenario of the wood sellers in such a way that it "*makes their wood selling practice, and their other practices, perfectly intelligible.*" (Canfield 1975: 470). In this contribution I contrast both views relating to the scenario of the wood sellers and I examine the role of the concept of *consistency* for our not being able to understand the alien wood sellers. First I examine briefly the background of the debate and Wittgenstein's account of mathematical necessity.

2 The Source of Necessity of Mathematical Truths

What is the source of the necessity of the statement that 5 + 7 equals 12? If we admit that there is a necessity involved, the possibility of an alternative answer to the question for the sum of 5 and 7 is not possible. But how are we supposed to account for the impossibility or the unintelligibility of such a claim?

In the philosophy of mathematics, it is common to contrast Platonism with constructivist accounts of constructivism like conventionalism (Cf. Dummett 1959: 324). Advocates of Platonism claim that abstract mathematical objects exist independently of us and that mathematical truths are to be discovered. Given that, the claim that 5 + 7 equals 13 is simply wrong, since "5 + 7" and "13" do not refer to the same object. A person assuming this to be true either misunderstands the instructions or ignores the facts. This entails, too, that there are no alternative ways of calculating. Conventionalists, on the other hand, acknowledge the possibility of alternative ways of calculating, inferring, and counting. For them the key source of the supposed necessity is convention, i. e. a decision made that determines what counts as addition, inference etc. it is a matter of decision which set of conventions one chooses and which necessities hold relative to a particular set of conventions. They maintain that if one understands our meaning of "5", "+", "7", "=", and "12", one will necessarily come to the conclusion that "12" is the sum of 5 and 7. This is, so to speak, built into the meaning of these expressions. And if one does not come up with the answer 12, then one does not understand what the expressions mean relative to the conventions in effect. It would be a contradiction if one accepts the conventions determining the meaning of these expressions and nevertheless writes down a number other than 12.

In contrast to Dummett, Stroud holds that for Wittgenstein neither of these alternatives are an option. Given this, Stroud argues that Wittgenstein faces a dilemma, which can be presented as follows: Against Platonism, he must show that there are alternative ways of calculating, adding, inferring etc. Against conventionalism, he must show that, in spite of alternatives, there is something necessary in our mathematical and logical practices. In other words, Wittgenstein needs to show that there is

> *the possibility of ways of counting, inferring, calculating, and so forth, different from ours, but which do not imply that our doing these things as we do is solely a result of abiding by, or having adopted, certain more or less arbitrary conventions to which there are clear and intelligible alternatives.* (Stroud 1965: 510)

Counting, adding, inferring etc. might have been done in a different way, for the way we are performing these activities depends on contingent facts. This is supposed to rule out Platonism. However, acknowledging the contingency of the ways these activities are carried out, does not mean that we understand alternative ways of counting, adding, inferring etc. The argument against conventionalism runs as follows: Conventionalists must assume that one could follow a different set of conventions. For, as Stroud rightly points out, the possibility to follow a particular set of conventions implies that one could follow a different set of conventions:

> One thing implied by saying that we have adopted, or are following, a convention is that there are alternatives which we could adopt in its place. (Stroud 1965: 509)

Conventionalists have to assume that we have *our* conventions for counting, adding, inferring and others may have *their* conventions to perform these activities. But, and this is the crucial point, if we cannot establish a connection between their conventions to count, add, or infer, then we have no reason why we should assume that they are counting, adding, inferring, or, following conventions at all. The appeal to their mere non-verbal behaviour as an expression of following conventions which is deemed to contrast with *our* way to do these activities is not sufficient. So if there is no counterpart to *our* conventions, it does not make sense to speak of *our* or *their* conventions at all, and, as a consequence, that a mathematical statement is true in virtue of conventions. Thus if one wants to evade conventionalism, one must show that we do not understand alternative answers to what we consider to be a necessary truth. This represents for Wittgenstein a safe path that avoids both Platonism and conventionalism. In RFM IV, §29 Wittgenstein writes:

> So much is clear: when someone says: "If you follow the *rule*, it *must* be like this", he has not any *clear* concept of what experience would correspond to the opposite.
> Or again: he has not any clear concept of what it would be like for it to be otherwise. And this is very important. (RFM IV, §29)

Stroud takes this up and asks –

> The solution to this dilemma is to be found in the explanation of why we do not have any clear concept of the opposite in the case of logical necessity, and why Wittgenstein speaks of our not having a *clear* concept here. How could we have any concept at all? (Stroud 1965: 511)

Now, and this is the crucial point: how can we show that we do not have a *clear* concept of what *different ways of calculating* might be? In other words, even if we admit that it might be possible that we calculate, count, infer etc. differently, we have to admit, too, that we do not understand what someone means if he or she actually does it. Since I do not assume that there is a general method of deciding whether we understand a particular statement or not, I now turn to the discussion of the practice of selling wood and alternative ways of determining the amount of wood. Could we communicate with people selling wood relative to the area covered by the pile of wood irrespective of the height of the pile and find out what they mean by "more"? Could we tell them what we mean by "more"? How can we show convincingly that calculating, counting, and measuring in this alien way is not intelligible?

3 Ways of Selling Wood

In RFM I, §143 Wittgenstein describes the practice of selling wood as follows:

> People pile up logs and sell them, the piles are measured with a ruler, the measurements of length, breadth and height multiplied together, and what comes out is the number of pence which have to be asked and given. They do not know 'why' it happens like this; they simply do it like this: that is how it is done. (RFM I, §143)

He rightly points out in §148 that we *do* make use of other methods of measuring wood and selling it accordingly as well. For instance, we measure and sell wood by the weight or by the time that it takes to fell the timber. These methods are perfectly comprehensible and intelligible. Wittgenstein writes:

> Those people–we should say–sell timber by cubic measure–but are they right in doing so? Wouldn't it be more correct to sell it by weight–or by the time that it took to fell the timber–or by the labour of feeling measured by the age and strength of the woodsman? And why should

they not hand it over for a price which is independent of all this: each buyer pays the same however much he takes (they have found it possible to live like that). And is there anything to be said against simply giving the wood away? (RFM I, §148)

Then he asks us to imagine a group of people selling wood at a price relative to the area covered by the pile of wood irrespective of the height of the pile.

> Very well; but what if they piled the timber in heaps of arbitrary, varying hight and then sold it at a price proportionate to the area covered by the piles?
> And what if they even justified this with the words: "Of course, if you buy more timber, you must pay more"? (RFM I, §149)

At first I want to point out that the transition words "Very well" at the beginning of §149 suggest that there is a contrast to be drawn between the methods of selling woods discussed earlier and the following one. Even if the quantity of woods and its relation to the price can be determined in manifold ways, the professed method introduced in §148 differs in an important respect.

Against the course of behaviour of the wood sellers in §148 advocates of Platonism would counter that they are simply wrong or insane. A pile 2 meters in length multiplied by 2 meters in height multiplied by 2 meters in width equals the same amount of wood of a pile in 1 meter length multiplied by 4 meters in hight multiplied by 2 meters in width, viz. 8 cubic metres of wood. Consequently, it is not more timber if the logs are arranged differently. Platonists would maintain that the wood sellers are ignorant or insane if they believe that they sell *more* wood if the logs are rearranged. In contrast, conventionalists have two options. First, they might hold that since the meaning of "counting", "calculating", "measuring", "inferring", "selling", "more" is as it is, the way the alien tribe wants to sell wood conflicts with what we mean by these expressions. Secondly, they follow an alternative set of conventions.

Here, according to Stroud, Wittgenstein argues that there is a third option and that allows him to escape the dilemma discussed in section 2. The appeal to the example serves to show that our way of calculating, inferring, and measuring is contingent and for this reason alternative ways of performing these acts are possible. This is part of the attack on Platonism. However, in undermining Platonism, it seems that we are dragged towards conventionalism. But if conventionalism was true, we would be committed to the view that alternative ways of calculating, inferring, and measuring are intelligible. Precisely this seems to be the moot point. Do we understand the wood sellers in claiming that "Now it is more wood and you have to pay more!" after they have rearranged a pile of logs so that they cover a larger area? Could they comprehend their way of measuring wood by arguing that they have adopted different conventions?

4 The Aberrant Wood Sellers

Could we find out what the wood sellers mean by "Now it is more wood" and could we convince them that they should not sell wood the way it appears they are selling wood? I take the criterion for this to be that we are able to retell or paraphrase what they intend to mean by uttering these words. This goes hand in hand with the question of whether we could teach them that they are not selling *more* wood in case an amount of wood is rearranged in such a way that it covers a larger area.

In §150 Wittgenstein discusses an attempt to teach them the use of "more wood" and "less wood" and he comes quickly to the conclusion that it is futile. From this he draws the conclusion that *"they have a quite different system of payment from us"*:

> How could I shew them that–as I should say–you don't really buy more wood if you buy a pile covering a bigger area? –I should, for instance, take a pile which was small by their ideas and, by laying the logs around, change it into a 'big' one. This *might* convince them–but perhaps they would say: "Yes, now it's a *lot* of wood and costs more"–and that would be the end of the matter.–We should presumably say in this case: they simply do not mean the same by "a lot of wood" and "a little wood" as we do; and they have a quite different system of payment from us. (RFM I, §150)

If we try to understand an utterance of theirs that "Now it is more wood", we might assume that they use "more" in the way we do. This could be justified by the fact that they demand more money if the wood is piled covering a larger area and that they mistakenly think that it is more wood after the rearrangement.

By "more" we mean that the quantity of an item is larger compared to another quantity of it or compared to another kind of the item. It indicates a two-place relation that is transitive and asymmetric. If pile A is more wood than pile B, and pile B is more wood than pile C, then pile A is also more wood than pile C. The law of transitivity manifests itself in the comparisons we make and it requires the specification of the amounts of arbitrary items we compare. It is dubious that we would have the concept of quantity or the concept of comparison of quantities without such a law. We compare, for instance, one litre of milk in one bucket and two litres of water in another bucket and we judge that there is *more* water than milk available. And we would also say that 200 shillings are *more* than 100 shillings. In the case of wood, if we add a row of logs to a stacked pile of wood, then the pile is higher and we would say that it is *more* wood than before. We would also say that a pile of wood, which is 2 metres wide, 2 metres deep and 2 metres high is *not* more wood than a pile which is 1 metre wide, 4 metres deep and 2 metres high.

It does not matter how the logs are arranged – in both cases the pile is 8 cubic metres.

Precisely this seems to be denied by the wood sellers. They indicate that the way the logs are arranged is an essential element for the question of whether more wood is sold or not. If a few logs are stacked on, say, 2 square meters and then the pile is rearranged so that these logs cover 4 square meters, then for them it is *more* wood and they charge a higher price. More money needs to be paid.

One might now say that they are simply wrong in claiming that it is more wood, which does not amount to an unintelligibility. But since the evaluation of a statement as being false presupposes its understanding, we have to assume that they believe that a mere redeployment could make *more* wood. We do so because, initially, we have no reason not to believe that the statement is unintelligible. As Stroud writes:

> I think the initial intelligibility and strength of Wittgenstein's examples derive from their being severely isolated or restricted. We think we can understand and accept them as representing genuine alternatives only because the wider-reaching consequences of counting, calculating, and so forth, in these deviant ways are not brought out explicitly. When we try to trace out the implications of behaving like that consistently and quite generally, our understanding of the alleged possibilities diminishes. (Stroud 1965: 512)

If one wants to show that the behaviour of the alien wood sellers is unintelligible, I hold that two factors need to be addressed. Firstly, we do not understand the wood sellers and their statement because we treat the statement in isolation. By that I do not mean the attempt to understand the utterance completely independent from other utterances, but that the use of the expressions as part of a statement is otherwise restricted. Secondly, we do not understand the wood sellers because if we trace out the implications of the statement, then sooner or later it dawns on us that we do not understand them. We end up with a form of life that we simply cannot grasp. This twofold line of reasoning needs to be seen in light of the argument by Stroud and the counter-argument by Canfield, which might be outlined as follows. After the alien wood sellers have rearranged the pile, they claim that it is now more wood. Stroud argues that if they believe this, then –

> [...] these people think of themselves as shrinking when they shift from standing on both feet to standing on one [...] (Stroud 1965: 512)

Against this line of thought Canfield objects that they might use "more" in different ways. To wit, they use the expression "more" the way we use it and as described earlier, but they use it in the aberrant way in the case of wood and only in the case of wood. He asks:

> Why assume that their practice with respect to selling piles of wood caries over to talk about their own size? (Canfield 1975: 472)

In other words, the idea is not so much that we are expected to understand a statement completely independent from other utterances, but that the expression "more" functions differently with regard to the item, or kind of it. In this case, the alien wood sellers might use "more" the way we use it in all cases except if amounts of wood are compared.

We do not make a difference between, say, wood or iron. If we add a row of logs to a pile of wood, then it is *more* wood than before and if we add a row of iron bars to a pile of iron, then it is *more* iron than before. We also compare piles of wood with piles of iron and sometimes we even compare piles of wood with amounts of liquids, In short, the use of "more" *does* carry over, even if not unrestrictedly. But for the sake of the argument, I grant Canfield the point that the wood sellers restrict the deviant use of "more" to wood and not to their own size. But even then, as Stroud points out –

> Surely they would have to believe that a one-by-six-inch board all of a sudden increased in size or quantity when it was turned from resting on its one-inch edge to resting on its six-inch side. (Stroud 1965: 512)

If their use of "more" is restricted to the comparing of amounts of wood, then they must believe that "more" toothpicks are produced out of one log if the position of the log is changed from its smaller edge to the longer one. One might now still ask why we could not restrict the aberrant use of "more" to boards of wood and not to toothpicks, even if they are also made of wood. Or, as Canfield ponders:

> Must we say then that for them the quantity of wood changes when the wood is taken out of a pile and made into a house? No. We may say rather that for them exists no transition from talk about quantity-of-wood-in-a-pile to talk about quantity-of-wood-in-a-house; no way of comparing the two quantities. (Canfield 1975: 473)

According to our standards they contradict themselves if they say that more wood is built into the house than was available when it was stapled as a pile or that the amount of toothpicks produced out of one board depends on the original position of the board. For us, it is the same amount of wood and we would contradict ourselves if we said that it is the same amount of wood and it is not the same amount of wood. If they treat this contradiction as unimportant or are unimpressed by it, as suggested by Canfield, then we could not paraphrase what they mean, which comes down to the assumed unintelligibility.[2]

We might ask them whether their use of "more" is applicable only to wood stapled in piles and not if wood is in the form of toothpicks and if wood is used in a house. In order to avoid the contradiction, they might come up with further criteria for the restriction of the strange use of "more". But if they proceed with this strategy, then, we have to assume that they use the expression for a singular case only and "more" could not be used on other occasions. But how could we find out what this could be? This means the statement containing the expression is unintelligible because it is taken to be a single and unconnected utterance and we would have no chance at paraphrasing what they mean by "more". We would lack criteria that determine whether we have grasped the use of "more" correctly or not.

The question of criteria for the correct application brings me now to the second line of reasoning, viz. the question of whether we could consistently, i. e. without contradiction, expand Wittgenstein's scenario in such a way that their deeds of the wood sellers do not turn out to be unintelligible. In contrast to Stroud, Canfield is optimistic that we could succeed in such an undertaking:

> Consistently with Wittgenstein's description, we can introduce various cultural surroundings for the wood sellers, countless different customs and practices that they might have and that connect up with, or fail to connect up with, their curious manner of selling piles of wood. And we can do this in a way that makes their wood selling practice, and their other practices, perfectly intelligible. (Canfield 1975: 470)

However, Canfield does not spend much time for spelling this out, so the moot question is whether we are indeed able tell such a story. As a side note, Canfield does not discuss primarily the determination of the amount of wood, but the practice of selling wood, or more generally, the exchange of wood.[3] In any case, I follow Wittgenstein and Stroud and the discussion of the question whether there is more or less wood available depending on the position of wood. In order to settle this question, we, as potential interpreters need to use "more" and observe their reactions to the way we use it. Only then we can find out whether we have understood their use of "more". To begin, it is clear that we must grasp, at least in broad strokes, how they use their the expression "more" and what practices it interdepends with. For a start, in order to rule out isolated cases as discussed

---

[2] This does not mean that we cannot communicate by using a paradox or a contradiction, cf. Williams 1964, but this is not meant here.
[3] Similarly, Baker and Hacker (Baker/Hacker 2009: 330).

before, we could make a pile of iron rods, rearrange them so that the they cover a larger area and then ask them whether now they think it is "more" iron. If they assent, this could be evidence that they do not make a distinction between wood and iron with respect to the application of "more". Since they ask a higher price if the logs cover a larger area, we can justly assume that they believe that it is more wood. As a second test, we could produce toothpicks out of two logs of wood. The first log is in upright position and the second one is laid down at the beginning of the production. If they really believe that it is more wood if the log lies laterally, then they have to say that more toothpicks have been produced (provided that toothpicks are counted). Similarly, they have to say that the volume of the logs changes depending on its position. If their use of "more" and their way to measure things were not consistent at all, as suggested by Canfield, it would be impossible to find out what they mean by it.

I now assume that they use the word consistently, i. e. that its use carries over from wood to their own size, to money and to the number of toothpicks produced. They use "more" consistently, so that they claim wholeheartedly that one can produce more toothpicks out of a log if its position at the beginning of the production covers a larger area. And they have to affirm that their size changes between the shifts from standing on two feet to standing on one. We could now try to come up, as Canfield suggests, with a story which somehow explains the shrinking.

In trying to do so, we would have to project ourselves into such a world, which entails that we would have to give up our familiar beliefs, or as Stroud writes our "familiar world":

> The reason for this progressive decrease in intelligibility, I think, is that the attempt to get a clearer understanding of what it would be like to be one of these people and to live in their world inevitably leads us to abandon more and more of our own familiar world and the ways of thinking about it upon which our understanding rests. The more successful we are in projecting ourselves into such a world, the less we will have left in terms of which we can find it intelligible. (Stroud 1965: 512 f.)

But why do we have to abandon more and more of our own familiar world, as Stroud suggests? By familiar world I understand the set of beliefs one holds to be true at a particular time. By interpreting the alien wood sellers, we would need to ascribe them the belief that it is more wood if the same amount of wood is arranged in such a way that it covers a larger area. In order to ascribe them this belief, we first need to understand what the content of this belief would be, i. e. we need to understand what the world would be like if the belief were true. This means that we must think what our set of beliefs would be if we adopted the belief. Without doubt, if one thinks about adopting the belief that it is more wood if the logs are rearranged, the set of beliefs would be inconsistent. As a consequence, one would have to choose between either giving up some of the original beliefs or rejecting the adoption of the new belief. In the latter case, one could not ascribe to them the belief that it is *more* wood after one pile is rearranged so that it covers a larger area. In other words, one does not know what they mean.[4] Alternatively, one might give up some of the beliefs one has so that she could integrate the new belief consistently in the existing set of belief. But the requirement of consistency would then impose also that she gives up the view that *more* indicates a transitive relation between two amounts on an item of a similar kind. And by that also the understanding of identity, equality and what the comparison of two amounts consists in. These concepts make sense only as parts of a larger framework of consistent beliefs and a substantial change would entail a number of further changes. Whether or not such a metamorphosis is feasible, the transmuted person perhaps would understand the wood sellers, if they speak at all – but I doubt that we would understand the transformed person. By the same token, we have to say that we do not understand the practice of selling wood as described in RFM I, §149.

Bibliography

Baker, G. P. and Hacker, P. M. S. (2009): *Wittgenstein: Rules, Grammar and Necessity. Essays and Exegesis of §§185 – 242*. Oxford: Wiley-Blackwell.
Canfield, John V. (1975): "Anthropological Science Fiction and Logical Necessity." In: *Canadian Journal of Philosophy*. Vol. 4, No. 3, 467 – 479.
Dummett, Michael (1959): "Wittgenstein's Philosophy of Mathematics." In: *The Philosophical Review*. Vol. 68, No. 3, 324 – 348.
Stroud, Barry (1965): "Wittgenstein and Logical Necessity." In: *The Philosophical Review*. Vol. 74, No. 4, 504 – 518.
Williams, Bernard (1964): "Tertullian's Paradox." In: *Philosophy as a Humanistic Discipline*. Princeton and Oxford: Princeton University Press, 3 – 21.
Ludwig Wittgenstein (1956): *Remarks on the Foundations of Mathematics*. Von Wright, G.H.; Rhees, R. (eds.); Anscombe, G.E.M. (ed. and trans.). 3rd edition 1978. Oxford: Basil Blackwell.

---

[4] This conclusion might appear to be too swift. For one might counter that we often ascribe beliefs to other people we do not assume to be true and we still understand them. But again, in order to judge a belief to be false, we need to understand it.

Susan Edwards-McKie

On the Infinite, In-Potentia: Discovery of the Hidden Revision of *Philosophical Investigations* and Its Relation to TS 209 Through the Eyes of Wittgensteinian Mathematics

Abstract: I shall build on my paper "Following a Rule without the Platonic Equivalent: Wittgenstein's Intentionality and Generality" (in *The Philosophy of Perception and Observation: Contributions to the 40th International Wittgenstein Symposium*, 2017) which explored the relation of the iterative operation to the potential infinite. Firstly, focussing on the principle of contextuality, I look at similarities and differences between Wittgenstein and Frege, which harmonize in interesting ways with the Dedekind cut and the actual infinite when viewed from the Fregean standpoint, but form a distinctly non-Dedekind paradigm when viewed from Wittgenstein's standpoint. I shall consider the principle of composition through Frege's critical question to Wittgenstein: "What cements things together?" with questions of range, part and whole. Wittgenstein's idea that it is the Eigenschaft of "5" to be the Gegenstand of the rule "3 + 2 = 5" is contrasted with Frege's Platonic work in "Der Gedanke". Questions of the role of the Tractarian *Gegenstand* in developing rules of iteration, compositionality and use, and McGuinness' and Pears' retranslation of Sachverhalte from "atomic fact" to that which is in-potentia (state of affairs) is briefly highlighted. Lastly, I provide a *Nachlass* discovery which suggests Wittgenstein continued to work on the highly mathematical TS 222, which later becomes *Remarks on the Foundations of Mathematics*, later than hitherto thought by scholars, precisely in the areas we have considered in the previous sections.

1 Introduction

The present paper is written as a companion piece to three others: "Wittgenstein's Solution to Einstein's Problem: Calibration Across Systems", an inaugural HAPP lecture, University of Oxford, on Wittgenstein and Physics, 2014, forthcoming Springer, 2018; "Wittgenstein's Wager: Mathematics, Culture and Human Action", invited lecture and publication on decision models for justice and public policy issues,

---

Susan Edwards-McKie, Darwin College, Cambridge (UK), susan.edwards-mckie@cantab.net

Krakow, Poland, 2014, Academia Verlag, 2017; "Wittgenstein's Wager: Quantum Decision Theory Revisited", invited contribution on decision modelling for high risk systems, based on Wittgensteinian philosophy of science and mathematics, Austrian National Defence Academy, Vienna, 2015; forthcoming 2019.

For Wittgenstein's projected text *Philosophische Bemerkungen* for Cambridge University Press in 1938, I have established that it is a text embedded in other typescripts, principally in TS 239 and its mathematical companion TS 221. Four forms of triangulation were met which have stood as benchmarks for the various candidates put forward as this text, the criteria extracted from the 1938 *Vorwort* (TS 225) and in response to concerns that no text had been discovered which fitted the numbering of the Rhees translation (TS 226). In order to appreciate what this revision can tell us I shall contextualise Wittgenstein's work by briefly reviewing the scientific and mathematical currents of thought in the 1929–39 period, arguing for a strong line of development within his own work between the 1930 TS 209 (published as *Philosophical Remarks*) discussed with Littlewood during the Trinity College fellowship considerations and the Hidden Revision 1938–39 composition proposed for CUP (published as a part of *Philosophical Investigations* and *Remarks on the Foundations of Mathematics, Part I*). These texts were a unity in 1938–39, and taken together – with the subtle shifts of meaning and emphases that the renumberings and the highlighted passages show alongside considerations of the themes of their lineage from the earlier TS 209 – offer us an incomparable view of Wittgenstein's philosophy of the infinite.

2 Discovery and Intellectual Placing of the Hidden Revision: *Philosophische Bemerkungen*, 1938, Overview

Northrop's famous 1946 essay on Leibniz and Einstein opens by praising Ludwig Wittgenstein as one among those "intimately acquainted with the concepts of mathematics and mathematical physics" who could fathom the "epistemological requirements of the mathematical character of space" (Northrop 1946: 444). Of particular relevance for the theme of this symposium, and the topic on which I have been asked to speak, is that during the 1937–39 period when Wittgenstein worked on the Hidden Revision of *Philosophical Investigations* his continued interest in infinity, and space and time is evidenced in his lectures, manuscripts and typescripts. Indeed, we find characterisations of space as potentially infinite and generality as a direction rather than an extension embedded across his *Nachlass*, crystalised in his

TS 209 composition for Trinity College, Cambridge, in his mathematically important essay *"Unendliche Möglichkeit"*, in the catalogued entries of his Whewell's Court short lectures: "Achilles and the Tortoise" and "Absolutely Determinate", and strongly developed in the *Philosophische Bemerkungen* proposed to CUP in 1938 which I have elsewhere called the Hidden Revision – and shall continue with this nomenclature – because the numbers were literally erased.

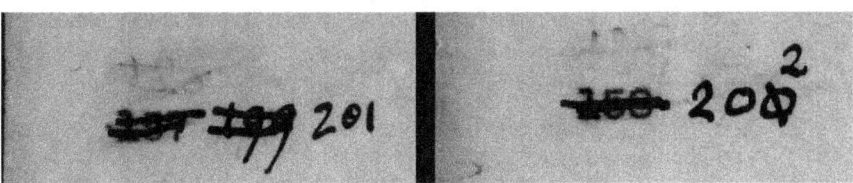

Fig. 1: Wittgenstein's Revisions: original TS 220: 157/8; HR 188/189; 1st ink 199/200; 2nd ink 201/202; Enhanced images copyright S. Edwards-McKie.

In the Hidden Revision 1938–39 remarks 1–192 were established as the Hidden Revision of TS 220 and remarks 193–316 were established as a continuation in internal remark numbering as well as the more obvious pagination across these two typescripts. By using colour filters and enhancement of the opening page of TS 239 I was able to establish that the typed *"Untersuchungen"* had been crossed through and *"Bemerkungen"* written in above to provide the title *Philosophische Bemerkungen* for its text. This exemplar, which precisely fits the Rhees translation numbering, has been much sought after within the Wittgenstein research community with von Wright, McGuinness, Pichler and Schulte expressing perplexity over the issue (von Wright 1979; McGuinness 2000, and see note 20:280). In the reconstruction the remark numbers in Wittgenstein's hand correlate faithfully with the Rhees translation remark numbering not only in remark numbers but the original intra-text numbers in parentheses of both TS 220 and TS 226 are crossed out and emended with the Hidden Revision numbers, which indicates that Wittgenstein worked on this for some time.

In addition, upon scrutiny not only is a consecutive renumbering visible, but there also appears to be a beginning system of cross-referenced numbering in Wittgenstein's hand placed beside and referencing other current Hidden Revision remarks: *"ausserdem/ die Nummern solcher Bemerkungen/tragen, die zu ihr in wichtigen Beziehungen/ stehen"* (MS 117: 123, draft of *Vorwort*), a process of writing which is carried through to the two inked revisions also shown in the above enhanced photo. This revision and its early highlighted remarks of 1938/39 are concerned with logical compulsion and with divisibility of groups, with the importance of not thinking in terms of the *essence* of a particular form when considering the possibilities of division.

The discovery of the Hidden Revision provides new and sustained evidentiary support for those scholars who conjectured that TS 220/221 were to be a unity in some form as the first version of the *Philosophical Investigations* of 1938 (von Wright 1979, KgE, Pichler 2004, Baker & Hacker 2005). The work does not support Rothhaupt 2010 that the projected CUP *Philosophische Bemerkungen* 1938 was the 'virtual' text of MSS114/115 and the Grosse Format (MS140), which was later typed by Waismann. The physical evidence of the HR was undiscovered earlier and is game-changing, and the intellectual developments in science and mathematics of the period and Wittgenstein's own development from the ideas of TS 209 to HR 1938–1939 make intellectual sense of the physical data.

While the standard dating of the close of Wittgenstein's work on the philosophy of mathematics is 1944, the terminus – of what became this elaborate revision – is, however, not conclusive. In addition, our full understanding of this text which became RFM I is complicated further by our loss of certainty of the reordering of the fragments of TS 222. My recent establishing through letters in Trinity that the hand of the numbers of the rearrangement is, in fact, that of von Wright rather than Wittgenstein requires further research and reappraisal when approaching the 'final' text. This finding does not affect the accuracy and efficacy of the Hidden Revision; indeed, we need to approach this text with fresh eyes.

The Hidden Revision even more fully distances Wittgenstein's work from the Fregean echo of "Der Gedanke" of 1918 (Frege 1918). Number for Frege is a *property* of logical/ formal concepts transmuted into objects, and *each* instantiation/extension is coded to the form. I have some sympathy with the work by Reck 2005 and others to rehabilitate Frege not as a naïve Platonist but as a non-ontological Platonist to the extent of agreeing that Frege's third realm is not modelled on a physical realm as is so often the case in naïve Platonism. However, the inclusion and exclusion principle that concept formation and extension entails in the Fregean logical system does operate and function as a law of the excluded middle in creating something very close to natural kinds; perhaps we could meaningfully call them natural logical kinds. Thus the first parameter that Reck creates to firmly exclude Frege from naïve Platonism does not appear to exclude him from a certain cosmogony with its epistemological roots.

For Wittgenstein number is only the *mark* of a concept which can be seen in signs, such as the notation of a series, and he characterised thinking itself as a *technique* of operating with signs. Indeed, as Sluga 1989: 116 points out contra Frege, for Wittgenstein thinking is mobilising signs as symbols in a variety of ways and even when we consider imaging and imagination Wittgenstein would not say that there is a private owner of the idea/image as Frege insisted. In a sense, we are what we do.

This can be understood as Wittgenstein's developing of a strong non-essentialism, which gathered pace after the typing of TS 221 and prefigured his selected crossing out of remarks from his sustained mathematical revision of 221 into TS 222. The crossed through remarks of the Hidden Revision, done in at least two distinct phases, fall along the lines that there is not an internal *property* which regulates division, or inversely *compositionality*. His selection complements the selection of remarks which were highlighted in the Hidden Revision of 1938–39. By the time of writing the materials for TS 221, 1937–38, the role of the propositions of logic was not that of laws of thought which bring out the essence of human thinking, such as Frege's law of identity about 'what men take for true' claims: "It is impossible for human beings [...] to recognize an object as different from itself" (RFM: 132). Relatedly for Frege, the individual members of the class exhibit an identity, which taken together constitute a type or 'natural' kind. In tandem with a move away from natural kinds cosmogony in a deterministic cosmology, Wittgenstein had edged 'essence' into *technique* of thinking (RFM: 133), which resonated with his *Denksystems*, a concept which emerged in the 1937–38 period with his highly complex technique of composition.

Much in keeping with Wittgenstein's idea of the relation of the potentially infinite to that of space is his criticism of Ramsey's conception of infinity as that which "presupposes that we were given the actual infinite and not merely the unlimited possibility of going on" (PR: 173; TS 209: 90; MS 106: 115). In a related vein to the remarks of 1930, TS 209 he writes in 1937, MS 117:

> In his fundamental law Russell seems to be saying of a proposition: 'It already follows – all I still have to do, is to infer it'. Thus Frege somewhere says that the straight line which connects any two points is really already there before we draw it; and it is the same when we say that the transitions, say in the series +2, have really already been made before we make them orally or in writing – as it were tracing them. (RFM: 21; HR: 200)

In TS 209 space and time are seen as potentially infinite, and Wittgenstein clearly distanced his work from that of set theory: "Time contains the possibility of all the future *now*. But all that of itself implies that time isn't infinite in the sense of the primitive conception of an infinite set. And so for space" (PR: 140). He cautions that one cannot speak of the *whole* infinite number series, as if it were an extension: "The infinite number series is only the infinite possibility of finite series of numbers" and "that the sign themselves contain the possibility and not the reality of their repetition" (PR: 144). He argues that set theory tries to express what can only be shown: "And when (as in set theory) it tries to express their possibility, i.e. when it confuses them with their reality, we ought to cut it down to size" (PR: 144). TS 209 makes the important connections that infinity is a direction, not a number or extension (PR: 142; TS 209: 64; MS 105: 145/149), and the refined conception of the infinite as a direction has been made in such a way that generality is entwined with the potential infinite and the mathematical operation rather than the actual infinite and the function.

3 *Extensionalitätsauffassung* Criticised: The Space of Human Movement Is Infinite.

As Rhees emphasised in a 1964 letter to von Wright and Anscombe concerning TS 209, the generality of a variable or proof was to be distinguished from the generality of the general proposition as Russell and Frege conceived it. In addition, in what McGuinness has argued is a Wittgenstein dictation to Ramsey of the notes for the paper on infinity which Wittgenstein delivered to the Aristotelian Society in Nottingham in 1929, we find: "Infinite *possibility* is represented by a variable whose place can be filled in infinitely many ways [...]" (McGuinness 2006). By 1935 he was re-exploring the idea of infinity as a property of space, rehearsed in MS 149 through one of the precursor drawings to the 1937 Cosmic Fragment drawings of the constructions of infinite series (MS 178e) (Edwards-McKie 2015b).

This MS 149 drawing is intellectually aligned with passages on space, time and infinite divisibility in the 1929–30 TS 209/ *Philosophical Remarks*. At remark 138: "Space has no extension, only spatial objects are extended, but infinity is a property of space [...] And the same goes for time." And at 139:

> How about infinite divisibility? Let's remember that there's a point to saying we can conceive of any finite number of parts but not of an infinite number; but that this is precisely what constitutes infinite divisibility. Now 'any' doesn't mean here that we can conceive of the *sum total of all* divisions (which we can't for there's no such thing). But that there is the *variable* 'divisibility' (i.e. the concept of divisibility) which *sets no limit* to actual divisibility; and that constitutes its infinity.

He concludes: "If you say that space is infinitely divisible, then strictly speaking that means: space isn't made up of individual things (parts)". Related points were developed more fully as applicable to space and infinite series in the mathematically important short essay "*Unendliche Möglichkeit*" (TS 215), traditionally dated 1932. This essay successfully draws together several mathematical issues. Importantly, space is characterised not as itself extended, but that "*Der Raum gibt der Wirklichkeit eine unendliche Gelegenheit der/ Teilung.*" (TS 215: 19).

In speaking specifically of TS 209, which the trustees referred to as Moore's Volume and later published as *Philosophical Remarks*, in the same letter in the early new year of 1964 Rhees argued that the *Remarks* (TS 209) was a different discussion from that of the *Big Typescript*, that the different ordering of the passages created a significantly different book. In the *Remarks*, he pointed out, Wittgenstein more prominently used discussions of the irrationals to highlight his deep concerns of the *Extensionalitätsauffassung* in mathematics.

Because we do not have the complete TS 208 from which 209 was made it is impossible to fully reconstruct the extent of their differences, but we can see the edges of the clippings in TS 209 to get some idea of which manuscript arrangements had already been made in the synopsis pre-cut which Wittgenstein gave to Russell, referred to by Russell in his letter to Trinity Council as a "bulky typescript, *Philosophische Bemerkungen*" (McGuinness,183; letter 8.5.1930). There are very many clippings in TS 209 pasted into a ledger, with some of the material uncut, but it is obvious that an overall reordering has taken place. Much of the material on infinity had been grouped together in TS 208, but we can see that even here there is further refinement. As a particularly good example of what was achieved in TS 209, let us briefly look at the sections on infinite divisibility and primary time, focusing on PR Remark 140/ TS 209: 63.

The manuscript sources are MS 106: 29 – 35 and MS 108: 105, but there is also an inked emendation which gives us a key to what the re-ordering and change of emphasis Rhees was certain was important to preserve was about. The typescript page 63, where this remark falls, is constructed of five pasted clippings. In TS 209 the passage originally read: "Is primary time infinite? That is, is it an infinite possibility?" (This has echoes of discussions with Ramsey.) It continues: "Even if it is only filled out as far as memory extends, that in no way implies that it is finite. It is infinite in the same sense as three-dimensional visual space (*Gesichteraum*) is infinite [...]" . In the revised text we have "*Gesicht und Bewegungsraum*" inked into the typescript in Wittgenstein's hand, thus he is including both seeing and moving as related to the potentially infinite. He then adds, from MS 105, a one-line remark, clipped and placed for specific effect and shift of argument: "The space of human movement is infinite in the same way as time."

Once human movement is emphasized, *constructive* possibility takes on a stronger mathematical-symbolic role, as in constructive proof and the seeing involved in complementing and conceptually expanding the operation of iteration. This moves generality firmly away from a disjunctive set interpretation. We are rapidly leaving fixed natural kinds cosmogony with a guided development of a series as replacement. Rather, as reiterated in a 1935 entry, and prefiguring the highlighted remarks of the Hidden Revision: "A certain symbolism will easily go with a certain aspect of looking at a thing" (MS 148: 22).

4 History of Production

Two scientific congresses framed Wittgenstein's return to working academic philosophy in 1929 with his production of TS 209 and fellowship at Trinity, and his production of *Philosophische Bemerkungen* 1938 (TS 220/221) as he returned to Cambridge seeking support from Trinity and the Professorship in Philosophy. The Second Conference of the Epistemology of the Exact Sciences, jointly sponsored by the Vienna Circle and the Berlin Circle, was held at Königsberg in 1930, and a similar congress in Copenhagen in 1936 focused on causality and quantum physics. Remarkably, the first established the strands of mathematical development for the 20th century with the important papers of Gödel's First Incompleteness Theorem, Carnap's on logicism, Heyting's on intuitionism, Von Neumann's on Hilbert's formalism, and finally, Waismann's presentation "The Nature of Mathematics: Wittgenstein's Standpoint", the latter an account which distinguished between set theoretic totalities and Wittgensteinian systems, essentially between the actual, completed infinite and the potential infinite. I have argued elsewhere for a systems approach to Wittgenstein's philosophy (Edwards-McKie 2015a; see also Sluga 2010), and I am suggesting in this paper that issues of causation and action at a distance were foundational for him in the 1938/39 compositions and lectures on mathematics, with the pivot-concepts of the infinite, space and time, divisibility, action at a distance and indeterminacy also pivot-concepts in the 1929–30 period.

MS 121, begun 26.4.1938 is a critique of the diagonal method as a platonising of Cantor's transfinite numbers, with many of the Whewell's Court lectures of this time period a criticism of Russell ideas of causation and judgment (Munz 2010: 88). The idea that systems cannot be extended in a transfinite way is brought out in the 12.7.1938 entry of this MS, with Wittgenstein voicing the absurdity that one infinity could be greater than another. That there can be embedded systems, intersecting systems, quantum systems is perfectly allowable. The demarcation around a particular system functions as a zone of potential action, movement, use which can interact with other systems, such as in the motley of mathematics he described. There is a problem with closure, but one which he is at pains throughout his work not to 'solve'. And if we try to represent our fuzzy border by drawing two concentric circles with the unsmooth 'boundary' in between these two limits we have only created an infinite regress if we try to reach a formal definition/line.

The Hidden Revision, with its renumbering decisively drawing together the typed version of the MS 142 manuscript breakthrough with questions about the foundations of mathematics in the companion typescript and then highlighting issues of divisibility and action at a distance parallels the developing philosophy of mathematics and physics of the first quarter of the 20th century, this time using his systemic, holistic argument to counter Einstein causality. Whether the holism, which both Einstein and Wittgenstein embraced, is envisaged as causal, as Einstein, Podolsky and Rosen argued (EPR Paradox 1935) or systemically, as Wittgenstein held, remains an entrenched research issue in contemporary physics and cosmology.

In addition, the focus on the mathematical materials of TS 221 was influenced by concerns to counter Turing's set theoretic argument. As a representative remark of the conflicting strands of Cambridge thought in the 1930s A. G. D. Watson, a physicist and advisor on mathematical issues to both Sraffa and Wittgenstein, writes:

> [...] all the remarkable problems and discoveries of the Foundations of Mathematics, the paradoxes of the theory of aggregates, Russell's theory of types, with its axiom of reducibility, Cantor's arithmetic of transfinite numbers, with its insoluble problems such as the 'continuum problem', the problems connected with functions in extension and the multiplicative axiom – all these merely express in one way or another the well-known difficulties which arise when we attempt to treat an infinite process as completed. (Watson 1938: 450)

In the 1938 Whewell's Court lectures, when talking about the two seeds, the ideas of indeterminacy, indeterminism and action at a distance surface from a variety of manuscript sources and earlier conversations with Waismann. In 30.29.1929, when attempting to describe visual space as purely Euclidean Wittgenstein found that he had to supplement it with, in Waismann's term, the "indeterminacy-factor" (McGuinness, 1979, 55). In his 1938 lecture it is addressed more directly: "The idea of action at a distance shocked scientists. This idea [of action at a distance/of indeterminacy] revolutionized science." (Smythies, Box 2, Lecture V, Trinity; WCL: 18, lecture 20.5.1938). Historically, we can see that these issues are precisely those which figure prominently in defining and developing various strands of Viennese modernism, indeed of modernism itself, with questions of uncertainty pushing towards post-modernism.

5 The Cambridge Connection

Denis Paul criticised publication of *Philosophical Remarks*, and Rhees' judgment to do so (Paul 2007: 17 –18; quoted in Rothhaupt 2010: 54) claiming that it was a poor introduction to Wittgenstein's thought as the selection and reordering of the remarks of the source TS 209 combined contradictory positions and were arranged in such a way which masked Wittgenstein's development of ideas more chronologically. Rhees' 1964 letter to von Wright and Anscombe argues, amongst

Tab. 1: Nachlass and Intellectual Synergies/Timeline (Copyright S. Edwards-McKie)

"Some Remarks on Logical Form"	Congress of the Epistemology of Exact Sciences 1930	C Series Note-Books	MS 142	Church/Turing *On Computable Numbers*	MS 178e 'Cosmic Fragment'	MS 117: 1–97	TS 220 The Moore TS	Whewell's Court Lectures	MS 117: 97–110	Vorwort *Philo-sophische Bemerkun-gen*	TS 221	TS 221	Hidden Revision
	Gödel's Incompleteness Proof	EPR Paradox 1935	first part of PI	Copenhagen Congress						CUP 1938		TS 226 Rhees translation	TS 220 + TS 221: 138–204
	TS 213 "Unendliche Möglichkeit"												MS 117: 1–97 = TS 221: 138–204
						MS 117: 127–148							
1929	1930–33	1933–36	1936–38	1936	1937 August	1937 September	1936–38	1938 Spring [Summer Term]	1938	1938 August	1938 Autumn [Michelmas term]	1938–39	1938–
precursor to entanglement ideas	property of space internal rule	C Series: space infinite division intervals		causality quantum physics Entscheidungs-problem	Härte des logischen Zwangs unendliche Reihe	Cont. a series internal properties of a series		infinite similarity action-at-a-distance Zeno Paradoxes indeterminism	Critique: Cantor set theory trans-finite number Diagonal method		Critique: Gödel Incompl Theorem Can a machine think?	Rhees late summer 1938– Michel. Term TS221 Michel. Term 1938–Lent term 1939	Cross-references: diagrammatic constructive proofs divisions of groups action-at-a-distance
systemic thinking	infinite as unending	mathematic aspect perception in constructive proofs construction of polygons			passen (fit) gleich (similarity)	"How we measure is what we measure". diagrammatic constructive Proofs Alternative divisions of groups aspect perception							
logical aspects of measurement	possibility: Unendliche Möglichkeit												

many substantial mathematical points, that the method used for this typescript was an internal relating of certain key features, and that when the remarks are not in this sort of relation it signals that a different view is being put forward by using the same material in a different way. In any case, after reflection and with great pressure of the 1929 Trinity fellowship funding upon him, he put it in *that order*.

Given that we have very few texts indeed which we are certain are the considered re-arrangement or intended connecting of manuscripts and typescripts by Wittgenstein himself, those of which certainty can be established need to be carefully studied from the point of view of the production process, and the meaning that can be gathered when considering the remarks in a particular combination. Sluga's exploration (Sluga 1989) of the reflective writing process suggests a lack of closure as a richness and an inevitability of philosophy, and with it a discrediting of the Fregean picture of thinking. In Wittgenstein's case it is interesting to note that periods of high level pressure produced a definitive result: the *Tractatus* during extreme war time conditions; TS 209 under extreme financial and intellectual need; the Hidden Revision under financial, intellectual and existential need.

In 1929, at a much more junior level, there had been ill feeling and misunderstandings in terms of the procedures and submission needed for the PhD. In a letter to Ramsey, who was Wittgenstein's supervisor, Wittgenstein voiced his dismay at Ramsey's cavalier attitude to his securing his PhD.: "I can't understand how, being my supervisor and even – as I thought – to some extent my friend having been very good to me you couldn't care two pins whether I got my degree or not. So much so that you didn't even think of telling Braithwaite that you had told me my book would count as a dissertation" (McGuinness & von Wright, 1980, 261: letter Frühjahr 1929). With the PhD. secured 18 June 1929, Wittgenstein approached Trinity for short term research funding in June 1929, and by March 1930 Wittgenstein was in Vienna composing the synopsis (TS 208) that he had been asked for as part of the application process for a longer term five-year fellowship funding at Trinity, which he was granted December 1930.

However, this fellowship stage of funding had been delayed when Russell, in his assessment letter 8.5.1930 to Trinity Council had praised Wittgenstein's theories as "novel, very original, and indubitably important", but felt bound to state "Whether or not they are true, I don't know" (McGuinness 2012: 183). TS 209 is Wittgenstein, and in very deep ways, Wittgenstein at his very best. From his cumbersome earlier synoptic TS 208 left with Russell, he rearranged, cut dross, refined, took risks, and argued his intellectual case to Littlewood.

Fully immersed in the Cambridge environment, and having fallen foul of it before at the BA level – an incident which left Wittgenstein and Moore no longer on speaking terms for many years – it is impossible to think that Wittgenstein was not keenly aware that as TS 209 in 1929 had forced a focus which gained him a first fellowship in 1930, the deeper level thinking in the philosophy of science and mathematics that he had reached through the Hidden Revision synthesis of 1938–39 was the order of the day for 1939. He had worked hard on it, and asked Rhees to begin translation of it for the proposed publication. Thus he fell back on what he always considered his profound calling in life: the philosophy of mathematics in its broadest and most theoretical terms, a currency which Trinity understood very well.

In 1929 Littlewood's comment that he had thought that Wittgenstein would not have developed and would have been "living on old capital", was, after discussing the ideas developed in TS 209 with Wittgenstein, corrected to:

> The idea about old capital is entirely groundless. W. wrote a book once before, & I mean it literally when I say that I see no reason whatever why he should not write another, & perhaps more important book. (McGuinness 2012: 186; Letter Littlewood, 1.6.30).

In 1938–39 Wittgenstein repeated this performance, even more under fire. He was elected to the Chair of Philosophy in February 1939.

6 A Developing Systeme: A Reflexive and Entangled Relation

We can accommodate "grasping the word in a flash", seeing a mathematical point from the Indian mathematician's point of view (an obvious reference to Ramanujan), without the future development already present in the grasping because a thinking/thought system does not have to be understood as a totality before we can get started within the system. In set theory one cannot get away from the beginning: the empty set necessary as generator. This point of view of Wittgenstein is sustained as a critique of Gödel's objections to contradiction and his thesis of unprovability. In a nutshell: In-potentia is to be distinguished from absolute and incomplete. Yet this *Systeme* does not reduce to a traditional hidden variable cosmology. One would say: I am at home in the infinitely rich system with its features of entanglement because I simply have no choice. It is where I find myself. Illusion is thinking that I am somehow 'above' it or that there is an internal variable which holds the system in place for me.

Even in the *Tractatus* all we have in terms of any sort of essence is that the *possibility* (*Möglichkeit*) of combination/concatenation of the *Gegenstände* like links in a chain is essential to the formation of the *Sachverhalte* – a consideration which informed the important retranslation of "Sachverhalt" from "atomic fact" to "state of affairs", the latter registering the aspect of in-potentia that is central to this notion (TLP; see also Edwards-McKie 2017b interview with Brian McGuinness). That Wittgenstein came to hold more strongly and clearly that there is not an internal property which regulates division and compositionality is a crucial intersection with Frege, and with Wittgenstein's sustained concern over Frege's early criticism of what he considered to be Wittgenstein's failed account of the unity of the proposition: what holds the elements of the proposition together? Wittgenstein would have to say in response: *In der Frage liegt ein Fehler*. Frege, with some humour, asks in his 28.7.1919 letter to Wittgenstein whether it might be something like gravitational force on planets (De Pelligrin 2011: 53).

The complexities of this are great. Since it could be argued that if Wittgenstein's conception of space as variable divisibility (as above) fulfils the two important Parmenidean criteria of Being which are indivisibility and homogeneity, then a hidden variable cosmology would seem to be the natural heir to this way of thinking: that I can and do step in the same river twice. I have argued elsewhere that even though Wittgenstein's work is not in contradiction to an Einsteinian general relativity with the cosmological constant in place, that his philosophy of time so impacts on his philosophy of space that it cannot be reconciled to the traditional view. What my discovery and study of the Hidden Revision has shown is that systemic time necessarily informs our conception of space as that in which human movement *defines* time: they are in a reflexive, entangled relation.

This is where action-at-a distance enters (specifically in the Hidden Revision highlighted extra numberings, stated in the neighbouring passages of TS 221: 195 – 196; HR: Remarks 301, 304; later published as RFM 62 and 65) as a counter to a cosmology of essentialism and local causal realism. Looking at diagrammatic proofs of division: "Suppose someone now asked: 'What does the action at a distance of the picture consist in?' – In the fact that I apply it." (HR: 304; RFM: 65). It is necessary that it be brought into a system or we might erroneously think a proposition could be true essentially. Or correlatively erroneously, we might think that there must be a mental process behind the meaning, a real sign behind our ordinary signs, platonic numbers behind our constructed series or an axiomatically or causally realised hidden variable behind our world.

It is of profound philosophical interest that in both Wittgenstein's *Big Typescript* (TS 213: Section 41) and in Rhees' *Philosophical Grammar* (PR: 81; PG: 39) – in which Rhees chose to publish Wittgenstein's revision of the *Big Typescript* in the virtual text MSS114/115 complex – Wittgenstein's use of action-at-a-distance is one in which action-at-a-distance is seen in contrast to *Anwendung* (application) while in the later discussion in the Hidden Revision, which is a much wider assemblage of interrelated strands, application *constitutes* action-at-a-distance. This shifts the paradigm from Newtonian, which is within the Fregean dialogue, to that of the entangled quantum.

Acknowledgement

I wish to thank Gabriele Mras, Paul Weingartner and Bernhard Ritter for the invitation to speak, and I wish to thank the Master and Fellows of Trinity College, Cambridge for their kind permission to reproduce images from the Wittgenstein Collection.

Bibliography

Baker, G.P and Hacker, P.M.S. (2005): *Wittgenstein. Understanding and Meaning: Volume 1 of an Analytical Commentary on the Philosophical Investigations*. Part II, 2nd Edition, Oxford: Blackwell.

Edwards-McKie, Susan (2012): "Some remarks arising from Hacking's 'Leibniz and Descartes: Proof and Eternal Truths'." In: *Contributions to the 34th Wittgenstein Symposium*. Kirchberg, Austria.

Edwards-McKie, Susan (2014a): "Looking through a microscope. Philosophische Bemerkungen 1938: the hidden revision which fits the 1938 Vorwort, the Rhees translation and what this tells us about Wittgenstein's philosophy of mathematics and cosmology." In: *Contributions to the 36th Wittgenstein Symposium*. Kirchberg, Austria.

Edwards-McKie, Susan (2014b): "Wittgenstein's solution to Einstein's problem: calibration across systems." An inaugural lecture HAPP, University of Oxford. Available St Cross College, Oxford website and WAB, University of Bergen website; forthcoming Springer 2018.

Edwards-McKie, Susan (2015a): "Quantum Decision Theory Revisited." Paper presented Austrian National Defence Academy, Vienna; forthcoming 2019.

Edwards-McKie, Susan (2015b): The Cosmic Fragment: *Härte des Logischen Zwangs* and *Unendliche Möglichkeit*. Nachlass discoveries and Wittgenstein's conception of generality and the infinite, Wittgenstein Studien, Band 6, 51 – 81.

Edwards-McKie, Susan (2017a): "Wittgenstein's Wager: Mathematics, Culture and Human Action." In: *Wittgenstein, Philosopher of Cultures*. Humphries, Carl; Walter Schweidler (eds.). Sankt Augustin, Germany: Academia Verlag; paper presented symposium *Wittgenstein Philosopher of Cultures*, Krakow, Poland.

Edwards-McKie, Susan (2017b): "A Tapestry: Susan Edwards-McKie interviews Professor Dr. B. F. McGuiness on the occasion of his 90th birthday." In: *Nordic Wittgenstein Review*. Vol. 6 No. 2, 85 – 90.

Einstein A., Podolsky B. and Rosen N. (1935): "Can Quantum-Mechanical Description of Physical Reality Be Considered Complete?" In: *Physics. Review*, Vol. 47, No. 10, 777 – 780.

Frege, Gottlob, (1918): "Der Gedanke." In: *Beiträge zur Philosophie des deutschen Idealismus*. "Thoughts." In: *Logical Investigations*. Geach, P.T. (ed.), Geach, P.T.; Stoothoff, R.H. (trans.). Oxford: Blackwell, 1977. 1 – 30.

McGuinness, Brian (2000): "Manuscripts and works of the 1930s." In: *Approaches to Wittgenstein*. London/New York: Routledge, 270 – 286.
McGuinness, Brian (2006): "Wittgenstein and Ramsey." In: *Cambridge and Vienna. Frank P. Ramsey and the Vienna Circle*. Galavotti, Maria Carla (ed.). Dordrecht: Springer, 19 – 28.
McGuinness, Brian (ed.) (2012): *Wittgenstein in Cambridge: Letters and Documents, 1911-1951*. Oxford: Blackwell Publishing Ltd.
Munz, Volker A. (2010): "The Whewell's Court Lectures: A Sketch of a Project." In: *Wittgenstein after his Nachlass*. Venturinha, Nuno (ed.). Basingstoke: Palgrave Macmillan.
von Neumann, J. (1930): "The Formalist Foundations of Mathematics." In: *Philosophy of Mathematics. Selected Readings*. Benacerraf, R.; Putman, H. (eds.), 1964. Englewood Cliffs, NJ: Prentice-Hall, 50 – 54.
Northrop, F. S. C. (1946): "Leibniz's Theory of Space." In: *Journal of the History of Ideas*. Vol. 7, No. 4, Leibniz Tercentenary Issue, 422 – 446.
Paul, Denis (2007): *Wittgenstein's Progress 1929-1951*. Bergen: Publications from the Wittgenstein Archives at the University of Bergen, No. 19.
De Pelligrin, Enzo (ed.) (2011): "Frege-Wittgenstein Correspondence." In: *Interactive Wittgenstein, Essays in Memory of Georg Henrik von Wright*. Floyd, Juliet (trans.). Dordrecht: Springer. 1 – 107
Pichler, Alois (2004): *Wittgensteins Philosophische Untersuchungen: Von Buch Zum Album*. Amsterdam/New York: Rodopi.
Reck, Erich (2005): "Frege on Numbers: Beyond the Platonist Picture." In: *The Harvard Review of Philosophy*. Vol. 13, No. 2.
Rothhaupt, Josef (2010): "Wittgenstein at Work: Creation, Selection, and Composition of 'Remarks'." In: *Wittgenstein after his Nachlass*. Venturinha, Nuno (ed.). Basingstoke: Palgrave Macmillan, 51 – 63.
Sluga, Hans (1989): "Thinking as Writing." In: *Wittgenstein in Focus – Brennpunkt: Wittgenstein*. McGuinness, Brian; Haller, Rudolf (eds.). Amsterdam, Atlanta: Rodopi, 115 – 141.
Sluga, Hans (2010): "Our grammar lacks surveyability." In: *Language and World*. Munz, Volker; Puhl, Klaus; Wang, Joseph (eds.). Frankfurt: Ontos, 185 – 204.
Schulte, Joachim (KgE): "Einleitung." In: *Ludwig Wittgenstein, Philosophische Untersuchungen: Kritisch-genetische Edition*. Schulte, Joachim (ed.). Frankfurt am Main: Suhrkamp, 12 – 50.
Turing, Alan (1936): "On Computable Numbers, with an Application to the Entscheidungsproblem." In: *Proceedings of the London Mathematical Society*. (Series 2), Vol. 42, 230 – 265.
Waismann, Friedrich (1979): "Totality and System." In: *Ludwig Wittgenstein and the Vienna Circle*. McGuinness, Brian (ed.). Oxford: Basil Blackwell.
Watson, A.G.D. (1938): "Mathematics and Its Foundations." In: *Mind*. Vol. 47, 440 – 451.
Wittgenstein, Ludwig (MS): *The Wittgenstein Papers*. Trinity College, Cambridge (UK).
Wittgenstein, Ludwig (TS): *The Wittgenstein Papers*. Trinity College, Cambridge (UK).
Wittgenstein, Ludwig (1956): *Remarks on the Foundation of Mathematics*. von Wright, G.H.; Rhees, Rush; Anscombe, Elizabeth (eds.). Oxford: Basil Blackwell.
Wittgenstein, Ludwig (1961): *Tractatus Logico-Philosophicus*. Pears, David F.; McGuinness, Brian F. (eds.). London: Routledge & Kegan Paul.
Wittgenstein, Ludwig (1975): *Philosophical Remarks*. Rhees, Rush (ed.). Oxford: Basil Blackwell.
Wittgenstein, Ludwig (1978): *Philosophical Grammar*. Rhees, Rush (ed.), Anthony Kenny (trans.). Berkeley/Los Angeles: University of California Press.

Wittgenstein, Ludwig (1986): "The Nature of Mathematics. Wittgenstein's Standpoint." In: *Ludwig Wittgenstein: Critical Assessments*. Shanker, Stuart (ed.), Vol. 3, London, Sydney and Dover, NH: Croom Helm.

Wittgenstein, Ludwig (KgE): *Philosophische Untersuchungen. Kritisch-genetische Edition*. Schulte, Joachim (ed.). Frankfurt: Suhrkamp.

von Wright, Georg H. (1979): "The Origin and Composition of Wittgenstein's *Investigations*." In: *Wittgenstein: Sources and Perspectives*. Luckhardt, C.G. (ed.). Hassocks/Ithaca: Branch Line, 138 – 160.

Wittgenstein, Ludwig (WCL): *Wittgenstein's Whewell's Court Lectures: Cambridge, 1938–1941, from the Notes Taken by Yorick Smythies*. Munz, Volker A.; Ritter, Bernhard (eds.) Chichester (UK): Wiley Blackwell, 2017.

Richard Heinrich
Incomplete Pictures and Specific Forms: Wittgenstein Around 1930

Abstract: Wittgenstein, when he distanced himself from some of the characteristic positions held in the TLP, did not in the first place give up the picture theory, neither did he give up the concept of elementary propositions as such. What he abandoned first was the understanding of elementary propositions as logically independent, and for a short period he thought of possibilities to adapt the picture theory accordingly. The peculiar concept of an incomplete picture stands for one such attempt. It is not primarily meant to single out a special kind of pictures, but rather to illustrate what he wants to say about a special kind of propositions – elementary propositions containing variables. The paper describes changes brought about by these considerations with respect to the TLP, in particular concerning Wittgenstein's view of generality and a re-evaluation of the concept of a specific form.

1 Introduction

Already in the autumn of 1914, when Wittgenstein initially sets out his view of propositions as pictures, the notion of an "unvollständige Abbildung eines Sachverhalts" ("incomplete portrayal of a situation", NB: 9e) makes its first appearance; the notebook-entry from June 16, 1915 contains the very expression "incomplete picture" (NB: 61e) in the context of a remark that will still figure in TLP 5.156. But only much later and for a short period around 1930, when rethinking his earlier positions concerning generality, elementary propositions, and logical form, Wittgenstein seems to have given the notion independent weight. These arguments are outlined in chapters VIII and (especially) IX of PR. In the following I will concentrate on a few paragraphs in the notes taken from Wittgenstein's conversations with Moritz Schlick and Friedrich Waismann at that time which are directly related to the passages in PR. My reasons for doing so are first, that here, at times prompted by questions of his interlocutors, Wittgenstein presents his views in a more differentiated way; second that continuity as well as divergence with regard to the TLP are being more clearly spelled out; and third that certain consequences concerning the notion of a picture as such become better visible. It is this last point where my interest in the subject mainly lies. It seems that in 1929–1930 Wittgenstein focused

Richard Heinrich, University of Vienna, richard.heinrich@univie.ac.at
DOI 10.1515/9783110657883-28

on the notion of an incomplete picture primarily as a possible means to clarify theoretical possibilities resulting from his abandonment of the thesis of mutual independence of elementary propositions.[1] Naturally in this way the question comes up whether – and if, in which sense – one could assertively speak of the existence and nature of incomplete pictures; and what kind of support the answer could lend to the conception of an elementary proposition containing variables. In this context Wittgenstein discusses situations where we ordinarily and unhesitatingly speak of the incompleteness of pictures. In the end, the arguments put forward in this direction do not yield much of a theoretically relevant result. They are retrospectively illuminating, on the other hand, as to the fragility and inherent ambiguities in the relationship of picture and logical form established in the TLP. In fact, they can be seen as a step in the dismantling of the Tractarian picture-theory. Wittgenstein continued to speak of propositions as pictures for quite some time, though; but when he included in BT (1933) the insights won around 1930 concerning the nature of generality, he no longer used the notion of the proposition as a picture. The single occurrence of the expression "general picture" in the opening sentence of section 70 (BT: 241e) is barely more than a reminder of the way he had confronted the issue three years earlier.

2 Generality

One aspect of Wittgenstein's statements in WVC is that he wanted to explain his attitude towards generality as put forward in the TLP, first of all the slogan that generality has to be kept separate from the truth-function:

> I dissociate the concept all from truth-functions. Frege and Russell introduced generality in association with logical product or logical sum. This made it difficult to understand the propositions '$(\exists x).fx$' and '$(x).fx$' in which both ideas are embedded. (TLP 5.521)

In the course of the meeting with Schlick and Waismann on December 22nd 1929, he sketched an example which could already have been used in the TLP to illustrate this point. To yield the meaning of the expression "all" as used in "All men in this room are wearing trousers", the (supposedly) underlying conjunction ("Professor Schlick is wearing trousers, Wittgenstein is wearing trousers, Waismann is wearing trousers…") is not sufficient and has to be complemented by a clause of the kind:

---

[1] On the relationship between the picture theory and the thesis of mutual independence of elementary propositions in the TLP (cf. Ricketts 1996: 84 – 85).

"and nobody else is in the room", thereby – via the generality of "nobody else" – leading to a circle. What then is the alternative understanding of generality?

In TLP, in the remarks following immediately upon 5.521, Wittgenstein says:

> What is peculiar to the generality-sign is first, that it indicates a logical prototype, and secondly, that it gives prominence to constants. (TLP 5.522)
> The generality symbol occurs as an argument. (TLP 5.523)

In 5.501 Wittgenstein describes, amongst others, how a set of propositions can be characterized in a general way by using "a function fx whose values for all values of x are the propositions to be described". Remarks 5.101 and 5.5, taken together, say that from such a set the logical sum of its members ("embedded" in the symbol $(\exists x).fx$) could be built via "successive applications [...] of the operation $(-----T)(\xi, ...)$". The point is that generality comes into the picture *before and independently of* the application of the operation $(-----T)(\xi, ...)$, in the form of the function "fx". Thomas Ricketts has given a more detailed picture of this approach:

> We can then form a sentence-function from an elementary sentence by converting any name or predicate in it into a variable-expression. We can use these elementary-sentence-functions to stipulate values of sentence-functions to serve as bases for an application of a truth-operation that yields a truth-function of those values. (Ricketts 2013: 130)

It is the priority of (the expression of) the constant form[2] common to all values of the function "fx" which tells us where generality has to be looked for when "dissociated [...] from truth-functions", namely the position of the argument.

In TLP Wittgenstein doesn't give examples of how to obtain propositions like "All men in this room are wearing trousers" or "I met a man"[3] in this way from elementary-sentence-functions[4]. That is the issue he confronts now, December 1929. His first example is the proposition "I see a square and in it a circle" or, as he also puts it, "There is a circle in the square". And he says, of course we do not want this to be understood as meaning: "Either this circle is in the square, or this circle ... or ...", where the various circles would be distinguished by diameter and position. The generality of "a circle" is not to be understood after the model of the

---

2 "Wittgenstein's idea in a sense reverses the conventional way of conceiving of variables: replacing a constant with a variable serves to make prominent not what has been removed from the proposition but what remains." (Potter 2009: 179).
3 Russell's example in Russell 1905: 481.
4 For the possibility and general methods to achieve this (cf. Geach 1981, Fogelin 1982, Varga von Kibed 2001, and Ricketts 2013).

logical sum. In what follows Wittgenstein obviously sees himself going beyond what he had proposed or implied, concerning the same issue, in TLP:

> It is clear that this is not an enumeration but something entirely different. I think that here there is a kind of proposition of which I used to have no idea and which corresponds roughly to what I want to call an incomplete picture. [...] The point is that in all such cases there is what I now want to call an elementary proposition that is an incomplete picture. (WVC: 39)

The "kind of proposition of which I used to have no idea" is *elementary propositions*: The word "incomplete" does not mean that the sentence "The circle is in the square" is not sufficiently analyzed; it has a determinate sense.

What strikes one in the formulation "[...] what I now want to call an elementary proposition that is an incomplete picture" is a certain ambiguity as to the substance (and direction) of the argument – if argument it is meant to be at all. If Wittgenstein wanted to express his surprise at the turning up of a hitherto overlooked kind of proposition, and his intention was to explain the phenomenon via the notion of an incomplete picture, it seems odd for him to assume from the outset the identity of explanans and explanandum ("that *is* an incomplete picture") – unless he had in mind something of the following kind: As every elementary proposition is a picture, an elementary proposition with a free variable has to be an incomplete picture. But this still leaves him with the task to show that there are indeed incomplete pictures (and to characterize them in a general way) – or else to resort to a different explanation. Now, as already noted, reflections on incomplete pictures came up in Wittgenstein's thought as soon as the picture theory itself. But when he at that earlier time pondered over the possibility of propositions being incomplete pictures, he never considered the incompleteness as contributing to (or characteristic of) the identity of the picture (or proposition) as such:

> If a proposition tells us something, then it must be a picture of reality just as it is, and a complete picture at that.– There will, of course, also be something that it does *not* say – but what it does say it says completely and it must be susceptible of SHARP definition.
> So a proposition may indeed be an incomplete picture of a certain fact, but it is ALWAYS *a complete picture*. [Cf. 5.156.] (NB: 61e)

That means that simply taking up the earlier notion of a proposition which is "an incomplete picture of a certain fact" certainly cannot do the job of explaining the recently discovered species of elementary propositions which are (characteristically) *incomplete pictures of what they are pictures of*. (If it were not so, how could he speak of something of which he "used to have no idea"?) So if there is here a question of argument at all, it must be the question, taken in a stronger (if yet vague) sense: Is there (and what is) an incomplete picture?

An answer – rather disappointing at first sight – can be found a few paragraphs later when Wittgenstein says: "The incompleteness of a picture consists in the occurrence of variables in a proposition" (WVC: 40). But what prima facie looks like blatantly begging the question can perhaps be read as an invitation to approach it from the other side, the side of a closer examination of the notion of an elementary proposition containing a free variable – and only then to state the expectations to be set in the incomplete picture.

3 Elementary Propositions

Wittgenstein gives an example to this effect (i.e. where incompleteness is examined exclusively from the perspective of the proposition) immediately following upon the recognition of the new species of propositions:

> I have seen two substances of the same colour. Then one might think this meant 'Both were green, or both were blue, or …'. It is clear to all of us that it cannot mean that. After all, we cannot produce such an enumeration. Whereas the following is the case: 'We saw a substance of the colour x and another substance of the colour x'. The point is that the Russellian analysis is not correct. (WVC: 39)

Referring back to the proposition "We saw a substance of the colour x" he then says a paragraph later:

> I mean, the right expression does not convey '$(\exists x).\varphi x$', but 'φx'.[5] […]
> 'φx' is hence a proper proposition, not merely preparatory for a proposition. Now I believe that certain data may be left out of an elementary proposition. A proposition is then an incomplete portrait of a state of affairs. WVC: 40

Waismann's notes from December 22nd contain a list of logical expressions relevant in the context of Wittgenstein's "men in trousers" (example WVC: 44), where the difference just mentioned becomes visible in the comparison of two lines: "'φx' = 'There is someone in the room'", whereas "'$(\exists x).\varphi x$' = '$\varphi a \vee \varphi b \vee \varphi c \ldots$'". On the one hand this can still be read as an illustration of TLP 5.521 (the separation of generality from the truth-function) and TLP 5.523 ("The generality symbol occurs as an argument"). The key the TLP provided for the understanding (and the possible consequences) of this program is the link between variable and constant form, established in TLP 4.1271: "For every variable represents a constant form that all

---

[5] "Wie lautet der richtige Ausdruck des Satzes? Ich meine: der Ausdruck lautet nicht: '$(\exists x).\varphi x$', sondern 'φx'" (WW III: 40).

its values possess, and this can be regarded as a formal property of those values."
But when, on the other hand, the expression "φx" is seen at the same time as
the representation of a constant form (i.e. a sentence-variable) *and* the correct
rendering of the (closed) sentence "There is someone in the room", the "kind of
proposition of which I used to have no idea" begins to take shape. The original
connection between the function *fx* and general propositions is now restated in the
form: "There must be incomplete elementary propositions from whose application
the concept of generality derives" (PR: 87).

When Wittgenstein directly confronts the question of elementary propositions
on January 2nd 1930 he recalls two ideas as characteristic of his earlier thinking:
First, that elementary propositions consist of simple signs (standing in for simple objects) in immediate connection, "without any help from logical constants"
(WVC: 74); and secondly, the mutual independence of elementary propositions.
And he says that he does no longer hold on to the second assumption, but still
adheres to the first. A detail noteworthy here is that when he speaks of elementary
propositions as consisting of simple signs he stresses the (negative) aspect that
they do not contain logical constants; this seems reasonable because he cannot
rely on the sole criterion of the immediate connection of simple signs if he wants
to count as elementary propositions also those "where certain data may be left out"
(WVC: 40) – and the immediate connection is therefore broken; but it is sufficient
that no signs of any other kind than simple ones – in particular no logical constants
– may occur.

Now, among the motives he gives for abandoning the thesis of the mutual
independence of elementary propositions, this one is salient: "What was wrong
about my conception was that I believed that the syntax of logical constants could
be laid down without paying attention to the inner connection of propositions"
(WVC: 74). The positive side of this diagnosis is expressed in PR: 109: "[…] there
are rules for the truth-functions which also deal with the elementary parts of
the proposition". This, in turn, has to be seen in the light of the essential role
truth-functions have to play in language:

> True-false, and the truth functions, go with the representation of reality by propositions. […]
> We could say: a proposition is that to which the truth functions may be applied. The truth
> functions are essential to language. (PR: 113)

What Wittgenstein has in mind when he speaks of "rules for the truth functions
which also deal with the elementary parts of the proposition" is that elementary
propositions can have something in common on the level of their respective constituents, and that the application of truth-functions to elementary propositions
has to be adapted to formal restrictions imposed by connections of this kind, as for

instance the rule that no specific coordinate (meaning some such sub-sentential connectedness) must be determined twice over – a point made clear in PR: 111:

> In my old conception of an elementary proposition there was no determination of the value of a co-ordinate; although my remark that a coloured body is in a colour-space, etc., should have put me straight on to this.
> A co-ordinate of reality may only be determined *once*.

One way to describe the idea could be to say that now he allows for inferential dependencies between elementary propositions, in contrast to TLP 5.134: "One elementary proposition cannot be deduced from another". But this would in a certain sense be too weak an interpretation. For it does not preclude a strategy relying on an antecedent classification of types of relations from which special rules of compatibility and deduction could be derived. That is not the direction intended by Wittgenstein.[6] He remained unflinchingly true to a principle already expressed in TLP 5.554: "It would be completely arbitrary to give any specific form". In 1929 this reads:

> Now I think that there is one principle governing the whole domain of elementary propositions, and this principle states that one cannot foresee the form of elementary propositions. It is just ridiculous to think that we could make do with the ordinary structure of our everyday language, with subject-predicate, with dual relations, and so forth. Real numbers or something similar to real numbers can appear in elementary propositions, and this fact alone proves how completely different elementary propositions can be from all other propositions. (WVC: 42)

While in the context of the TLP this principle meant above all that the concept of a specific form is fundamentally alien to that of logical form, here it means that given some specific coordinate there are no apriori restrictions as to the ascription of different coordinates to reality.[7] And from that Wittgenstein draws the conclusion that the given elementary proposition as such exhibits the complete structure by which it is co-ordinated with other propositions via its specific form. This is not a *consequence* of an antecedently established law or rule like the following:

> Whatever colour I see, I can represent each of them by mentioning the four elementary colours red, yellow, blue, green, and adding how this particular colour is to be generated from the elementary colours. (WVC: 42)

---

[6] Cf. WVC: 182: "[…] regarding these questions we cannot proceed by assuming from the very beginning, as Carnap does, that the elementary propositions consist of two-place relations etc."
[7] WVC: 91: "[…] a proposition can be varied in as many dimensions as there are constants occurring in it."

Rather it represents its complement on equal footing: "One elementary proposition describes all the colours in space" (WVC: 41). The inferences specifically possible in color-space are based on *internal* formal relationships. In a short argument with Waisman on January 5th, 1930, Wittgenstein makes that clear:

> WAISMANN: You use the word 'compare'. But when I compare a proposition with reality I know that the azalea is red and from this I *infer* that it is not blue, nor green, nor yellow. What I see is nothing less than a state of affairs. But I never see that the azalea is not blue.
> WITTGENSTEIN: I do not see red: rather, I see *that the azalea is red*. In this sense I also see that it is not blue. It is not that a conclusion is drawn consequential upon what is seen: no – the conclusion is known immediately as part of the seeing. (WVC: 87)

The word Wittgenstein uses to address those (respective) internal formal determinations, particularly when seen as restrictions on the application of truth-functions, is "syntax": "[...] the rules for the logical constants form only a part of a more comprehensive syntax about which I did not yet know anything at that time" (WVC: 74). It appears most notably when he states the consequences of his new insights for his earlier view of the mutual independence of elementary propositions: "In this way syntax draws together the propositions that make one determination" (PR: 113).

But once the striking affinity is noticed between a remark like "One elementary proposition describes all the colours in space" (WVC: 41) on the one hand, and TLP 3.42: "A proposition can determine only one place in logical space: nevertheless the whole of logical space must already be given by it" on the other hand, it becomes obvious that Wittgenstein is in fact changing – in the sense of loosening – his concept of logical form as such. The logical form of a proposition is now determined by its truth-functional complexity (based on its true-false polarity) *as restricted by* its coordination with other propositions on the sub-sentential level. This is strongly suggested by a remark from January 2nd, 1930: "Every proposition is part of a system of propositions that is laid against reality like a yardstick. (Logical space)" (WVC: 76); it is also expressed – the other way round, as it were – when Wittgenstein in response to a question by Schlick says that syntax and with it logic, in some sense, is empirical (WVC: 76f.). This is due to logic's now comprising dependencies of kinds that cannot be foreseen. It is important that this broader understanding of logic preserves the difference between truth-functional dependencies and specific forms. "The truth functions are essential to language" (PR: 113), whereas the variety of the underlying systems of syntactical rules cannot be anticipated. So the critical divergence from TLP 5.554 ("It would be completely arbitrary to give any specific form") seems just to be the insistence that some such specific form has necessarily to be defined lest the logical form of the proposition remains undetermined. The impact of this requirement becomes visible, however, when one realizes that from it follows that logical form can no longer be regarded as independent of contingent determinations, logic no longer as "transcendental" (TLP 6.13: "Logic is transcendental"). Wittgenstein makes the point in an argument with Schlick:

> SCHLICK: Is there not a feeling that the logical constants (the truth functions) are something more essential than the particular rules of syntax, that for instance the possibility or constructing a logical product '$p.q$' is more general, more comprehensive as it were, than the rules of syntax according to which red and blue cannot be in the same place? For the former rule does not contain anything about colour and place.
> WITTGENSTEIN: I do not think that there is a difference here. The rules for logical products, etc., cannot be severed from other rules of syntax. Both belong to the method of depicting the world. (WVC: 80f.)

Now is there in this train of thought something that makes it necessary that elementary propositions be considered as incomplete in the sense of "leaving something out"? One could take the proposition "A is red" as being true in case A (supposed to be a simple object) is related in certain definite ways to a number of (yet unspecified) simple objects reached at in a process of analysis on Tractarian lines, that is: by resolving complex expressions into descriptions until one arrives at elementary propositions consisting of simple signs. Such an analysis in some sense makes the form "redness" explicit; but it fails to provide an explanation of the truth-functional relationship of "A is red" to "A is green" and vice versa. Anyhow, this requirement of complete analysis is the only way to conceive, on Tractarian grounds, the sense of "A is red" as determinate. If instead "A is red" is a legitimate application of the maxim "One elementary proposition describes all the colours in space" (WVC: 41) the existence of internal connections with "A is green" can be made intelligible along the lines sketched above; but obviously the proposition leaves something open in the same way as "There is a circle in the square" or "I met a man" leave some determinations out (and contain free variables): it does not say *which color* A has. Instead, by ascribing the whole system of color-relationships to reality like a measuring rod, it provides the means to determine this color to variable degrees of exactness. But all this holds in exactly the same ways for the proposition "A is green", which is the reason why Wittgenstein at one point (PR: 113) says: "[...] for 'A is green', the proposition 'A is red' is not, so to speak, another proposition – and that strictly is what the syntax fixes – but another form of the same proposition".[8] This should be read in the light of the azalea-example: Of

---

8 Which is, in turn, the reason why the conjunction "A is green and A is red" must not be construed – it would violate the general syntactic rule that "A co-ordinate of reality may only be determined *once*" (PR: 111).
That "for 'A is green', the proposition 'A is red' is not, so to speak, another proposition" is the

course, when the position of A with regard to the color-circle is determined to a sufficient degree we will see that A – for instance – is red; but again – and as the azalea example is intended to convey –, this *is* inseparably and immediately and at the same time the perception of the falsehood of "A is green".

In the conversation on December 22nd, when Wittgenstein says that "One elementary proposition describes all the colours in space", he follows that up with the remark: "Perhaps the way things are is that all incomplete descriptions – all incomplete propositions with gaps – link together to form a complete elementary proposition." It might be tempting to read this as the concession that ultimately only the one "complete elementary proposition" *really is* an elementary proposition with a determinate sense. But that is not Wittgenstein's position. He insisted – and claimed it as the intellectual advancement he had made in the years before 1930[9] – that propositions like "A is green" *are* elementary propositions and *do have* a perfectly determinate sense in spite of the fact that they leave some determinations out and therefore contain free variables. With regard to the proposition "A is red" in the above example this means that it must be supposed to contain already in its present form a description exact enough to decide its truth or falsity – notwithstanding the fact that the exact colour of A is not determined.[10]

It is this systematic point where Wittgenstein, still understanding propositions as pictures, may have expected to find support from a usage of his notion of an incomplete picture. It would require a shift of accent which can be brought out by means of a contrast. Remark TLP 5.156 reads: "A proposition may well be an incomplete picture of a certain situation, but it is always a complete picture of something." What had to be examined now, in contrast, is the notion of an incomplete picture with a definite sense which need not by necessity be complete in some

---

counterpart – under the terms of the wider understanding of logical form – of TLP 4.0621: "The propositions 'p' and '$\neg p$' have opposite sense, but there corresponds to them one and the same reality."

9 In a conversation with Waismann on December 9th, 1931 (WVC: 182) he said: "[…] I did think that the elementary propositions could be specified at a later date. Only in recent years have I broken away from that mistake", and he called this the "dangerous" mistake pervading the TLP.
10 That a proposition which leaves something open – i.e. a general proposition – must exhibit by itself the form in which specifications can be made is a constant principle in Wittgenstein's thought from the early notebooks until at least the mid-thirties, independently of far-reaching changes in his conception of analysis. Cf. a note from June 17th 1915: "In other words the proposition must be completely articulated. Everything that its sense has in common with another sense must be contained separately in the proposition. If generalizations occur, then the forms of the particular cases must be manifest" (NB: 63e) and "The multiplicity a general proposition anticipates for its possible particular cases has to be located in the grammatical rules for its terms. What isn't located in these rules isn't anticipated." (BT: 242e).

other respect, as it is roughly stated in PR, remark 87: "This incomplete picture is, if we compare it with reality, right or wrong: depending on whether or not reality agrees with what can be read off from the picture" (PR: 115). To really confront this issue means to engage in reflections on our understanding of (and manner of speaking about) pictures – or else the circle in the reciprocal explanation of incomplete picture and general elementary proposition cannot be broken.

4 Pictures

In one passage in Waismann's notes from December 22nd 1929 (WVC: 39f.) Wittgenstein is reported as elaborating, in a way typical of this irresoluteness, on the "circle-in-the-square" example. He starts with the *description* of "a state of affairs which consists in there being a circle of a specific size at a specific point of the square" and calls this "a complete picture"; then he imagines himself replacing the numerical values which identify center and diameter *in the sentence describing the state of affairs* by variables; finally, in the third and last clause of the sentence, he says: "[…] and then I shall get an incomplete picture". Now from beginning to the end of this scenario the distinction between sentence and picture is purely nominal: a specific description is a complete picture, a description made incomplete (by replacing numbers by variables) is an incomplete picture. The point of calling the sentence a picture consists solely in the suggestion of a certain (if unspecified) way the description relates to (or: has to be compared with) the state of affairs. A picture-theory to this end is contained in TLP, but it does not take into account the role of specific forms. Now, at the time of his conversations with Schlick and Waisman, Wittgenstein thinks that to compare a proposition like "A is red" to a certain state of affairs presupposes an *internal* association with a model which exhibits the basic syntactical rules relevant for the determination of the sense of the proposition:

> When I built language up by using a coordinate system for representing a state of affairs in space, I introduced into language an element which it doesn't normally use. This device is surely permissible. And it shows the connection between language and reality. The written sign without coordinate system is senseless. Mustn't we then use something similar for representing colours? (PR: 79)

A model in this sense (like the color-octahedron) is internally linked to the proposition (is part of the symbolism) *and* is located "in the same space" as the state of affairs:

> If I say something is three feet long, then that presupposes that somehow or other I am given the foot length. In fact it is given by a description: in such and such a place there is a rod one foot long. The 'such and such a place' indirectly describes a method for getting there; otherwise the specification is senseless. The place name 'London' only has a sense if it is possible to *try to find* London. (PR: 79)

The role a map plays in helping "to get there" makes explicit the syntactic rules holding for a sentence like "the rod is in that place" – and a sketch of the situation on paper can be seen as a manifestation of the pictorial character of the proposition:

> You can draw a plan from a description. You can translate a description into a plan.
>   The rules of translation here are not essentially different from the rules for translating from one verbal language into another. (PR: 63)

> If you think of propositions as instructions for making models, their pictorial nature becomes even clearer. (PR: 57)

One way to bring out the difference with respect to TLP is to say that instead of just *interpreting* propositions as pictorial models (esp. TLP 4.01) now real pictures (in the wide sense of models) are *exhibiting* the specific form of propositions. If this rough sketch of a complicated change in Wittgenstein's view of pictures is read into his account of the "circle-in-the-square" example, it still does not yield a satisfactory explanation of the relation between incomplete picture and general proposition. This is so because for the move from the given proposition to the general form there is no parallel under the pictorial aspect; there is no hint at what an incomplete picture is – i.e. what the specifically pictorial manifestation of the form of a general proposition is. That he somehow sensed this deficit in his account seems to be the reason why Wittgenstein here abruptly resorts to a banal scenario where the incompleteness of the picture in question can be taken for granted according to commonsense:

> Imagine a portrait in which I have left out the mouth, then this can mean two things; first, the mouth is white like the blank paper; second, the picture is always correct whatever the mouth is like. (WVC: 40)

But it seems problematic whether under this description either of the two readings is eligible to argue against the position that the picture is still essentially a complete picture[11] – or, positively: to argue for the picture to be, in its incompleteness, a picture in its own right. Regarding the first line of Wittgenstein's alternative, if

---

[11] In the sense of TLP 5.156: "A proposition may well be an incomplete picture of a certain situation, but it is always a complete picture of something."

in a (painted) portrait one wants to convey the message that the mouth is white, the obvious means would be to paint the mouth with white pigment; that the omission in itself, without further comment, should suggest the whiteness of the mouth is not plausible. In any case, Wittgenstein obviously votes for his second alternative, that "the picture is always correct whatever the mouth is like" – because he continues with the already quoted words: "The incompleteness of a picture consists in the occurrence of variables in a proposition" (WVC: 40). This is indeed essential incompleteness – but not conveyed by the picture as such; it can only be said by a limine interpreting the picture as a proposition with a free variable. For the question how the change from complete to incomplete state affects the picture as a picture this yields no additional information. It is odd, however, that Wittgenstein does not take into consideration the most obvious interpretation of the "portrait with the mouth left out" – namely that the painting simply is unfinished. Under this assumption the portrait would still be of the person intended, and at the same time the absence of a mouth would justify the ascription of incompleteness; a disadvantage only lies in the fact that the portrait itself, when considered as unfinished, is not what it is intended to be; it is *not yet the portrait*, one would like to say. Anyhow, this case points in the direction of a more interesting rewriting of the whole scenario. In one short remark Wittgenstein presents a simplified and as it were reversed variant of the case of the unfinished picture which now helps to bring out the vital point. Of a complex picture, a square containing two circles, he says: "If I leave out one symbol, I still get a picture" (WVC: 40). That is right: The result is an incomplete picture which is still a picture in every relevant aspect. What makes this description appropriate is that it focuses on *the move* from the given picture to its incomplete variant. Only with respect to this transition as a whole – and comprising three phases (picture given – one symbol removed – incomplete picture) – do we speak of the incomplete picture.

The deficiency of Wittgenstein's "portrait with the mouth left out" lies in the fact that it takes this decisive transition as already made – relegating it as it were to a pre-historical status. But then the scenario cannot make visible exactly how the incomplete picture differentiates itself from the complete one – therefore the need to fall back on the difference between the respective propositions.

The difference of the "two circles in a square" scenario with respect to the "unfinished portrait" scenario lies elsewhere, notably in the fact that in the comprehensive description of the "two circles in a square" case the two pictures are being held against each other and it can therefore be *seen* in which ways the one is a richer portrayal of the same reality. Here the objection suggests itself that the same effect can be produced by an expansion of the "unfinished portrait" variant: Let the painter finish her work, and a description of the whole process will allow for the comment that what at the earlier stage was incomplete can now be seen in its complete form. This is correct. The difference still remaining is that Wittgenstein simply does not treat this case, and that in any case it would have had a serious strategic disadvantage for him. This has to do with the most obvious problem presented by the "two circles in a square" case itself.

For is this not again just another confirmation of the fact that there are, once and for all, no incomplete pictures – no pictures *incomplete in themselves*? That a picture can only be called incomplete with respect to a different picture *it is not*? And that all talk of incomplete pictures (like in Wittgenstein's scenarios of the missing mouth and the two circles) is justified only insofar as it takes into account ("respects") this fact and has to abstain from all existential commitment?

Now the decisive point for Wittgenstein is that the *talk of incomplete pictures* is justified and can be characterized by a description of the kind of the example with the two squares; and not that there is a general formal trait that distinguishes intrinsically between complete and incomplete pictures. In short: It is at least as important for Wittgenstein that there are no incomplete pictures as it is important that there are incomplete pictures. What the "two circles in a square" example conveys is that if a picture is given and is made incomplete by deleting one of its elements, we will still have a picture in the full sense as the result. That is exactly what is needed when the goal is to support the idea that there can be an elementary proposition with a free variable which is "a proper proposition, not merely preparatory for a proposition" (WVC: 40) *from the side of the idea that propositions are pictures*. The enormous strategic advantage of the "two circles in a square" case over the "unfinished portrait" case (which seems only to reverse the direction) lies in its being a smart travesty of the central message in §9 of Frege's *Begriffsschrift*: If we delete from an expression one sign (thinking of it as replaceable by other signs), the resulting expression is fundamentally different from the one we started with; if we set out with a proposition ("beurteilbarer Inhalt"), the result will be a function (a concept), and not a proposition. But if it is assumed that propositions are essentially pictures, and it can be shown that the result of deleting an element of a picture is still a picture in the full sense – then this need not be so. In the "two circles in a square" scenario the direction of the argument is right: It reveals the one real, tangible feature in our dealings with pictures which lends support to the idea of elementary propositions as incomplete pictures. And so Wittgenstein completes the example with the following words which make unequivocally clear what he thought could be gained by a more concentrated consideration of incomplete pictures:

> If I leave out one symbol, I still get a picture – contrary to the ordinary conception of things according to which I get only a preliminary to a proposition by omitting a part of a proposition (WVC: 40)

But to all this applies a proviso: The argument is completely dependent on the assumption that propositions are pictures. For this there are, around 1930, no justifications independent of the picture-theory of the TLP which will soon lose its attraction (which does not mean at all that the notion of a picture loses interest) for Wittgenstein. And it has to be added that at the very time of this experimenting with incomplete pictures Wittgenstein had other and much more forward-looking visions of the consequences to be drawn from the abandonment of his thesis of the mutual independence of elementary propositions. In this sense the incomplete pictures of 1929–1930 are the farewell performance of the picture-theory of TLP – deserving to be referred to by the dictum "A picture held us captive" (PI: 53).

Bibliography

Fogelin, Robert (1982): "Wittgenstein's Operator N." In: *Analysis*. Vol. 42, 124 – 127.
Geach, Peter (1981): "Wittgenstein's Operator N." In: *Analysis*. Vol. 41, 168 – 170.
Potter, Michael (2009): *Wittgenstein's Notes on Logic*. Oxford: Oxford University Press.
Potter, Michael (2013): "Wittgenstein's pre-Tractatus manuscripts: a new appraisal." In: *Wittgenstein's Tractatus. History and Interpretation*. Sullivan, Peter; Potter, Michael (eds.). Oxford: Oxford University Press, 13 – 39.
Ricketts, Thomas (1996): "Pictures, logic, and the limits of sense in Wittgenstein's Tractatus." In: *The Cambridge Companion to Wittgenstein*. Sluga, Hans; Stern, D. G. (eds.). Cambridge: Cambridge University Press, 59 – 99.
Ricketts, Thomas (2013): "Logical segmentation and generality in Wittgenstein's Tractatus." In: *Wittgenstein's Tractatus. History and Interpretation*. Sullivan, Peter; Potter, Michael (eds.). Oxford: Oxford University Press, 125 – 142.
Russell, Bertrand (1905): "On Denoting." In: *Mind, New Series*. Vol. 14, No. 56, 479 – 493.
Varga von Kibed, Matthias (2001): "Variablen im Tractatus." In: *Tractatus Logico-Philosophicus, Klassiker Auslegen*. Vossenkuhl, Wilhelm (ed.). Berlin: Akademie Verlag, 209 – 229.
Wittgenstein, Ludwig (1961): *Notebooks 1914–1916*. von Wright, G.H.; Anscombe, G.E.M. (eds.). New York: Harper and Bros. Quotations by page numbers.
Wittgenstein, Ludwig (1967): *Schriften 3. Wittgenstein und der Wiener Kreis. Von Friedrich Waismann*. McGuinness, Brian (ed.). Frankfurt a.M.: Suhrkamp. Quotation by page number.
Wittgenstein, Ludwig (1974): *Tractatus logico-philosophicus*. Pears, David; McGuiness, Brian (trans.). London, New York: Routledge. References by numbers of remarks.
Wittgenstein, Ludwig (1975): *Philosophical Remarks*. Rhees, Rush (ed.). Oxford: Blackwell. Quotations by page numbers.
Wittgenstein, Ludwig (1979): *Ludwig Wittgenstein and the Vienna Circle*. McGuiness, Brian (ed.). Oxford: Blackwell. Quotations by page numbers.
Wittgenstein, Ludwig (2005): *The Big Typescript: TS 213*. Luckhardt, C. Grant; Aue, Maximilian (ed. and trans.). Oxford: Blackwell. Quotations by page numbers.
Wittgenstein, Ludwig (2009): *Philosophical Investigations*. Hacker, P.M.S.; Schulte, Joachim (eds.). Oxford: Wiley-Blackwell. Quotation by page number.

Oliver Feldmann
„Man kann die Menschen nicht zum Guten führen" – Zur Logik des moralischen Urteils bei Wittgenstein und Hegel

Abstract: Die in jüngerer Zeit beliebt gewordenen Versuche, Wittgenstein und Hegel in ein Naheverhältnis zu bringen, sind müßig. Die beiden könnten kaum gegensätzlicher sein, und Wittgenstein hat diesen Gegensatz auch selbst festgehalten mit seinem Pochen auf den Standpunkt der „Verschiedenheit" gegenüber jenem der Allgemeinheit bei Hegel. Dennoch gibt es einen zentralen Punkt, an dem die beiden Denker sich begegnen – und das sogar zweifach. Es ist der Gedanke der ‚Grenze', an dem Wittgenstein sich immer wieder entlang bewegt. Und der bei Hegel stets Kritik kennzeichnet. Wittgensteins Diktum, dass „das Gute" „außerhalb des Tatsachenraums" liege, welcher ob der Zufälligkeit seines „So-Seins" alles sinnvolle Sprechen über „Höheres" verunmögliche, findet seinen Widerhall in dem Urteil Hegels über Kant: „Die vollendete Moralität muss ein Jenseits bleiben; denn die Moralität setzt die Verschiedenheit des besonderen und allgemeinen Willens voraus." Während Wittgenstein glaubt, alles eigentlich „Wichtige" sei genau dort anzusiedeln und unzugänglich, ergeben sich bei Hegel interessante Einsichten über die Verwandtschaft von Moral und Heuchelei und den notwendigen Idealismus der Moralphilosophie.

> *Denn die Art, wie man lebt, ist so verschieden von der Art, wie man leben sollte, dass, wer sich nach dieser richtet statt nach jener, sich eher ins Verderben stürzt, als für seine Erhaltung sorgt.*
>
> – Niccolò Machiavelli, *Der Fürst*, XV

1

Der Titel bzw. das Thema dieses Artikels mag ein wenig befremdlich anmuten.[1] Auch wenn gegenteilige Behauptungen derzeit etwas en vogue sind – die Berüh-

---

1 Die zitierte Stelle bei Wittgenstein lautet vollständig: „Man kann die Menschen nicht zum Guten

Oliver Feldmann, Vienna University of Business and Economics, oliver.feldmann@wu.ac.at

DOI 10.1515/9783110657883-29

rungspunkte zwischen diesen zwei Denkern, Hegel und Wittgenstein, erscheinen in der Tat recht spärlich. Und, vor allem: Wittgenstein hat sich zum Thema Ethik, wie man weiß, kaum, und wenn, dann mehr mahnend und abwehrend geäußert. Dennoch: Es wird sich, wie ich hoffe, zeigen, dass aus dem Vergleich der beiden Erkenntnisse zu gewinnen sind über den Status und die Funktionsweise der Moral bzw. des moralischen Bewusstseins; und folglich auch über Wittgensteins – ich nehme gleich vorweg: unentschiedene Stellung dazu.

Wittgenstein selbst hat ja die Differenz, eigentlich muss man besser von einem *Gegensatz* sprechen, zwischen seiner Philosophie und derjenigen Hegels in einem Gespräch mit Drury im Herbst 1948 so festgehalten:

> Nein, mit Hegel könnte ich vermutlich nichts anfangen. Mir scheint, Hegel will immer sagen, dass Dinge, die verschieden aussehen, in Wirklichkeit gleich sind, während es mir um den Nachweis geht, dass Dinge, die gleich aussehen, in Wirklichkeit verschieden sind. (Rhees 1987: 217)

Ob Wittgenstein sich direkt mit Texten Hegels auseinandergesetzt hat, ist nicht überliefert – und eher unwahrscheinlich. Sicherlich hat er in Cambridge via Russell und andere Bekanntschaft mit einigen von Hegels Gedanken gemacht. Im Vorwort zu seinem *Tractatus* schreibt er jedenfalls – nahezu – so, als ob er bei Hegel in die Schule gegangen wäre.

Hören wir zunächst Hegel. Dieser hatte bekanntlich eine hohe Meinung von Kant. Und dennoch, oder gerade deshalb, ging er mit ihm immer wieder besonders scharf ins Gericht. Von der „Leere" der Kantischen Bestimmungen ist Mal ums Mal die Rede, vom armseligen Formalismus seiner Moralphilosophie, vom „Barbarischen" seiner Terminologie…; und immer wieder griff Hegel Kant an wegen seiner Untersuchung des *Erkenntnisvermögens* und der Frage nach dessen *Leistungsfähigkeit*. Die mittlerweile in der geistesgeschichtlichen Rezeption allseits in den Rang einer ‚philosophischen Grundfrage' aufgestiegene Problemstellung „Was kann ich wissen?" hielt Hegel für albern; die darin beschlossene Suche nach einer „Schranke" oder „Grenze der Erkenntnis" für widersprüchlich und falsch. In seiner ‚Großen Logik' heißt es hierzu:

> Es pflegt zuerst *viel* auf die Schranken des Denkens, der Vernunft usf. gehalten zu werden, und es wird behauptet, es *könne* über die Schranke *nicht* hinausgegangen werden. In dieser Behauptung liegt die Bewusstlosigkeit, dass darin selbst, dass etwas als Schranke bestimmt ist, darüber bereits hinausgegangen ist. Denn eine Bestimmtheit, Grenze ist als Schranke

---

führen; man kann sie nur irgendwohin führen. Das Gute liegt außerhalb des Tatsachenraums." (VB: 454). An selber Stelle liest man: „Wenn etwas gut ist, so ist es auch göttlich. Damit ist seltsamerweise meine Ethik zusammengefasst."

nur bestimmt im Gegensatz gegen sein Anderes überhaupt als gegen sein *Unbeschränktes* [...] (WdL I: 145)

Wer dem Denken, nicht dem individuellen, sondern dem Denken als solchem, eine Schranke/Grenze zuspricht, der kennt diesen ‚Grenzpfahl' offensichtlich – und ist somit „darüber hinaus".

Wie klingt diese Überlegung bei Wittgenstein? Zum besseren Verständnis seines *Tractatus* stellt er dem Text folgende Lesehilfe voran:

> Das Buch will also dem Denken eine Grenze ziehen, oder vielmehr – nicht dem Denken, sondern dem Ausdruck der Gedanken: Denn um dem Denken eine Grenze zu ziehen, müssten wir beide Seiten dieser Grenze denken können (wir müssten also denken können, was sich nicht denken lässt.) Die Grenze wird also nur in der Sprache gezogen werden können und was jenseits der Grenze liegt, wird einfach Unsinn sein. (TLP, Vorwort)

„*Nahezu in die Schule gegangen*" schrieb ich vorhin. Die Parallele ist unübersehbar. Wittgenstein ist offensichtlich, aus welcher Quelle auch immer, vertraut mit dieser Hegelschen Kritik an der Widersprüchlichkeit der Vorstellung von der Endlichkeit, der Beschränktheit des menschlichen Geistes. Er versteht und akzeptiert diese Kritik – und will dennoch das Anliegen, das hierin kritisiert ist, aufrecht erhalten können: dem Denken eine Grenze ziehen – aber eben nicht im Denken, das wäre auch in seinen Augen ein unhaltbarer Widerspruch, sondern in dessen *Ausdruck*, in der *Sprache*.

Man könnte den Ausspruch Nestroys, den Wittgenstein als Motto seinen *Philosophischen Untersuchungen* voranstellt, an dieser Stelle heranziehen und vom Auseinanderfallen von Schein und Sein in puncto „Fortschritt" sprechen. Denn in der Tat liegt hier so etwas wie ein kleines Lehrstück in Sachen – verschlungener – philosophischer Fortschritt vor. In zweifacher Hinsicht: Zum einen methodisch: Ein als widerlegt anerkannter Standpunkt wird am Leben gehalten durch eine Zusatz-Annahme, eine Perspektiven-Verschiebung. So, als würde dieser widersprüchliche Gedanke von der Endlichkeit des Denkens dadurch richtiger, dass man ihm, dem Denken, *außerhalb* seiner und *physisch* die Limitation vorsetze. Zum anderen, inhaltlich, in dieser verschobenen Sichtweise, eine interessante Auskunft über das, was *Linguistic Turn* seinem Wesen nach ist.

Ziehen wir zur Erläuterung dieses Punktes eine andere Feststellung Wittgensteins aus dem *Tractatus* heran:

> Im Satz drückt sich der Gedanke sinnlich wahrnehmbar aus. (TLP 3.1)

Auch bei Hegel ist die Sprache das „Dasein" (PhG: 478) oder die „Existenz des Geistes" (VGP III: 106 – 107). (In TLP 4.002 z.B. bestimmt Wittgenstein sie als

„Kleid", aber auch als „Verkleidung" des Gedankens.) Dennoch würde Hegel Wittgenstein – und durchaus zu Recht! – in dieser Auffassung der sinnlichen Wahrnehmbarkeit des Gedankens im Satz widersprechen. „Freilich", hätte er Wittgenstein zugerufen, „wird im Satz der Gedanke sinnlich wahrgenommen. Und doch zugleich nicht *als Gedanke*! Als Gedanke, in seinem Gedanken*inhalt*, muss er noch immer geistig erfasst, *nach*gedacht werden und wird nicht sinnlich wahrgenommen. Sinnlich wahrgenommen wird seine Gestalt/sein Kleid, die Zeichen und/oder Töne.[2]

„Wittgenstein glaubt aber, das Verhältnis so fassen zu können; und daraus ergibt sich, was vorhin benannt war als „interessante Auskunft" über den Charakter des *Linguistic Turn* in der Philosophie: Man hat es hier mit einer *empiristischen Wende* innerhalb der Erkenntnistheorie zu tun.[3] Die Sprache ist, wie Wittgenstein ganz richtig sagt, der „Ausdruck der Gedanken", ihr leibliches Gewand – und insofern, in seinen Augen, jener *Ort*, wo man des Gedankens *habhaft* werden könne, ein objektiver/realer *Hebel*, um ihm die gesuchte Grenze seiner Betätigung zu setzen.[4] (Später, in seinen PU §119, verwendet Wittgenstein, durchaus passend, die Redeweise von „Beulen", die sich „der Verstand beim Anrennen" gegen die Sprache bzw. deren Grenzen hole.[5])

2

Mit dieser Grenzziehung soll bei Wittgenstein geschieden werden zwischen der „Welt der Tatsachen" und dem, was wirklich wichtig ist. In Wittgensteins Worten,

---

[2] Dieser Unterschied ist Wittgenstein klarerweise auch bekannt; so heißt es bei ihm andernorts: „Die Klasse der Trios unterscheidet sich von der Zahl 3 ungefähr ebenso, wie sich ein Gehirnvorgang von einem Bewusstseinszustand unterscheidet. " (WWK: 222)

[3] Genauer muss man eigentlich sagen: Mit einer *Radikalisierung* des Empirismus jener Epistemologie, die nach den „Voraussetzungen" oder „Bedingungen " fragt, die Erkenntnis „möglich" machen sollen.

[4] Gabriele Mras hat in ihrem sehr lesenswerten Buch zu Frege *Wahrheit, Gedanke, Subjekt* darauf hingewiesen, dass die Idee von der Sprache als *dem* eigentlichen Aufklärungsobjekt der Philosophie ein durchgängiger Topos des 19. Jahrhunderts gewesen ist und insofern genau genommen weder bei Frege noch bei Wittgenstein wirklich Originalität für sich beanspruchen kann (siehe Mras 2001: 123 – 128).

[5] Auch bei Schopenhauer – Wittgenstein verweist auf ihn wiederholt, anerkennend wie kritisch, als einen seiner intellektuellen Bezugspunkte (etwa: VB: 476, Drury 1987: 218; vgl. auch Janik/Toulmin 1973, Magee 1997, Schroeder 2012 et al.) – findet sich dieses Bild einer physischen Eingeschlossenheit des Geistes: „Darum stoßen wir mit unserem Intellekt [...] überall an unauflösliche Probleme wie an die Mauer unseres Kerkers." (Schopenhauer 1920: 1456).

in seinem Brief an seinen Verleger Ludwig von Ficker:

> [...] der Sinn des Buches [*Tractatus logico-philosophicus*] ist ein ethischer." Und als ‚Schlüssel' dazu könne man ansehen: [...] das „Werk bestehe aus zwei Teilen: Aus dem, der hier vorliegt, und aus alledem, was ich nicht geschrieben habe. Und gerade dieser zweite Teil ist der Wichtige. Es wird nämlich das Ethische durch mein Buch gleichsam von innen her begrenzt, und ich bin überzeugt, dass es, streng, *nur* so zu begrenzen ist. Kurz, ich glaube: Alles das, was viele heute schwefeln, habe ich in meinem Buch festgelegt, indem ich darüber schweige. (B: 96f.)

Hin und wieder freilich hat Wittgenstein dieses sein Schweige-Gebot durchbrochen; folgen wir ihm darin. Am ausführlichsten tut er dies in seinem *Vortrag über Ethik* aus dem Jahr 1929. Der Standpunkt ist noch ganz derselbe wie zur Zeit seines *Tractatus*. Wittgenstein beginnt mit einem Zitat Moores „Die Ethik ist die allgemeine Untersuchung dessen, was gut ist"[6], (er) unterscheidet dann zwischen dem „relativ Guten" auf der einen Seite, etwa dem ‚guten Pianisten', dem richtigen, ‚gut gewählten' Weg – also der in diesem „gut" gemeinten Entsprechung von Mittel und Zweck, der Angemessenheit gegenüber einem „vorher festgelegten Maßstab"; und dem „absolut Guten", dem eine *Notwendigkeit* innewohne, der man sich nicht entziehen könne, wie der „Zwangsgewalt eines absoluten Richters" (VE: 14). Und im Verlauf dieser Überlegungen fällt jener oft zitierte Satz, den er selbst zurückhaltend und dessen Absurdität damit etwas abmildernd als „Metapher" bezeichnet:

> Wäre jemand imstande, ein Buch über Ethik zu schreiben, das wirklich ein Buch über Ethik wäre, so würde dieses Buch mit einem Knall sämtliche anderen Bücher auf der Welt vernichten. (VE: 13)

Man sieht: Anders als die im selben Jahr entstandene „Programmschrift" des „Vereines Ernst Mach", also des Wiener Kreises, der explizit erklärt, dem praktischen „*Leben dienen*" zu wollen und können[7] und gleichzeitig – und im Widerspruch zu diesem seinem Selbstverständnis – moralische Fragen für eine Angelegenheit des

---

6 Es lohnt sich, hier darauf hinzuweisen, dass Wittgenstein – bei aller Abneigung gegen dessen Detailarbeit – nicht nur im Ausgangspunkt recht nah bei Moore bleibt: „[...] weil diese Frage, wie ‚gut' zu definieren ist, die fundamentalste Frage der ganzen Ethik ist. [...] der *einzige* einfache Gegenstand des Denkens, der der Ethik eigentümlich ist. Seine Definition ist deshalb der entscheidende Punkt bei der Definition der Ethik; [...] Wenn ich gefragt werde: ‚Was ist gut?', so lautet meine Antwort, dass gut gut ist, und damit ist die Sache erledigt. Oder wenn man mich fragt: ‚Wie ist gut zu definieren?', so ist meine Antwort, dass es nicht definiert werden kann [...]." (Moore 1970: 34 – 41)

7 „Die wissenschaftliche Weltauffassung dient dem Leben und das Leben nimmt sie auf." (WWWK: 29 – 30) – Ein Idealismus, den das *politische* Leben Österreichs bekanntlich allerdings bald barsch und gewaltsam zurückwies.

je individuellen Standpunkts erklärt und die Metaphysik für etwas derer „man sich bald [...] schämen wird" (Schlick 1976: 16), ist Wittgenstein der Auffassung, dass alle Wissenschaft letztlich verblassen müsse und würde, wenn dieser ethische Bereich dem menschlichen Denken zugänglich wäre.

Aber dem ist eben nicht so; er liegt für Wittgenstein jenseits der Grenze des Denk- bzw. Sagbaren.

Wie steht es nun um diese Notwendigkeit, diese richterliche Zwangsgewalt des Ethischen, deren Fehlen Wittgenstein in seinem *Lecture on Ethics* beklagt?

In der Tat: Es kann sie nicht geben. Aber das Argument kann nicht lauten: Weil sie „außerhalb des Tatsachenraums" liegt, wie Wittgenstein immer wieder betont. Das *will* es, das Ethische, das moralische Urteil ja gerade: „außerhalb des Tatsachenraums" liegen, den Fakten *entgegen* stehen. Das moralische Urteil ist ja *Kritik*; es drückt – ganz absichtsvoll – *Einwände* gegen die vorhandene Welt, gegen die Tatsachen aus. Es ist also recht verquer, der Ethik, dem moralischen Urteil – wie berechtigt oder unangebracht es im Einzelfall auch sein mag – vorzuhalten, dass es in der gegenwärtigen Welt der Tatsachen keine Entsprechung finde.

Vom moralischen Standpunkt aus müsste man sogar sagen, dass diese Zurückweisung *zynisch* ist. Das ist ja gerade der *Ausgangspunkt* des moralischen Urteils – wie gesagt: wie berechtigt auch immer –, dass es mit den „Fakten", mit den tatsächlich vorliegenden Handlungsweisen nicht einverstanden ist und hier Änderung verlangt. Das moralische Urteil hat seine sprachliche Heimat im Modalverb ‚Sollen'. Und das ist klarerweise dem ‚Ist'-Zustand entgegengesetzt.

3

Was ist nun der Maßstab dieses Sollens?

Im Vorwort zu seinen *Philosophischen Bemerkungen* notiert Wittgenstein – unvorsichtigerweise, muss man sagen –, es sei „in gutem Willen geschrieben". Von welchem Standpunkt aus, Wittgenstein ist mit ihm offensichtlich zumindest vertraut, kann von einem „guten Willen" gesprochen werden? Und was *will* ein solcher „guter Wille" erreichen?

Nehmen wir uns vor der Beantwortung dieser Fragen noch einmal kurz Wittgensteins Überlegung zur (fehlenden) Notwendigkeit des Ethischen bzw. zur „richterlichen Zwangsgewalt" vor. Selbstverständlich *kann* kein einziges moralisches Urteil eine derartige Zwangsgewalt für sich in Anspruch nehmen. Dem moralischen Gebot fehlt notwendig die Verbindlichkeit. Um es als Paradox auszudrücken: Notwendig fehlt ihm die Notwendigkeit. Das zweite, psychologische, Lebenselement der Moral (ihr erstes, sprachliches, war das Sollen) ist nämlich die *Freiwilligkeit*. Moral ist eben kein *Recht* (kann allenfalls zu einem werden, aber dann ist es eben nicht mehr ‚bloß' Moral).[8] Sie spricht von Handlungs-Alternativen, die es gibt/gegeben hätte/geben sollte. Sie ist darin also auch notwendig *idealistisch*. Während das Sollen des Rechts sanktionsbewehrt ist und gerade damit für das sorgt, was gerne euphemistisch „Rechtssicherheit" genannt wird, will das moralische Urteil appellieren/überzeugen/zumindest aber überreden.

Hierzu lesen wir bei Wittgenstein:

> Der erste Gedanke bei der Aufstellung eines allgemeinen ethischen Gesetzes von der Form „Du sollst …" ist: „Und was, wenn ich es nicht tue?" (TB, 30.07.16 = TLP 6.422)

Dieser Idealismus der Moral ist ja auch zugleich Grundlage für den *schlechten Ruf*, den sie – neben ihrem guten – in der Öffentlichkeit hat. Und heute, interessanterweise, mehr noch als in früheren Jahren. Wo etwa, wie hierzulande, die Ausdrücke „Gutmensch" und „Tugendterror" anerkannte und durchgesetzte Injurien darstellen, ist das, was ich den *Doppelcharakter der Moral* nennen möchte, mehr als offenkundig. Das moralische Votum stellt sich *absichtsvoll* neben und über den „Tatsachenraum" und will – in der Regel auch im Fall des (Selbst)Lobs – auf diesen handlungsanleitend und insofern verändernd einwirken. Das moralische Urteil ist keine „Tatsache" – will aber gerne eine werden.

Dass es deshalb, wie Wittgenstein meint, „eine Bedingung der Welt sein" müsse, (ebenso) „wie die Logik" (TB, 24.07.16)[9], wird man bezweifeln müssen. Denn dafür ist das moralische Urteil gerade aufgrund seines soeben dargestellten Doppelcharakters viel zu uneindeutig.

---

8 Gerhard Polt, der kleinere Nachfahre des großen Karl Valentin, sorgt in einer Szene damit für einen Lacher, dass er den über den Tisch gezogenen Kunden eines Autohauses beteuern lässt, er habe den Prozess gegen dieses Autohaus vor Gericht „gewonnen – (Pause) moralisch".

9 Hier zeigt sich recht klar – Janik 1985, Glock 2013 u.a. weisen zu Recht darauf hin – der Einfluss Otto Weiningers auf Wittgensteins damalige Gedankenwelt. Logik und Ethik haben für beide denselben, geradezu existenziellen, Status. So heißt es bei Weininger: „Logik und Ethik aber sind im Grunde nur eines und dasselbe – Pflicht gegen sich selbst. […] Alle Ethik ist nur nach den Gesetzen der Logik möglich, alle Logik ist zugleich ethisches Gesetz." (Weininger 1903: 200) In seinem alltäglichen Leben war Wittgenstein Mal ums Mal bestrebt, auch die kleinsten Verrichtungen als Vollzug einer solchen „Pflicht gegen sich selbst" zu exekutieren: „Engelmann, for example, reports that his (W.'s) decision to cast off his necktie was as deliberate as that to dispose of his fortune. In every situation he appears to have been confronted with a decision involving duty to himself." (Janik 1985: 67)

4

Gefragt war also nach dem Maßstab des moralischen Sollens. Oder anders gesagt: Wie lautet der Standpunkt, von dem aus die moralische Beurteilung der jeweiligen Handlungen und Tatsachen stattfindet? Trotz seiner Behauptung in der *Lecture on Ethics* und anderswo

> Die Ethik ist, sofern sie überhaupt etwas ist, übernatürlich, und unsere Worte werden nur Fakten ausdrücken; (VE: 13)

hat Wittgenstein hiervon offensichtlich doch einigermaßen klare Vorstellungen. Immerhin konfrontiert er Freunde und Bekannte im Laufe der Jahre immer wieder (für die Zuhörer mitunter offensichtlich recht leidvoll!) mit „Geständnissen" und „Beichten" über sein bislang misslungenes Leben und dem Versprechen, sich „bessern" zu wollen. In seinem Ethik-Vortrag formuliert er

> In der Ethik geh(t) es darum, [...] zu untersuchen, was das Leben lebenswert macht (VE: 10f.).

– um dann schließlich rückblickend – so die letzte von ihm überlieferte Äußerung – ausrichten zu lassen, dass sein Leben durchaus darunter falle und „ein wunderbares" gewesen sei. Und auch eine Empfehlung an Anhänger der sowjetischen Revolution in England hielt Wittgenstein bereit: Sie sollten, statt sich politisch zu engagieren, lieber „gutherzig" sein. „Seien Sie einfach gut zu den anderen." (Rhees 1987: 48)

Ein Beispiel, mit dem Wittgenstein in seinem Vortrag seine Überlegungen illustrieren will, gibt Auskunft darüber, dass er diesen Maßstab des moralischen Urteils sehr wohl kennt – auch, wenn er ihn zugleich der Unzugänglichkeit bezeichnen will:

> Angenommen, ich könnte Tennis spielen, und einer von Ihnen beobachtete mich beim Spiel und sagte: „Na, Sie spielen aber ziemlich schlecht", und ferner angenommen, ich erwiderte: „Das weiß ich, ich spiele schlecht, aber ich will gar nicht besser spielen", dann bliebe dem anderen gar nichts anderes übrig, als zu antworten: „Schon recht, dann ist ja alles in Ordnung." Aber denken wir uns, ich hätte einen von Ihnen aberwitzig angelogen, und nun käme er auf mich zu und sagte: „Sie benehmen sich abscheulich." Wenn ich darauf erwiderte, „Ich weiß, dass ich mich schlecht benehme, aber ich will mich gar nicht besser benehmen", könnte der andere dann antworten: „Schon recht, dann ist ja alles in Ordnung"? Nein, das ginge bestimmt nicht, sondern er würde sagen: „Na, dann *sollten* Sie sich aber besser benehmen wollen." Hier haben wir es mit einem absoluten Werturteil zu tun [...]. (VE: 11)

„Das ginge *bestimmt* nicht" – „Certainly(!) not" im Englischen. Und zwei Seiten später spricht Wittgenstein von dem Gefühl der „Scham", das sich einstellen müsse, wo gegen das wahrhaft Richtige verstoßen werde. (So es dies denn gäbe.) Auch moralische Gefühle und „Bewusstseinszustände" – Reue, Heuchelei, Empörung... – sind Wittgenstein also bekannt. *Moralische Tatsachen dieser Welt* wird man – entgegen seinem Diktum! – sagen müssen.

Woraus speisen sie sich? Was ist das *Ge-* oder *Ver*bot, die *Grenze* – ziehen wir das Wort wieder herbei, denn es passt hier sehr gut –, gegen die jeweils verstoßen wird? Wittgenstein wird uns hier offensichtlich nicht helfen. Für ihn gilt auf engem Raum gleichermaßen, dass es die Ethik auch ohne menschliches Lebewesen gäbe –

> Kann es eine Ethik geben, wenn es außer mir kein Lebewesen gibt? Wenn die Ethik etwas Grundlegendes sein soll: ja! (TB, 02.08.16) –

wie, dass der menschliche Wille „Träger von Gut und Böse" (TB, 21.07.16) und der „Träger der Ethik" (TB, 04.08.16) sei. Und zusammenfassend, auch im Tagebuch (TB, 02.08.16) und durchaus zutreffend: „Die völlige Unklarheit aller dieser Sätze ist mir bewusst."

Fragen wir also erneut bei Hegel nach. Wieder in Auseinandersetzung mit Kant schreibt er, in der Kantischen Moralphilosophie, wie überhaupt im moralischen Bewusstsein, das Hegel wiederholt als „*ganzes Nest von Widersprüchen*" kennzeichnet (PhG: 453 und VGP III: 371), werde stets „postuliert", dass „*der besondere Wille dem allgemeinen gemäß sein*" müsse. Jedoch:

> Die vollendete Moralität muss ein Jenseits bleiben; denn die Moralität setzt die Verschiedenheit des besonderen und allgemeinen Willens voraus. (VGP III: 369)

Der oben dargestellte Idealismus der Moral spiegele sich notwendig in der Moralphilosophie wieder. Für diese gelte: „Das absolute Gut bleibt Sollen ohne Objektivität" (VGP III: 372); und der Standpunkt des Sollens sei somit ein immerwährender, ein „unendlicher Progress" (VGP III: 369). – Warum ist das so?

Nehmen wir drei kleine, letzte Hegel-Stellen dazu – und es wird sich erhellen, wie man *sinnvoll* davon reden kann, dass das „gute", das „gelungene Leben", notwendig „transzendent" bleibt, wie Wittgenstein sich ausdrückt.

Die Moral ist und will sein Verpflichtung des individuellen Willens auf den übergeordneten Standpunkt eines *WIR*s, eines Kollektivs, *gegen* (nehmen wir den Fall der moralischen *Kritik*; für das Lob, insbesondere den Standpunkt der Selbstgerechtigkeit, gilt, mutatis mutandis, das Analoge) den je aktuell vorliegenden Willensinhalt. In den Worten Hegels von vorhin: „die Moralität setzt die Verschiedenheit des besonderen und allgemeinen Willens voraus". Das heißt, sie hat gerade in dieser „Verschiedenheit" – und sagen wir dazu der Ehrlichkeit halber besser

„*Gegensatz*", denn sonst wäre das Sollen des moralischen Imperativs, das Pochen auf die Pflicht, kaum vonnöten – ihren Ausgangspunkt und *Grundlage*.

Der/die moralisch so Ermahnte *soll* von seinem/ihrem „besonderen Willen" Abstand nehmen zugunsten des „höheren", allgemeinen, im eingeklagten, ideellen „Wir" vorgebrachten Interesses.

In Hegels Diktion:

> Die Pflicht ist ein *Sollen* gegen(!) den besonderen Willen, gegen die selbstsüchtige Begierde und das willkürliche Interesse gekehrt; dem Willen, insofern er in seiner Beweglichkeit sich vom Wahrhaften(!) isolieren kann, wird dieses als ein Sollen vorgehalten. (WdL I: 47)

Ob das gelingt, diese „Beweglichkeit" zuungunsten der „selbstsüchtigen Begierde ", steht dahin – bzw. in der Einschätzung (-> „Freiwilligkeit"!) des so angesprochenen Subjekts, ob der eventuell in Aussicht stehende Lohn der Tugend sich messen lassen kann mit dem Verlust aus dem erheischten Abstandnehmen von der vorgängigen Begierde.

Aber es kommt noch schlimmer:

> Für die Niederträchtigkeit ist allein die Moralität als Beziehung zur Tugend möglich (A: 545)

heißt es bereits beim frühen Hegel. Es geht beispielsweise um die *Heuchelei*. Also die *moralische Lüge*. „Können Hunde Schmerzen heucheln?", fragt sich Wittgenstein in seinen *Philosophischen Untersuchungen* (§250). Und meint damit „simulieren", also „schauspielern" – aber eben auch *Mitleid* – ein weiteres moralisches Gefühl, also wieder eine moralische Tatsache! – evozieren.

Sie können es wohl nicht. Aber Menschen machen es, *gerade weil* sie moralisch sind, täglich, das Heucheln. Die Akkommodation des eigenen Interesses an einen übergeordneten Standpunkt zur Verkleidung (!) eben dieses Interesses ist offensichtlich eine Ausgeburt eben jener Ausgangslage, die zugleich die Moral generiert. Hier haben wir also einen echten Fall von „Familienähnlichkeit". Heuchelei und Moralität sind ganz offensichtlich Geschwister.

Ein letztes Mal Hegel:

> Das Subjekt ist [...] von der Vernunft des Willens unterschieden und fähig, sich das Allgemeine selbst zu einem Besonderen und damit zu einem Scheine zu machen. (Enz III §509)

Bibliography

Carnap, Rudolf; Hahn, Hans; Neurath, Otto (1929): *Wissenschaftliche Weltauffassung. Der Wiener Kreis*. Wien: Artur Wolf.

Drury, Maurice O. (1987): *Gespräche mit Wittgenstein*. In: Rhees 1987, 142 – 235.
Glock, Hans-Johann (1996): *A Wittgenstein Dictionary*. Oxford: Blackwell.
Hegel, G.W.F. (A): „Aphorismen." In: *Jenaer Schriften*. Frankfurt a. M.: Suhrkamp, 1970, 540 – 567.
Hegel, G.W.F. (PhG): *Phänomenologie des Geistes*. Frankfurt a. M.: Suhrkamp, 1970.
Hegel, G.W.F. (WdL I): *Wissenschaft der Logik I*. Frankfurt a. M.: Suhrkamp, 1969.
Hegel, G.W.F. (Enz III): *Enzyklopädie der Wissenschaften III*. Frankfurt a. M.: Suhrkamp, 1970.
Hegel, G.W.F. (VGP III): *Vorlesungen über die Geschichte der Philosophie III*. Frankfurt a. M.: Suhrkamp, 1971.
Janik, Allan; Toulmin, Stephen (1973): *Wittgenstein's Vienna*. New York: Simon and Schuster.
Janik, Allan (1985): *Essays on Wittgenstein and Weininger*. Amsterdam: Rodopi.
Machiavelli, Niccolò (2001): *Der Fürst*. Von Oppeln-Bronikowski, Friedrich (Übers.). Frankfurt a. M., Leipzig: Insel.
Magee, Bryan (1997): *The Philosophy of Schopenhauer*. Revised and Enlarged Edition. Oxford: Oxford University Press.
Moore, G. E. (1970): *Principia Ethica*. Stuttgart: Reclam.
Mras, Gabriele (2001): *Wahrheit, Gedanke, Subjekt. Ein Essay zu Frege*. Wien: Passagen.
Pascal, Fania (1987): „Meine Erinnerungen an Wittgenstein." In: Rhees 1987, 35 – 83.
Rhees, Rush (1987): *Ludwig Wittgenstein: Porträts und Gespräche*. Frankfurt: Suhrkamp.
Schlick, Moritz (1976): „Vorrede." In: Waismann 1976, 11 – 23.
Schopenhauer, Arthur (1920): *Die Welt als Wille und Vorstellung*. Leipzig: Insel.
Schroeder, Severin (2012): „Schopenhauer's Influence on Wittgenstein." In: Vandenabeele, Bart (Hrsg.): *A Companion to Schopenhauer*. Oxford: Blackwell, 367 – 384.
Waismann, Friedrich (1976): *Logik, Sprache, Philosophie*. Stuttgart: Reclam.
Weininger, Otto (1903): *Geschlecht und Charakter: eine prinzipielle Untersuchung*. Wien: Braumüller.
Wittgenstein, Ludwig (1998): *Logisch-philosophische Abhandlung. Tractatus Logico-philosophicus*. Kritische Edition. McGuinness, Brian; Schulte, Joachim (Hrsg.). Frankfurt a. M.: Suhrkamp.
Wittgenstein, Ludwig (1984): *Tagebücher 1914–1916*. In: *Werkausgabe*; Bd. 1. Frankfurt a. M.: Suhrkamp, 87 – 233.
Wittgenstein, Ludwig (2001): *Philosophische Untersuchungen. Kritisch-genetische Edition*. Schulte, Joachim (Hrsg.). Frankfurt a. M.: Suhrkamp.
Wittgenstein, Ludwig (1984): *Philosophische Bemerkungen*. Rhees, Rush (Hrsg.). Frankfurt a. M.: Suhrkamp.
Wittgenstein, Ludwig (1984): *Ludwig Wittgenstein und der Wiener Kreis. Gespräche, aufgezeichnet von Friedrich Waisman* (= *Werkausgabe*; Bd. 3). McGuinness, Brian (Hrsg.). Frankfurt a. M.: Suhrkamp.
Wittgenstein, Ludwig (1984): *Vermischte Bemerkungen*. von Wright, G. H. (Hrsg.). In: *Werkausgabe*; Bd. 8. Frankfurt a. M.: Suhrkamp, 445 – 573.
Wittgenstein, Ludwig (1984): *Über Gewissheit*. In: *Werkausgabe*; Bd. 8. Anscombe, G. E. M.; von Wright, G. H. (Hrsg.). Frankfurt a. M.: Suhrkamp, 113 – 257.
Wittgenstein, Ludwig (1989): „Vortrag über Ethik." In: *Vortrag über Ethik und andere kleine Schriften*. Schulte, Joachim (Hrsg.). Frankfurt a. M.: Suhrkamp, 9 – 19.
Wittgenstein, Ludwig (1980): *Briefwechsel mit B. Russell, G. E. Moore, J. M. Keynes, F. P. Ramsey, W. Eccles, P. Engelmann, L. v. Ficker*. McGuinness, Brian; von Wright, G. H. (Hrsg.). Frankfurt a. M.: Suhrkamp.

Esther Ramharter
Der Status mathematischer und religiöser Sätze bei Wittgenstein

Abstract: Mathematical and religious propositions would seem, at first glance, to be located at opposite ends of the spectrum of certainty: the former being considered as the very paradigm of certainty, the latter as highly doubtful and questionable. In the philosophy of Ludwig Wittgenstein, however, these two sorts of propositions tend – from an epistemological point of view – to converge. Both can be characterized as hinge propositions: propositions on which large parts of our language and beliefs rest. Both types of propositions are normative. On this view, the initial assumption of radical difference is shown to be misguided. For with both mathematical and religious propositions their normative and foundational status is central, perhaps even characteristic. To get a clearer picture of the similarities and differences between them it is useful, as I will try to show, to distinguish internally different types of each of those kinds of propositions.

1 Einleitung

Von einem üblichen Standpunkt aus gesehen, befinden sich mathematische und religiöse Sätze an gegenüberliegenden Enden eines Spektrums: Mathematische Sätze zählen zu den sichersten überhaupt, religiöse Sätze gelten dagegen als höchst kontroversiell und bezweifelbar. Wittgensteinianer könnten allerdings dazu tendieren, jene starke Ähnlichkeit in den Vordergrund zu rücken, die darin besteht, dass religiöse wie mathematische Überzeugungen fundamental für Lebensform und Sprache sind, dass sie zu den festesten Überzeugungen von Menschen zählen (vgl. LRB I: 54). Das Ziel dieses Aufsatzes besteht darin, eine differenziertere Sicht auf Ähnlichkeiten und Unterschiede zu entwickeln.

Ein allgemeiner Vergleich zwischen Mathematik und Religion würde sich mit folgendem Bedenken konfrontiert sehen: Während wir gewöhnlich Mathematik als aus Sätzen bestehend erachten, denken wir bei Religion an Erfahrungen, Einstellungen und mentale Zustände, die nicht notwendigerweise in Sätzen ausgedrückt werden können. Einerseits jedoch erwägt Wittgenstein sehr wohl die Möglichkeit, dass es eine Mathematik ohne Sätze geben könne (BGM I, §144; IV, §§15 – 16), und andererseits lassen sich auch Bedenken dagegen äußern, dass es religiöse

---

Esther Ramharter, University of Vienna, esther.ramharter@univie.ac.at

Überzeugungen geben könnte, die prinzipiell unausdrückbar sind. Gleichwie, in der Folge werde ich mich jedenfalls auf Sätze beschränken.

2 Einige Arten mathematischer Sätze

In der Literatur zu Wittgenstein ist häufig von „mathematischen Sätzen", ohne weitere Spezifizierung, die Rede. Demgegenüber scheint es mir wichtig und lohnend, verschiedene Arten von mathematischen Sätzen zu unterscheiden. Die folgende Tabelle listet vier Typen von mathematischen Sätzen auf, die auch für Wittgenstein relevant sind. Mein Anliegen besteht weder darin, scharfe Trennungen zu ermöglichen, noch möglichst feine Differenzierungen zu schaffen, sondern ich möchte nur zeigen, dass es sehr verschiedene Phänomene gibt, die jeweils unter „Satz" fallen. Ich verwende für die allgemeine Charakterisierung Ausdrücke, die nicht von Wittgenstein stammen, obwohl alle Unterscheidungen von Wittgensteinschen Überlegungen herrühren. Die Wittgensteinschen Termini, die mir am passendsten erscheinen, setze ich nur in Klammern, da ich mich nicht verbindlich darauf festlegen möchte, dass Wittgenstein diese Termini – nur oder ausschließlich oder zu jeder Zeit – entsprechend gebraucht hat (am Beispiel „grammatischer Satz" kann man sehen, dass eine solche Festlegung nicht möglich wäre).

Beispiele		Allgemeine („formale") Charakterisierung
(M1)	$2+2=4$	basaler Satz
		(– empirischer Satz zu einer Regel verhärtet[1]
		– Formulierung eines Paradigmas[2]
		– Angel-Satz/hinge proposition[3])
(M2)	$25 \times 25 = 625$	technischer[4] Satz
		(– grammatischer[5] Satz, hinge proposition)
(M3)	„Es gibt keine größte Kardinalzahl",	theoretischer Satz
	$2^{\aleph_0} > \aleph_0$	(– grammatischer Satz, keine hinge proposition)
(M4)	Auswahlaxiom	Postulat

---

[1] Siehe BGM I, §165; VI, §§22 – 23.
[2] Siehe BGM I, §§62 – 67; III, §28.
[3] Siehe ÜG §341, §655. Zu hinge propositions gibt es nicht nur sehr viel Literatur, es hat sich eine eigene Disziplin – hinge epistemology – entwickelt. Siehe Coliva et al. 2016, Kusch 2016, Schönbaumsfeld 2016; für einen Zusammenhang mit Religion Pritchard 2000, Wright 2004.
[4] In ungefähr meinem Sinn verwendet Wittgenstein das Wort „technisch" in BGM VI, §2.
[5] Vgl. BGM I, §128. Eine Auswahl an Literatur zu grammatischen Sätzen bei Wittgenstein: Aidun 1981, Baker et al. 1985, Glock 2013, Schroeder 2009.

Die genannten Sätze weisen nicht nur eine steigende Komplexität (ihrer Beweise) auf, sondern sie (ihre Rechtfertigungen) sind auch von verschiedener Natur.

Sätze der Art (M1) werden von Wittgenstein selbst als durch Bilder vermittelte Paradigmen, als zu Regeln verhärtete Erfahrungen beschrieben. Ich unterscheide sie von den Sätzen der Gruppe (M2) dadurch, dass sie in dem Sinn keine grammatischen Sätze sind, als sie keine „verselbständigte" Sprache, Grammatik, Algorithmik,... brauchen. Während sich 2 + 2 = 4 als „verhärtete Erfahrung" auffassen lässt, ist das bei 25 × 25 = 625 wohl kaum der Fall. Bei Multiplikationen mit größeren Zahlen ist es eine bestimmte Technik, ein Algorithmus, eine übersichtliche Struktur,..., was sich in der Praxis bewährt hat, nicht direkt der Zusammenhang, der ausgedrückt wird. (Wir bestehen auf der Richtigkeit von 25 × 25 = 625 nicht deswegen, weil wir in zahlreichen Fällen beobachtet haben, dass sich irgendwelche Gegenstände entsprechend verhalten – obwohl wir natürlich tatsächlich in grundlegende Schwierigkeiten kämen, wenn wir mit Umständen konfrontiert wären, die der Rechnung widersprechen.[6]) Natürlich ist der Übergang zwischen (M1) und (M2) aber fließend (auch 2 + 2 = 4 hat eine Technik des Zählens, zumindest mit kleinen Zahlen, im Hintergrund).[7]

Ein Argument, 2 + 2 = 4 doch – entgegen meiner Festlegung – als grammatischen Satz zu klassifizieren, ließe sich auf Basis folgender Stelle bei Wittgenstein formulieren: „Einen Satz als unerschütterlich gewiß anzuerkennen – will ich sagen – heißt, ihn als grammatische Regel zu verwenden: dadurch entzieht man ihn der Ungewißheit." (BGM III, §39) In einem gewissen Sinn ist es durchaus angebracht, hier von einer grammatischen Regel zu sprechen, aber es handelt sich dann um eine ganz basale Regel, die Grundlage einer Grammatik ist, sich nicht aus einer Grammatik ergibt. Ein Vergleich mit Axiomen der Mathematik hilft hier: Sind sie beweisbar (d.h. sind sie ein Ergebnis der „Grammatik" der Mathematik)? In einem trivialen Sinn schon, denn das Axiom kann unter Verweis auf es selbst „bewiesen" werden, und das System wäre nicht es selbst, wenn das Axiom fehlte. In einem gehaltvolleren Sinn von „Beweis" würde man hier allerdings nicht von Beweis sprechen. – Und in einem analogen Sinn möchte ich 2 + 2 = 4 als einen basalen, aber nicht grammatischen Satz bezeichnen.

Sätze vom Typ (M3) bezeichne ich als „technische" Sätze; sie sind grammatische Sätze wieder in dem Sinn, dass ihre Anerkennung aus der „Grammatik" – in diesem Fall: dem mathematischen Formalismus –, ohne direkte Bezugnahme auf Erfahrung resultiert. Es handelt sich allerdings nicht um hinge propositions, da

---

6 Siehe BGM I, §37.
7 Dennoch scheint mir die Nähe von „2 + 2 = 4" zu „Das ist meine Hand" größer als zu einer komplexen mathematischen Rechnung und meine Unterscheidung somit gerechtfertigt.

ihre Verwerfung – aus welchem Grund auch immer – unsere Sprache und deren Verankerung in einer Lebensform nicht erschüttern würde. (M4) schließlich umfasst Sätze, die wir als Postulate nehmen, die man also annehmen oder (ebenso gut) ablehnen kann.[8]

3 Einige Arten religiöser Sätze

Wie bei den mathematischen Sätzen legt es sich auch bei den religiösen nahe, verschiedene Typen zu unterscheiden. Die folgende Aufstellung enthält Beispiele, deren allgemeine Charakterisierungen analog zu den entsprechenden Beispielen aus der Mathematik sind:[9]

(R1) „Gott ist mit mir"
(R2) Glaubensbekenntnisse, die Zehn Gebote, grundlegende Regeln, die eine religiöse Person befolgt, Sätze der Bibel, ...
(R3) theologische Folgerungen
(R4) Es wird ein Jüngstes Gericht geben.

Überzeugungen vom Typ (R1) hat Wittgenstein wohl im Sinn, wenn er schreibt:

> Das Leben kann zum Glauben an Gott erziehen. Und es sind auch *Erfahrungen*, die dies tun; aber nicht Visionen, oder sonstige Sinneserfahrungen, die uns die ‚Existenz dieses Wesens' zeigen, sondern z.B. Leiden verschiedener Art. Und sie zeigen uns Gott nicht wie ein Sinneseindruck einen Gegenstand, noch lassen sie ihn *vermuten*. Erfahrungen, Gedanken, – das Leben kann uns diesen Begriff aufzwingen. (VB: 571)

In Bezug auf den Beispielsatz für (R1) „Gott ist mit mir" könnte man einwenden, dass er für jemand, der den Satz behauptet, nicht basal für das Leben sein muss (keine hinge proposition sein muss). Eine solche Verwendung des Satzes ist durchaus möglich – wenn etwa jemand unreflektiert Glaubensaussagen seiner Kind-

---

[8] Wenn Wittgenstein sagt, dass der Sinn eines mathematischen Satzes seine Verifikation ist PB: 166f., kann er wohl nur Sätze vom Typ (M2) oder (M3) meinen. Ich erwähne das, weil es zeigt, dass meine Differenzierung Konsequenzen hat.
[9] Dass die Unterscheidung aus der Mathematik kommt und sich möglicherweise nicht gleichermaßen aufdrängt, wenn man sich ursprünglich mit Religion beschäftigt, sehe ich als einen Vorzug, nicht als einen Mangel des Vergleichs, da man so eine neue Perspektive auf religiöse Sätze gewinnen kann.

heit übernimmt, jedoch jederzeit ohne Widerstand davon abzugehen vermag – das wäre aber dann ein anderer Gebrauch als mit (R1) gemeint ist.

(R2) subsumiert solche Redeweisen, die einer gewissen Technik oder eines umfangreicheren, ritualisierten Kontextes bedürfen – so wie etwa die Psalmen, die als Formeln konzipiert, aus gebräuchlichen, vielseitig verwendbaren Bausteinen zusammengesetzt sind oder das Beten eines Rosenkranzes oder die Formulierung eines Glaubensbekenntnisses. Solche Texte und Gebete beinhalten Aussagen über Gott,[10] aber solche, die nicht – wie jene vom Typ (R1) – allein stehen können, nicht das Ergebnis individueller zu Gewissheit verhärteter Erfahrung(en) sind,[11] sondern in einen fixen oder gar formelhaften Zusammenhang eingebunden.

(R3) meint theologische Folgerungen, die nicht fest in der Lebenspraxis des Religiösen verankert sind, (R4) umfasst jene Sätze, die Theologen wie Gläubige nur als Postulate ansehen würden, die sich also nicht beweisen, nicht widerlegen lassen, die man annehmen, aber auch ablehnen kann.

Zum letzten Beispiel bedarf es einer Klarstellung. Der Satz „Es wird ein Jüngstes Gericht geben" kann auf verschiedene Weise verstanden werden,[12] z.B.:

(a) als „Ich werde danach beurteilt werden, was ich in meinem Leben getan habe"
(b) als „Eines Tages wird sich auf der Welt etwas ereignen, was wir ‚Jüngstes Gericht' nennen werden"

Wenn (a) gemeint ist, so handelt es sich einen Satz vom Typ (R2), wenn (b) gemeint ist, liegt ein Satz vom Typ (R3) oder (R4) vor. Wittgenstein diskutiert in den *Lectures on Religious Beliefs* die Frage, ob er diesem Satz widersprechen würde oder ob seine Distanzierung davon anderer Art wäre (LRB I: 53 – 59). Die Bedeutung, die er diesem Satz gibt, changiert in seinen Bemerkungen, er tendiert aber sicher eher dazu, den Satz im Sinn von (a) zu verstehen. Ich werde den Satz in der Folge dagegen im Sinne von (b) interpretieren, so, dass es sich dabei um ein Postulat handelt, das man – auch als Gläubiger – akzeptieren oder auch ablehnen kann, dass der Satz also in die Kategorie (R4) fällt.

---

[10] Wittgenstein erwähnt auch die Überzeugung von Katholiken, dass „eine Oblate unter gewissen Umständen ihr Wesen gänzlich ändert" (ÜG §239). Vgl. auch VB: 494.
[11] Einen Extremfall stellen die Versuche von Mystikern, ihre Erlebnisse zu beschreiben, dar.
[12] Wittgenstein erwägt u.a. die Bedeutung „Particles will rejoin in a thousand of years, and there will be a Resurrection of you." (LRB I: 53)

4 Vergleich zwischen mathematischen und religiösen Sätzen – Ähnlichkeiten

Es lassen sich also durchaus Verwandtschaften im Status religiöser und mathematischer Sätze feststellen.[13] Hinsichtlich des Status der Sätze ergibt sich folgende Tabelle von Entsprechungen – wobei sowohl für die Mathematik als auch für die Religion fließende Übergänge zwischen den verschiedenen Typen bestehen.

	Mathematik	Religion	
(1)	$2 + 2 = 4$	„Gott ist mit mir"	basaler Satz (hinge proposition, grundlegend für die Sprache)
(2)	$25 \times 25 = 625$	Bekenntnisse, Regeln, ritualisierte Gebete	technischer Satz (hinge proposition, grammatisch/regelgeleitet.
(3)	$2^{\aleph_0} > \aleph_0$	theologische Folgerungen	theoretischer Satz (systematische Implikationen, grammatisch)
(4)	Auswahlaxiom	„Es wird ein Jüngstes Gericht geben"	Postulat

Bei näherer Betrachtung dieser Tabelle zeigt sich, dass wir wohl geneigt sind, die Ähnlichkeit in den Fällen (3) und (4) zu akzeptieren, sehr viel weniger dagegen jene in (1) und (2). Niemand wird sich daran stoßen, wenn – in der Theologie ebenso wie in der Mathematik – aus vorausgesetzten Sätzen Schlussfolgerungen gezogen werden bzw. Postulate – wenn sie auch als solche deklariert sind – aufgestellt werden.

Hinter Sätzen der Art (1) und (2) stehen feste Überzeugungen – sowohl im Fall der Mathematik als auch der Religion. Wittgenstein bemerkt:

> Der feste Glaube. (An eine Verheißung z. B.) Ist er weniger sicher als die Überzeugung von einer mathematischen Wahrheit? – Aber werden dadurch die Sprachspiele ähnlicher! (VB: 554)

---

13 Ähnlichkeiten, die ich hier nicht ausführlicher thematisieren werde, sind aus einer Wittgensteinschen Perspektive etwa auch: Für beide Bereiche spielt Übersichtlichkeit eine Rolle (für die Mathematik vgl. BGM I, §154; III, §1, für die Religion RoF: 9e), für beide haben Bilder eine bedeutungsstiftende Funktion (für die Mathematik vgl. BGM I, §§62 – 67, für die Religion LRB III: 71 – 72), der Umgang mit Widersprüchen ist in beiden ein neuralgischer Punkt (für die Mathematik siehe BGM III, §§80 – 81; IV, §§55 – 60, für die Religion LRB I: 75). In allen diesen Fällen würde eine Differenzierung von verschiedenen Arten von Sätzen wie die hier vorgeschlagene fruchtbar sein.

Man könnte also sagen, die konstatierte Ähnlichkeit sei bloß oberflächlich – die Sprachspiele ähneln sich wenig. Das mag von einem gewissen Standpunkt seine Berechtigung haben, von einem anderen aus gesehen, bedeutet es aber doch eine wesentliche Gemeinsamkeit von Sätzen, wenn sie hinge propositions sind oder sonstwie eine fundamentale Rolle spielen. Mir scheint weiters, dass man sich mit der Auskunft, dass die Sprachspiele dennoch verschieden sind, jedenfalls nicht begnügen muss. Im nächsten Abschnitt werde ich zu zeigen versuchen, *inwiefern* die Sprachspiele sich unterscheiden. Schließlich sei noch eine Gemeinsamkeit zwischen mathematischen und religiösen Sätzen genannt, die zunächst für einen Unterschied gehalten werden könnte: Man könnte meinen, dass es sich bei religiösen Sätzen um Sollen-Sätze, Normen, etc. handelt, während mathematische Sätze Aussagen sind; Wittgenstein jedoch hält mehrfach fest, dass auch mathematische Sätze ausdrücken, was bei einer Rechnung herauskommen *soll* (siehe z.B. BGM I, §154; III, §9, 28).

5 Unterschiede zwischen religiösen und mathematischen Sätzen gleicher Art

Der Unterschied zwischen (M1) und (M2) auf der einen Seite und (R1) und (R2) auf der anderen lässt sich schon auf relativ formaler Ebene festmachen: Zwar fungieren alle diese Sätze, *wenn sie akzeptiert werden*, als fundamentale Sätze (hinge propositions), während es jedoch bei (R1) und (R2) möglich ist, dass ein Satz für jemand fundamental (eine hinge proposition) ist und für jemand (anderen) nicht (möglicherweise sogar das Gegenteil einer hinge proposition ist),[14] ist das bei (M1) und (M2) nicht der Fall. Wir würden solche Sätze schlicht nicht Mathematik nennen.[15] Entsprechend nennt man den Übergang von einer Überzeugung zur anderen im Fall von (R1) und (R2) eine Bekehrung, im Fall von (M1) und (M2) spricht man – je nach Richtung – von Heilung bzw. Ausschluss aus der Sprechergemeinschaft. (Wenn jemand, trotz Übereinstimmung in allen Voraussetzungen und Hintergründen auf 2 + 2 = 5 besteht, könnten wir nicht mehr weiterreden mit ihm.)

Der grundlegendste Unterschied zwischen religiösen und mathematischen Sätzen besteht aber meines Erachtens in einer Art von Rechtfertigung, die es im religiösen Kontext gibt, im mathematischen dagegen nicht: Es handelt sich um die

---

[14] Vgl. ÜG §107.
[15] Vgl. BGM III, §§75 – 76; VI, §21. Die Unumstößlichkeit von Mathematik beschreibt Wittgenstein auch so: „Es bricht kein Streit darüber aus (etwa zwischen Mathematikern), ob der Regel gemäß vorgegangen wurde." (PU §240)

Berufung auf ein individuelles Eingreifen, eine individuelle Offenbarung: „Gott hat gestern zu mir gesagt...", „Gott schenkt mir Glück",... Wittgenstein widmet sich solchen Sätzen in seinen philosophischen Überlegungen nicht sehr ausführlich, es findet sich aber etwa folgender Tagebucheintrag vom 25. 11. 1936:

> Heute ließ Gott mir einfallen – denn anders kann ich's nicht sagen - daß ich den Leuten hier im Ort ein Geständnis meiner Missetaten machen sollte. Und ich sagte, ich könne nicht! Ich will nicht obwohl ich soll. Ich traue mich nicht einmal der Anna Rebni und dem Arne Draegni zu gestehen. So ist mir gezeigt worden daß ich ein Wicht bin. Nicht lange ehe mir das einfiel sagte ich mir ich wäre bereit mich kreuzigen zu lassen. (TB: 70)

Weiters kann man einen Wortwechsel zwischen Wittgenstein und Yorick Smythies, einem seiner Lieblingsstudenten, in den *Vorlesungen über religiösen Glauben* so verstehen, dass es dabei um die Anerkennung einer solchen Möglichkeit für den religiösen Menschen geht. Ich gebe den Dialog in einer gekürzten Version, die ich von Kusch 2011 übernehme, wieder:

> [Wittgenstein:] „God's eye sees everything" – I want to say of this that it uses a picture. ... We associate a particular use with a picture.
>
> Smythies: This isn't all he does – associate a use with a picture.
>
> Wittgenstein: Rubbish. I meant: what conclusions are you going to draw? etc. Are eyebrows going to be talked of, in connection with the Eye of God?... If I say he used a picture, I don't want to say anything he himself wouldn't say. ... The whole weight may be in the picture. ... I'm merely making a grammatical remark ... (LRB III: 71, Kusch 2011: 44)

Kusch argumentiert nun, gegen einige andere Autoren,[16] dass Smythies' Einspruch sich auf genau jenes individuelle Geschehen zwischen Gott und dem Gläubigen bezieht: „[O]n my reading, Smythies insisted that a view that reduces religion to the use of pictures misses its most important aspect: the relationship between the religious believer and God." (Kusch 2011: 47) Ich stimme mit Kusch in diesem Punkt überein. Er setzt fort:

> Wittgenstein's response to this criticism was to say that anyone who draws the contrast between the two ideas (of the pictures and of the relationship) in this way must assume that the pictures are of little weight. On Wittgenstein's rendering of the role of pictures and narratives in religion they do not stand in the way of a relationship with God. On the contrary, they are essential to that relationship. (Kusch 2011: 47)

---

[16] Kusch 2011 wendet sich gegen die Interpretationen von Diamond 2005, Putnam 1992, Schönbaumsfeld 2007.

Auch darin stimme ich mit Kusch überein, dass Wittgensteins Antwort auf Smythies festhält, dass wir, um über Gott sprechen zu können, auf Bilder angewiesen sind und dass ein Ignorieren der Unabdinglichkeit der Bilder irregeleitet wäre. Meine Auffassung unterscheidet sich allerdings von jener Kuschs darin, dass Wittgenstein meines Erachtens Smythies' Punkt nicht trifft. Smythies ist durch seinen Einwurf keineswegs darauf festgelegt zu bestreiten, dass Bilder notwendig sind, um religiöse Begriffe wie „Gott" verstehen bzw. verwenden zu können. Um zu Gott sprechen zu können, braucht der religiöse Mensch Bilder – das kann Smythies ohne weiteres zugestehen. Wenn Kusch meint, Wittgenstein weist Smythies Kommentar zu Recht als „rubbish" zurück, dann braucht es ein Argument, dass die Bilder hier nicht nur notwendig sind, sondern auch ausreichend – dass dann *nichts mehr fehlt* („This isn't all he does" muss falsch sein). Das trifft zwar tatsächlich zu, wenn der Religiöse über Gott spricht („Gott ist gut") und auch wenn er zu Gott spricht, aber nicht wenn er *„wegen"* Gott spricht, also wenn er etwas auf Gott zurückführen möchte. Wenn jemand sagt, dass er gestern seinen Freund getroffen hat, dann assoziiert er nicht nur einen Namen mit einem Bild. Denselben Anspruch stellt ein religiöser Mensch. Er möchte Aussagen dadurch *rechtfertigen* können, dass für ihn ein gewisser Zusammenhang mit Gott besteht.

Kusch scheint diesen Einwand zu sehen, denn er beruft sich auf Johann Georg Hamann[17] – den Wittgenstein gelesen haben könnte –, um sagen zu können:

> On Wittgenstein's rendering of the role of pictures and narratives in religion they do not stand in the way of a relationship with God. On the contrary, they are essential to that relationship. [...] For Hamann the holy scripture is not just a report on God's deeds, it is first and foremost a divine action towards us. We understand the bible only because God enables us to do so; and the text and our reaction to it are of one piece. In other words, for Hamann the bible is a weighty picture because it is the picture through which God relates to us, and we to him. (Kusch 2011: 47)

Auf diese Weise wird der Unterschied, den ich gemacht habe, nivelliert: Ein Bericht über Gott wird dasselbe wie eine individuelle Erfahrung mit Gott. Was genau aber sagt Hamann?

> [J]a ich bekenne, daß dieses Wort Gottes eben so große Wunder an der Seele eines frommen Christen, er mag einfältig oder gelehrt seyn, thut als diejenigen die in demselben erzählt werden, daß also der Verstand dieses Buchs und der Glaube an den Inhalt desselben durch nichts anders zu erreichen ist als durch denselben Geist, der die Verfasser desselben getrieben, daß seine unaussprechlichen Seufzer die er in unserm Herzen schafft mit den unausdrücklichen

---

[17] Der Bezug zu Hamann wird dadurch motiviert, dass es sehr plausibel scheint, dass Wittgenstein eine Aussage über Luther von Hamann übernommen hat.

> Bildern einer Natur sind, die in der heiligen Schrift mit einem größern Reichthum als aller Saamen der ganzen Natur und ihrer Reiche, aufgeschüttet sind. (Hamann 1950: 43)

Mag sein, dass Gott durch die Schrift handelt, aber keineswegs lässt sich diesem Zitat entnehmen, dass das die einzige Art wäre, auf die er handeln kann. Smythies' Bedenken wären allenfalls dann ausgeräumt – oder ausräumbar –, wenn Gott ausschließlich durch Sprache mit Menschen in Beziehung treten kann. Offenbar kennt aber auch Hamann andere Arten, wie Gott sich zeigen kann. Um willkürlich eines von vielen Beispielen herauszugreifen:

> Ich bin in Riga dem Ehebruch sehr nahe gewesen, ich habe Versuchungen des Fleisches und Blutes sowohl als des Witzes und Herzens gehabt und Gott hat mich gnädig bisher selbst von den Schlingen der Huren, ich möchte sagen, durch ein Wunder behütet. (Hamann 1950: 18)

Zudem gibt es überhaupt keine Hinweise, warum Wittgenstein in dieser Angelegenheit ausgerechnet Hamanns Sichtweise übernommen haben sollte. Wenn Wittgenstein seine Auffassung auf Lektüre zurückgehen sollte, kämen ebenso gut Augustinus, Karl Barth, Johannes Calvin, J. B. S. Haldane, Sören Kierkegaard, Martin Luther, Leo Tolstoi, u.v.a.[18] in Frage, die sehr verschiedene Vorstellungen haben und die Wittgenstein ebenfalls gelesen hat. Mehr noch, Wittgensteins explizite Bezugnahme auf Hamann (BEE: 183, 67f.) legt eher nahe, dass er dessen Aussage kritisch gegenübersteht. (Man könnte jetzt noch argumentieren, dass nicht Wittgenstein Hamanns Auffassung übernimmt, sondern dass er Smythies unterstellt, er habe Hamanns Auffassung – das aber scheint mir nun endgültig zu sehr Spekulation zu sein.)

Anzumerken bleibt ferner noch, dass Kusch nicht den gesamten, zugegebenermaßen etwas wirren Dialog zitiert (vielleicht haben mehrere Personen gleichzeitig gesprochen?). Die ersten Auslassungszeichen stehen für Sätze, in denen es um Vorbereitungen für das Leben nach dem Tod geht. Es ist also keineswegs so klar, dass Wittgenstein und Smythies an „God's eye sees everything" anknüpfen[19] – sie könnten sich auch auf eine Praxis beziehen, die sich nicht auf das Assoziieren mit Bildern beschränken lässt.[20]

---

18 Siehe Biesenbach 2011: 1 – 3.
19 Auch dass Smythies das Pronomen „he" verwendet, weist darauf hin, dass er an die Sätze über den Mann, der Vorbereitungen für seinen Tod trifft, anknüpft.
20 Man könnte noch einwenden, Wittgenstein gehe es aber eben um sprachliche Beschreibungen, und Smythies protestiere in diesem Zusammenhang. Dann hätte Wittgensteins Reaktion aber fairerweise statt in „rubbish" darin bestehen müssen, dass er Smythies erklärt, dessen Anliegen interessiere ihn im Moment nicht.

Was Smythies also, meines Erachtens von Wittgenstein missverstanden oder zumindest ignoriert, moniert, ist die Möglichkeit einer Rechtfertigung (von Sätzen, Handlungen, Überzeugungen,...) unter Berufung auf Gott. Zwar reicht es, um zu Gott zu sprechen (zu meinen[21]) und um Geschichten über Gott zu erzählen, dass man eine passende Sprache hat – und deren Möglichkeit beruht auf Bildern –, aber der Anspruch, etwas auf Gott zurückführen zu können – wie man etwas auf ein Zusammentreffen mit einem anderen Menschen zurückführt – geht darüber hinaus. Wenn der Religiöse sagt „Gott macht, dass es mir gut geht", dann stellt er nicht eine allgemeine Aussage über Gott wie „Gott ist gut" zur Disposition, sondern beansprucht, dass ein bestimmter Zusammenhang zwischen etwas – nämlich Gott – und ihm *tatsächlich besteht*. Ebendiese Art der Rechtfertigung scheint mir charakteristisch für gewisse religiöse Sätze, der Mathematik sind sie dagegen fremd.[22] In der Mathematik haben Rechtfertigungen nie etwas Individuelles – nicht nur ist keine Rechtfertigung unter Berufung auf Gott möglich, auch die Berufung auf eine andere Person vermag niemals einen mathematischen Satz zu rechtfertigen.[23]

Um genauer zu sein: Kuschs Erklärung geht für (gewisse) Sätze der Art (R2) möglicherweise auf, aber nicht für (alle) Sätze der Art (R1). Wenn jemand Sätze äußert, die ihre Rechtfertigung und Anerkennung allein aus sprachlichen Zusammenhängen ziehen – wie es auch für mathematische Sätze der Art (M2) zutrifft –, dann ist Bilder zu assoziieren möglicherweise „all he does". Wenn jemand aber Gott als Rechtfertigung für die Gültigkeit von Sätzen heranzieht, dann beansprucht er – ob legitimer oder nicht legitimer Weise sei dahingestellt – etwas darüber hinaus, etwas, das es bei mathematischen Sätzen nicht gibt.

Abschließend sei noch erwähnt, dass sich für die religiösen Sätze eine – vermeintliche – Spannung zeigt: Religiöse Sätze vom Typ (R1) und (R2) sind zum einen fundamental, Welt-verändernd, das Leben in seiner Gesamtheit betreffend, zum anderen sind sie individuell bestimmt. (Wittgenstein drückt eine ähnliche Spannung einmal aus: „Die Welt des Glücklichen ist eine andere als die des Unglücklichen." TLP 6.43)

---

21 Aus der Sicht des Nicht-Religiösen spricht der Religiöse ins Nichts.
22 Wittgenstein stellt das in BGM I, §106 fest: Man glaubt ein mathematisches Resultat nicht „auf die Versicherung eines Andern hin" – man kann das Ergebnis einer Rechnung von einem anderen „annehmen", aber man kann es nicht „glauben".
23 Natürlich berufen sich Mathematiker auf andere Papers, aber diese werden nie als individuelle Äußerungen verstanden. Die Anerkennung der Leistung des Autors bedeutet nicht, dass er die rechtfertigende Instanz ist.

6 Fazit

Die Ähnlichkeit gewisser mathematischer und religiöser Sätze – nämlich jener vom Typ (3) und (4) – beruht darauf, dass diese Sätze unspezifisch für ihr Feld sind, dass sie also die jeweiligen Gemeinsamkeiten *mit allen Sätzen einer bestimmten Art* – auch solchen, die weder mathematisch, noch religiös sind – aufweisen. Es handelt sich dabei um *ihrer Rolle nach* unkontroversielle Sätze. (Natürlich müssen Postulate etwa inhaltlich keineswegs unkontroversiell sein, aber als Postulate sind sie es.)

Bemerkenswerter ist die Ähnlichkeit von fundamentalen mathematischen und religiösen Sätzen – Sätzen vom Typ (1) und (2). Sie teilen in verschiedener Hinsicht einen besonderen Status: Sie sind, so sie als zutreffend erachtet werden, besonders grundlegend für Sprache und Lebensvollzug, mit besonders festen Überzeugungen verbunden, – direkt oder indirekt – als Verhärtung von Erfahrung, durch Identifizieren einer Regel, eines Musters entstanden. Während allerdings mathematische Sätze für alle Menschen unserer Kultur diesen Status haben, gibt es bei den religiösen Sätzen Menschen, für die diese Sätze fundamental (hinge propositions) sind, und Menschen, die sie nicht einmal für zutreffend halten (für die unter Umständen sogar die Negationen hinge propositions sind).

In Sätzen vom Typ (1) kann weiters ein Spezifikum des Religiösen zum Ausdruck kommen: die Berufung auf ein Individuelles. (Sie sind gleichzeitig individuell und allgemein, grundlegend.) In der Mathematik dagegen gibt es keine Rechtfertigung der Art: „*X* hat gesagt".

Bibliography

Aidun, Debra (1981): „Wittgenstein on Grammatical Propositions." In: *The Southern Journal of Philosophy*. Vol. 18, No. 2, 141 – 148.

Baker, Gordon P. und Hacker, Peter M.S. (1985): *Wittgenstein: Rules, Grammar and Necessity*. Oxford: Blackwell.

Biesenbach, Hans (2011). *Anspielungen und Zitate im Werk Ludwig Wittgensteins*. Bergen: Publications from the Wittgenstein Archives at the University Bergen.

Coliva, Annalisa und Moyal-Sharrock, Danièle (Hrsg.) (2016): *Hinge Epistemology*. Leiden: Brill.

Diamond, Cora (2005): „Wittgenstein on Religious Belief: The Gulfs Between Us." In: *Religion and Wittgenstein's Legacy*. Phillips, Dewi Zephania und von der Ruhr, Mario (Hrsg.). Aldershot: Ashgate, 99 – 137.

Glock, Hans-Johann (2013): „Necessary Truth and Grammatical Propositions." In: *Phenomenology as Grammar*. Padilla Gálvez, Jesús (Hrsg.). Heusenstamm: Ontos Verlag, 63 – 76.

Hamann, Johann Georg (1950): „Gedanken über meinen Lebenslauf." In: *Sämtliche Werke*. Historisch-kritische Ausgabe, Josef Nadler (Hrsg.), Vol. 2. Wien: Herder, 9 – 54.

Kusch, Martin (2011): „Disagreement and Picture in Wittgenstein's ‚Lectures on Religious Belief'." In: *Image and Imaging in Philosophy, Science and the Arts*. Vol. 1. Heusenstamm: Ontos Verlag, 35 – 58.
Kusch, Martin (2016): „Wittgenstein on Mathematics and Certainties." In: *International Journal for the Study of Skepticism*. Vol. 6, No. 2–3, 120 – 142.
Pritchard, Duncan (2000): „Is ‚God Exists' a ‚Hinge Proposition' of Religious Belief?" In: *International Journal for Philosophy of Religion*. Vol. 47, No. 3, 129 – 140.
Putnam, Hilary (1992): *Renewing Philosophy*. Cambridge, MA: Harvard University Press.
Ramharter, Esther und Weiberg, Anja (2006): *Die Härte des logischen Muss. Wittgensteins Bemerkungen über die Grundlagen der Mathematik*. Berlin: Parerga.
Schönbaumsfeld, Genia (2007): *A Confusion of the Spheres. Kierkegaard and Wittgenstein on Philosophy and Religion*. Oxford: Oxford University Press.
Schönbaumsfeld, Genia (2016): „‚Hinge Propositions' and the ‚Logical' Exclusion of Doubt." In: *International Journal for the Study of Skepticism*. Vol. 6, No. 2–3, 165 – 181.
Schroeder, Severin (2009): „Analytic Truths and Grammatical Propositions." In: *Wittgenstein and Analytic Philosophy*. Glock, Hans-Johann und Hyman, John (Hrsg.). Oxford: Oxford University Press.
Wittgenstein, Ludwig (1966): „Lectures on Religious Belief." In: *Lectures & Conversations on Aesthetics, Psychology and Religious Belief*, Cyril Barrett (Hrsg.). Berkeley/Los Angeles: University of California Press, 53 – 72.
Wittgenstein, Ludwig (1998–2000): *Wittgenstein's Nachlass: The Bergen Electronic Edition*. Oxford: Oxford University Press.
Wittgenstein, Ludwig (1984): *Bemerkungen über die Grundlagen der Mathematik, Werkausgabe Bd. 6*. Frankfurt am Main: Suhrkamp.
Wittgenstein, Ludwig (1984): *Philosophische Bemerkungen, Werkausgabe Bd. 2*. Frankfurt am Main: Suhrkamp.
Wittgenstein, Ludwig (1991): *Bemerkungen über Frazers Golden Bough/Remarks on Frazer's Golden Bough*. Denton: The Brynmill Press.
Wittgenstein, Ludwig (1984): „Tagebücher 1914–1916." In: *Tractatus logico-philosophicus, Tagebücher 1914–1916, Philosophische Untersuchungen, Werkausgabe Bd. 1*. Frankfurt am Main: Suhrkamp.
Wittgenstein, Ludwig (1984): *Tractatus logico-philosophicus*. In: *Tractatus logico-philosophicus, Tagebücher 1914–1916, Philosophische Untersuchungen, Werkausgabe Bd. 1*. Frankfurt am Main: Suhrkamp.
Wittgenstein, Ludwig (1984): *Philosophische Untersuchungen*. In: *Tractatus logico-philosophicus, Tagebücher 1914–1916, Philosophische Untersuchungen, Werkausgabe Bd. 1*. Frankfurt am Main: Suhrkamp.
Wittgenstein, Ludwig (1989): „Über Gewissheit." In: *Bemerkungen über die Farben. Über Gewißheit. Zettel. Vermischte Bemerkungen, Werkausgabe Bd. 8*. Frankfurt am Main: Suhrkamp.
Wittgenstein, Ludwig (1999): „Vermischte Bemerkungen." In: *Bemerkungen über die Farben. Über Gewißheit. Zettel. Vermischte Bemerkungen, Werkausgabe Bd. 8*. Frankfurt am Main: Suhrkamp.
Wright, Crispin James Garth (2004): „Hinge Propositions and the Serenity Prayer." In: *Knowledge and Belief*. Löffler, Winfried und Weingartner, Paul (Hrsg.). Wien: Hölder-Pichler-Tempsky, 287 – 306.

Richard Raatzsch
Gutes Sehen

Abstract: The argument turns around Wittgenstein's observation according to which we observe in order to see what we would not see if we did not observe. Looked at in tis way, seeing is essentially embedded in some kind of activity. This idea does not only offer a solution to the old dispute about whether or not seeing is something active. This idea also extends itself naturally to the question of whether or not goodness can be seen. Essential steps in this extension are: to note that there are standards for activities, to remind oneself that a part of these standards often concerns the ability to recognize, f. i.: see, the goodness of something with which the activity in question is internally connected, and to accept that this in turn defines standards for seeing something, or someone. Since the form of the argument is rather against Wittgenstein's way of arguing, the end of argument is a methodological recall action.

1 Einleitung

1. Im Folgenden soll folgende These erwogen werden:

> Man kann Gutes nur sehen, wenn man gut sehen kann; und gut sehen kann man allein dann, wenn man Gutes sehen kann.

Um das Verständnis des Kommenden zu erleichtern, hier einige wichtige Bestandteile des Arguments:

- Ein Jegliches ist mehr oder weniger gut (oder eben schlecht, böse, ...).
- Dabei kann es *auf verschiedene Weise* gut sein.
- Diese Weise kann das, was ist, *bestimmen*, ist dem, was gut ist, *nicht äußerlich*; es ist die Sache selbst, soweit sie ihrem *Begriff* entspricht (s.o.).
- Sehen gehört zum Handeln.
- Die Maßstäbe guten Handelns ruhen in dem, wie man handelt.
- Gutes Sehen ist insofern ein *allgemeines* Vermögen. Wertblindheit ist ein *Mangel*.
- Nur der hat den Mangel nicht, der Gutes sehen kann.

Richard Raatzsch, EBS Universität für Wirtschaft und Recht, Deutschland, richard.raatzsch@ebs.edu

2. Im Folgenden wird vorausgesetzt, dass das Wort „gut" nicht in (wenigstens) zwei völlig verschiedenen Bedeutungen oder nur in höchstens einem Fall überhaupt bedeutungsvoll gebraucht wird. Damit scheint jede Antwort auf die Frage, ob, und ggf. wie, das *moralisch* Gute zu sehen ist, auch intern mit der Frage verbunden zu sein, wie es um die Erkenntnis alles *anderen* Guten bestellt ist, vorausgesetzt, die Frage seiner Sichtbarkeit betrifft die Natur des Guten.

Die Autonomie des moralisch Guten besteht dann darin, dass es in anderer Weise verschieden von anderen Formen des Guten ist, als diese (oder doch sehr viele von ihnen) es untereinander sind. So wie ein Handballtorwart in anderer Weise von einem Kreisspieler verschieden ist, als dieser sich von einem Centerspieler unterscheidet. Somit fiele auch das moralisch Gute unter die Eingangsthese.

3. In einer Hinsicht ist die Sichtbarkeit des moralisch Guten keine offene Frage: so reden wir eben. Und man *schließt* (häufig, meistens, immer?) aus dem, was man sieht, nicht auf die moralische Qualität des Gesehenen. Man erkennt es (häufig, ...) unmittelbar; man sieht es.

Dass wir ausdrücklich von moralischer *Blindheit* (dem Gerechtigkeits*sinn*; *moral sense*) reden, kann natürlich rein metaphorisch sein, eine *Redeweise*, wie die herrschende Meinung besagt. – Das Folgende treibt in dieser Hinsicht stromauf; ist also eher altmodisch. Warum auch soll das wirkliche Rätsel nicht lauten: Wie kann die Tatsache zum Wert, das Natürliche zum Moralischen, die Welt zu Gottes Güte usw. hinzutreten?, statt umgekehrt?

2 Sehen und Gesehenes

4. In einem Sinn des Ausdrucks „gutes Sehen" gibt es so etwas wie Sehtests für gutes (resp.: normales) Sehen. – Kommt hierbei auch etwas (Gutes als) Gesehenes vor?

Nehmen wir Sehtests für Führerscheine. Die „guten Augen und die klare Sicht", denen ein solcher Test gilt, *passen*, gewissermaßen, zum (gewöhnlichen) Straßenverkehr. (Autos fahren zwar auch dann, wenn man vor lauter Nebel keine drei Meter weit sehen kann. Aber das hilft einem bei einem Sehtest nicht, bei dem herauskommt, dass man nur drei Meter weit sehen kann.) Dem entspricht, dass bei anderen Tests gerade Leute gut abschneiden, die bei Führerscheinsehtests schlecht abschneiden, etwa Kurzsichtige. Gewöhnlich gilt Kurzsichtigkeit als *Defekt*. Aber es gibt auch eine ganze Reihe verschiedener Tätigkeiten, die mit dem bloßen Auge *besser* von Kurzsichtigen ausgeführt werden (können) als von andern, etwa feinmechanische Arbeiten. An diesen gemessen, wäre eher Normalsichtigkeit ein Defekt. Gut sehen würde hier also der, welcher sonst schlecht sieht. Das zeigt, dass Kurzsichtigkeit nicht an sich schlecht ist, soweit „an sich" bedeutet „unter allen denkbaren Bedingungen". Allgemein gesprochen, eine Art des Sehens kann (besser oder schlechter) zu einer Art von Sichtbarem *passen* (als eine andere).

In einem andern Sinn dagegen ist das, was bei Führerscheintests gemessen wird, *gar keine* Art des Sehens. Sondern der Begriff „Art des Sehens", der durch diesen Fall exemplifiziert wird, gehört eher zu dem, was man „physiologische Betrachtungsweise des Sehens" nennen könnte (vgl. Abschnitt IV). Die philosophische Betrachtungsweise zeigt sich dort, wo man etwa sagt, wie Heidegger (Heidegger 1949: 33), Sehen entdecke immer Farben (wie Hören immer Töne). – Das haben wir nicht etwa durch empirische Untersuchungen herausgefunden; wir haben es, genau genommen, *überhaupt nicht* herausgefunden. Wir haben uns nur dessen *erinnert*, wie wir über Farben und Töne, Sehen und Hören reden. Wir haben uns auf unsere *Begriffe* besonnen.

Aber nicht nur die *Art* der Wahrnehmung, das Sehen im Unterschied zum Hören, und die des Wahrgenommenen, die Farben im Unterschied zu den Tönen, „passen (in dieser Weise) zueinander". Auch innerhalb dieser Arten finden wir ein Passen. Denn wenn jemand sagt, er habe gesehen, dass *p*, und ein anderer erwidert, das könne er nicht gesehen haben, da es nicht der Fall war, dass *p*, kann der Erste zwar erwidern: „Nun, da frage ich mich, was ich dann gesehen habe."; hier jedoch könnte die Antwort eben auch lauten: „*q!*", und wenn „Gar nichts!" die Antwort wäre, was es durchaus auch könnte, mangelte es dieser Antwort so lange an völliger Verständlichkeit, bis sie ergänzt werden würde um einen Satz wie etwa: „Es war wohl nur ein Schatten", oder auch: „Ich hatte nur einen Sinneseindruck." Jede dieser Antworten exemplifiziert das „Passen": die ersten beiden Antworten unmittelbar, wenn auch auf je verschiedene Weise, die dritte vermittelt über eine nähere Bestimmung des Terminus „sehen" derart, dass von einem Unterschied zwischen Sehen und Gesehenem entweder gar nicht mehr die Rede ist oder in einer Weise, dass beides einander nicht mehr gegenübersteht.

Dass Sehen *immer* Farben entdeckt, kann mindestens zweierlei bedeuten:

– dass man (eigentlich) *nur* Farben sieht, oder
– dass man immer *auch* Farben sieht.

Selbst wenn man erklärte, Formen seien schlicht Grenzen von Farbflächen, verwenden wir nicht immer eine solche Erklärung, wenn es darum geht, jemanden zu lehren, was ein Kreis ist. Ganz zu schweigen von dem Fall, in dem es etwa um Freude geht. Kurz, man kann einen Kreis, die Freude im Gesicht eines Andern u. v. m. unmittelbar sehen, und muss nicht auf sie aus dem schließen, was man „wirklich" sieht, seien dies Farben oder was auch immer.

Das wirft auch schon ein Licht auf die These, dass man, wenn man etwas sieht, immer auch Farben sieht. Welche Farbe hat denn die Freude? Selbst wenn sie eine Farbe hätte („Grün wie die Hoffnung, rot wie die Liebe."), hätte sie diese auf eine ihr eigene Weise. Man sähe sie nicht so, wie man das Rot einer roten Rübe sieht. Die rote Rübe sieht im Dunkeln schwarz aus, oder ist im Dunkeln schwarz. Was aber entspräche dem im Fall der Freude?

Alle diese Fälle liefern Beispiele eines „Zueinander-Passens" von Sehen und Gesehenem; Farben liefern nur eines unter anderen. (Wir kommen auf diesen Punkt unten noch einmal zurück.) Allgemein gesprochen, Sehen und Gesehenes können auf zahllose Arten zueinander passen, je nachdem, wovon jeweils die Rede ist.

5. Wenn man mehr oder Anderes als Farben sehen kann, dann auch verschiedene *Ganzheiten*, *Arten*, von Sichtbarem. Farbige Formen könnten etwa eine Art bilden, Farben und der Wechsel ihres Sättigungsgrades eine andere, etwas wie ein Pulsieren der Farbe, ohne dass dabei überhaupt Formen im Spiel sein müssten. Aber daraus, dass jemand farbige Formen sehen kann, folgt nicht schon, dass er auch jenes Pulsieren wahrnehmen kann; und umgekehrt. Vielleicht sieht jemand überhaupt nur etwas, wenn sich etwas ändert, wie beim Pulsieren. Der Satz

> Wer sieht, wie farbige Formen pulsieren, sieht, weil es farbige Formen sind, die pulsieren, auch farbige Formen

ist nicht an sich wahr. „Farbige Formen pulsieren sehen" muss genauso wenig auf einfache Weise kompositional sein, wie es Ausdrücke sein müssen, in denen das Wort „gut" vorkommt. Wo zwei Menschen nur jeweils eine von beiden Ganzheiten sehen können, wäre nichts natürlicher, als zu sagen, dass ihre *Sehvermögen* verschiedener Art sind. („In einem solchen Fall" wohlgemerkt, einem für bestimmte Zwecke konstruierten Fall.) Diese verschiedenen Sehvermögen wären selbst, oder ihre Ausübung resultierte dann, in zwei *Arten von Sehen*. – Je vielfältiger der Art nach das Sichtbare ist, umso vielfältiger sind auch die Arten des Sehens.

Ähnliches zeigt die Ansicht, Sehen entspreche dem (oder falle unter das), was man traditionell auch „Anschauung" nennt, wenn man außerdem glaubt, die Anschauung könne nicht „blind" sein, sondern sei „begrifflich verfasst", und schließlich die Auffassung vertritt, unsere Begriffe, soweit sie in Anschauungen „eingehen", ließen sich in verschiedene Gruppen einteilen. Unter diesen Voraussetzungen gäbe es offensichtlich auch verschiedene Arten von Anschauungen, entsprechend den Gruppen von Begriffen, die sie kennzeichnen. Wenn nun Anschauungen jeweils ein Ganzes aus Anschauen und Angeschautem sind, so dass Anschauen das (m. o. w. passende) Gegenstück zum Angeschauten sein kann, muss es, wenn es verschiedene Arten von Anschauen geben können soll, auch verschiedene Arten von Angeschautem geben können, wenn Anschauen und Angeschautes zwar verschieden sind, aber zugleich in dem Sinne intern zusammengehören, als die Anwendung jener Begriffe, welche in den sprachlichen Ausdruck eingehen, dessen Verständnis uns sagt, was angeschaut wurde, das Anschauen wesentlich charakterisiert. Ausgedrückt in der Terminologie von „Sehen" und „Sichtbares" ergibt dies: Jede Art von Sichtbarem erfüllt, gewissermaßen, eine Form des Sehens, und umgekehrt. Soweit aber auch jeweils alle Arten des Sichtbaren einerseits und alle Formen des Sehens andererseits zusammengehören, „zueinander passen", steht einer Vielfalt an Formen des Sehens eine solche des Sichtbaren gegenüber, wobei es zwischen einzelnen Elementen wiederum Entsprechungen gibt. (Hier ist der Ausdruck „eine Vielfalt von zu einer Einheit verbundenem X" erkennbar ein Notbehelf. Es ist ja gerade die Einheit, was jene Vielfalt von X zu einer Vielfalt von X, statt einer von Y, macht. Hier ist also etwas begriffliche Nachsicht auf Seiten des Lesers gefragt.)

Was Sehen und Sichtbares angeht, ist beides wichtig: Entsprechung auf der Ebene der Vielfalt und auf der Ebene ihrer jeweiligen Elemente. Denn natürlich könnte es eine Vielfalt geben, der keine andere entspricht, und eine Vielfalt, der zwar eine andere entspricht, aber in keiner (auf der Ebene ihrer „Elemente") geordneten Weise. Das aber war unser Ausgangspunkt: dass es zwischen einzelnen „Elementen" Entsprechungen gibt. Nur weil diejenigen „Elemente", die jeweils auf der einen, und diejenigen, die jeweils auf der andern Seite stehen, unter jeweils einen Begriff fallen, d. h. eine Einheit bilden, haben wir Entsprechungen auf der Ebene der Vielfalt und der ihrer „Elemente".

6. Das ist bemerkenswert. Denn wenn wir auf unsere gewöhnlichen Sätze schauen, zeigt sich zunächst reine Gleichmacherei. Man sagt, man sehe einen *Tisch*, wie man sagt, man sehe einen *Film*. Wir sehen einen *Berg*, seinen *Schatten*, ein *Nachbild* des Schattens eines Berges. Soweit es um die Form des Satzes geht, wie er dasteht, legt das nahe, dass man einfach Verschiedenes sieht. Man tut dasselbe, soweit Sehen ein Tun ist, nur jeweils in Bezug auf verschiedene Gegenstände; oder es widerfährt einem dasselbe, wenn Sehen ein Widerfahrnis ist, nur ausgehend von unterschiedlichen Dingen. In der Sprechweise des vorangehenden Paragraphen: einer Vielfalt von Gesehenem steht eine einzige Sache (ein Tun oder ein „Erleiden") gegenüber. Aber hier sind offensichtlich einige Unterschiede nicht im Blick: einen Tisch aus verschiedenen Perspektiven sehen; einen Film aus dieser oder jener Perspektive sehen; sehen, wie ein Beweis geführt wurde; ein Haus von innen sehen. Selbst wenn nicht alle diese Unterschiede wichtig sind, müssen wir uns, wenn wir nicht von vornherein ausschließen können, dass es zumindest manche sind, alle diese Hinsichten ausdrücklich offen halten. Genau dazu fordert unsere

These vom Zueinander-Passen des Sehens und des Gesehenen auf. Sie dringt darauf, die Art des Sehens derart näher zu bestimmen, dass der Sinn des Wortes „sehen" mit dem des Ausdrucks für das Gesehene intern verbunden ist.

Diese Forderung ist radikal. Denn sie scheint die *Einheit* des Begriffs zu opfern. – Nun stimmt es zwar, dass „Sehen" nicht in dem Sinne mehrdeutig ist wie „Bank". Aber damit ist nicht schon über die *Art* der Einheit alles dessen entschieden, was unter einen Begriff fällt. Sehen muss weder immer gleich sein, egal, was gesehen wird, noch muss es etwas sein, das, wenn es das Sehen von X ist, nichts mit dem zu tun hat, was man „Sehen von Y" nennt oder nennen könnte. (Hier bietet sich offenbar Wittgensteins Konzept der Familienähnlichkeit an.) Das heißt aber nicht, das Gute und sein Sehen stünden in der Art von Beziehung zueinander, in der etwa Tische und ihr Sehen zueinander stehen – also in einer Beziehung, die u. a. mit einem (bestimmten) Begriff der Perspektive verbunden ist. Auch diese Angleichung kann wesentliche Unterschiede verschleiern. Wir müssen also auf das Gute und sein Sehen selbst eingehen.

3 Das Gute und seine Sichtbarkeit

7. Wenn wir davon ausgehen, wie wir das Wort „gut„ verwenden, dann kann kein Zweifel daran bestehen, dass es eine *Formenvielfalt* des Guten gibt. Hier ist ganz kurzer Ausschnitt aus seiner Inventarliste:

- gute Messer, Kellen, Hobel usw. – kurz: gute Werkzeuge,
- gute Köche, Maurer, Tischler usw. – oder: gute Handwerker,
- gute Arbeit,
- gutes Essen, gute Häuser, gute Tische usw. – also: gute Arbeitsprodukte,
- gute Hunde, Kühe, Pferde usw. – allgemein: gute Haustiere,
- gute Tierzüchter,
- gute Augen, Herzen, Bänder und Sehnen usw. – d. h.: gute Organe,
- gute Gesundheit,
- ein gutes Leben.

Genau genommen, gibt es nicht die Inventarliste, aber manche Listen bieten sich durchaus an. Auch gibt es Zusammenhänge zwischen einzelnen Mitgliedern verschiedener Reihen: gute Messer, gute Köche, gute Arbeit, gutes Essen gehören zusammen – die gute Gesundheit schließt sich hier an und vielleicht steht auch ein gutes Leben in dieser Linie. (In einer anderen Darstellungsweise wäre dies also eine Reihe.) Wir haben somit, wie schon beim Sehen, beides zugleich: Vielfalt und Einheit.

Wenn gesagt wurde, dass vielleicht auch der Begriff des guten Lebens in die Reihe gehört, die mit dem guten Messer beginnt und sich über den guten Koch fortsetzt, dann sollte das „vielleicht" andeuten, dass die Unterschiede zwischen den Kriterien, die bei der Beurteilung von etwas als gut ins Spiel kommen, nicht einfach nur in dem Sinne verschieden sind, wie ein Messer und eine Schere einerseits und ein Messer und das Kochen andererseits verschieden sind, sondern auf eine noch andere Weise verschieden sein können. Anders gesagt, es ist nicht offensichtlich unsinnig, zu sagen, ein gutes Leben habe nichts mit guter Gesundheit, zu tun, selbst wenn es tatsächlich nicht richtig sein sollte.

Entspricht dieser Vielfalt des Guten auch tatsächlich eine des *Sehens*?

8. Einer Vielfalt des Guten kann überhaupt nur dann eine des Sehens entsprechen, wenn zumindest manches Gute sichtbar ist. Wie steht es also um die Sichtbarkeit (der Formen) des Guten?

Nehmen wir Messer. Angenommen, wir wissen, wann ein Messer (dieser oder jener Art) ein gutes Messer ist. Wenn wir nun ein (solches) Messer sehen – *schließen* wir dann auf seine Güte? *Manchmal* ja; etwa dann, wenn Zweifel „in der Luft liegen". Aber wir tun es nicht immer, nicht in der Regel. Wäre es anders, könnten wir dann einen Fehler der Art machen, wie ihn ein Arzt macht, der zwar weiß, dass rote Flecken dieser Art Masern bedeuten, solche roten Flecken auch sieht, aber trotzdem nicht auf Masern schließt? Wenn man von roten Flecken auf Masern schließt, dann schließt man jedenfalls von einer Sache auf eine andere. In diesem Sinne aber sind ein Messer und seine Güte kein Zweierlei. Es gibt nicht dort das Messer und hier sein Gutes, so dass, wenn etwa alle Messer zerstört wären, es immerhin noch ihr Gutes geben könnte. – Muss man das Gute an einem Messer dann nicht einfach sehen können?

9. Man könnte einwenden, dass man deshalb nicht wirklich sehen kann, ob ein Messer gut ist, weil ein Messer, das gut *aussieht*, nicht gut sein muss. Das Kriterium für die Güte eines Messers ist nicht sein Aussehen, es ist sein *Funktionieren*.

Richtig ist, dass wir die Art, auf die ein Messer funktioniert, oder: sich gebrauchen lässt, dem Messer, wenn es einfach vor uns liegt, nicht in jedem Fall ansehen können. Angenommen, ein Messer muss für einen bestimmten Zweck besonders elastisch sein. Aber wenn es nur auf dem Tisch herumliegt, im Unterschied dazu, dass es gerade gebraucht wird, ist es häufig schwer, wenn nicht gar unmöglich, zu sehen, ob ein Messer elastisch ist. Wenn man es dagegen zu biegen versucht, kann man es durchaus sehen. (Wie auch nicht, wenn man es doch zu biegen versuchen kann, *um zu sehen*, ob es elastisch ist!)

Es können auch weitere Kriterien des Guten ins Spiel kommen, deren Vorliegen nicht durch die Betrachtung eines (einmaligen) Gebrauchs ermittelt werden kann, wie etwa das Kriterium der Haltbarkeit oder der Zahl der Benutzungen, über die hin ein Messer (gut) zu gebrauchen sein muss, um ein gutes Messer zu sein. Aber dies alles spricht höchstens dafür, dass es schwierig sein kann, zu sehen, ob ein Messer ein gutes Messer ist, aber es reicht nicht für die These, dass man dies *niemals* sehen könnte. Und warum auch sollte man den Begriff des Gebrauchs von dem des Sichtbaren völlig trennen wollen, wenn das Vormachen des Umgangs mit einem Werkzeug ein Standardfall des Lehrens seines Gebrauchs sein kann? „Wenn du wissen willst, wie man es macht, sieh' zu!" Und wie man es macht, ist, wie man es richtig macht – wie man es gut macht. Und hierin liegt schon inbegriffen, dass man nicht nur über das Werkzeug und den Umgang mit ihm, sondern auch über die Güte des Handelnden als solchem durch Hinsehen Aufschluss erlangen kann. Ebenso sehen, um noch eine dritte Form des Guten zu erwähnen, Pferdehalter, ob ein Pferd ein gutes Pferd ist oder nicht. Natürlich kann ein Pferd, das gut aussieht, einen verborgenen Mangel haben, der verhindert, dass es ein gutes Pferd ist. Aber – der Mangel ist dann eben *verborgen*, das heißt: nicht *gleich*, nicht *auf den ersten Blick zu sehen*, aber eben nicht: unter *keinen* Umständen sichtbar. Wenn nichts von dem verborgen ist, worauf es für die Güte ankommt, sollte man sie dann nicht sehen können? Auch ein Pferd kann nur wie ein gutes Pferd aussehen, wenn man, im Prinzip, auch sehen kann, dass es gut ist.

4 Das Sehorgan

10. *Womit* aber ist das Gute zu sehen, soweit es sichtbar ist? – „Mit dem Auge!", möchte man natürlich sagen. Nur, was an den wissenschaftlichen Untersuchungen des Auges hat mit dem Sehen des Guten in dem Sinne zu tun, in dem vieles an solchen Untersuchungen etwas mit dem Sehen von Farben zu tun hat? Selbst wenn das Sehen nicht mit dem zusammenfällt, was man kausal untersuchen kann, bleibt immer noch, dass auf Nichts auch nichts supervenieren kann. – Entgegen dem Anschein, geht es hier nicht um ein, sondern um zwei Dinge: um das *Organ* und um die *Struktur* des Sehens.

11. Beginnen wir mit der Frage nach der Struktur. Die allgemeine Form des Einwandes, der sich auf Untersuchungen der kausalen Mechanismen des Farbsehens beruft, ist die folgende. Vorausgesetzt, man sieht, dass etwas gut ist, sieht man eigentlich *Zweierlei*:

(a) dasjenige, was gut ist, und
(b) dass es gut ist (oder auch: das Gute an dem, was gut ist).

Soweit Farbiges betroffen ist, so die Idee, sagen uns u. a. physiologische Untersuchungen, wie wir dieses sehen (können). Soweit uns physiologische Untersuchungen über das Sehen Aufschluss geben, geben sie uns aber gerade keinen, wenn es um das Sehen von Gutem geht.

Dieses Argument lässt das Gute bestenfalls als einen Zusatz zu dem erscheinen, was gut ist und für sich bestehen können soll. Man kann also, im Prinzip, etwas sehen, ohne zu sehen, ob es gut ist. (Soweit „etwas sehen" bedeutet „etwas kennen", kann man das Fragliche also kennen, ohne seine Güte zu kennen.) Entgegen dem Anschein, den unsere Sprechweise uns nahe legt, versteht sich dies alles jedoch nicht von selbst. – Bevor wir jedoch hierauf näher eingehen, werfen wir einen Blick auf das Organ. Lassen wir die Frage beiseite, ob wirklich immer ein Organ im Spiel ist, wenn man etwas sieht. Die Frage ist, ob, selbst wenn, oder soweit, es stimmt, dass wir ohne Augen nicht sehen können, dies bedeutet, dass sog. kausale Mechanismen des Sehens den Begriff des Sichtbaren *beschränken*.

Offensichtlich kann man „das Sehen" nur dann physiologisch untersuchen, wenn das zu Untersuchende unabhängig von der Untersuchung als Sehen bestimmt ist. Also ist der Begriff des Auges als Sehorgan zunächst einmal gerade kein physiologischer Begriff. Wie aber ist das Auge als Sehorgan unmittelbar bestimmt? – Unter anderem durch seinen Bezug zum *Sichtbaren*. Dass Sehen immer Farbe entdeckt, heißt auch, dass, wenn es das Auge ist, was diese entdeckt, das Auge durch den Begriff der Farbe *bestimmt* ist. Wenn wir mit dem Auge sehen, dann ist ein Auge das, womit wir Farben sehen. Zu sagen, man könne nur das sehen, was das Auge zu sehen erlaubt, ist aber unsinnig, wenn man das Auge nach dem *Gesehenen* bestimmt. Man könnte höchstens sagen, beide sind dem jeweils Andern das Maß – eine Art prästabilierter Harmonie.

Zwar spricht auch der Physiologe vom Auge als Sehorgan. Aber was interessiert ihn dabei? – Die physiologische Beschaffenheit des Auges, sein Verhalten unter verschiedenen Bedingungen, unter denen wir sagen, dass wir dies oder jenes sehen (können). Der Physiologe setzt, wie gesagt, unsern gewöhnlichen Begriff des Sehens voraus – den Begriff, den es schon gab, als noch niemand eine Ahnung von Physiologie oder physikalischer Optik hatte.

Angenommen, aus dem Keller dringt ein seltsames Geräusch, und ich steige hinab, um nachzusehen, woher es kommt. Zurückkehrend werde ich gefragt, was ich gesehen habe. Ich antworte: „Nichts, das Licht ging nicht an, ich muss noch mal runter, mit der Taschenlampe." – Kann man hier sagen, ich hätte nicht gesehen, wie man sagen kann, ich habe nichts gesehen? Vielleicht nicht; aber selbst dann ist dieser Fall von dem verschieden, in dem ich bei hellem Tageslicht in der Schublade nachsehe, was drin ist, und auf die Frage, was ich gesehen habe, antworte: „Nichts." Insofern scheint jenes „Nichts!" aus dem ersten Fall die Grenze des Sehens zu markieren. Dagegen ist ein physiologischer Zustand als solcher so gut wie jeder andere. Und wieso sollten unsere Augen, wenn wir nicht(s) sehen, nicht in einem bestimmten physiologischen Zustand sein? (Wie ja auch eine Saite mit der Frequenz 10000 schwingen kann und vielleicht gar unser Trommelfell mit ihr, auch wenn wir nichts hören.)

12. Die Idee, dass wir mit dem Auge sehen, ist daher so gefährlich, wie sie natürlich ist. Sie ist gefährlich, insofern sie uns einen gewissermaßen physiologischen Begriff des Sehens nahe legt. Ein solcher lässt es rätselhaft erscheinen, dass wir Dinge sehen können, die mit dem Licht im physikalischen Sinn nichts zu tun haben. Aber diese *physiologische* Betrachtungsweise des Sehens verzerrt die *begriffliche* Lage. Entzerrt, ist nichts mehr rätselhaft an der Tatsache, dass im Auge nichts geschieht, das physiologisch derart verschieden ist, wie es verschieden ist, wenn wir einmal Farben und einmal Freude, einmal Freude und einmal das Böse sehen, so dass man, wenn man weiß, was im Auge geschieht, darauf schließen kann, was sein Besitzer gerade sieht.

5 Gutes und gutes Sehen

13. Jemand kann in dem Sinn ein Messer sehen, ohne zu wissen, was ein Messer ist, als er später, wenn er es weiß, zu Recht sagen darf, er habe *damals* ein Messer gesehen. Ja, man könnte auch sagen, Sehen bestehe im *primitivsten* Fall darin, auf optische Reize derart zu reagieren, dass zwischen Reizwandel und Reaktionswandel m. o. w. systematische Korrelationen bestehen. Aber dadurch, dass es primitiv ist, deutet es begrifflich schon über sich selbst hinaus: auf das *normale* Sehen. Das bedeutet jedoch nicht, dass jenes in diesem als *Element* enthalten wäre.

Dieser Fall ähnelt dem, in welchem man sagt, dass man Gegenstände durch ein verzerrendes oder sonst wie störendes Medium sieht, obwohl sie unerkennbar sind. Frage: Wann sagt man das? Antwort: Man sagt es, wenn man schon weiß, oder doch Gründe hat für die Überzeugung, dass es sich um Dinge handelt, die man (normal) sehen würde, läge zwischen ihnen und uns *kein* störendes Medium. – Wer aber kann so etwas sagen? Nur der, der sich hierin auskennt.

Zwar kann jemand ein Messer sehen, ohne zu wissen, was ein Messer ist; aber das kann er nur dank der Tatsache, dass es jemanden (besser: viele) gibt, der (die) es weiß (wissen). *X* zu sehen, ohne zu wissen, was *X* ist, ist ein parasitärer Fall des

Sehens von X. Aber es wäre Unsinn, zu sagen, in jedem Normalen stecke etwas Parasitäres.

Das deutet auch schon darauf hin, was es heißt, zu wissen, was X ist. – Es heißt, kurz gesagt, gewisse Dinge mit X tun zu können. Der Punkt, auf den es jetzt ankommt, ist also der Zusammenhang des Sehens mit bestimmten Handlungen. Ihr Beherrschen kennzeichnet auch den Sehenden. Nicht nur besteht zwischen dem Sehen und dem Gesehenen eine Art von „Entsprechung", sondern diese erstreckt sich bis hinab zum Sehenden, vermittelt über die Einbettung des Sehens in ein bestimmtes Handeln.

Vom Sehen zu sagen, es sei in dieses oder jenes Handeln eingebettet, soll zunächst einmal dem Sehen selbst den Charakter des Handelns absprechen. Natürlich kann man sich vorstellen, dass Menschen sich gegenseitig beschreiben, was sie sehen. Aber das nimmt der Aufforderung, jemand solle sagen, was er sieht – und zwar rein das, was er sieht, nicht das, was er beobachtet, was ihm in die Augen springt o. ä., nichts von ihrer Seltsamkeit. Es ist völlig unklar, wozu jene Aufforderung eigentlich auffordert, was also als ihre Befolgung zu gelten hat. Natürlich kann man (oft, in der Regel) beschreiben, was man sieht. Man beschreibt, was man zu sehen gelernt hat, und zwar im Zusammenhang dieses oder jenes Tuns. Das macht das Sehen nicht schon zu einem *Handeln*; aber, weil nicht dieses, auch nicht einfach zu einem *Widerfahrnis*.

Zwar wird bei Sehtests für Führerscheine unser Sehvermögen geprüft, und zwar an Fällen seiner Ausübung. Aber dieses Vermögen ist nicht von der Art des Vermögens, Aufgaben der Höheren Mathematik zu lösen. Sehen können heißt nicht, eine Fähigkeit *beherrschen*. Wer einen Sehtest nicht besteht, bekommt keine zweite Chance, verbunden mit der Aufforderung, vorher etwas zu üben oder es überhaupt erst mal zu lernen. Viel eher schickt man ihn zum Augenarzt. Insofern ähnelt die Sehfähigkeit der Fähigkeit eines (normalen, gesunden) Fußes, gewissem seitlichem Druck zu widerstehen, also nicht umzuknicken. Man nimmt sich Sehen auch nicht vor, und jemandes Augen können einem alle möglichen Überraschungen bereiten.

Was der, welcher in der Lage ist, auf bestimmte Reize so zu reagieren, dass wir sagen würden, er sehe, allein dadurch nicht auch schon kann, ist dies: *hin*sehen, *nach-*, *zu-*, sich *um*sehen usw. Sehen einerseits und Hin-, Zusehen usw. andererseits sind von verschiedener Art. Zu sagen, jemand habe gesehen, dass das Messer auf dem Tisch liegt, heißt nicht dasselbe, wie zu sagen, er habe dies durch Nachsehen herausgefunden.

Eine Form von Hin-, Nach- und Zusehen ist das Beobachten. Wenn „[m]an beobachtet, um zu sehen, was man nicht sähe, wenn man nicht beobachtet[e]", wie Wittgenstein (BÜF, Teil III, §326) sagt, sieht man dann, *indem* man beobachtet, oder *dadurch, dass* man es tut? Sicher sieht man nicht einfach *auf Grund dessen, dass* man beobachtet, oder *in Folge von ...*, wenn man bei diesen Wendungen an

Fälle wie den denkt, dass man Kopfschmerzen hat auf Grund dessen, dass man zu viel Sonnenschein ausgesetzt war. Auf *diese* Weise sind Beobachten und Sehen nicht verschieden. Eher schon sind beide auf eine begriffliche Weise verbunden. Denn dass man beobachtet, um zu sehen, was man nicht sähe, wenn man nicht beobachtete, deutet darauf hin, dass man, sozusagen, nicht einfach nur sieht, wenn man etwas sieht. Wenn Beobachten ein Tun ist, dann hat man also damit, dass man sagt, Sehen sei ein Widerfahrnis, nicht etwas *Falsches* gesagt, sondern etwas von wenig Interesse. Denn man hat etwas Wesentliches in Bezug auf das Sehen weggelassen: *seine Einbettung in ein Tun* (wie es das Beobachten ist). Dann aber fängt das Problem bereits mit dem Stellen der Frage, ob Sehen ein Tun oder ein Widerfahrnis, passiv oder aktiv ist, an, statt erst mit der Suche nach einer Antwort. Denn die Frage nach der Natur des Sehens setzt in dieser Form voraus, dass man das Wesentliche am Sehen durch die Betrachtung des Sehens *für sich* erkennen kann. Es könnte aber durchaus sein, dass das, was man auf diese Weise herausbekommt, nicht mehr ist, als das, was man dadurch herausbekommt, dass man beim Sehen ausschließlich an etwas denkt, was ein Sehtest für Führerscheine testet.

Dass man beobachtet, um zu sehen, was man nicht sähe, wenn man nicht beobachtete, bedeutet natürlich nicht, dass man beim Beobachten, oder auch: während man beobachtet, nur das sehen kann, was zu sehen man das Beobachten anstellte. (Man kann beim Beobachten von Tieren sehen, wie ein Flugzeug abstürzt – eine *Ablenkung*, die „*während* des Beobachtens" geschieht, und möglich ist, weil es so etwas wie ein Beobachten des Flugverkehrs geben könnte.)

Das Gleiche zeigt sich auch in der Form, in der man dafür zur Verantwortung gezogen werden kann, etwas gesehen zu haben, was man nicht hätte sehen dürfen. Wer etwas sieht, was er nicht sehen soll, indem er hinsieht, wo er nicht hinzusehen hat, kann sich nicht damit rausreden, dass nicht (schon) das Hinsehen, sondern (erst) das Sehen der Grund dafür sei, dass er jetzt weiß, was er besser nicht wüsste (oder dies die Form seines Wissens ist), das Sehen aber, im Unterschied zum Hinsehen, etwas ist, das ihm widerfährt. Es stimmt, dass es ihm widerfährt; aber hier handelt es sich, sozusagen, um eine aktive Passivität, ein getanes Widerfahrnis.

14. Wenn wir davon ausgehen, dass „die natürliche Heimat des Begriffs des Sehens" sich in der Weise beschreiben lässt, wie es Wittgensteins Beobachtung andeutet, erscheint die Idee in einem neuen Licht, wonach das, was der, welcher das Messer sah, ohne zu wissen, was ein Messer ist, schließlich erwarb, ein bestimmtes *Verständnis* dessen war, was er zuerst gesehen hatte, und nicht die *Fähigkeit, es zu sehen*. Soweit dies zu der Idee zurückführt, dass wir eigentlich nur bestimmte Farbkomplexe (oder gar nur „Lichtmuster") sehen, die wir dann in bestimmter Weise deuten, selbst wenn uns dies als solches nicht bewusst werden sollte – „Sein Auge

verändert sich nicht, nur sein Gehirn, soweit sich in diesem Verständnisfähigkeiten materialisieren." – reicht es hin, daran zu erinnern, dass wir jemandem, der ein Messer beschreiben kann, das vor ihm liegt, diese Fähigkeit nicht absprechen würden, wenn sich herausstellen sollte, dass er nicht in der Lage ist, die „in der gleichen Situation vorliegenden manifesten Farbkomplexe" zu beschreiben. Unsere Kriterien dafür, dass jemand jene Fähigkeit hat oder sie ausübt, sind in dieser Hinsicht *unabhängig* von den Kriterien für die andere Fähigkeit und deren Ausübung. Soweit jemandes Vermögen, solche Beschreibungen zu geben, Kriterium dafür ist, was er sieht oder zu sehen vermag, kann man also jenes sehen, ohne dieses sehen können zu müssen.

An dieser Stelle sollte der noch fehlende Schritt beim Erwägen unserer These offensichtlich sein: Handlungen gehen einher mit *Standards der Korrektheit*. Solchen Standards kann man mehr oder weniger gerecht werden. Man kann eine Handlung mehr oder weniger gut ausüben. Wer eine Handlung (in der Regel) gut ausübt, *beherrscht* sie. Dazu gehört, zumindest in einigen Fällen, zu erkennen, wie es um den Wert der Dinge, die ins Handeln eingehen, bestellt ist. Nur ein guter Koch, und was ihm in Hinsicht auf Messer verwandt ist, sieht, alles in allem, ob ein (Küchen-) Messer ein gutes Messer ist oder nicht. Umgekehrt erkennt man, ob jemand ein guter Koch ist, auch daran, ob er die Güte eines solchen Messers sehen kann. Zugleich ist dieses Sehen des Guten in dem Sinne etwas, was über den Sehenden, wie er hier und jetzt beschaffen ist, hinausgeht, also etwas zeitlich Ausgedehntes und Gemeinschaftliches, als nicht alles, was jemand zu sehen glaubt, das ist, was er wirklich sieht. Dass jemand glaubt, ein gutes Messer zu sehen, bedeutet nicht, dass das, was er sieht, wirklich ein gutes Messer ist. Dass jemand, der gut im Sehen ist, glaubt, ein gutes Messer zu sehen, bedeutet also nur insofern, dass das, was er sieht, wirklich ein gutes Messer ist, als sein Gutsein darin besteht, jenem Maßstab zu verkörpern, dessen Existenz nicht allein davon abhängt, dass er ihn verkörpert. Insofern geht sein Gutsein über ihn hinaus, und könnte man fast sagen, er sei insofern gut, als er am Guten teilhat, welches als solches nicht mit seinem Gutsein zusammenfällt, insoweit von ihm unabhängig ist. Man könnte also fast sagen:

Das Gute erkennt sich, in der Verschiedenheit seiner Formen, selbst.

Zwar kommt außerhalb so oder so bestimmter Situationen der Satz „Ich habe etwas Gutes gesehen„ dem Satz gleich „Ich habe etwas gesehen"; und dieser Satz sagt, außer eben in wohlbestimmten Situationen, so gut wie gar nichts. Dass er gar nichts sagt, macht ihn aber nicht einfach bedeutungslos. Sondern aus dem vielen, welches er sagen könnte, greift er nicht auf das „Gemeinte" zu. Er sagt insofern nichts, als er zu viel sagt. Denn das Gute kommt in einer solchen Vielfalt daher, wie es Dinge, Zustände, Eigenschaften, Ereignisse usw. gibt, von denen man sagen kann, sie seien gut. (Und wovon sagen wir das nicht alles!)

15. Aber selbst wenn der Kontext klar macht, um welches Gute es geht, sieht man das Gute nicht so, wie man sieht, dass etwas grün ist. Das Gute ist keine solche Eigenschaft, soweit es überhaupt eine Eigenschaft ist. Wenn es aber auch nicht außerhalb ihrer ist, muss es gewissermaßen die Sache selbst sein.

Damit kann der Wertblinde nicht einfach in Hinsicht auf Werte defizitär, aber sonst völlig in Ordnung, sein. Sondern sein Problem ist so beschaffen, wie es beschaffen sein muss, damit es Sinn macht, dass wir ihm seine Art der Blindheit damit zu überwinden helfen, dass wir ihm die Lage der Dinge schildern, erklären, wie man mit den Dingen umgeht, wozu sie dienen usw. Und genau das tun wir auch!

Wie reagieren wir zum Beispiel auf jemanden, der den moralischen Wert eines Geschehens nicht erkennt? Wir lenken etwa sein Augenmerk auf die Folgen der Geschehnisse für Andere, auf die Motive, aus denen heraus gehandelt wurde, berufen uns auf die allgemein anerkannten Verhaltensmaßstäbe u. a. m. – alles m. o. w. sichtbare Dinge. Unter diesen ist nichts, von dem wir sagen würden, es sei das Gute an diesem Tun. Aber dennoch sagen wir ihm alles, was in Bezug auf das Gute eines Tuns zu sagen ist. (Ähnlich erklären wir jemandem, was ein gutes Messer ist, indem wir ihm Messer zeigen, ihren Gebrauch vorführen usw. – darunter ist vieles m. o. w. sichtbar. Aber es ist nichts dabei, von dem wir sagen würden, es sei das Gute an diesem Messer, obwohl nichts fehlen muss.)

Die Besonderheit des Moralischen zeigt sich auch darin, dass wir zuweilen über Menschen, die Verwerfliches tun, sagen, sie wüssten nicht, was sie tun, statt: sie wüssten nicht, dass das, was sie tun, verwerflich ist. Und doch wollen wir nicht sagen, jene handelten unfreiwillig – wie ich unfreiwillig eine Explosion verursachen kann, wenn ich, bevor ich das Licht anschalte, nicht bemerke, dass Gas im Zimmer ist. Den Unwissenden, den Wertblinden zeichnet eine andere Art von Defekt aus, ein Defekt, der insofern *grundsätzlicher* ist, als er ihren Status als Handelnde selbst betrifft. Sie gleichen eher tollwütigen Hunden, die auch nicht einfach freiwillig beißen, dennoch aber bestraft werden. (Aber nicht müssen – was die Aufforderung, ihnen zu vergeben, bedeutsam macht.) Was ihnen fehlt, ist das, was, wenn sie es hätten, ihnen erlauben würde, ein ganz anderes Leben zu führen, und sich in einem solchen Leben, auch zeigte, und zwar ein Leben, das selbst dann ein ganz anderes wäre, wenn es sich in nichts von dem vorherigen unterscheiden würde. Es wäre ein Leben, welches in einem anderen *Geist* geführt würde, und daran ändert auch die Tatsache nichts, dass ein Leben, welches in diesem (neuen, oder überhaupt einem) Geist geführt wird, häufig doch anders aussieht.

6 Rückrufaktion?

16. Mit der vorstehenden Betrachtung war ausdrücklich das Ziel verbunden, eine gewisse These zu *erwägen*, nicht, sie zu *begründen*. Das lässt dem Zweifel ausdrücklich Raum: Wie steht es um die Möglichkeit der Täuschung bei der Wahrnehmung von Gutem? Was ist, wenn etwas in mehr als einer Hinsicht einem Maßstab des Guten unterliegt, ein Bankraub etwa, eine üble Sache, brillant ausgeführt wird? Natürlich kann man sagen, dass man ihn so, aber auch anders betrachten kann. Aber das heißt nicht, dass man ihn doppelt sieht. Was schließlich ist, wenn zwei Wertsysteme aufeinander stoßen – in welchem Sinne kann man hier sagen, das Gute nicht sehen zu können sei ein Defekt?

Das Hauptbedenken aber ist von anderer Art. Auf die Idee des Sehens des Guten kamen wir *ursprünglich* durch eine Art Ausschlussverfahren: Wenn wir nicht auf das Gute schließen, wie könnten wir *sonst* von ihm wissen, soweit wir von ihm wissen können, *außer* durch das Wahrnehmen, etwa durch das Sehen des Guten. Zwar ist ein Ausschlussverfahren oft nützlich, in unserem Fall geht es aber mit der Suche nach einer *Begründung* unserer Sprechweise einher. Eine solche scheint nötig zu sein, eben weil sich sowohl etwas dagegen sträubt, zu sagen, wir würden aus dem Sehen einer Sache auf ihre Güte schließen, als auch gegen die These, wir würden das Gute sehen. Wie aber könnte es eine Begründung geben – und dann auch einer bedürfen – wenn wir uns nur auf die normale Sprache berufen können, also alle Argumente philosophisch gleichwertig sind?[1]

Bibliographie

Heidegger, Martin (1949): *Sein und Zeit. Erste Hälfte*. 6. Aufl., Tübingen: Neomarius.
Wittgenstein, Ludwig (1984): *Werkausgabe Band 8. Bemerkungen über die Farben / Über Gewißheit / Zettel / Vermischte Bemerkungen*. Frankfurt am Main: Suhrkamp, 7 – 112.

---

[1] Das Vorstehende ist eine Skizze des Skeletts eines größeren Ganzen gleichen Titels. Dort werden auch die Positionen, auf die hier nur – sei es kritisch, sei es konstruktiv – angespielt wird, namentlich genannt und diskutiert. Dass die hier vorliegenden Überlegungen Wittgenstein weit mehr zu verdanken haben als nur die Bemerkung aus *Bemerkungen über die Farben* sollte offensichtlich sein. Mein Dank gilt Raymond Geuss (Cambridge), Fabian Freyenhagen (Essex) und Werner Wolff (Berlin).

Timm Lampert
Wittgenstein's Conjecture

Abstract: In two letters to Russell from 1913, Wittgenstein conjectured that first-order logic is decidable. His conjecture was based on his conviction that a decision procedure amounts to an equivalence transformation that converts initial formulas into ideal symbols of a proper notation that provides criteria for deciding the logical properties of the initial formulas. According to Wittgenstein, logical properties are formal properties that are decidable on the basis of pure manipulations of symbols. This understanding of logical properties (such as provability or logical truth/falsehood) is independent of and prior to any interpretation or application of logic. Wittgenstein's conception of logic is incompatible with the undecidability proof of Church and Turing from 1936. Thus, Wittgenstein's conjecture and his understanding of logic appear to be refuted. This paper argues that Wittgenstein did not drew this conclusion and it explains why he never withdrew his conjecture.

1 The Conjecture

In a November 1913 letter to Russell, Wittgenstein conjectured that first-order logic (FOL) is decidable, stating that "there is one Method of proving or disproving all logical prop[osition]s" (CL: 54). He illustrated this by means of his ab-notation for propositional logic and conjectured that it "must also apply" to the "Theory of app[arent] var[iable]s", i.e., FOL (CL: 54). NL: 95f. presents the following examples of the simplest quantified expressions:

$$\forall x \varphi x : \quad a - \forall x - a - \varphi x - b - \exists x - b \qquad (1)$$

$$\exists x \varphi x : \quad a - \exists x - a - \varphi x - b - \forall x - b \qquad (2)$$

Wittgenstein emphasized that his conjecture depended not on the ab-notation itself (cf. CL: 52) but rather on his conviction that "logical truth" and "logical falsehood" are identifiable by symbolic properties of their representations in a proper notation.

For propositional logic, the well-known method of truth tables may serve as an illustration. While in the "old notation" (NL: 93d), i.e., "Russell's method of symbolising" (NL: 102c), tautologies (= logically true formulas) have no common formal property that can serve as their identity criterion, the method of truth tables

---
Timm Lampert, Humboldt-Universität, Berlin, lampertt@staff.hu-berlin.de

DOI 10.1515/9783110657883-32

makes it possible to identify tautologies by the common property that the value "False" does not occur in the column of the main sentential connective.

Wittgenstein, however, envisaged his *ab*-notation, rather than truth tables, as a decision procedure. The main reason for this is that the *ab*-notation is designed to be generalised to the whole realm of FOL. The *ab*-notation does not merely combine truth values of different *types* of propositional functions; instead, it combines all poles of different *tokens* that occur in the initial formula. Thus, it does not presume any internal dependencies of partial expressions. This becomes important as soon as first-order formulas are considered and internal relations become more complex (cf. Lampert 2017a for details). The reference to poles of tokens is the reason why a propositional tautology is identified in the *ab*-notation by the property that its outermost *b*-pole is connected to opposite innermost poles of one and the same propositional variable; cf. figure 1.[1]

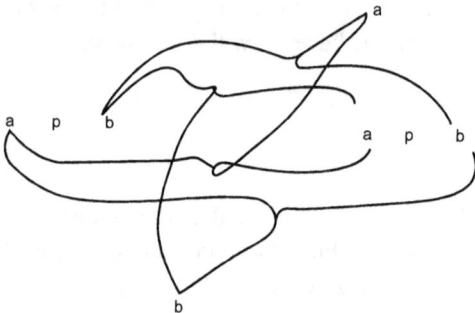

Fig. 1: Wittgenstein's *ab*-diagram of $p \equiv p$ (CL: 57)

Wittgenstein was unclear on how to realize his understanding of a decision procedure for identity, but he had "NO doubt that it must be possible to find such a notation" (CL: 60) for the whole realm of quantified formulas.

Wittgenstein envisaged a method of proving and disproving logical formulas that would differ from both (i) an automated proof search within a calculus of inference rules and (ii) an automated search of models and counter-models with finite domains. In case (i), the search may go on forever if the initial formula is not provable. In case (ii), the search is restricted to finite models or counter-models and, therefore, does not yield a decision for formulas with only infinite models or counter-models. Such automated search methods do not support the intuition

---

[1] Republished with permission of John Wiley and Sons Inc; permission conveyed through Copyright Clearance Center, Inc.

that FOL is decidable. In fact, modern logic engines do not decide formulas that are neither theorems nor contradictions and have only infinite models or countermodels.[2]

According to Wittgenstein, however, a decision procedure is an equivalence transformation from formulas written in a conventional notation that does not provide criteria for identifying logical properties into ideal symbols of a proper notation that does provide such criteria. Such a method is essentially the same in the cases of both proving and disproving a formula, and it is independent of any semantics or model theory that inevitably faces the problem of infinite domains once quantification becomes involved.

Of course, without being in possession of such a procedure, one can question whether it exists. Nevertheless, given that an automated proof search can be conceived as a reduction to expressions of the form $A \vee \neg A$ or $A \wedge \neg A$ in the case of theorems or contradictions, one might understand the conviction or intuition that it is possible to fully algorithmize an equivalence transformation that also identifies non-theorems or satisfiable formulas by means of syntactic criteria. In addition to truth tables and the ab-notation for propositional logic, the translation of propositional formulas into disjunctive normal forms (DNFs) provides another immediate and straightforward illustration of a decision procedure of this kind: an initial formula is contradictory iff each disjunct contains both an atomic formula A and its negation $\neg A$. It is rather simple to extend this kind of decision procedure to fragments of FOL that are known to be decidable; cf. Lampert 2017a for Wittgenstein's ab-notation and Lampert 2017b for an equivalence procedure that applies to all FOL formulas that are reducible to FOLDNFs (= DNFs of anti-prenex FOL formulas in negation normal form) that do not contain $\vee$ within the scope of universal quantifiers. In this latter case, a formula is contradictory iff each disjunct contains a "unifiable pair of literals". This, in turn, can be decided by generating subformulas containing only two literals from a disjunct and deciding upon their contradictoriness; cf. Lampert 2017b for details.

Every attempt to generalize decision procedures for fragments of FOL is confronted with the problem that any complete calculus for FOL implies a rule that increases complexity. The problem is to identify criteria for non-provability if complexity increases during the course of an automated proof search. However, this does not necessarily mean that identity criteria are missing in more complex cases. A natural idea for a termination criterion in the case of non-provability in more complex cases that is consistent with Wittgenstein's views on induction in his

---

[2] Cf., e.g., SYO635+1.p, SYO636+1.p, SYO637+1.p and SYO638+1.p from the TPTP library: http://tptp.cs.miami.edu/~tptp/cgi-bin/SystemOnTPTP.

middle period (cf. section 3, page 524) is to determine the impossibility of finding a proof along proof paths by detecting loops ("visible recursions") in the proof search given suitable equivalence transformations.

Thus, Wittgenstein's remarks in his letters to Russell have the ingredients of a significant conjecture: (i) a well-defined problem (the "Entscheidungsproblem"); (ii) a method for its solution (equivalence transformation to reduce formulas to ideal symbols); (iii) paradigmatic, partial solutions (the ab-notation for propositional logic and fragments of FOL, or, more generally, the reduction to FOLDNFs of fragments of FOL); and (iv) the conviction that partial solutions can be generalized because the decision problem is nothing but the problem of defining an algorithm for symbolic manipulation such that one is left with symbols that provide criteria for deciding the logical properties of the initial formulas.

According to Wittgenstein, a distinctive feature of formal properties, such as properties of mathematical or logical expression, is that they can be represented and identified by symbolic properties of a proper notation. In this respect, they differ from "material properties", which can be represented by propositional functions within a logical symbolism (cf. TLP 4.126). The "confusion" between material and formal properties "pervades the whole of traditional [i.e. old, T. L.] logic" (TLP 4.126) since it treats formal properties as material ones and represents them by propositional functions. The failure to distinguish between material and formal properties is the fundamental mistake of what Wittgenstein calls the "old" logic (TLP 4.126), i.e., mathematical logic in the vein of Frege and Russell, as opposed to his "new" logic.

It has often been noted (cf., e.g., the editor's comment in CL: 52[3]; Landini 2007: 112 – 118,[4]; and Potter 2009: 181 – 183) that Wittgenstein's conjecture and its implied understanding of logic are refuted by the proofs of Church and Turing from 1936. From the perspective of modern mathematics and mathematical logic, the undecidability proof of FOL is the most obvious and stringent objection to Wittgenstein's understanding of logic and logical proof. In the following, I neither dispute this objection nor argue for or even discuss the systematic value of Wittgenstein's point of view. Instead, I argue that Wittgenstein never relinquished his early conviction that the formal properties of logic and mathematics are not properly representable *within* a logical symbolism. Instead, he acknowledges only symbolic identity criteria of an ideal notation as providing the proper representa-

---

[3] "It is [...] interesting that he [Wittgenstein] was looking for a decision method for the whole realm of logical truth. This problem, as we now know, cannot be solved."

[4] "The undecidability of quantification theory is a significant blow to Wittgenstein's conception of logic. [...] it undermines Wittgenstein's hope of finding a notation in which all and only logical equivalents have one and the same representation."

tion of formal properties. This is the basic foundation of Wittgenstein's philosophy of logic and mathematics throughout his life and, as I will show, the reason why he never withdrew his conjecture.

2 The *Tractatus*

Given the significance that Wittgenstein attributed to the decision problem in 1913 and to his *ab*-notation as the framework for solving this problem, it seems surprising that the *Tractatus* does not contain any hint regarding how to decide first-order formulas. Instead, Wittgenstein explicitly restricted his method to "cases where no generality-sign occurs" (TLP 6.1203). However, this does not mean that he doubted the feasibility of defining a decision procedure for FOL.

First of all, Wittgenstein was never interested in engaging in the logical business of working out a decision procedure. Furthermore, the intention of the *Tractatus* was not to present the technical details of a decision procedure for logic. Remark 6.1203 was inserted into TS 202 on a separate sheet as late as 1919, and TS 203 and TS 204 do not invoke it. This remark does not extend significantly beyond Wittgenstein's *ab*-notation from 1913. It serves merely to illustrate Wittgenstein's idea regarding the identification of tautologies by means of a decision criterion for a proper notation in the simple case of propositional logic. It is this idea of deciding logical properties based on criteria for a proper notation that is essential to the logic of the *Tractatus*, not the details of its realization in more complex cases. The fact that Wittgenstein restricted the illustration of his method to propositional logic in TLP 6.1203 does not mean that he no longer believed that it could be generalized. He simply did not bother to do so. Moreover, he never maintained that his *ab*-notation is necessary to do this work. TLP 5.1311 and 6.1201 illustrate his idea of identifying logical properties by means of equivalence transformations without alluding to the *ab*-notation, and they relate to formulas of propositional logic as well as to quantified formulas. His conviction is based on the general idea of a decision procedure in the form of an equivalence transformation for converting initial formulas into ideal symbols, not on any specific method of realizing this idea.

Whereas Wittgenstein referred to the logic of the *Principia Mathematica* (PM) in his 1913 letter to Russell, the logic of the *Tractatus* deviates in some respects from the usual understanding of FOL as elaborated in PM. First of all, Wittgenstein rejected the use of identity as a relation between objects in the *Tractatus* (TLP 5.5301). Consequently, he did not permit the use of the identity sign as a primitive symbol in logic (TLP 5.53). Therefore, pure identity statements such as $\forall x \, x = x$

do not fall within the scope of FOL according to the *Tractatus* (TLP 5.534). Nevertheless, logic implies quantifiers and the power to express indefinite quantifiers such as "some" as well as definite quantifiers such as "exactly one" according to the *Tractatus*. Russell showed how to express quantified expressions of this sort within FOL with identity. In doing so, he presumed the usual inclusive reading of quantifiers: $\exists x \exists y (Fx \wedge Fy)$, thus, means "At least one x and at least one (*the same or different*) y such that x is F and y is F", which is equivalent to $\exists x Fx$. Therefore, to express "There are at least two objects that satisfy F", Russell was obliged to introduce identity due to the inclusive reading of quantifiers: $\exists x \exists y (Fx \wedge Fy \wedge x \neq y)$. By contrast, to apply his understanding of logic to identity, Wittgenstein deviated from the usual inclusive reading of bound variables in favour of an exclusive reading that permits the elimination of identity from a proper notation of logic (TLP 5.531ff.). According to his exclusive reading, $\exists x \exists y (Fx \wedge Fy)$ reads "At least one x and at least one *different* y such that x is F and y is F" (= "At least two objects satisfy F"). Thus, $\exists x \exists y (Fx \wedge Fy)$ is not equivalent to $\exists x Fx$ according to the exclusive reading. By his exclusive reading, Wittgenstein wanted to eliminate the need to invoke identity to express quantified concepts such as "at least two" or "exactly one".

This reductionism represents a significant difference with respect to Wittgenstein's attempt to find some proper notation for identity within his *ab*-notation from 1913. Ultimately, his solution to the problem of representing identity in a proper notation of logic was to abandon it in favour of a different reading of quantifiers. Wittgenstein even went one step further in the *Tractatus* by also advocating for a reductive analysis of quantification (TLP 5.52 and 6 – 6.01). In the *Tractatus*, quantified expressions are analysed as truth functions of atomic propositions, which opens the door for problems of infinity in logic. This is a radical difference with respect to his original *ab*-notation, in which quantifiers are not reduced; cf. NL: 95f. and formulas (1) and (2) on page 515. Later, Wittgenstein called his reduction of quantification to propositional logic "his biggest mistake of the *Tractatus*" (Wright 1982: 152, cf. PG II, §8) and returned to a non-reductive analysis of quantified formulas in which quantifiers are accepted as "primitive" (cf. VW: 165), as is the case in the *ab*-notation for quantification (cf. NL: 95f.).

The logic of the *Tractatus* differs from the usual FOL. The technical details of *Tractarian* logic are still a subject of discussion; cf. Rogers/Wehmeier 2012, Weiss 2017, and Lampert/Säbel 2017. However, it can be stated that Wittgenstein's reductive analysis of identity and quantification in the *Tractatus* is motivated by his analysis of propositions as truth functions of bipolar atomic propositions. By contrast, his conjecture is based on the general conviction that formal properties such as theoremhood or refutability must be decidable in any proper system of logic because they are inherent properties of the structure of propositions that should be revealed by a proper notation. This conviction is independent of the philosophically motivated analysis of propositions. The analysis of quantified expressions as truth functions of atomic propositions does not imply that a decision procedure must remove quantifiers, nor does an exclusive reading of quantifiers imply that Wittgenstein no longer assumed the usual, inclusive FOL to be decidable. One should distinguish the peculiarities of a *Tractarian* conception of logic from Wittgenstein's general claim that any proper system of logic must enable the identification of logical properties by syntactic criteria during the course of equivalence transformations. This claim applies to the usual calculus of FOL in the vein of Frege and Russell as well as to the specific exclusive and/or reductive *Tractarian* conception of logic.

Wittgenstein's conjecture is independent of the inclusive or exclusive reading of the quantifiers. This is evident from the fact that he did not consider these peculiarities when discussing the similarities and differences between his proof method and the axiomatic proof method. He called the axiomatic proof conception of Frege and Russell the "old procedure" (MN: 109e) or the "old conception of logic" (TLP 6.125). This conception relies on "primitive propositions" (TLP 5.43, 6.127f.), i.e., axioms, and "laws of inference" (TLP 5.132) to derive "only tautologies [...] from tautologies" (TLP 6.126, cf. MN: 109e). Wittgenstein accepted the axiomatic method as a method that makes it possible "to give in advance a description of all 'true' logical propositions" (TLP 6.125). In modern terminology, one might say that he did not question that traditional FOL is correct and complete. Given his critique of identity, one should restrict this claim to FOL without identity. However, what mattered to him in discussing the two different proof conceptions was, first and foremost, that the axiomatic proof method was "not at all essential" (TLP 6.126, cf. TLP 5.132). He denied the ability of such a proof conception to justify the logical truth of the derived tautologies: it neither explains why axioms are tautologies (cf. TLP 6.127f) nor why laws of inference preserve logical truth (TLP 5.131). The axiomatic method requires semantic or extra-logical evidence (cf. 6.1271), which Wittgenstein wanted to make superfluous with his "new" proof conception.

The key idea of this proof conception is to identify the logical properties of propositions by criteria related to their proper notation. A proof in accordance with Wittgenstein's proof conception does not consist of inferring a proposition from other propositions. Instead, it consists of converting a proposition into some equivalent ideal expression that allows one to identify its logical properties. That is why "every proposition is its own proof" (TLP 6.165). This does not merely imply that "all the propositions of logic are of equal status" since "it is not the case that some of them are essentially primitive propositions and others are derived propositions" (TLP 6.127, cf. 5.43). Rather, it additionally means that Wittgenstein did not restrict his proof conception only to "logical propositions" (tautologies). Instead, he referred to "*every* proposition"(see TLP 6.165 above; emphasis mine) and stated "that we can actually do without logical propositions; for in a suitable notation we can in fact recognize the formal properties of the propositions by mere inspection of the proposition themselves" (TLP 6.122).

Since his proof conception applies to "every proposition", i.e., every logical formula, and enables the identification of their logical properties, a proof procedure implies a decision procedure, according to Wittgenstein's proof conception. This can be seen from the following: if the mechanical transformation of an initial formula into its ideal expression in a proper notation does not result in the representative expression for all tautologies, this outcome is sufficient to decide that the initial formula is not a tautology. According to Wittgenstein, the logical *properties* of each logical formula are independent of and prior to its internal *relations* to other propositions (cf. TLP 5.131, 6.12). A proof is a reduction of initial formulas to their ideal representative expressions through equivalence transformations. It is this proof *method*, not the outcome or the specific understanding of quantification and identity, that constitutes the crucial difference between Wittgenstein's "new" logic and the "old logic" of Frege and Russell.

The *Tractatus* does not abandon the conviction that in any proper system of FOL, it must be possible to decide the logical properties of formulas by converting the initial formulas into a proper notation. Hence, the idea that motivated Wittgenstein's conjecture as stated in his letter to Russell 1913 is still prominent in the *Tractatus*.

3 The Middle Period: Before 1936

In his middle period, instead of working out his conception of a "new" logic in more detail, Wittgenstein was much more interested in applying his algorithmic understanding of logic to mathematics.

The idea of applying his understanding of logical proofs in terms of equivalence transformations to mathematics was already Wittgenstein's primary concern regarding his conception of mathematics in the *Tractatus*. He called mathematics "a logical method" (6.2, cf. 6.234). In doing so, he argued not for a logicist reduction of mathematics to logic but for analogous conceptions of proof in both logic and mathematics. Similar to a logical proof, a mathematical proof involves manipulating symbols with the aim of identifying mathematical properties by means of the properties of the resulting ideal expressions. Whereas the propositions of logic are tautologies, the propositions of mathematics are equations (TLP 6.22). In both cases, these propositions are meaningless. Wittgenstein called the method of logic for combining propositions into tautologies the "zero-method" (TLP 6.121); it involves identifying the properties of logical implication and logical equivalence by relating single propositions to senseless tautologies. Likewise, Wittgenstein attributed to mathematics "the method of substitution" (TLP 6.24), which involves combining mathematical terms into meaningless (nonsensical) propositions. However, he emphasized that the application of these methods is not necessary for identifying the properties of logical implication/equivalence or mathematical identity since the internal relations of the related expressions follow from their internal (formal) properties (TLP 6.122, 6.126d, 6.1265, 6.23 – 6.2323).

As in logic, "intuition" ("Anschauung", not "self-evidence") is needed in mathematics to solve mathematical problems (TLP 6.233), namely, the intuition that "the process of *calculation* serves to bring about" (TLP 6.2331). This intuition refers to the expressions resulting from logical or mathematical equivalence transformations. Whereas Wittgenstein's *ab*-notation illustrates this intuition in the case of logic, he also conceived of a specific notation for the case of arithmetic. His Ω-notation is designed to support the application of the method of substitution to reduce both sides of an arithmetic equation to identical ideal symbols. TLP 6.241 illustrates this process in the case of the proof of $2 \cdot 2 = 4$, in which both $2 \cdot 2$ and 4 are reduced to $\Omega'\Omega'\Omega'\Omega'x$.[5]

Wittgenstein seamlessly evolved from his *Tractarian* proof conception to the work of his middle period. Since he abandoned the *Tractarian* reductive analysis of quantification in his middle period, he strengthened his claim that the axiomatic proof method of Frege and Russell and his own notation both "achieve the same result" (WVC: 92) in logic. He focused not on any difference in the results within FOL but on the irrelevance of the axiomatic method. According to Wittgenstein, this method is not essential since it is similarly possible to identify tautologies "in my notation" (WVC: 92). As in the *Tractatus*, he concluded that tautologies are "indeed quite irrelevant" since his proof method applies to any proposition and is not restricted to inferring theorems from axioms. For this reason, he did not differentiate between proof and decision methods: "That inference is *a priori* means only that syntax decides whether an inference is correct or not. Tautologies are only one way of showing what is syntactical" (WVC: 92).

Yet, in his middle period, Wittgenstein was primarily interested in applying his algorithmic proof conception to mathematics. He thus took up the challenge

---

5 Cf. the following URL for a full computer program that implements Wittgenstein's Ω-notation for the whole realm of rational numbers:

http://www2.cms.hu-berlin.de/newlogic/webMathematica/Logic/q-decide.jsp

Cf. the "Introduction" for a description of the program.

to the *Tractatus* issued by Ramsey, asking how Wittgenstein's "account can be supposed to cover the whole of mathematics" (Ramsey 1923: 475). In no way did this make him doubt his conviction of the decidability of formal properties. Instead, he began by drawing the same analogies he had drawn in the *Tractatus*: "Logic and mathematics are not *based on* axioms [...] [t]he idea that they are involves the error of treating the intuitiveness, the self-evidence, of the fundamental propositions as a criterion for correctness in logic" (PG: 297). Instead, it is intuition ("Anschauung") rather than intuitiveness ("self-evidence") that is relevant; "intuition ["Anschauung"] of symbols" (WVC: 219) is not related to a belief in the truth of axioms, as intuitiveness is. What Wittgenstein had said regarding the logical proof of propositions in TLP 6.1265, he then explicitly related to mathematical propositions in *Philosophical Remarks*: "the completely analysed mathematical proposition is its own proof". (PR: 192) A mathematical proof is "an analysis of the mathematical proposition" (PR: 179) rather than a logical derivation from axioms. As in the *Tractatus*, he compared "the method of tautologies" to the "the proof of an equation", stating that both "[make] evident the agreement between two structures".

In Wittgenstein's middle period, he extended his idea of proving mathematical properties by reducing them to symbolic properties of a proper notation to more sophisticated areas of mathematics, such as induction, impossibility proofs and analysis. He rejected any axiomatic or extensional (set-theoretical) understanding of infinity as well as meta-mathematical impossibility proofs in favour of his algorithmic view. In this view, the intent is to reduce any sort of infinite regression, such as in the case of approximating real numbers, to a "visible recursion" in a proper notation (cf. PR: 187, 243, and, e.g., PR, XIV – XVIII, PG II, §32). He positioned this view in opposition to "arithmetic experiments", in which sequences are generated without making manifest the laws governing their construction (PR: 235; cf. TLP 6.2331). In contrast, he sought only a progression to infinity in accordance with a "recognizable law" (PR: 235). This idea can be traced back to the *Tractarian* concept of operations that generate forms in mathematics or logic through iterative application (TLP 5.25 – 5.254).

In his middle period, Wittgenstein repeatedly challenged the view that propositions concerning formal properties have any decisive mathematical or logical meaning independent of the possibility of deciding them either by means of a given decision procedure or by inventing one. In contrast to the case of Hilbert's formalism, Wittgenstein did not understand symbolic manipulation as something that requires additional interpretation. Instead, it is the decision procedure itself that gives meaning to the formal properties and entities in question. In this respect, Wittgenstein advocated for an algorithmic analysis of meaning in pure mathematics and logic. One might object that we do understand statements about logical formulas or mathematical statements without being able to decide them and that much higher work in mathematics goes beyond Wittgenstein's narrow understanding thereof. The essential point of Wittgenstein's view, however, is not the extent of understanding logic or mathematics in the case that no decision procedure is available. Rather, the crucial point is that he adheres to an ideal of a most decisive and precise understanding of logical or mathematical properties that is based on nothing but equivalence transformation within a symbolism and, thus, does not require any reference to entities outside of that pure symbolism. According to Wittgenstein, no higher mathematics or meta-mathematics can threaten this ideal. Instead, mathematics should always strive for the reduction of its concepts and proofs to the realm of decidability. In a case of conflict, this ideal is the standard of proof and cannot be questioned by proof methods that do not adhere to this ideal.

In the *Tractatus*, Wittgenstein was not content to define logical truth on the basis of general validity (TLP 6.1231). In his view, a full understanding of a logical property implies the ability to identify it through symbolic manipulation. He held this view also for mathematical equations, as he explicitly stated in his middle period:

> We cannot *understand* the equation unless we recognize the connection between its two sides.
>
> Undecidability presupposes that there is, so to speak, a subterranean connection between the two sides; that the bridge *cannot* be made with symbols.
>
> A connection between symbols which exists but cannot be represented by symbolic transformations is a thought that cannot be thought. If the connection is there, then it must be possible to see it. (PR: 212f.)

It was Wittgenstein's belief that pure mathematics and logic are sciences that concern nothing beyond the finite, rule-guided manipulation of signs, thus ruling out the possibility of undecidability:

> Of course, if mathematics were the natural science of infinite extensions of which we can never have exhaustive knowledge, then a question that was in principle undecidable would certainly be conceivable. (PR: 213)

The all-important point is that Wittgenstein's conviction of decidability is rooted in his general understanding of propositions concerning formal properties. This analysis is incompatible with a set theoretical representation of formal properties or any consideration of their decidability within meta-mathematics or any other analysis that goes beyond pure symbolic equivalence transformations.

To understand Wittgenstein's later reaction to undecidability proofs (see the next section), one must consider that his understanding of logic and mathematics is diametrical to the emergence of mathematical logic and the efforts to lay down foundations of mathematics that go beyond what is computable. Before Wittgenstein was ever confronted with Turing's negative solution to the decision problem in 1937, he had already ruled out the possibility of a negative meta-mathematical answer to the decision problem on the basis of his algorithmic understanding of logic:

> Logic isn't metamathematics either; that is, work within the logical calculus can't bring to light essential truths *about* mathematics. Cf. here the "decision problem" and similar topics in modern mathematical logic. (PG: 297)

At the same time, he did not expect any progress to be made in mathematics from the solution to the decision problem since he held that logic and mathematics are analogous in their algorithmic methods but autonomous in their languages, concepts and calculi. It was for that reason that he rejected any logical formalization of mathematics as the basis of the axiomatic method when applied beyond pure logic to mathematics and in undecidability proofs. Whereas many mathematicians feared that a positive solution to the decision problem threatened to make all mathematical questions mechanically solvable without the need for any further human ingenuity, Wittgenstein regarded the decision problem not as a "leading problem" but as a "problem of mathematics like any other" (cf. WA 3: 268i, from MS 110: 189). There is no evidence that this remark from 1931 was directed against Wittgenstein's early conjecture (against Floyd 2005: 95). Wittgenstein questioned not the solvability of the Entscheidungsproblem but rather its importance to mathematics and its foundations. According to Wittgenstein, his conjecture is like any other conjecture in need of an algorithmic solution.

4 Wittgenstein's Reaction to Undecidability Proofs: After 1936/7

Wittgenstein was in close contact with Turing in the late thirties. He was one of the first to read Turing's undecidability proof of 1937.[6] According to Floyd, Turing's proof must have "struck" Wittgenstein (Floyd 2016: 30). Indeed, this would have been the most reasonable reaction if Wittgenstein had accepted the proof and its underlying method. Alternatively, he could have stuck to his convictions and turned

---

[6] Cf. Turing's letter to his mother from 11th February 1937, in which he mentions Wittgenstein as the second outside King's College to whom he already had send a copy. Cf. AMT/K/1/54, Turing Digital Archive (http://www.turingarchive.org/browse.php/K/1/54) and Floyd 2016: 9, footnote 3.

them against the undecidability results of the thirties. One may deny that this is a profound reaction. Yet, given Wittgenstein's convictions and his algorithmic understanding of logic and mathematics, he hardly had any choice unless he was ready to radically change his views and suddenly embrace the application of the axiomatic method in mathematics and meta-mathematics. However, although Wittgenstein changed many of his views throughout his life, there is no evidence that he ever abandoned his critique of the logical formalization of mathematics and meta-mathematics, which characterizes the axiomatic method. This is due to his alternative algorithmic conception of proof that lays at the heart of his philosophy of mathematics throughout his life. Wittgenstein refined this conception in his later philosophy by embedding it into a pragmatic and cultural context that placed focus on a "surveyable representation" ("übersichtliche Darstellung") rather than on mechanical decision procedures. However, this is better understood as a further development in considering the foundations of an algorithmic proof conception, rather than a renunciation of it.

Unfortunately, Wittgenstein did not explicitly discuss Turing's undecidability proof of FOL. Instead, he discussed Gödel's undecidability proof for axiomatized arithmetic theories (such as Peano Arithmetic, PA) in remarks that largely stem from 1937 to 1939.[7] In the following, I will first consider this discussion and then apply it to the decision problem and Turing's proof. I do not deny that there are crucial differences between Turing's and Gödel's undecidability proofs. However, I will argue that these differences do not matter from Wittgenstein's point of view since both similarly question his algorithmic proof conception, and Wittgenstein reacts to both of them with a fundamental critique of the underlying axiomatic method.

4.1 Wittgenstein's Reaction to Gödel's Undecidability Proof

Gödel's undecidability proof proves that there exists at least one formula G in the language of PA (henceforth denoted by L_A) such that neither G nor $\neg G$ is provable from the axioms of PA. If Gödel had proven this result by providing a decision method for provability in PA, this would be in line with Wittgenstein's own proof conception. His paradigm for acceptable, algorithmic proofs of unprovability is manifested in the algebraic proofs of the unsolvability of certain problems within

---

[7] Wittgenstein delivered his "Lectures on Gödel" (WCL: 50 – 57) in the Eastern Term 1938. This lecture begins with his sketch of Gödel's proof from MS 117 (written end of 1937), for a critique of this proof sketch see Lampert 2006. RFM I, appendix I has parallels to WCL. The remarks on Gödel in RFM V, §§18f. are from 1941.

Euclidean geometry, such as the problem of angle trisection with a straightedge and compass (cf. RFM I, appendix I, §14). Such proofs of unprovability are part of a decision procedure that distinguishes between possible and impossible constructions on the basis of their algebraic representations: the angles that can be constructed with a straightedge and compass are those and only those that are representable by algebraic equations that can be solved with nested square roots.[8] This fits with Wittgenstein's algorithmic conception of proof in terms of a finite transformation of the problem into a representation in some notation that allows one to decide the initial question based on properties of the resulting expressions.

However, Gödel's proof is not of this sort. Instead, it rests on the representation of a formal property, namely, PA-provability, in L_A, i.e., a language that is based on FOL supplemented with constants for numbers and arithmetic functions. This means that provability is expressed by a certain open formula (abbreviated by $\exists y Byx$, according to Gödel's definition 46) in L_A iff, for all Gödel numbers n of L_A-propositions, n is provable iff $\exists y By\bar{n}$ is true according to the intended interpretation of L_A.[9] According to Wittgenstein's proof conception, any intent to represent a formal property, such as provability, by an open formula (propositional function) must be founded on confusion between material and formal properties, which is the fundamental mistake of mathematical logic. In contrast to Gödel, Wittgenstein claimed that formal properties can only be "shown", i.e., identified through a decision procedure; they cannot be "said", i.e., expressed within the formal language to which they apply.

Wittgenstein rejected the application of the axiomatic method in Gödel's undecidability proof of his formula G. He did not do so by referring to the relevant proof of the representability of recursive functions within L_A (cf. theorems V and VII in Gödel 1931: 186; theorem 13.4 in Smith 2007: 109; and Lampert 2018b for detailed discussions). Instead, he was aware that he was instead "bypass[ing]" (RFM V, §17, last sentence) Gödel's proof since he was discussing not the details of the proof but rather what could be taken as a *"forcible reason* for giving up the search for

---

8 For an implementation of the Kronecker algorithm that allows to decide whether the respective algebraic equations are solvable with nested square roots, cf.

http://www2.cms.hu-berlin.de/newlogic/webMathematica/Logic/k-decide.jsp

9 $\bar{n}$ is the expression for the number n in L_A. Under the presumption of the *expressibility* (or definability) of provability, Gödel then proves that the property of provability cannot be *captured* in PA. This means that it is not true for all n that either $\exists y By\bar{n}$ or its negation is provable from the axioms of PA. Cf. Smith 2007: 34f., for the definitions of representing (expressing, defining) and capturing properties within PA.

a proof" (RFM I, appendix I, §14). For Wittgenstein, this was a question of what counts as a "criterion of (un)provability" (cf. RFM I, appendix I, §14 – 16, and V, §18f.). According to his algorithmic proof conception, a criterion for a formal property must be a decision criterion in terms of some property of ideal symbols. This is why the proof of the impossibility of trisecting an angle with a straightedge and compass counts as a criterion for giving up the search for such a construction (RFM I, appendix I, §14). By contrast, the criterion for a "forcible reason" to give up the search for a decision procedure is not satisfied by meta-mathematical undecidability proofs since they are based on the representation of a formal property by a propositional function within the formal language itself. According to Wittgenstein, undecidability proofs reduce the possibility to represent provability as a propositional function to absurdity, not the assumption of a decision procedure that is independent of such a representation. Indeed, the verdict regarding the representation of formal properties by propositional functions had lain at the heart of Wittgenstein's critique of mathematical logic since the beginning (cf. TLP 4.126).

One reason why Wittgenstein thought that formal properties are not representable by propositional functions is that he rejected the possibility of self-referential representations within a formalism based on FOL (cf. TLP 3.332f). He distinguished operations from functions and considered that it is only with operations that self-application comes into play (TLP 5.25f). However, the application of operations is a part of symbolic manipulation and is not something that is expressible by functions within a logical symbolism. Undecidability proofs, meanwhile, rest on diagonalization and, thus, on a formula that is intended to represent that the formula itself does (not) have a certain property. Gödel's formula G, for example, is intended to represent the property of unprovability of the formula G itself. On this basis, he proved that G cannot be captured in PA.[10] This proof method gives priority to semantics (representation) over syntax (capturing). It is only this priority that makes it possible to prove meta-mathematically that an algorithmic proof conception is limited. Such reasoning cannot convince an advocate of the algorithmic proof conception since such an advocate instead places priority on syntax. In the case of conflict, said advocate would deny the definability of the formal property in question. Thus, given G were provable from the axioms of PA,

---

[10] Note that Gödel's syntactic proof presumes the representation of "y is a proof of x", Def. 45, and "x is provable", Def. 46, in Gödel 1931: 186. Claiming that Gödel's "syntactic" proof does not rely on the representability of arithmetic and meta-mathematical properties (and, in this respect, on semantics) demonstrates a misunderstanding of this proof. Gödel never maintained that. Instead, he made it clear that his syntactic version of the proof is based on the expressibility of provability within L_A; cf. Gödel 1931: 176. In contrast to Gödel's so-called semantic proof, his so-called syntactic proof presumes only the consistency of PA, not its correctness.

the diagonal case would simply turn out to be such a case of conflict. Therefore, Wittgenstein would not infer that PA is inconsistent but instead would deny that G, in fact, represents its own unprovability (RFM I, appendix I, §8, §10). This is also why Wittgenstein could not accept Gödel's undecidability proof as an proof of incompleteness.

Wittgenstein analysed undecidability proofs as proofs by contradiction (cf. RFM I, appendix I, §14, and cf. PI §125 below). In the case of Gödel's undecidability proof, he mainly considered the contradiction as one between a supposed proof of G and the fact that G represents its own unprovability (RFM I, appendix I, §8, §10f.). However, his rejection also applies to the so-called syntactic version of Gödel's proof since this version also relies on the assumption that the formal property of provability can be represented within L_A, which involves self-referential interpretations in the diagonal case. No proof of contradiction can be a compelling reason to give up the search for a decision procedure since an advocate of the algorithmic proof conception questions the assumption of representability for the formal property in question.

Wittgenstein compared the contradiction arising in an undecidability proof to a paradox (RFM I, appendix I, §12f, §19). According to Wittgenstein's analysis, so-called semantic paradoxes, such as the Liar paradox, as well as paradoxes of mathematical logic, such as Russell's paradox, rely on the representation of formal properties by propositional functions (cf. TLP 3.33 – 3.334; WVC: 121; and PR: 207f.). The problem lies not with the specific properties (semantic properties vs. set-theoretical properties) but with the analysis of self-reference as something that is expressible by propositional functions and thus capable of being represented in a symbolism based on FOL. The distinction between meta- and object-language is not sufficient to prevent paradoxes, according to Wittgenstein's analysis. Instead, it is the distinction between formal and material properties that must be considered. This distinction comprises both semantic paradoxes and the paradoxes of mathematical logic. It even applies to arithmetic properties and their meta-mathematical correlates. For Wittgenstein, the arithmetic and meta-mathematical interpretations in the language of L_A were not an "absolutely uncontroversial part of mathematics" (Wang 1987: 49; however, cf. also Gödel 1931: 149, footnote 14) but rather the outcome of the fundamental mistake of mathematical logic, namely, the assertion that formal properties of mathematics and meta-mathematics can be expressed by propositional functions. Wittgenstein's algorithmic proof conception rules out such a possibility since it maintains that formal properties can be expressed only by symbolic properties of a proper notation. Wittgenstein believed in an algorithmic proof conception as the standard for a rigorous proof that can never be affected by any underlying intended interpretations of a logical symbolism to represent any properties, since such an interpretation necessarily extends beyond the realm of mere symbolic manipulations.

4.2 Wittgenstein's Reaction to the Undecidability of FOL: PI §125

Turing's undecidability proof for FOL differs from Gödel's proof in many respects. It rests not on recursive functions but on Turing machines; it refers not to L_A and PA but directly to FOL, and it proves without referring to axioms that the property of provability (or, likewise, logical validity) of FOL formulas cannot be decided. However, in terms of Wittgenstein's attitude regarding undecidability proofs, these differences do not matter since Turing's proof similarly rests on the representation of formal properties within the language of FOL. In the case of Turing, these properties are properties of his machines (such as printing a 0 or, in more modern versions of the proof, halting). Thus, Wittgenstein's critique applies to Lemma 2 in Turing 1936/7: 262. This lemma is proven by referring to the intended interpretations of logical formalizations of the configurations and instructions of Turing machines. In the diagonal case, the provability of a logical formalization of the behaviour of machines involving a decision machine for FOL must be interpreted as a statement about the behaviour of that same machine. However, it is possible to define machines involving FOL such that their behaviour contradicts the intended interpretation. According to Wittgenstein's critique of the method of logical formalization, one is not obliged to infer that no decision machine for FOL exists. Instead, he would reject the interpretation that the provability of the logical formalization is correlated to the behaviour of the formalized machine in the diagonal case. According to Wittgenstein, a logical formalization cannot fully express and capture computation and what is computable. [11]

The fact that Wittgenstein applied his fundamental critique of the axiomatic method and its application within undecidability proofs to the undecidability proofs of FOL is made clear by PI §125. This passage follows up on the remark about "the leading problem of mathematical logic" from 1931 (see above, page 526) and refers to Ramsey's phrasing of the decision problem. PI §125 is an echo

---

[11] This statement does not in any way question the Turing thesis. Turing's thesis merely states that computability can be reduced to what Turing machines can compute. One can simultaneously accept this and reject the logical formalization of Turing machines. The same applies to the Church thesis: what is questioned is not the reduction of computability to recursive functions but the logical formalization of those recursive functions.

of Wittgenstein's critique of the axiomatic method and its application in meta-mathematical undecidability proofs:

> It is not the business of philosophy to resolve a contradiction by means of a mathematical or logico-mathematical discovery, but to render surveyable the state of mathematics that troubles us - the state of affairs before the contradiction is resolved.
>
> Here the fundamental fact is that we lay down rules, a technique, for playing a game, and that then, when we follow the rules, things don't turned out as we had assumed. So that we are, as it were, entangled in our own rules.
>
> This entanglement in our rules is what we want to understand: that is, to survey.
>
> It throws light on our concept of meaning something. For in those cases, things turn out otherwise than we have meant, foreseen. That is just what we say, for example, when a contradiction appears: "That's not the way I meant it."

Even after Turing's proof, it was obvious to Wittgenstein that the Church-Turing theorem, like Gödel's theorem, rests on "a contradiction that needs to be resolved". This could be done by solving the decision problem. However, this would be a logico-mathematical endeavour. The business of philosophy is instead concerned with the state of mathematics before this discovery. It is to analyse the confusion between formal and material properties that underlies the axiomatic method and its acceptance of meta-mathematical criteria for undecidability. For Wittgenstein, it seemed out of the question that this method should rely on "rules, a technique" that would lead to a situation in which an advocate of the axiomatic method would face the discovery of a decision procedure and then be in need of a philosophical analysis that would explain why "things turn out otherwise" than foreseen.

Wittgenstein inserted §125 from TS 228, written in between 1945 and 1948, into the manuscript of PI, demonstrating that he held onto his conjecture throughout his life. He never accepted a proof method in meta-mathematics that relied on the logical formalization of formal properties. Instead, he adhered to an algorithmic understanding of proof that had already lain at the heart of his early philosophy.

5 Conclusion

From the perspective of modern mathematics and its foundations in mathematical logic, Wittgenstein's point of view might seem like the reactionary dream of an advocate of mathematics in the manner in which it was done before the foundational crisis. To say the least, Wittgenstein's point of view is in stark contrast to the development of modern mathematical logic. Wittgenstein's critical reaction to the emergence of mathematical logic was very general, and he deliberately neither discussed undecidability proofs in detail nor made the effort to solve the decision problem. However, if one wishes to take Wittgenstein's point of view seriously, this work must be done.

Bibliography

Boolos, George S.; Burgess, John P., Jeffrey, Richard C. (2002): *Computability and Logic.* Cambridge (UK): Cambridge University Press, 2002^4 (1974).

Fan, Zhao (2018): "One-Step, Two-Step: Different Diagonal Arguments." In: Mras, Gabriele M. et al (eds.): *Contributions to the 41st International Wittgenstein Symposium.* Kirchberg a. W., 43 – 45.

Floyd, Juliet (2005): "Wittgenstein's Philosophy of Logic and Mathematics." In: Shapiro, Stewart (ed.): *The Oxford Handbook of Philosophy of Mathematics and Logic.* Oxford: Oxford University Press, 75 – 128.

Floyd, Juliet (2012): "Wittgenstein's Diagonal Argument: A Variation on Cantor and Turing." In: Dybjer P. et al (eds): *Epistemology versus Ontology: Essays on the Philosophy and Foundations of Mathematics in Honour of Per Martin-Löf.* Dordrecht: Springer, 25 – 44.

Floyd, Juliet (2016): "Chains of Life: Turing, Lebensform, and the Emergence of Wittgenstein's Later Style." In: *Nordic Wittgenstein Review*, Vol. 5, No. 2.

Gödel, Kurt (1931): "Über formal unentscheidbare Sätze der *Principia mathematica* und verwandter Systeme I." In: *Monatshefte der Mathematik und Physik*, Vol. 38, 173 – 198.

Lampert, Timm (2006): "Wittgenstein's 'Notorious Paragraph' About the Gödel Theorem." In: *Contributions of the Austrian Wittgenstein Society 2006*, Kirchberg a. W., 168 – 171.

Lampert, Timm; Säbel, Markus (2017): "Wittgenstein's Elimination of Identity for Quantifier Free Logic", http://philsci-archive.pitt.edu/12974/ (preprint).

Lampert, Timm (2017a): "Wittgenstein's ab-Notation: An Iconic Proof Procedure." In: *History and Philosophy of Logic*, Vol. 38, No. 3, 239 – 262.

Lampert, Timm (2017b): "A Decision Procedure for Herbrand Formulae without Skolemization." https://arxiv.org/abs/1709.00191.

Lampert, Timm (2018a): "Iconic Logic and Ideal Diagrams: The Wittgensteinian Approach", in: Chapman, Peter et. al. (eds.), *Diagrammatic Representation and Inference*, Heidelberg: Springer, 624 – 639.

Lampert, Timm (2018b): "Wittgenstein and Gödel: An Attempt to Make 'Wittgenstein's Objection' Reasonable." In: *Philosophia Mathematica*, Vol. 26, No. 3, 2018, 324 – 345.

Landini, Gregory (2007): *Wittgenstein's Apprenticeship with Russell.* Cambridge (UK): Cambridge University Press.

Ramsey, Frank Plumpton (1923): "Critical Notices." In: *Mind*, Vol. XXXII, No. 128, 465 – 478.

Potter, Michael (2009): *Wittgenstein's Notes on Logic*, Oxford: Oxford University Press.

Rogers, Brian; Wehmeier, Kai (2012): "*Tractarian* First-Order Logic: Identity and the N-Operator." In: *Review of Symbolic Logic*, Vol. 5, No. 4, 538 – 573.

Smith, Peter (2007): *An Introduction to Gödel's Theorems*, Cambridge (UK): Cambridge University Press.

Turing, Alan (1936/7): "On Computable Numbers, With an Application to the Entscheidungsproblem." In:*Proceedings of the London Mathematical Society*, Vol. 2, No. 42, 230 – 265.

von Wright, Georg Henrik (1982): *Wittgenstein.* Oxford: Blackwell.

Wang, Hao (1987): *Reflections on Kurt Gödel*. Cambridge (MA): MIT Press
Weiss, Max (2017): "Logic in the *Tractatus*." In: *Review of Symbolic Logic*, Vol. 10, No. 1, 1 – 50.
Wittgenstein, Ludwig (NL): "Notes on Logic." In: *Notebooks 1914–1916*. Oxford: Blackwell 1979, 93 – 107.
Wittgenstein, Ludwig (MN): "Notes Dictated to G. E. Moore in Norway, April 1914." *Notebooks 1914–1916*. Oxford: Blackwell 1979, 108 – 119.
Wittgenstein, Ludwig (TLP): *Tractatus Logico-Philosophicus*. London: Routledge, 1994.
Wittgenstein, Ludwig (WVC): *Wittgenstein and the Vienna Circle*. Oxford: Basil Blackwell, 1979.
Wittgenstein, Ludwig (PR): *Philosophical Remarks*. Chicago: Chicago Press, 1975.
Philosophical Grammar. Rhees, Rush (ed.); Kenny, Anthony (transl.); Blackwell, Malden (MA), etc., 1980^2 (1974).
Wittgenstein, Ludwig (PI): *Philosophical Investigations*. 4th edition, Chichester (UK): Wiley Blackwell, 2009.
Wittgenstein, Ludwig (RFM): *Remarks on the Foundations of Mathematics*. Cambridge (MA): M.I.T. Press, 1967.
Wittgenstein, Ludwig (RPP): *Remarks on the Philosophy of Psychology*. Oxford: Basil Blackwell, 1991.
Wittgenstein, Ludwig (WA 3): *Wiener Ausgabe*; Vol. 3, Vienna: Springer, 1995.
Wittgenstein, Ludwig (CL): *Cambridge Letters*. Oxford: Blackwell, 1997.
Wittgenstein, Ludwig (VW): *The Voices of Wittgenstein*. London: Routledge, 2003.
Wittgenstein, Ludwig (WCL): *Wittgenstein's Whewell's Court Lectures: Cambridge, 1938–1941*. Munz, Volker; Ritter, Bernhard (eds). Chichester (UK): Wiley Blackwell, 2017.

Index of Names

Ackermann, Wilhelm   134, 250, 256–257, 298, 307
Anscombe, G. E. M.   170, 383, 439, 446, 449, 471
Aristotle   IX, 103, 145, 191
Barwise, Jon   119, 421
Ben-Yami, Hanoch   105
Bencivenga, Ermanno   107
Berlin, Brent   202
Bernays, Paul   249–252, 254–257, 260, 297–310
Billock, Vincent   203–205
Birkhoff, Garret   386–387
Bishop, Errett   96
Black, Max   420, 427
Blackburn, Simon   370
Boghossian, Paul   369–370
Boole, George   IX, 133–134, 136–137, 139–141, 144, 147–148, 277, 344, 387, 393
Boolos, G. S.   77, 216–217, 220
Bourbaki / Bourbakists   327–328, 334–335, 337–339, 387
Bradley, F. H.   377
Brouwer, L. E. J.   X, 51, 267, 297, 301–304, 309
Brown, D. J.   73
Canfield, J. V.   429–430, 435–438
Cantor, Georg   IX, 67, 219, 224, 250, 275, 288–289, 291–292, 347, 448–450
Carnap, Rudolf   91, 276, 390, 416–417, 448, 463
Carroll, Lewis   370–371
Chierchia, Gennaro   7–8, 10–11, 21–22
Chudnoff, Elijah   370
Church, Alonzo   95, 97, 135, 254, 279–280, 283, 331, 336, 450, 515, 518, 531
Coniglio, M. E.   71, 79, 81, 83
Curry, H. B.   315
Davidson, Donald   362
De Morgan, Augustus   IX, 144–145, 147, 348
de Swaart, Harrie   92, 102

Dedekind, Richard   123, 127–128, 250, 343, 347–353
Drury, Maurice   474, 476
Dummett, Michael   156, 170, 199, 300, 376, 430–431
Einstein, Albert   441–442, 449
Epstein, R. L.   73–74, 76–77, 79
Euler, Leonhard   140–142, 144–145
Figallo, Martin   83
Fine, Kit   34, 56–57, 129–130, 390, 418, 420–423, 426
Floyd, Juliet   263–278, 280, 282–286, 288–290, 292, 526, 533
Frascolla, Pasquale   416, 420
Frege, Gottlob   IX–X, 95, 118, 133–140, 148–150, 153–157, 159, 162, 169–181, 183–190, 192–194, 215–217, 220–221, 250, 328, 343, 345, 349, 375–380, 382–383, 387, 423, 429, 441, 444–446, 453, 458, 470, 476, 518, 521–523
Gauss, C. F.   94
Geach, Peter   375, 459
Gentzen, Gerhard   71–74, 77, 97, 192, 298, 305, 309, 316, 324
Glivenko, Valery   71–74, 77
Glock, Hans-Johann   362, 410, 413, 479, 486
Gödel, Kurt   X, 71–74, 77, 96, 102, 161, 257, 260, 272–274, 279–281, 283, 286–287, 292, 297–299, 302, 304–310, 313, 323, 333–334, 354, 387, 448, 450, 452, 527–532
Goethe, J. W. von   202
Grice, H. P.   50
Haack, Susan   39
Hamann, J. G.   493–494
Harburger, Walter   91
Hegel, G. W. F.   473–476, 481–482
Heidegger, Martin   501
Hempel, C. G.   417
Henkin, Leon   119–120, 122, 274
Herbrand, Jacques   260, 279, 297, 306–309, 352

Hering, Ewald 203
Heyting, Arend 448
Hilbert, David IX–X, 72, 91, 93–94, 103, 134, 249–252, 254–258, 260, 267, 279, 297–306, 308–309, 314, 330, 334, 349, 448, 524
Hintikka, Jaakko 117, 120–125, 128–129, 199–200
Höfler, Alois 201
Horty, John 153–154, 158–160, 162–163
Howard, William 315
Hsieh, Po-Jang 203
Husserl, Edmund 103, 155, 174, 197–198, 201
Janik, Alan 476, 479
Jerábek, Emil 46–47, 50, 79
Jeshion, Robin 192–194
Jourdain, Philip 170, 378, 382
Kant, Immanuel 40, 92, 136–137, 139, 145, 217, 251–252, 255, 299, 303, 473–474, 481
Kay, Paul 202
Kitcher, Philip 332
Kleene, Stephen 92, 162, 254, 279–280, 283, 308, 351–352
Klein, Felix 57, 198, 202
Köhler, Wolfgang 361, 367–368, 371
Kolmogorov, A. N. 71–74, 77
Kreisel, Georg 255, 290
Kripke, Saul 51, 166–167, 263, 330, 332, 359, 362, 364, 368–370, 390
Kronecker, Leopold 57, 251, 302, 347, 528
Kuhn, Thomas 49
Kunen, Kenneth 327–328
Kusch, Martin 486, 492–495
Landman, Fred 10, 17, 19–20, 25–33
Leibniz, G. W. 3–4, 12, 134–135, 139–141, 147, 185–186, 277, 393, 417, 442
Lewis, C. I. 71, 73,
Lewis, David 417–418, 422
Link, Godehard 3–20, 23–25, 27, 29, 31, 33–34
Littlewood, J. E. 442, 451–452
Löb, M. H. 274
Lovecraft, H. P. 205
Łukasiewicz, Jan 42–44, 78–79, 93, 97, 165, 387

Machiavelli, Niccolò 473
Malmnäs, P.-E. 76–77
Mancosu, Paolo 57, 218–220, 224, 252, 324
Martin-Löf, Per 189, 192–194, 306
McGuinness, Brian 360, 371, 389, 441, 443, 446–447, 449, 451–453
Moore, G. E. 272, 378–379, 381, 447, 450–451, 477
Moreira, A. P. R. 71, 81–83, 85
Mras, Gabriele 144, 203, 476
Negro, Antonio 419–420, 425–426
Nestroy, Johann 475
Northrop, F. S. C. 442
Nozick, Robert 56
Ogden, C. K. 362, 380–381, 383, 389
Padro, Romina 370
Paul of Venice / Paulus Venetus 207
Peano, Giuseppe IX, 134, 148, 277, 333, 343–355, 377, 387
Peirce, C. S. 134, 144–145, 147–148, 352–353, 381, 387
Petrus Hispanus 96
Plato IX, 114, 271–272, 300
Poincaré, Henri 250–251
Post, Emil 79, 281, 387, 390
Potter, Michael 177, 459, 518
Prawitz, Dag 73–74, 76–77, 316
Putnam, Hilary 133–135, 148, 203, 264–265, 298, 313, 332, 335, 492
Quine, W. V. 96, 133–135, 142, 300–301, 387
Ramsey, F. P. 267, 271, 275, 375, 380–382, 419–420, 427, 445–447, 451, 524, 531
Reichenbach, Hans 41
Rhees, Rush 272, 442–443, 446–447, 449–450, 452–453, 474, 480
Richards, I. A. 362, 381
Ritter, Bernhard 204, 291
Rothstein, Susan 6–10, 13, 16–18, 21, 23–24, 27, 29–30, 33
Russell, Bertrand IX–X, 21, 134, 148–150, 155, 170, 187, 215–216, 250–251, 275–276, 278, 285, 307, 328–329, 353, 362, 375, 377–378, 380, 382, 387, 389, 392, 397, 400, 445–449, 451, 458–459, 474, 515, 518–523

Ryle, Gilbert 103, 418
Sandu, Gabriel 117, 121–122, 129–130, 199–200
Schlick, Moritz 146, 203, 360, 457–458, 464–465, 467, 478
Schopenhauer, Arthur 476
Schröder, Ernst IX, 133–134, 136, 140, 144, 147–148, 344, 347, 351–352
Shapiro, Steward 21, 185, 221, 334, 371
Sieg, Wilfried 249–251, 257–258, 279, 281, 298, 303
Skolem, A. T. 120, 252
Sloman, Aaron 156, 167
Sluga, Hans 444, 448, 451
Smythies, Yorick 449, 492–495
Stone, Marshall 392
Strawson, P. F. 5
Stroud, Barry 429–433, 435–438
Suszko, Roman 47–48, 73
Tarski, Alfred 57, 64, 74, 77, 86, 91, 93, 96, 98–99, 102, 118, 197–199, 385, 387, 391, 393
Troeltsch, Ernest 91
Turing, Alan X, 95, 255, 263–292, 366, 449–450, 515, 526–527, 531
Vasiliev, Nikolay 197, 206–207
Venn, John 141–145
Vienna Circle / Wiener Kreis 448, 477
von Neumann, Johann 260, 307–309, 448
von Wright, G. H. 443–444, 446, 449, 451, 520
Waismann, Friedrich 267, 359–371, 444, 448–449, 457–458, 461, 464, 466–467
Watson, Alister 270, 272–273, 275, 279, 288, 290, 449
Weingartner, Paul 50, 412
Weininger, Otto 479
Weyl, Hermann 267, 297–298, 301–302, 304
Whitehead, A. N. 13–15, 134, 148, 151, 250–251, 275, 389
Wigner, Eugene 56
Williamson, Timothy 189
Wittgenstein, Ludwig X–XI, 40, 117, 133, 136, 150–151, 169–172, 174–181, 184, 187, 189, 197–199, 201–208, 254, 263–278, 281–283, 285, 287–292, 328–332, 336–337, 339, 359–370, 375, 377–383, 387, 389–393, 397–398, 403, 410, 412–414, 416–418, 420, 423, 425–426, 429–435, 437, 441–449, 451–454, 457–471, 473–481, 485–495, 499, 504, 509–510, 515–533
Wójcicki, Ryszard 73–77
Wolff, Michael 133, 137–139
Wolniewicz, Bogusław 398–401, 404–409, 413
Woodward, James 56
Wright, Crispin 215–217, 369–371, 486
Yablo, Stephen 166, 415, 417–418, 421, 423, 426
Yaqub, Alladin 92, 94
Zach, Richard 249, 256
Zermelo, Ernst 219–220, 226–227

Index of Subjects

a priori / apriority   40, 70, 146, 150, 193, 201, 203, 252, 523
ab-notation   515–520, 523
abstraction principle   X, 215, 217, 219–221, 224–225, 231–232, 234–237, 239, 241
Achilles and the Tortoise   370, 443
alternative logics / logical systems   40, 50
ambiguity, 24–25, 331, 336, 345, 367, 377–378, 458, 460
analytic / analyticity   X, 37, 146, 185, 203, 217, 280, 314, 362, 420
arithmetic   X, 68, 99, 102, 140, 149, 161, 163, 166, 185, 188, 217, 236, 250–251, 258, 277, 301, 303, 305, 308–309, 315, 329–330, 333, 343–345, 349, 351–352, 354, 393, 449, 523, 529
classical ~   72–73
intuitionistic ~   72–73, 303, 308–309
first-order ~   297, 299, 305
foundations of / for ~   XI, 347, 352
second-order ~   102, 215–216, 306
algorithmic proof   523, 527, 529–530
arithmetical language   154, 159, 161
arithmetical operation   343–344, 353
arithmetization   96, 103, 161, 302
Aspect-Restriction Strategy   17, 26, 29, 31
assertion   45, 169–180, 183–184, 187, 189–191, 194, 199–200, 206
assertion sign   X, 170, 175
atomic individual   4–5
atomic proposition   520–521
autological   286
axiom of choice / Choice Axiom   78, 94–95
axiom of reducibility   449
Basic Law V / BL V   187–188, 215–217, 219–221, 224–225, 232–234, 239–241
Basic Laws of Arithmetic / Grundgesetze der Arithmetik   134, 175, 181, 187–188, 379
Bedeutung (reference)   153, 179, 378
begging the question   461
Begriffsschrift   IX, XI, 133–135, 171, 173, 179, 185–187, 345, 375–376, 379, 382–383
Begriffsschriftsätze / propositions of *Begriffsschrift*   175, 180, 185
bijection / bijective   27, 45, 83, 240
Bild   476, 487, 490, 493, 495 (cf. 'picture' in Wittgenstein)
bivalence / bivalent   41–42, 45, 47, 49–51, 58, 300
Boole-Schröder algebra   134
Boolean   7, 134–136, 139, 141, 144, 222–223, 323, 392–393, 406, 423
~ algebra   136, 385, 391–393, 406
Bourbakian   328–329, 334–335
calculus   X, 74, 93, 107, 109, 134, 136, 141, 144, 147, 149, 267, 271, 279, 313, 318, 320–321, 362, 516–517, 521, 526
~ logic   135, 137, 139–140, 148
predicate ~   92, 102
propositional ~   72–73, 79, 91–93, 97 (cf. propositional logic)
typed lambda ~   313, 318–322
cardinal numbers   366
cardinality   66–67, 98–99, 199, 219, 232–234, 388, 407–408
causality   359–362, 367, 448–450
cause and reason   359, 363, 365–366, 368–369, 371
certainty   351, 444, 485
chain of reasons   359, 366–368, 370
Church-Turing thesis / theorem   323, 532
circular / circularity   37–39, 91, 250, 285, 343, 370
classical propositional logic / CPL   38, 41, 47, 72–73, 76–77, 79, 348, 385–388, 390–393
rules / principles of ~   38–41, 49, 72, 277, 385–386
colo(u)r   68, 176–178, 180, 197–198, 204, 201–209, 270, 422–423, 443, 461, 463–467
~ concept / term   201–203, 206
~ octahedron   201–202, 467
logic of ~s   201, 204

common sense   266, 268, 271
completeness   86, 91–92, 94, 100,
    314–315, 322, 387
~ theorem   99–102, 313, 387
compositionality   95, 159, 197, 200, 208,
    441, 445, 453
comprehension   X, 221–223, 230–231,
    236, 238, 241
~ assumption / hypothesis   226–227, 229,
    238
~ axiom   102, 216
~ principle   230, 241
concept formation   256, 444
conceptual content   375–377, 379–380
conditional   32, 42, 100, 160, 178, 165,
    304, 316, 416, 427
consequence operation / operator   46–47,
    71, 74, 86, 92, 388
consequence relation   48, 71, 75–76, 388
consistency   71, 91, 94, 241, 249–252,
    255, 258, 260, 277, 286, 299, 305–310,
    314–315, 322–323, 419, 426, 430, 439,
    529
~ of (classical) arithmetic   72–73, 250
~ proof   250–251, 257–258, 260, 299,
    305–308, 310, 314–315, 323
constructive proof   73, 447
Constructive Type Theory / CTT   183,
    189–191, 194
constructivism / constructivist   96, 102,
    189, 199, 251, 300, 430
content of a judgement / assertible /
    judgable ~   171–174, 179–181, 186,
    188, 376
content / sense containment   419–421,
    425–427
contingent   167, 379, 398–399, 401, 406,
    409, 417, 431, 433, 465
continuum   94, 273, 304, 449
contradiction   72, 124, 201, 206–207,
    216–218, 224, 228, 230, 242, 245–246,
    274, 277–278, 284, 286, 290, 328, 393,
    406, 426, 430, 436–437, 452–453, 517,
    530, 532
convention   148, 291, 430–431, 433
conventionalism   429–431, 433
countable   3–4, 7, 11, 15–16, 21, 23, 25,
    405, 408, 411
countable sets   411
counterfactual   57–58, 61–63
criterion / criteria   64–66, 92, 137, 155,
    202, 205, 207, 216, 235, 250, 300, 323,
    348, 380, 434, 437, 462, 515, 517–519,
    521, 524, 529
culture   268, 270
cumulative / cumulativity   9, 18, 20, 22, 79,
    85
Curry-Howard isomorphism   315, 317,
    319–321, 323
decidability / decidable   77, 92, 94–95, 97,
    99–100, 102, 515, 517, 520, 524–525
decision problem   278, 518–519, 526–527,
    531–532
deduction   39–41, 92, 97–98, 101, 138,
    254, 256, 258–259, 307, 313–317,
    320–322, 354, 380, 463
deduction theorem   43, 92, 98, 101
deductive logic   40–41, 277
deductive system   82, 98
definite description   176, 178, 307, 378
degree of invariance   55, 58, 61–62, 65, 68
denotation   4, 7, 10–14, 22, 65, 175
dependence   97, 102, 117–129, 166, 209,
    280
Dependence Logic   117, 128–130
dependence / dependency among quantifiers
    117–118, 122–123, 125, 130
mutual ~   117, 125, 128
determinism   361
diagonal argument / argumentation   263,
    266, 275, 278–279, 287, 290–292
~ method   288, 290, 448
~ proof   283, 289
diagonalization   278, 529
disjunction   44, 46–47, 65, 92, 101, 120,
    139–140, 145–147, 223, 377, 423–424,
    426
divisiveness   10
evident / (unmittelbar) einleuchtend   XI, 37,
    39, 97, 183–184, 186–188, 192–194,
    303, 524
elementary proposition / sentence   150,
    201, 267, 387, 390, 457–467, 470–471
independence of ~   458, 462, 464, 471

Index of Subjects — 541

elimination   109–110, 302, 314–315, 317, 320, 354, 520
empty concept / predicate   113, 235
Entscheidungsproblem / decision problem   263–264, 271, 278–281, 283, 518–519, 526–527, 531–533
epistemic   37, 39–41, 49–50, 55, 121, 183, 185–194
epistemology / Erkenntnistheorie   37, 188, 255, 359, 364, 369, 476, 486
equality   345, 398, 439
equivalence   42, 46, 48, 51, 57, 78, 98, 101–102, 110–111, 125, 218, 221, 233, 235, 313, 344–345, 348–349, 393, 398–399, 517, 523
~ relation   78, 219–220, 231–236, 393, 423
~ transformation   515, 517–519, 522–523, 525
essence / Wesen   184, 202, 279, 444–445, 488–489, 452
essentialism   453
Ethik   474, 477–481
Buch über ~   477
Euclidean geometry   202, 528
Euler circles   140–141, 147
Euler diagram   133, 136, 140–141, 144, 147
everyday discourse / language   206, 265, 463
existence   82, 105–108, 110–113, 117, 119, 142, 148–149, 215–216, 219–221, 301–303, 353, 397, 458, 465
~ of a set   119, 216
existential import   105, 107, 142
explicit definition   253, 256
extensional point of view / extensionalist   273, 291
Extensionalitätsauffassung   446–447
extensionality   221, 224, 226, 233, 235
Farbe / Färbung   376, 502, 507 (cf. colour)
field of knowledge   58, 64, 67
finite domain   99–100, 516
finitism / finiter Standpunkt   249, 251–252, 254–257, 260, 309–310
finitist / finitistic proof   250, 309
first-level concept / predicate   93, 106, 217, 238
first-order arithmetic   297, 299, 305
first-order (predicate) logic   91, 93, 99–100, 106, 123, 125–126, 279, 301, 307, 387–388, 391, 515
form of a proposition   267, 464, 468
formal language   74–75, 334, 382, 528–529
formal system   93, 107, 251–252, 257–258, 266, 269, 273, 279–281, 286–287, 380
formality   66, 197–199, 207–208
formalization   X, 101–102, 105, 250, 257, 279, 286, 327, 337–338, 526–527, 531–532
foundation   X, 37, 134, 149, 251, 264–265, 271, 277, 297, 309, 327–329, 335–337, 339, 347–348, 351, 368, 519, 527, 532
~ for / of mathematics   93, 217, 220, 263, 272–274, 299–300, 328, 339, 343, 448–449, 526
~ of logic   183, 264, 281, 291
foundational reductions   315
Foundations of Arithmetic / Die Grundlagen der Arithmetik   185, 215–216, 249
free algebra   74–75, 81, 387, 391
free logic   105–112, 114
Frege point   375
general form of proposition   267, 468
generality   X, 70, 146–150, 267, 269, 442, 446–447, 457–459, 461–462, 519
geometric proposition   252
geometrical   94, 201, 203, 303
geometry   IX, 57, 94, 198, 202–203, 206, 255, 300, 305, 351, 528
Gödel-Malcev theorem   102
grammatischer Satz   486–487, 490
green   197, 201, 203–205, 461, 463–466
Grenze   473–476, 478, 481, 508
Group-Forming Strategy   17–19, 21, 23, 25–26, 30–31
guessing game   359, 364, 366, 368
guise   17, 26–33, 146, 278
Gute, das (absolut) / goodness   473, 477, 481, 499, 504–507, 511–513
Herbrand's theorem   297, 306, 308
Hilbert's program   250–251, 299, 305, 314
hinge proposition / Angel-Satz   485–488,

490–491, 496
homomorphism   5, 197
horizon   327, 337–339
horizontal, the   171, 173–176, 178, 180–181 (cf. assertion sign)
Hume's Principle / HP   215–217, 219, 241
hyperintensional   17, 29–30, 51
idealism   300
Idealismus   454, 477, 479, 481
identity criterion / criteria   216, 333, 515, 517–518
identity sign   154, 519
identity statement   154, 176, 519
IF-Logic   200
implication   38, 42, 44, 47–48, 97–98, 101, 149, 187, 241, 273, 276, 344–346, 350, 354, 369, 371, 435, 523
analytic ~   420
deductive ~   77–78
material ~   X, 42
strict ~   73
implicit knowledge   157, 190
impossibility   130, 219, 302, 338, 430, 518, 524, 529
incomplete picture   457–458, 460–461, 466–471
incompleteness   260, 272, 274, 290, 297, 313, 333, 458, 460–461, 468–469, 530
~ theorem(s)   102, 287, 299, 308, 323
inconsistency   31, 150, 215, 217, 219–221, 224, 241, 285
inconsistent abstraction principle   217, 220–221, 235, 241
independence   117, 122, 195, 350–351, 360, 397 (cf. elementary proposition)
Independence-Friendly Logic   123, 200
indeterminism   449–450
induction   39–41, 95–96, 225, 251, 256, 307–309, 344, 346–347, 353, 517, 524
inductive proof / proof by induction   352–353, 355
complete ~   250, 347
mathematical ~   94, 249, 254, 258, 260, 308, 352
inference rule / rule of ~   X, 94, 97, 101, 183, 186, 191–194, 316, 320, 343, 354, 369, 516

infinite domain   517
infinite regress   37, 158, 179, 284, 448
infinite totality / the actual ~   225, 301, 441, 445, 448
infinity   123, 128, 218, 273, 301, 348, 366–367, 442, 445–448, 520, 524
~ axioms (Dedekind's)   348
intensional / intensionality   17, 19, 25–28, 34, 44, 47–49
internal property   197, 207–208, 445, 453
internal relation   150–151, 190, 198, 201, 208, 268, 363, 516, 522–523
internal structure   153–154, 156, 161
interrelations between logics / logic systems   71, 74, 86
intuitionism   X, 72, 102, 189, 297, 302–306, 448
intuitionistic   73, 101–102, 286, 297, 299, 302, 309–310, 315
~ arithmetic   72–73, 303, 308–309
~ (propositional) logic   50–51, 71, 93, 101, 304
intuitionists / ~logician   102, 286
invariance   55–58, 60–70, 198, 202, 209
invariant   55–64, 66, 197–198
judg(e)ment   XI, 72, 141–150, 169–181, 183–194, 197, 319–320, 371, 375–376, 381, 448–449
judg(e)ment stroke   169–175, 179–181, 183–188, 190, 194 (cf. the vertical)
justification / Rechtfertigung   15, 37–41, 49–50, 103, 183–185, 190–194, 346, 368, 471, 491, 495
justification regress   39–40, 103
Kantian   209, 297, 420
knowledge   X, 58, 63–64, 67, 95, 121, 135, 150, 154–155, 157–158, 188–191, 194, 215, 525
language(-)game   117, 172, 199, 202, 205–208, 266, 268–269, 281, 289, 364
lattice   94, 392, 397–398, 400–401, 404–413
hybrid ~   397, 411–413
~ theory   386–387
law   X, 44, 47, 55–56, 58, 61–65, 67–69, 73, 95–96, 149, 170, 184–188, 206, 216, 253, 278, 284, 286, 289–292,

300–302, 304, 347–348, 364, 393, 434, 444–445, 463, 524
arithmetical ~/ ~ of arithmetic   185–186, 188, 302
~ of contradiction   206, 393
~ of excluded middle   X, 47, 72, 278, 284, 286, 290, 292, 300–302, 444
~ of identity   61–62, 65, 67
~ of inference   521
logical ~/ ~ of logic   44, 55, 58, 65, 185–187, 206, 290
~ of nature   56, 68
~ of thought   146, 393, 445
~ of truth   184
limits of computability   323
limits of logic   290
Linguistic Turn   475–476
logical atomism   387, 397–398, 400
logical consequence   45, 55, 65, 313, 385, 418–419, 425–426
logical constant   56, 64–65, 91, 94, 122, 415, 423, 462, 464–465
logical form   146, 150, 321, 457–458, 463–464, 466
logical pluralism   40, 50
logical proposition   204, 380, 521–522
logical truth   42, 45, 55–56, 65, 369, 515, 518, 521, 525
logicism   X, 66, 149, 186, 216–217, 221, 251, 304, 329, 448
logicist   185, 329, 522
Löwenheim-Skolem theorem   313, 323
Löwenheim-Skolem-Tarski theorem   99
machine as symbol   263, 269, 273
many-valued logic   X, 48, 78, 93, 102 (cf. multi-valued logic; three-valued logic)
Markov principle   102
mass quantities   5, 12–13, 15
mathematical entities   333–335
mathematical practice / practice of mathematics   117, 323–324, 332, 335–337, 431
mathematical propositions / mathematische Sätze   267, 485–486, 488, 491, 495–496, 524
maximally invariant properties   62–64, 66
Maßstab des (moralischen) Sollens   478, 480
Maßstab des Guten / guten Handelns   477, 499, 513
meaning preservation   41, 45–47, 423
measure / measurement   15, 24, 60, 432, 438, 450
mereological   3–4, 6, 13–14, 418
~ relation   3, 6–8, 11–13, 15–16, 21–23, 33–34
mereology   3, 8, 11, 34
meta-induction   40–41
metalanguage / meta-language   39, 48–51, 80, 92, 279, 382
meta-logical proof   38, 51–52
metalogic / meta-logic   37–38, 51–52, 91–103, 206–207
metamathematics / meta-mathematics   91–95, 102–103, 280, 297, 308–309, 333–335, 337, 339, 525–527, 530, 532
metaphor   155, 274, 363, 420, 427
metaphysics   IX, 5, 68, 94, 185
mixed colo(u)r   197–198, 201, 203, 208
modal logic   44, 71, 76, 93, 100–101, 390
modern logic   117, 133–135, 137–139, 385, 517
modus ponens   38–39, 97, 192, 369–371
moralisches Urteil   473, 475, 477–481 (cf. Werturteil)
moralischer Wert   512
multi-valued logic   37, 40, 42–43, 47
multiple relation theory of judgment   381
natural deduction   97, 221, 313–317, 320–322
natural language   8, 15, 34, 56, 105–106, 129, 171
natural number   69, 98–99, 102, 159, 216, 218–219, 236, 251, 297, 319, 323, 329, 333, 344–345, 347–348, 351, 353, 355, 387, 399, 405
necessity   X, 55, 57–58, 61–65, 68–69, 100, 143, 149, 201, 345, 417, 429–430, 466
degree of ~   58, 61–62, 64–65
logical ~   X, 432
metaphysical ~   55, 68
types of ~   65, 69
negation   44, 64–65, 97–98, 101, 138,

142–143, 146, 149, 159, 206–207, 274, 285–286, 290, 302, 377, 412–413, 417–418, 423–424, 517
double ~ 51, 72–73
negative fact 176, 413
negative free logic 105–106, 109–112, 114
neo-Kantian / Neokantians 91, 297
neologicism / neologicist X, 215, 217, 220
nominalism 300–301
non-classical (propositional) logic 37, 40–41, 47–51, 85, 93, 101, 314
non-extensionalist 273
normal form 121, 313, 317–318, 321-322, 517
normative 170, 183, 188, 193–194, 271, 281, 485
notion of a class 136
noun 4, 6–10, 15, 19, 22–23, 27, 29, 34, 172, 181
count ~ 4, 6–7, 9–10, 12, 14–15, 22
mass ~ 4, 6–11, 13, 21–22, 25
~ phrase 8, 15–17, 19, 26–30, 33–34, 174
numeral 251, 254, 258–259, 303, 306, 329–330, 334–335
numerical 67, 99, 199, 257–258, 467
object language 39, 98, 127, 129, 382
obviousness 186–187, 192–194, 380
Occam's Razor 5
one-domain analysis 7–8, 16, 21, 23, 34 (cf. two-domain analysis)
one-to-one correlation / correspondence 216, 218, 231
one-to-one mapping 219, 407
ontology 5, 197–198, 201–202, 206–207, 397–398, 402, 412, 414
operation 12, 18–20, 30, 38–39, 46–47, 92, 140, 147, 198, 223, 263, 267, 273, 282, 303, 318, 343–344, 350–353, 355, 366, 392, 413, 424, 441, 446–447, 459, 524, 529
optimality 37, 40–41, 49–50
ordinary language 92, 95, 263, 349, 378 (cf. natural language; everyday language)
paraconsistent logic X, 48, 50, 86, 102
paradigm 133, 137, 139, 141, 147–148, 323, 327, 352, 426, 441, 454, 485–486, 527

~ shift 141, 453–454
paradox X, 50, 57, 136, 149, 153, 158, 167, 218, 237, 285, 301, 437, 449–450, 478, 530
Carroll's ~ 370–371
Grelling ~ 286
Richard's ~ 308
Russell / Russell's paradox ~ 21, 149, 187, 216, 235, 308, 530
semantic ~ 153–154, 156, 158, 160–161, 167, 530
Zeno ~ 450
Liar ~ 161–163, 165, 308, 530
parthood 4–5, 7, 13–14, 29, 34, 424
part-whole 215, 217–221, 224–225, 232–236, 241
~ principle 218–219, 221, 234
Peano arithmetic 333, 527
Peano axioms 343, 345, 349, 351
petitio principii 103
philosophy of logic X, 153–154, 375–377
philosophy of mathematics X, 96, 215, 300, 323, 430, 452, 454
physics 5, 55–56, 63, 68–69, 103, 149, 202, 299, 337, 442, 448–450
pictorial 150, 421, 468
'picture' in Wittgenstein 150, 328–329, 332, 335–336, 379, 416, 421, 457–461, 466–471, 492–493
picture theory 379, 457–458, 460
Platonism 297, 299–303, 305, 308, 429–431, 433, 444
restricted ~ 301
possible world 38, 40, 100, 158, 390, 399, 401, 405, 409, 411, 415–417, 419–420, 424
Post-completeness 99–100
potential infinite / potentionally infinite 441–442, 445, 448
practice XI, 184, 199, 207, 332, 336, 338–339, 370, 430–432, 436–437, 439 (cf. mathematical practice)
predicate X, 4, 7, 10, 14–15, 17–18, 24, 27, 106, 112–114, 143, 155, 157, 161, 171, 179, 184, 188, 203, 221–222, 224, 226, 228–229, 238, 240, 245, 249, 306, 459, 463

Index of Subjects — 545

binary ~   225–226, 237, 245
distributive ~   9, 13–14, 18
~ logic   64, 79, 106, 133, 135, 137–138, 140, 148–149, 250, 306
~ calculus   92–93, 102, 106
existence ~   105–106, 108, 110–111, 113
identity ~   99–100, 110, 113
unary ~   105–106, 112–114, 161
truth ~   154, 161–163, 179
~ variable   222, 225, 229
primitive, 42, 97, 169, 193, 256, 260, 327, 330, 333, 351, 354, 519–520
~ expression, 159, 162, 335
~ language, 334–335
~ proposition   350–351, 521
~ recursion / recursive   254, 256–257
~ recursive arithmetic / PRA   249–250, 252
~ recursive function   249–250, 255–257, 260, 303, 308, 352
Principia Mathematica   134–135, 377, 389, 519
private language   263, 291
problem of circularity   38, 91
problem of universality   101
proof method / method of proof   282, 314, 324, 521–523, 525, 529, 532
proof-object   190–191
proof theory   94, 250, 257–258, 260, 297, 306, 386
propositional content   417–418, 420–421, 424–425, 427
propositional function   516, 528–530
propositional logic   38, 41, 46–47, 72, 74, 83, 97, 140, 148, 253, 348–349, 385–386, 391, 515, 517, 519–520 (cf. calculus)
provability   102, 221, 240, 274, 515, 517, 527–531 (cf. unprovability)
psychologism / psychologistic   170–171, 183–185, 192, 194, 263, 378
psychology   188, 202, 208, 367, 380
quantification   IX, 93, 106–107, 112, 114, 143, 199–200, 222, 226, 229, 234, 302, 304, 517–518, 520, 522
multiple ~/ quantifier   X, 117
Quantified Argument Calculus / Quarc   105–114

quantifier   X, 65, 99–100, 105–108, 117–130, 136–137, 140, 148–150, 158, 161–162, 199–200, 221–223, 229, 245, 252, 256, 307, 377, 517, 520–521
binary ~   198, 208
existential ~   65, 100, 106, 119–120, 124, 130, 254
particular ~   105–106, 423
Lindström binary ~   199
Mostowski ~   199
universal ~   65, 120, 130, 161, 223
quantifier-free   119, 158, 252, 256
Quarc and free logic   105–107, 111–112
real number   99, 273, 278, 297, 304, 402, 411, 463, 524
realism   300, 302, 323, 453
recursive definition   256, 343, 345, 347, 350–353
red   18–21, 24–25, 203–206, 463–467
reddish green   201, 203–204
reduction   47, 86, 256, 315, 317–318, 320–321, 323, 327–331, 335–337, 339, 520, 522, 525, 531
~ of mathematics to set theory   X, 327–329, 337, 517–518
reductionist / reductionism   263, 265, 520
reference   25–26, 95, 149, 153–155, 157, 159, 161, 167, 173–176, 180, 185, 252, 334–335, 378, 383, 385, 387, 450, 452, 516, 525, 530
self-reference   161
reflexive / reflexivity   100, 220, 423, 453
regress   37, 39–40, 158, 179, 284, 370, 448 (cf. justification regress)
religious proposition / religiöser Satz   485, 488, 495
Remarks on the Foundations of Mathematics   328, 349, 365, 429, 441–442
rule-following / follow a rule   X, 263–264, 268–269, 271, 273, 289, 291–292, 363–364, 368–371
Satz   185, 187, 388–389, 416, 475–477, 486–491, 495, 501–503, 511
~system   266–267
scepticism   103, 361
science   IX, 4–5, 56–57, 68–69, 91, 95–96, 134, 146–150, 208, 265, 271, 299–300,

304, 323, 337, 352, 442, 444, 449, 452, 525
philosophy / philosophers of ~ 68, 266, 323, 442
second-order 119–120, 124, 127, 220–221, 232, 235, 245, 257
~ arithmetic 102, 215–216, 306
~ logic 101, 129, 215–216, 220–221, 229, 234, 236, 241
~ property 17, 28–29
seeing / sehen 186, 203, 291, 378–380, 447, 452, 464, 499–511, 513
self-evidence X, 183, 193–195, 380, 523–524
self-evident 183, 187–188, 192–194
~ judgement / proposition / truth 183, 188, 193–194
semantic proof 38–39, 529
semantics 4, 8, 14, 32, 47–50, 78–79, 85–86, 93–95, 125, 158–159, 199, 386, 392, 415–416, 418, 420–421, 424–427, 529
Fregean / Frege's ~ 167, 375, 378
Game-Theoretical ~ 120–121
hyperintensional ~ 28
Kit Fine's ~ 421
Kripke / Kripkean ~ 166, 390
Link's / Link-style ~ 5, 12–14
Tractarian ~ 416
truth-conditional ~ 416
two-domain analysis 7–8, 16 (cf. one-domain analysis)
sense / Sinn 95, 149, 153–156, 160, 167, 187, 278, 378, 380, 383, 423, 477, 488 (cf. reference) 470–471, 500
sequent calculus 107, 109
set of all sets 285
set-theoretical X, 119, 219–220, 224, 227, 524, 530
set theory XI, 94, 99, 148, 217–219, 301, 327–330, 336–337, 339, 347, 445, 450, 452
axiomatic ~ 327, 335, 337
Zermelo-Fraenkel ~ 323
sign and symbol 375, 379
similarity 34, 166, 177, 315, 382, 385, 387, 423, 441, 450, 485, 521
singular term 27, 105–108, 154, 157, 167, 329, 378
situation 143, 147, 219, 267, 277, 397–401, 404–406, 408–410, 412–413, 420–421, 457, 466, 468
elementary ~ 398–400, 404–406, 408, 413
independent ~ 409
lattice of elementary ~s 398, 400, 404
Skolem function 117, 120–121, 123, 125, 130
Sollen, das 478, 482
sortal distinction 7–8, 16, 19
standard model 161, 219
state of affairs / atomic fact / Sachverhalt 397, 410, 412–413, 416, 420–422, 441, 452–453, 461, 464, 467, 532
independence of ~ 397
structure of propositions / sentences 376, 520
structure of the world 150, 379
subject matter 68, 251, 377–378, 415, 418, 421–427
substitution, rule of 97, 253, 258
substitutivity 8, 26–28
subsumtion / falling under a concept 137, 145, 148–149, 216, 232
successor function / operation 333, 345–346, 348, 350–351, 355
surprises in logic 197, 207
syllogism / syllogistic / syllogistics X, 133, 137–143, 145–146
symbolism 97, 170, 349, 378, 391, 447, 467, 518, 525, 529–530
synthetic a priori 201, 203, 252, 255
Tatsache 476, 478–479, 481–482, 500, 508, 512
moralische ~ 479, 481–482
Tatsachenraum 473–474, 478–479
tautological / tautologous / tautology 99, 133, 138, 143–144, 149, 151, 178, 201, 234, 276, 278, 284–285, 306, 388, 420, 426, 515–516, 519, 521–524
technique 96, 140, 142, 200, 239, 264, 270–271, 273–274, 278, 289, 444–445, 532
theory of types 308, 449 (cf. Constructive

Type Theory)
third realm 185, 378, 381, 444
three-valued logic 42–43, 48–50, 106 (cf. many-valued logic; multi-valued logic)
token 375, 381, 516
tone 376
topological ontology 402
topology of situations 397, 404 (cf. Wittgenstein's topology)
Tractatus Logico-Philosophicus 133, 150, 169, 171, 175–176, 181, 189, 201, 207, 267, 269, 331, 375, 379, 381–382, 385–393, 397, 414–415, 451–452, 471, 474–475, 477, 519–520, 522–525
Tractatus logic 150, 519–520
traditional logic 139, 143
transcendental 40, 139, 150, 207, 465
transfinite 72, 225, 260, 309, 448–449
transitive 13, 100, 123, 127, 244, 306, 399, 423, 434, 439
transitivity, 228, 434
translation 37, 40–41, 43, 45–52, 71–86, 230, 260, 324, 434, 439, 468, 517
abstract contextual ~ 71, 81–82, 84
concept of ~ 71–73, 75, 79, 81, 83–84
conservative ~ 71, 76–79, 81–82, 85–86
contextual ~ 71–72, 76–77, 80–84, 86
hypercontextual ~ 83
universal ~ 40–41, 50
~ between logics 41, 71, 73, 75, 83
true-false polarity 464
truth conditions 157, 167, 178–179, 208, 422
truth function 140, 162, 267, 462, 464–465, 520–521
truth predicate 154, 161–163, 179,
truth table 38–39, 43–44, 48, 515–517
truth value IX, 38, 40, 42, 46–48, 138, 153–166, 169, 173–178, 180–181, 319, 385, 416, 516
truth-conditional content 160
truth function 159, 161, 267, 458–459, 461–462, 464

truth-functional / truth-functionality, 138, 140, 145, 147–148, 267, 416, 464–465
two-domain analysis 7–8 (cf. one-domain analysis)
type distinction 137, 139–140
type of logic 133, 140, 144
uncountable 3–4, 408
undecidability / undecidable 95, 272, 286, 299, 515, 518, 525–527, 529–532
~ proof 518, 525–527, 529–532
unintelligibility 429–430, 435, 437
unity of the proposition 453
universality of logic 98–99
unprovability 452, 527–530 (cf. provability)
unsurveyability / unsurveyable 331–332
vague / vagueness 22–23, 135, 172
validity 37–39, 42–43, 45, 192, 203, 279, 284, 307, 525, 531
variable X, 45–47, 74, 77, 80, 106, 117–121, 123–130, 161, 202, 222–223, 253–254, 256–258, 318–320, 389, 393, 446, 452–453, 457–459, 461, 465, 467, 469, 516
bound ~ 126, 309, 520
boxed ~ 45, 47
dependence between ~s 117, 119–120, 123, 125, 127–128
free ~ 126, 222, 223, 229, 234, 258, 307, 319, 460–461, 465–466, 469–470
object ~ 222, 229
predicate ~ 222, 225, 229
propositional / sentence ~/ ~for propositions 42–43, 46, 76, 80, 389, 462, 516
quantified ~ 118, 121, 126
Venn's diagram 142
vertical, the 170–171 (cf. judgement stroke)
visual space 202, 447, 449
ways of being true 415, 418, 420
Wert 361, 500, 511–512
~urteil 480 (cf. moralisches Urteil)
Wertblindheit 499, 512
Wittgenstein's topology 403, 405–407
Wolniewicz's lattices 398, 407–408

www.ingramcontent.com/pod-product-compliance
Lightning Source LLC
Chambersburg PA
CBHW050523300426
44113CB00012B/1930